# TIME SERIES, UNIT ROOTS, AND COINTEGRATION

**Phoebus Dhrymes**

*Columbia University*
*New York, New York*

**ACADEMIC PRESS**

San Diego   London   Boston   New York   Sydney   Tokyo   Toronto

Academic Press
*a division of Harcourt Brace & Company*
525 B Street, Suite 1900, San Diego, California 92101-4495, USA
http://www.apnet.com

Academic Press Limited
24-28 Oval Road, London NW1 7DX, UK
http://www.hbuk.co.uk/ap/

Library of Congress Card Catalog Number: 97-80318

International Standard Book Number: 0-12-214695-6

PRINTED IN THE UNITED STATES OF AMERICA
97  98  99  00  01  02  QW  9  8  7  6  5  4  3  2  1

*To P.C.B. Phillips,*
*whose work on integrated processes*
*infused clarity and depth into the subject*

# Contents

# Preface

This book deals with the theory of stationary and certain nonstationary processes, the latter as developed in the literature of econometrics from the late 1970s to the present. It is intended for advanced students and professionals. Thus, it deals with the taxonomy of stochastic sequences (Chapter 1), estimation and prediction for stationary processes (Chapter 2), the general problem of unit roots and $I(1)$ regressors (Chapter 3), and various aspects of cointegration (Chapters 4, 5, and 6). Because such material makes extensive use of topics in probability theory generally not easily accessible to students and professionals, I included three background chapters: Brownian motion (Chapter 7), stochastic integration (Chapter 8), and functional central limit theorems and ancillary topics (Chapter 9). A number of other novel features are worth noting.

Chapter 2 contains a discussion of various algorithms useful in the estimation of stationary processes, such as the Durbin-Levinson-Whittle and the innovations algorithms. It also discusses a number of model selection criteria such as various Akaike criteria and Bayesian information criteria. In addition, it covers models in state space form, and Kalman procedures. The latter include the fitering, prediction and fixed point smoothing procedures.

Chapters 4 through 6 deal with various issues of cointegration, including cointegration in vector autoregressions, autoregressive moving average models, and the classic case where the process underlying the cointegrated $I(1)$ process is a moving average. Chapter 6 includes a number of tables for use in testing various cointegration hypotheses. The tables are more extensive than those currently in the literature and are designed for conformity cointegration tests. However, as a special case, they contain tables appropriate for testing for cointegration (rank) when the test statistic reflects the cointegration hypothesis, i.e. for the likelihood ratio tests associated with the contribution of Johansen.

Chapter 9 is perhaps more extensive than it need be for the narrow purpose of serving as background. It offers, in as readable a manner as I could manage, material on the space $C(R_+, R^d)$, the space of continuous functions, $D(R_+, R^d)$, the space of discontinuous functions with certain properties, and Skorohod topology; it also includes a discussion of various types of mixing, strong and weak laws of large numbers, the law of the iterated logarithm for mixing processes and concludes by presenting an extensive collection of central limit and functional central limit theorems for stationary and/or mixing processes.

As an added help to the reader I included 22 pages of graphs depicting realizations from various types of time series: the standard i.i.d. sequence, linear deterministic trend plus an i.i.d. sequence, stationary sequences, integrated sequences that are cointegrated of various ranks and so on.

The book is written in a more or less uniform notational scheme, which facilitates the absorption of the literature it encompasses. Thus, it may be used as a main text in an advanced course in time series when cointegration is the main concern. In such a case Chapter 2, dealing with the estimation of stationary series, is useful chiefly because it introduces a method of estimation for autoregressive moving average models, which is very close to standard procedures in econometrics. This eases the discussion of cointegration in integrated autoregressive moving average models.

For a conventional course in time series, with limited coverage of integratred processes, the instructor may use this book as a reference and/or as a secondary textbook.

Finally I express my thanks to the many individuals and institutions that were helpful in the course of writing this volume. Chapter 9 was completed during my visit to the Institute for Advanced Studies in Vienna. Accordingly, I would like to thank that institution and its directorate for providing a congenial and supportive environment for my work on this book. I also acknowledge the very able research assistance of my student Dimitrios Thomakos; he programmed and prepared all the tables that appear in this volume, read the work in its entirety, and provided helpful suggestions. My former research student Dr. Julian Silk also read the work in its entirety and provided many editorial suggestions that improved the exposition in several chapters. Steve Durlauf, Cheng Hsiao, Ioannis Karatzas, Peter Robinson, and Aris Spanos read parts of the book and offered helpful comments.

To all I express my deep appreciation and enter the usual *caveat* that all remaining errors are entirely mine.

Last but not least I would be remiss if I failed to acknowledge my debt to the literature, and particularly to the work of Steve Durlauf, Rob Engle, Clive Granger, Soren Johansen, Bruce Hansen, Pierre Perron, Peter Phillips, Gregory Reinsel, and many others.

Phoebus J. Dhrymes
Bronxville, New York
August 1997

# Chapter 1

# Stochastic Sequences

## 1.1 Preliminaries

The term "time series" is generally, but not universally, defined in econometrics chiefly by contrast to the term "cross section" or "panel", and generally refers to the type of data being analyzed. Thus, for example, the methods employed in analyzing inventories held by a group of firms, or aggregately in the economy, is commonly referred to as time series analysis. The data in question are indexed in real time, so a natural ordering exists in terms of prior and subsequent observations. By contrast, cross section data, which refer to contemporaneous observations on a number of economic agents, are indexed by an integer index set, but there exists no natural ordering in which the observations may be arranged relative to the index. Panel data, on the other hand, refer to repeated observations over real time, on a number of economic agents. As such, they combine both cross section and time series features.

Most time series work in econometrics consists of processing data and obtaining inferences about underlying economic phenomena in the "time domain", i.e. by examining second (and lower) moment relationships in the data. Regression, as well as simultaneous equations methods, belong essentially to this variety of analysis, although there is nothing to prevent these methods from being employed in a cross section and/or panel context.

By the fifties economists had already become concerned with the intertemporal relationships between two or more series; perhaps the prototype of this approach lies in the cobweb model of agricultural supply, in which the previous history of prices (or only the last observed price) determines current supply.

Subsequent models of **distributed lags** sought to determine the time profile of the impact of a change in the economic environment on a given set of agents, or economic variables. A typical example is the early investment model, which sought to determine the manner in which previous changes in the demand for output evoked additions to the capital stock (investment). At roughly the same time, in the engineering control literature various linear prediction and control models were examined in connection with communications and other branches of engineering. In many other branches of science, similar models were explored. For example, in hydrology simple models of accretion and evaporation led to what economists would have called distributed lags, i.e. the representation of a given stock as the **integral** of previous net accretions, in this case the volume of water in a given lake.

In the seventies, the publication of *Time Series, Forecasting and Control* by Box and Jenkins (1970) popularized the analysis of time series data by essentially obtaining their autocorrelation and partial autocorrelation functions, and matching the shapes of such functions with orders of autoregression (AR) or moving average (MA) for the underlying series. Often this is referred to as analysis in the "time domain".

At roughly the same time, another strand of the literature sought to exploit the simpler representation of "time series" in the frequency domain. Thus, if one models an economic phenomenon that "unfolds in real time" by a stochastic sequence which has certain stationarity features, one may analyze it not in terms of its autocorrelation function, as in Box and Jenkins, but rather in terms of the **Fourier transform** of the autocorrelation function, which is called the **spectral density function**. Although this affords a considerably more convenient representation of the properties of stochastic sequences, it has had much less acceptance in econometric practice, quite possibly owing to the increased mathematical burden that this approach places on the investigator.

The proclivity to analyze economic data, primarily macro and financial data, in terms of time series time domain techniques was further aided by a growing dissatisfaction with structural (simultaneous equations)[1] models and the desire to employ simpler procedures that do not rely on " prior information". Unfortunately, it is not possible to approach the analysis of economic data *tabula rasa*, and what occurred is the substitution of one set of "prior information" by another.

More recently, it was increasingly felt that certain economic phenomena may not be properly representable as stationary sequences, and for this reason we have witnessed a new development of the literature in the direction of modeling through nonstationary sequences–the unit root approach, which leads directly into the theory of Brownian motion.

In what follows we give an integrated presentation of these aspects of the literature, including **cointegration**, after having provided a somewhat informal discussion of the theory of stationary sequences, and issues arising in the estimation of their parameters (from a time domain point of view). In that phase we introduce several algorithms that are not generally found in the literature of econometrics, although they have been well established in the literature of statistics. These algorithms provide initial consistent estimators that are important in dealing with nonlinear estimation problems, and otherwise make the estimation task considerably simpler. Ancillary topics that are important in understanding recent developments in the estimation and testing of **cointegrated** nonstationary sequences, such as Brownian motion, stochastic integration, and central limit theorems (CLT) including functional central limit theorems (FCLT), are given respectively in Chapters 7, 8, and 9. These chapters may be studied in their own right, for the tools they provide; if the reader is not particularly interested in such aspects, they may be

---

[1] Not to mention the fact that by the late seventies the research potential of the general linear structural (simultaneous) equations framework had been nearly exhausted.

omitted without loss of relevance and may be utilized only as references for
results invoked in the arguments of the main body of this work, contained in
Chapters 1 through 6.

Although we attempt to make the presentation as self contained as possible,
we shall refer extensively, when necessary, to the author's *Topics in Advanced
Econometrics: Probability Foundations*, (1989).

## 1.2   Stationary Processes

Because most applications in econometrics involve **vector processes** while,
due to the relative ease of presentation, most textbook discussions dwell on
**scalar processes**, we make a conscious attempt in this volume to discuss
adequately the properties of vector processes. Many results, however, could
be stated uniformly for both vector and scalar processes; in such cases the
context will make clear whether we deal with random variables, or random
vectors (r.v.). When a random vector is used, it will be understood that it
is a $q$-element **row** vector, unless otherwise indicated.

**Definition 1.** Let $T$ be a linear [2] index set and $\{X_t : t \in T\}$ a collection of
r.v. defined on the probability space $(\Omega, \mathcal{A}, \mathcal{P})$. The collection $\{X(t,\omega) :
t \in T, \omega \in \Omega\}$, also denoted by $\{X_t(\omega) : t \in T, \omega \in \Omega\}$, or $\{X_t : t \in T\}$,
is said to be a **stochastic process**.

**Remark 1.** Many authors distinguish between the terms **stochastic process** and **stochastic sequence**, respectively, according to whether $T$ is a
"continuous" or discrete set. For example, if $T = (-\pi, \pi)$ one would term
$\{X_t(\omega) : t \in T, \omega \in \Omega\}$ a **stochastic process**, while if, say $T = \mathcal{N}$,[3] one
would use the term **stochastic sequence**. We do not observe this distinc-
tion and instead use the two terms interchangeably. The context then would
make clear whether we are dealing with a continuously indexed family of r.v.
or one which is discretely indexed.

---

[2] This means that if $t_1, t_2 \in T$ then $t_1 + t_2 \in T$.

[3] The notation $\mathcal{N}$, denotes the so called integer lattice on $R = (-\infty, \infty)$, i.e. it is the
set $\{n : n = 0, \pm 1, \pm 2, \ldots\}$. We may also use the notation $\mathcal{N}_0 = \{n : n = 0, 1, 2, \ldots\}$,
$\mathcal{N}_+ = \{n : n = 1, 2, \ldots\}$, and $\mathcal{N}_- = \{n : n = -1, -2, -3, \ldots\}$.

Moreover, in the mathematics literature, the entity of Definition 1 is occasionally referred to as a **time series**, although this is a term of declining popularity. In theoretical econometrics, on the other hand, the term remains popular, and the subject of this volume would be characterized as time series analysis.

**Definition 2.** Let $\{X_n : n \in \mathcal{N}\}$ be a sequence of random variables defined on the probability space $(\Omega, \mathcal{A}, \mathcal{P})$, where $\mathcal{N} = \{0, \pm 1, \pm 2, \ldots\}$. The sequence is said to be **strictly stationary** or **stationary in the strict sense** if and only if for all $k, s \in \mathcal{N}$ and (appropriate) cylinder sets $B_{(k)} \in \mathcal{B}(R^\infty)$,

$$\mathcal{P}(A_1) = \mathcal{P}(A_2), \quad A_1 = \{\omega : (X_0(\omega), X_1(\omega), \ldots, X_k(\omega)) \in B_{(k)}\},$$

$$A_2 = \{\omega : (X_s(\omega), X_{s+1}(\omega), \ldots, X_{k+s}) \in B_{(k)}\}. \tag{1.1}$$

**Remark 2.** Notice that strict stationarity implies that the r.v. of the sequence are **identically distributed**. This is so since if we take $k = 1$, we conclude that for arbitrary $s$, $X_\tau$ and $X_{s+\tau}$ have the same probability characteristics, and hence the same distribution. In particular, if the r.v. possess moments, say to the $r^{th}$ order, then **for all** $s$,

$$E|X_s|^m = E|X_0|^m, \quad m \le r.$$

**Definition 3.** Let $\{X_n : n \in \mathcal{N}\}$ be a sequence of r.v. defined on the probability space $(\Omega, \mathcal{A}, \mathcal{P})$. The sequence is said to be **stationary in the wide sense**, or **covariance stationary**, if and only if

i. $EX_k = \mu$,[4] for all $k \in \mathcal{N}$;

ii. $\text{Cov}(X_{n+\tau}, X_n) = K(n + \tau, n) = K(\tau)$, for all $n$, $\tau \in \mathcal{N}$.

If $X_n$ is a **scalar** the function $K$ is said to be the **autocovariance** function and is denoted by $\kappa(\tau)$; if $X_n$ is a **vector** the function $K$ is said to be the **autocovariance kernel**.[5]

---

[4] For convenience, we shall always assume that $\mu = 0$, unless otherwise indicated.

[5] This term is not standard, but it is employed here in order to facilitate simplicity of exposition.

**Remark 3**. Matters are facilitated if we take all random variables to be **complex**, i.e. if we think of the scalar process $X_k$ as having the form $X_k = X_{k,1} + iX_{k,2}$, and of a vector process as having the form $X_{k\cdot} = X_{k,1\cdot} + iX_{k,2\cdot}$, where $i$ is the imaginary unit, obeying $i^2 = -1$; moreover, we place all our discussion in the context of the Hilbert space $\mathcal{H}^2(\Omega,\ \mathcal{A},\ \mathcal{P})$, or more concisely $\mathcal{H}^2(\mathcal{P})$, consisting of all r.v. possessing second moments, and defined over the probability space $(\Omega,\ \mathcal{A},\ \mathcal{P})$.[6] The usefulness of the space $\mathcal{H}^2(\mathcal{P})$ in this context is quite evident since we define the inner product on that space by

$$< x, y > = \int_\Omega x\bar{y} d\mathcal{P}, \quad x, y \in \mathcal{H}^2(\mathcal{P}).$$

In this, and particularly the next, chapter the usefulness of this framework in studying properties of, and issues relating to, weakly stationary sequences is amply demonstrated.

In the framework of Remark 3, the autocovariance function is defined by

$$\kappa : \mathcal{N} \longrightarrow \mathcal{C},$$

where $\mathcal{C}$ is the field of complex numbers; the autocovariance kernel is defined by

$$K : \mathcal{N} \longrightarrow \mathcal{C}^{q^2},$$

where $\mathcal{C}^m$ denotes the Cartesian product of $\mathcal{C}$ with itself $m$ times. Thus, $K(\tau)$ is a **complex matrix** and can be written as

$$K(\tau) = K_{(1)}(\tau) + iK_{(2)}(\tau).$$

Moreover, for both the autocovariance function and kernel, **note the conventions** involved; if we employ the standard notation $X_{n\cdot}$ so that the stochastic process in question is represented as a **row** vector, and it is assumed to have zero mean, the definition is

$$K(\tau) = EX'_{n+\tau\cdot}\overline{X}_{n\cdot};$$

---

[6] For a definition of Hilbert space, see Dhrymes (1989), p. 73.

thus, the conjugated process multiplies on the right,[7] and the argument of $K$ is the difference between the indices of the first and second terms of the product. In this context, then, we have

$$K(-\tau) = EX'_{n-\tau\cdot}\overline{X}_{n\cdot}, \quad \overline{K}(\tau) = E\overline{X}'_{n+\tau\cdot}X_{n\cdot} = [EX'_{n\cdot}\overline{X}_{n+\tau\cdot}]' = [K(-\tau)]'.$$

Finally, note that

$$
\begin{aligned}
K_{(1)}(\tau) &= E(X'_{n+\tau,1\cdot}X_{n,1\cdot} + X'_{n+\tau,2\cdot}X_{n,2\cdot}), \\
K_{(2)}(\tau) &= E(X'_{n+\tau,2\cdot}X_{n,1\cdot} - X'_{n+\tau,1\cdot}X_{n,2\cdot}).
\end{aligned}
$$

Since $K(\tau) + \overline{K}(\tau) = 2K_{(1)}(\tau)$, and by the preceding discussion $\overline{K}(\tau) = [K(-\tau)]'$, it follows that

$$K_{(1)}(-\tau) = K_{(1)}(\tau)', \quad K_{(2)}(-\tau) = -K_{(2)}(\tau)'.$$

For the scalar case we have the obvious result

$$\kappa(-\tau) = \overline{\kappa}(\tau),$$

and hence that

$$\kappa_{(1)}(-\tau) = \kappa_{(1)}(\tau), \quad \kappa_{(2)}(-\tau) = -\kappa_{(2)}(\tau).$$

In the context of the space $\mathcal{H}^2(\Omega, \mathcal{A}, \mathcal{P})$ we introduce, for scalar processes, the **inner product**

$$< X_m, X_n > = EX_m\overline{X}_n,$$

and for vector processes, the inner product[8]

$$< X_{m\cdot}, X_{n\cdot} > = EX_{m\cdot}\overline{X}'_{n\cdot}.$$

---

[7] This would be of no consequence in the autocovariance **function** due to the commutative property of (scalar) multiplication.

[8] Because of the definition of inner product in terms of the autocovariance function, and the assumption that $EX_n = 0$, **scalar** variables which are uncorrelated are said to be **orthogonal** in a Hilbert space context. Thus, the inner product is quite useful for **scalar sequences** in the sense that a zero inner product between two scalar r.v. in a Hilbert context means that they are **uncorrelated**. The inner product, however, is not quite as useful in multivariate (vector) sequences, since a zero inner product between two **vectors**, say

$$0 = < X_{m\cdot}, X_{n\cdot} > = EX'_{m\cdot}\overline{X}_{n\cdot} = \text{tr}[K(m-n)],$$

**need not imply** that $K(m-n) = 0$ and, thus, the two vectors **need not be uncorrelated**; for that **we require** that $K(m-n) = 0$. Evidently, if the latter condition holds, the inner product is zero, but the converse **is not true**.

Since a weakly stationary sequence is basically a sequence which is stationary in its first two moments, the mean function must be independent of the index, i.e. a constant; thus, it entails no loss of generality to take it to be zero, so that the inner product will coincide with the autocovariance function.

Since a Hilbert space is a normed space with an inner product, we shall complete the description of $\mathcal{H}^2(\mathcal{P})$ by defining the norm for scalar processes as

$$\| X_k \| = < X_k, X_k >^{1/2},$$

and for vector processes as

$$\| X_k \| = < X_k, X_k >^{1/2} = [\mathrm{tr} K(0)]^{1/2}.$$

We formalize the preceding discussion by restating a number of useful properties of the autocovariance function for scalar processes in the form of a proposition. We have,

**Proposition 1.** Let $X = \{X_n : n \in \mathcal{N}\}$ be a covariance stationary sequence in $\mathcal{H}^2(\mathcal{P})$, and $\kappa(\cdot)$ its autocovariance function. The following statements are true:

   i. $\kappa(0) \geq 0$;

  ii. $\overline{\kappa}(\tau) = \kappa(-\tau)$;

  iii. $\kappa(0) - (1/2)[\kappa(\tau) + \overline{\kappa}(\tau)] \geq 0$, in the sense that the difference is non-negative for every $\tau$;

  iv. the autocovariance function is positive semidefinite, in the sense that for an arbitrary sequence $s_j \in \mathcal{N}$, and complex (scalars) $z_j$, $j = 1, 2, \ldots, N$

$$\sum_{i=1}^{N} \sum_{j=1}^{N} z_i \overline{z}_j \kappa(s_i - s_j) \geq 0.$$

Proof: The proof of (statement) i is immediate since $\kappa(0) = (X_k, X_k) = E X_k \overline{X}_k$.

The proof of ii follows from the definition

$$\overline{\kappa}(\tau) = E \overline{X}_{n+\tau} X_n = E X_n \overline{X}_{n+\tau} = \kappa(-\tau).$$

As for iii we note that by definition, for any $\tau$,

$$
\begin{aligned}
0 \leq \text{Var}(X_{n+\tau} - X_n) &= 2\kappa(0) - \kappa(\tau) - \kappa(-\tau) \\
&= 2\{\kappa(0) - (1/2)[\kappa(\tau) + \overline{\kappa}(\tau)]\}.
\end{aligned}
$$

Finally, to prove iv let $Y = \sum_{j=1}^{N} z_j X_{s_j}$. Then,

$$
0 \leq \text{Var}(Y) = EY\overline{Y} = \sum_{i=1}^{N} \sum_{j=1}^{N} z_i \kappa(s_i - s_j)\overline{z}_j.
$$

q.e.d.

The analog for a vector process is given below.

**Proposition 2.** Let $X = \{X_{n\cdot} : n \in \mathcal{N}\}$ be a vector covariance stationary sequence in $\mathcal{H}^2(\mathcal{P})$, and $K(\cdot)$ its autocovariance kernel. The following statements are true:

   i. $K(0) \geq 0$, in the sense that the matrix is positive semidefinite;

   ii. $\overline{K}(\tau) = K(-\tau)'$;

  iii. $K(0) - (1/2)[K(\tau) + \overline{K}(\tau)'] \geq 0$, in the sense that the difference above is positive semidefinite for every $\tau$;

  iv. the autocovariance kernel is positive semidefinite, in the sense that for an arbitrary sequence $s_j \in \mathcal{N}$, and complex (column) vectors $z_j$, $j = 1, 2, \ldots, N$, $\sum_{i=1}^{N} \sum_{j=1}^{N} z_i' K(s_i - s_j)\overline{z}_j \geq 0$.

Proof: The proof of i is immediate by noting that

$$
K(0) = E(X_{n,1\cdot}' X_{n,1\cdot} + X_{n,2\cdot}' X_{n,2\cdot}) \geq 0.
$$

The proof of ii was given in the discussion above. As for iii we note that

$$
0 \leq E(X_{n+\tau\cdot} - X_{n\cdot})'(\overline{X}_{n+\tau\cdot} - \overline{X}_{n\cdot}) = 2K(0) - K(\tau) - \overline{K}(\tau)'
$$

$$
= 2\{K(0) - (1/2)[K(\tau) + \overline{K}(\tau)']\}.
$$

Finally, to prove iv let

$$
Y = \sum_{j=1}^{N} z_j' X_{s_j\cdot}'.
$$

Then,

$$0 \leq \text{Var}(Y) = EY\overline{Y} = \sum_{i=1}^{N}\sum_{j=1}^{N} z_i'[EX_{s_i.}'\overline{X}_{s_j.}]\overline{z}_j = \sum_{i=1}^{N}\sum_{j=1}^{N} z_i' K(s_i - s_j)\overline{z}_j.$$

q.e.d.

**Remark 4.** Note that the notation $\| \cdot \|$ when applied to **nonrandom** vectors or matrices means the square root of the inner product as defined on $n$-dimensional Euclidean space, $R^n$. Thus, for example, if $v \in R^n$, $\| v \|^2 = v'v = \sum_{j=1}^{n} v_j^2$; if $A$ is an $m \times q$ matrix then we may define $\| A \|^2 = [\text{vec}(A)]'\text{vec}(A)$. Other norms for a matrix are also possible such the largest root or the trace of $A'A$ or $AA'$.

We conclude this section by introducing some additional concepts.

**Definition 4.** Let $X = \{X_{n.} : n \in \mathcal{N}\}$ be a vector covariance stationary process with autocovariance kernel $K(\cdot)$. The **kernel generating function** is given by

$$G(z) = \sum_{\tau=-\infty}^{\infty} K(\tau)z^\tau, \quad \text{provided} \quad \sum_{\tau=-\infty}^{\infty} \| K(\tau) \| < \infty, \quad z \in \mathcal{C}.$$

The corresponding definition for scalar processes is

$$g(z) = \sum_{\tau=-\infty}^{\infty} \kappa(\tau)z^\tau, \quad \text{provided} \quad \sum_{\tau=-\infty}^{\infty} |\kappa(\tau)| < \infty, \quad z \in \mathcal{C}.$$

**Definition 5.** In the context of Definition 4, consider the process

$$Y_{n.}' = \sum_{j=-\infty}^{\infty} C_j X_{n-j.}',$$

where $Y_{n.}$ and $X_{n.}$ are, respectively, $s$- and $q$-element row vectors. The sequence $C = \{C_j : j \in \mathcal{N}, C_j \in \mathcal{C}^{sq}\}$, where $\mathcal{C}$ is the field of complex numbers, is said to be a **linear time invariant filter**; if $C_j = 0$, for $j > |N|$, it is said to be a finite (linear time invariant) filter.

In the scalar case the same applies, *mutatis mutandis*.

We often omit the terms in parentheses above and refer to such entities as simply filters, or finite filters.

The usefulness of the kernel generating function may be illustrated by a simple example, using the entities of Definition 5. To this end obtain the kernel of the $Y$-process as

$$K_y(\tau) = EY'_{n+\tau}.\overline{Y}_{n\cdot} = \sum_{j_1=-\infty}^{\infty} \sum_{j_2=-\infty}^{\infty} C_{j_1}[EX'_{n+\tau-j_1}.\overline{X}_{n-j_2\cdot}]\overline{C}'_{j_2}$$

$$= \sum_{j_1=-\infty}^{\infty} \sum_{j_2=-\infty}^{\infty} C_{j_1}K_x[\tau - (j_1 - j_2)]\overline{C}'_{j_2}.$$

The kernel generating function of the $Y$-process is thus

$$G_y(z) = \sum_{\tau=-\infty}^{\infty} \sum_{j_1=-\infty}^{\infty} \sum_{j_2=-\infty}^{\infty} [C_{j_1}z^{j_1}]\{K_x[\tau - (j_1 + j_2)]z^{\tau-(j_1-j_2)}\}[\overline{C}'_{j_2}z^{-j_2}].$$

Make the change in index, $s = \tau - (j_1 - j_2)$, and sum to obtain

$$G_y(z) = C(z)G_x(z)C^*(z^{-1}), \quad C(z) = \sum_{j=-\infty}^{\infty} C_j z^j, \quad C^*(z^{-1}) = \sum_{j=-\infty}^{\infty} \overline{C}'_j z^{-j}.$$

If the $C_j$ and the $X$-process **are both real**, and thus $z \in R$, the relation above simplifies to

$$G_y(z) = C(z)G_x(z)C(z^{-1})'.$$

This allows us to evaluate routinely the kernels $K_y(\tau)$, if we are given the kernels $K_x(\tau)$ and the filter $\{C_j : j \in \mathcal{N}\}$.

Because we often have occasion to deal with complex matrices it is useful to recall a number of important properties of such matrices.

**Definition 6.** Let $H$ be a square matrix of order $q$ and suppose $h_{ij} \in \mathcal{C}$; the matrix is said to be **Hermitian** if and only if $H = \overline{H}'$.

Let $A$ be a square matrix of order $q$ with $a_{ij} \in \mathcal{C}$; it is said to be a **unitary** matrix if and only if $\overline{A}'A = I_q$.

**Remark 5.** Over the field of complex numbers, $\mathcal{C}$, a **Hermitian** matrix is the analog of a **symmetric** matrix over the field of real numbers $R$. The properties ascribed to real symmetric matrices generally carry over to Hermitian matrices over the space $\mathcal{C}$.

**Assertion 1.** If $H$ is Hermitian as in Definition 6, its characteristic roots are real.

Let $H$ be Hermitian; then there exists a unitary matrix $A$ and a diagonal matrix $\Lambda = \operatorname{diag}(\lambda_1, \lambda_2, \ldots, \lambda_q)$ such that

$$HA = A\Lambda, \quad \text{or} \quad H = A\Lambda\overline{A}', \quad \text{or} \quad \overline{A}'HA = \Lambda.$$

Proof: Although such proofs are straightforward, they are a distraction and the reader is referred to Bellman (1960), pp. 24-25, or Gantmacher (1959), pp. 331-338.

**Remark 6.** In the context of Definition 6 and Assertion 1, item iv of Proposition 1 simply states that the **matrix** $[\kappa(s_i - s_j)]$ is a positive semidefinite Hermitian matrix; items ii, iii, and iv of Proposition 2 simply state the fact that the **block matrix** $[K(s_i - s_j)]$ is a positive semidefinite Hermitian matrix. The reader may be somewhat puzzled by the nature of this matrix, in that its appearance is rather different from covariance matrices usually encountered in econometrics. To disabuse the reader of this notion consider the GNLSEM (general nonlinear structural econometric model) with additive errors, see e.g. Dhrymes (1994b) Chapter 6,

$$y_{t\cdot} = g_{t\cdot}(\theta) + u_{t\cdot}, \quad t = 1, 2, \ldots, T,$$

and suppose that the vector $u_{t\cdot}$ represents a covariance stationary process. If we write all observations in the form of a **column** vector, the error terms may be written as

$$u = (u_{1\cdot}, u_{2\cdot}, \ldots, u_{T\cdot})', \quad \text{and} \quad \operatorname{Cov}(u) = [K(t - t')], \quad t, t' = 1, 2, \ldots, T.$$

Covariance stationarity implies that the main diagonal blocks (all) consist of the matrix $K(0)$, the first subdiagonal blocks (all) consist of the matrix $K(1)$, the second subdiagonal blocks (all) consist of the matrix $K(2)$, and so on. If the process is **real** the superdiagonal blocks correspond to the transpose of the corresponding subdiagonal blocks; if the process is complex then the superdiagonal blocks consist of the transpose of the complex conjugate of the corresponding subdiagonal blocks. Note, further, that stationarity implies that $\lim_{\tau \to \infty} K(\tau) = 0$.

A particularly simple elementary example is the covariance matrix of the first-order autoregressive error term in the context of the GLM (general linear model).

## 1.3  Spectral Representation

A complete derivation of the spectral representation of covariance stationary
processes is completely outside the scope of this volume. Nonetheless a brief
heuristic discussion of this issue may be quite useful. In this section, until
further notice, we shall be dealing with scalar processes so that r.v. refers
exclusively to random variables. Let $\{X_n : n \in \mathcal{N}\}$ be a sequence of r.v.
defined on the probability space $(\Omega, \mathcal{A}, \mathcal{P})$ and suppose

$$X_n = \sum_{s=-\infty}^{\infty} z_s e^{i\lambda_s n}, \quad \lambda_s \in (-\pi, \pi], \lambda_i \neq \lambda_j, \ i \neq j. \tag{1.2}$$

Suppose further that $\{z_s : s \in \mathcal{N}\}$ is a zero mean, square integrable sequence
of complex orthogonal r.v. such that

$$\sum_{s=-\infty}^{\infty} \sigma_s^2 < \infty, \quad \sigma_s^2 = E z_s \bar{z}_s.$$

The preceding implies that the series representation of $X_n$ converges (**absolutely**) **in mean square**. Now, if we define

$$Z(\lambda) = \sum_{\lambda_s \leq \lambda} z_s, \quad \Delta Z(\lambda_s) = Z(\lambda_s) - Z(\lambda_s-), \tag{1.3}$$

where the summation is over all indices $s$ such that $\lambda_s \leq \lambda$, we note that
$Z$ is a **stochastic process** with orthogonal increments,[9] defined for $\lambda \in$
$(-\pi, \pi]$. In view of Eq. (1.3) we may rewrite Eq. (1.2) as

$$X_n = \sum_{s=-\infty}^{\infty} e^{i\lambda_s n} \Delta Z(\lambda_s), \tag{1.4}$$

which looks very much like the approximating sums in the Riemann-Stieltjes
or the Lebesgue-Stieltjes integral. Unfortunately, the situation is far more
complex. Nonetheless, it may be shown, but not in this venue, that for **any**
**covariance stationary stochastic process**, $\{\xi_n : n \in \mathcal{N}\}$, there exists
an **orthogonal stochastic measure**, $Z$, defined on the Borel $\sigma$-algebra
$\mathcal{B}([-\pi, \pi])$ generated by $[-\pi, \pi]$, such that

$$\xi_n = \int_{-\pi}^{\pi} e^{i\lambda n} Z(d\lambda), \quad \text{or alternatively,} \quad \xi_n = \int_{-\pi}^{\pi} e^{i\lambda n} dZ. \tag{1.5}$$

---

[9] This is so since disjoint intervals contain a **different collection of the** $z_s$.

The alternative representation is seen to be quite compatible with the original, if we note that, without loss of generality or relevance, we may think of the **stochastic measure**, above,

$$\{Z(\lambda) : Z(\lambda,\omega) = Z(I_\lambda,\omega), \quad I_\lambda = (-\infty,\lambda] \quad \lambda \in [-\pi, \pi]\}$$

as a **stochastic point process** with orthogonal increments; for $I = (\lambda_1, \lambda_2]$, $\lambda_1 \le \lambda_2$, the (stochastic) measure of $I$, obeys $Z(I) = Z(I_{\lambda_2},\omega) - Z(I_{\lambda_1},\omega)$; thus, it is the **difference** of the stochastic point process evaluated at $\lambda_1$ and $\lambda_2$. This may be made more palatable if we note that

$$I_{\lambda_2} = (I_{\lambda_2} \cap I_{\lambda_1}) \cup (I_{\lambda_2} \cap \overline{I}_{\lambda_1}) = I_{\lambda_1} \cup I.$$

Since the two sets in the last member above are disjoint we have, by the additive properties of measures,

$$Z(I_{\lambda_2},\omega) = Z(I_{\lambda_1},\omega) + Z(I,\omega), \quad \text{or} \quad Z(I_{\lambda_2},\omega) - Z(I_{\lambda_1},\omega) = Z(I,\omega).$$

Moreover, it may be shown that we may represent the autocovariance function by

$$\kappa(\tau) = \int_{-\pi}^{\pi} e^{i\lambda\tau} F(d\lambda), \quad \text{or alternatively,} \quad \kappa(\tau) = \int_{-\pi}^{\pi} e^{i\lambda\tau} dF(\lambda), \quad (1.6)$$

where $F$ is the spectral measure (or distribution function), i.e. it is a nonnegative nondecreasing function defined over sets $I \in \mathcal{B}([-\pi, \pi])$ and such that **for any simple interval** $I = (\lambda_1, \lambda_2] \in [-\pi, \pi)$, $\lambda_2 \ge \lambda_1$,

$$F(I) = EZ(I)\overline{Z(I)}. \quad (1.7)$$

As was the case with the stochastic process $Z$ above, we may also regard $F$ as a point function,[10] in which case $F$ is defined, by $F(\lambda) = F(I_\lambda)$, $I_\lambda = (-\infty, \lambda]$. When $F$ is viewed as a measure, and $I = (\lambda_1, \lambda_2]$ we obtain $F(I) = F(I_{\lambda_2}) - F(I_{\lambda_1})$.

For notational ease, we employ the alternative representation of the two integrals in Eqs. (1.5) and (1.6). We can also establish, by the continuity of the point process, that

$$EdZ(\lambda)\overline{dZ(\theta)} = dF(\lambda), \quad \text{if} \quad \lambda = \theta,$$

---

[10] This is analogous to the cumulative distribution function.

and zero otherwise.

**Definition 7.** If the (point) function $F$, above, is **absolutely continuous** [11] over $(-\pi, \pi]$, its derivative $f$ is said to be the **spectrum**, or **spectral density function of the stochastic sequence**. Moreover, the spectral density function, $f$, and the autocovariance function, $\kappa(\tau)$, are a pair of Fourier transforms, i.e.

$$\kappa(\tau) = \int_{-\pi}^{\pi} e^{i\lambda\tau} f(\lambda)\, d\lambda, \quad f(\lambda) = \frac{1}{2\pi} \sum_{s=-\infty}^{\infty} e^{-i\lambda s} \kappa(s). \qquad (1.8)$$

An interesting interpretation of the spectrum becomes evident, [12] if we note that

$$\kappa(0) = \int_{-\pi}^{\pi} E\, dZ(\lambda)\overline{dZ(\lambda)} = \int_{-\pi}^{\pi} f(\lambda)\, d\lambda,$$

demonstrating that **the variance of a covariance stationary sequence** is merely the integral of contributions by (the variance of) the random amplitude $dZ$ in Eq. (1.5) corresponding to the sinusoids $e^{i\lambda n}$, or of the random amplitude of $(\Delta Z =)z_s$ in Eqs. (1.4) or (1.2), corresponding to the sinusoids $e^{i\lambda_s n}$.

The situation is entirely analogous if we are dealing with vector sequences; thus, if $X = \{X_{n\cdot} : n \in \mathcal{N}\}$ is a covariance stationary (**row**) **vector** sequence, there exist orthogonal stochastic measures, say $Z(\lambda) = (Z_1(\lambda), Z_2(\lambda), \ldots, Z_q(\lambda))'$, on the measurable space $(\psi, \mathcal{B}(\psi))$, where $\psi = [-\pi, \pi]$, such that for the $j^{th}$ component of $X_{n\cdot}$ we have

$$X_{nj} = \int_{-\pi}^{\pi} e^{i\lambda n} dZ_j(\lambda), \quad j = 1, 2, \ldots, q,$$

or, in vector form

$$X'_{n\cdot} = \int_{-\pi}^{\pi} e^{i\lambda n}\, dZ(\lambda). \qquad (1.9)$$

The interpretation of the (continuous parameter) vector stochastic process $Z(\lambda)$ is the same as in the scalar case, i.e. we may think of it as a (vector) measure $Z(\lambda) = Z(I_\lambda, \omega)$ where $I_\lambda = (-\infty, \lambda]$, as in the previous discussion.

---

[11] Absolute continuity in this context means that $F$ is the (indefinite) integral of its derivative.

[12] Notice the similarity between the spectrum and the autocovariance generating function defined in the previous section, especially when we take $z = e^{i\lambda}$.

For an interval $I = (\lambda_1, \lambda_2]$ with $\lambda_1 \leq \lambda_2$ and $\lambda_1, \lambda_2 \in [-\pi, \pi]$ we have the interpretation

$$Z(I) = Z(I_{\lambda_2}, \omega) - Z(I_{\lambda_1}, \omega). \qquad (1.10)$$

Again, as in the scalar case, we have

$$EdZ(\lambda)\overline{dZ(\theta)}' = dF(\lambda), \quad \text{if} \quad \lambda = \theta, \qquad (1.11)$$

and zero otherwise. The **matrix** $F$ is said to be the matrix of **spectral distribution functions** and, *mutatis mutandis*, has the same meaning and interpretation as given above for the scalar case. When the elements of $F$ are absolutely continuous, the matrix of derivatives

$$\frac{dF}{d\lambda} = \left[ \frac{dF_{ij}(\lambda)}{d\lambda} \right] = f(\lambda) = [f_{ij}(\lambda)], \quad i, j - 1, 2, \ldots, q \qquad (1.12)$$

is said to be the **spectral matrix**; the diagonal elements of such a matrix, namely $f_{ii}(\lambda)$, $i = 1, 2, \ldots q$, are said to be the **spectral density functions** or **spectra** (of the $q$ elements of the vector process $X_n.$); the off diagonal elements $f_{ij}(\lambda)$, for $i \neq j$, are the said to be the **cross spectral densities**, or simply the **cross spectra** of the components of the vector process.

As noted earlier, the elements of the matrix $F$, i.e. the $F_{ij}$ may be thought of as measures, and the **point function** $F_{ij}(\lambda)$ may be interepreted as the value assumed by the measure $F_{ij}$ on the set $I_\lambda = (-\infty, \lambda]$. For a particular set $I = (\lambda_1, \lambda_2]$ with $\lambda_1 \leq \lambda_2$ and $\lambda_1, \lambda_2 \in [-\pi, \pi]$ we have the interpretation $F(I) = F(\lambda_2) - F(\lambda_1)$.

Moreover, we have the $q^2$ autocovariance functions

$$K_{ij}(\tau) = EX_{n+\tau, i}\overline{X}_{nj}, \quad K(\tau) = [K_{ij}(\tau)] = EX'_{n+\tau.}\overline{X}_n. \qquad (1.13)$$

where, when viewed as a function of $\tau$, $K(\tau) = [K_{ij}(\tau)]$ is termed the **autocovariance kernel**. Occasionally, for fixed $\tau$, $K(\tau)$ is said to be the autocovariance matrix of **order** $\tau$, or of **lag** $\tau$. Notice that $K_{jj}(\tau)$ corresponds to the autocovariance function of a scalar process discussed earlier.

Again as in previous discussions, autocovariances and spectral (or cross spectral) densities form pairs of Fourier transforms, i.e.,

$$f_{rs}(\lambda) = \frac{1}{2\pi} \sum_{\tau=-\infty}^{\infty} e^{-i\lambda\tau} K_{rs}(\tau), \quad K_{rs}(\tau) = \int_{-\pi}^{\pi} e^{i\lambda\tau} f_{rs}(\lambda) \, d\lambda, \qquad (1.14)$$

where $r, s = 1, 2, 3, \ldots, q$, or in matrix notation

$$f(\lambda) = \frac{1}{2\pi} \sum_{\tau=-\infty}^{\infty} e^{-i\lambda\tau} K(\tau), \quad K(\tau) = \int_{-\pi}^{\pi} e^{i\lambda\tau} f(\lambda) \, d\lambda. \qquad (1.15)$$

Further discussion of the spectral representation and related aspects of vector processes will take us too far afield.

## 1.4 Taxonomy of Stationary Sequences

### 1.4.1 Scalar Sequences

In econometrics, certain types of sequences occur frequently, and it is useful, at this stage, to review them and their properties.

**Definition 8.** Let $\epsilon = \{\epsilon_t : t \in \mathcal{N}\}$ be a sequence of zero mean, finite variance, mutually uncorrelated, random variables defined on the probability space $(\Omega, \mathcal{A}, \mathcal{P})$. Such a sequence is generally termed **white noise**. [13]

This definition is useful when dealing with **covariance stationary** as distinct from **strictly stationary** processes; when dealing with the latter, and indeed in most of this volume, we shall define **white noise** to be a **sequence of independent identically distributed (i.i.d.)**, zero mean, finite variance random variables.

**Definition 9.** Let $\epsilon$ be a white noise sequence as in Definition 8, and define

$$u = \{u_n : u_n = \sum_{j=-\infty}^{\infty} a_j \epsilon_{n-j}, \quad n \in \mathcal{N}\}.$$

The sequence $u$ is said to be a (bilateral or two-sided) moving average (MA) process of infinite extent, denoted by $MA(\infty)$; if $a_j = 0$, $j < 0$, it is said to be a (unilateral or one-sided) moving average of infinite extent; if $a_j = 0$, $j > n$ it is said to be a moving average **of order** $n$, and is denoted by $MA(n)$.

---

[13] Often the additional requirement that the variance is unity is added to the definition of white noise.

**Remark 7**. Notice that for the MA process we cannot leave both the series coefficients and the variance of the $\epsilon$-sequence unrestricted. It is **common in econometrics to impose the normalization convention** $a_0 = 1$.

Notice further, in this as well as in later examples, that if white noise is defined in terms of mutually uncorrelated (orthogonal) random variables, the MA process is weakly stationary. If, on the other hand, it is defined in terms of i.i.d. random variables, the resulting MA process is **strictly stationary**.

**Definition 10**. Let $\epsilon$ be a white noise sequence as in Definition 8, and define

$$u = \{u_n : \sum_{j=-\infty}^{\infty} b_j u_{n-j} = \epsilon_n, \quad n \in \mathcal{N}\}.$$

The sequence $u$ is said to be a (bilateral) **autoregression** (AR) of infinite order and is denoted by $AR(\infty)$. If $b_j = 0, j < 0$, it is said to be a unilateral autoregression of infinite order; if $b_j = 0$ for $|j| > m$, it is said to be a finite autoregression of order $m$ and is denoted by $AR(m)$.

**Definition 11**. Let $\epsilon$ be a white noise sequence as in Definition 8, and define

$$u = \{u_n : \sum_{j=-\infty}^{\infty} b_j u_{n-j} = \sum_{s=-\infty}^{\infty} a_s \epsilon_{n-s}, \quad n \in \mathcal{N}\};$$

$u$ is said to be a bilateral **autoregressive moving average (ARMA) process of infinite order**, and is denoted by $ARMA(\infty, \infty)$. If $b_j = 0$, $j < 0$ and $a_s = 0$, $s < 0$, it is said to be a unilateral process of infinite order. If $b_j = 0$, $|j| > m$ and $a_s = 0$, $|s| > n$, it is said to be a finite autoregression-moving average of order $(m, n)$ and is denoted by $ARMA(m, n)$.

Even though the representation of processes in bilateral terms simplifies the expositional aspects of our discussion, the reader should note that such processes are **physically unrealizable**; in our universe "time" has a definite direction and it is not possible for a phenomenon at "time" $t$ to be affected by what happens at "time" $t + \tau$, $\tau > 0$. It may be affected, at least in economics, by what one might "expect" to happen at time $t + \tau$, but it **cannot** be affected by what **actually** happens at that time since what actually does happen at time $t + \tau$ is **not knowable** at time $t$. In all succeeding applications we shall deal with **physically realizable** systems, and thus only with **unilateral** $AR$ or $MA$ or $ARMA$ processes.

In the discussion above, we have dealt with entities that involve the summation of a countably infinite number of terms. The question, then, must arise as to whether such series converge, and if so in what mode. We address these issues next.

**Proposition 3.** Let $\xi = \{\xi_n. : n \in \mathcal{N}\}$ be a (vector) sequence such that $\xi'_{n.} = \sum_{j=-\infty}^{\infty} A_j X'_{n-j.}$, where $X = \{X_n. : n \in \mathcal{N}\}$ is a sequence of random vectors, or elements, defined on the probability space $(\Omega, \mathcal{A}, \mathcal{P})$ and such that $\sum_{j=-\infty}^{\infty} \| A_j \| < \infty$; the following statements are true:

i. if, in addition, $\sup_{n \in \mathcal{N}} E|X_n.| \leq m_1 < \infty$, the series converges absolutely to an a.c. finite r.v. with probability one;

ii. if, in addition, $\sup_{n \in \mathcal{N}} E|X_n.|^2 \leq m_2 < \infty$, the convergence in question is in quadratic mean ($L^2$).

Proof: Since $\{f_N : f_N(\omega) = \sum_{j=-N}^{N} \| A_j \| |X'_{n-j.}(\omega)|, N \in \mathcal{N}_+\}$ is a monotone nondecreasing sequence, it converges to $\psi_n = \sum_{j=-\infty}^{\infty} \| A_j \| |X'_{n-j.}|$ almost everywhere on $\Omega$, and the only question is whether it is a.c. finite. By the monotone convergence theorem [see Dhrymes (1989), p.44]

$$E\psi_n = \lim_{N \to \infty} E \sum_{j=-N}^{N} \| A_j \| |X'_{n-j.}|$$

$$\leq \lim_{N \to \infty} \sum_{j=-N}^{N} \| A_j \| \sup_{j \in \mathcal{N}} E|X'_{n-j.}| \leq m_1 \sum_{j=-\infty}^{\infty} \| A_j \| < \infty;$$

Since $|E\xi'_{n.}| \leq E\psi_n$, it follows immediately that $\psi_n$ and hence $\xi_n.$ are a.c. finite. For suppose not; then there exists a set $A = \{\omega : |\psi_n(\omega)| = \infty\}$ for which $\mathcal{P}(A) > 0$. Thus, with $\overline{A}$ the complement of $A$

$$E\psi_n = \int_{\overline{A}} \psi_n \, d\mathcal{P} + \int_A \psi_n \, d\mathcal{P} \geq \int_A \psi_n \, d\mathcal{P} = \infty,$$

which is a contradiction, thus completing the proof of i.

To prove ii we note that

$$0 \leq E \left\| \xi_n. - \sum_{j=-N}^{N} A_j X'_{n-j.} \right\|^2 \leq \sum_{j>|N|} \| A_j \|^2 E|X'_{n-j.}|^2$$

$$\leq m_2 \sum_{j>|N|} \| A_j \|^2 \leq m_2 \left( \sum_{j>|N|} \| A_j \| \right)^2 \to 0,$$

which completes the proof of $L^2$ convergence.

q.e.d

Another very important issue is whether, and under what circumstances, a given relationship may be inverted. For example, when can a $MA(n)$ be represented as an $AR(\infty)$; when does an $AR(m)$ have a representation as a $MA(\infty)$? The answer to all these questions is intimately related to the spectral density of the underlying sequence, as the following proposition makes clear.

**Proposition 4.** Let $X = \{X_n : n \in \mathcal{N}\}$ be a covariance stationary sequence in $\mathcal{H}^2(\mathcal{P})$, and $f_x(\lambda)$ its spectral density function; if $f_x > 0$ a.e. [almost everywhere with respect to Lebesgue measure, defined on $\mathcal{B}([-\pi, \pi])$], there exists a linear time invariant filter, $c = \{c_j : j \in \mathcal{N}\}$ and a white noise sequence, $WN(1)$, such that

$$X_n = \sum_{j=-\infty}^{\infty} c_j \epsilon_{n-j}, \qquad (1.16)$$

i.e. **every covariance stationary sequence with positive spectral density a.e. has a $MA(\infty)$ representation**.

Proof: Since $X$ is covariance stationary, there exists a stochastic measure, $Z$, such that the elements of the sequence can be represented as

$$X_n = \int_{-\pi}^{\pi} e^{i\lambda n}\, dZ_x, \quad EZ_x(I)\overline{Z}_x(I) = F_x(I),$$

where $I \in \mathcal{B}([-\pi, \pi])$, and $F_x$ is the spectral measure corresponding to the $X$-process. The spectral density, $f_x$, is the derivative of $F_x$ viewed as a point function on $[-\pi, \pi]$. Since, evidently, $\int_{-\pi}^{\pi} f_x(\lambda)\, d\lambda < \infty$ and moreover, $f_x > 0$ a.e., there exists a square integrable function $\phi(\lambda) \neq 0$ a.e. such that $f_x(\lambda) = (1/2\pi)|\phi(\lambda)|^2$. From the theory of Fourier series we have

$$\phi(\lambda) = \sum_{j=-\infty}^{\infty} c_j e^{-ij\lambda}, \quad c_j = \frac{1}{2\pi}\int_{-\pi}^{\pi} e^{ij\lambda}\phi(\lambda)\, d\lambda, \quad c_{-j} = \overline{c}_j, \qquad (1.17)$$

in the sense of $L^2$ convergence, i.e.

$$\int_{-\pi}^{\pi} \left| \phi(\lambda) - \frac{1}{2\pi}\sum_{j=-N}^{N} c_j e^{-ij\lambda} \right|^2 d\lambda \to \infty.$$

Define the function $\phi^*(\lambda) = [1/\phi(\lambda)]$, which is square integrable and well defined a.e., and the stochastic measure

$$Z^*(I) = \int_I \phi^*(\lambda) \, dZ_x(\lambda), \quad I \in \mathcal{B}([-\pi, \pi]). \tag{1.18}$$

For the particular interval $I = (a, b]$, $b \geq a$, we find

$$F^*(I) = EZ^*(I)\overline{Z}^*(I) = \frac{1}{2\pi} \int_a^b |\phi^*(\lambda)|^2 \, |\phi(\lambda)|^2 \, d\lambda = \frac{b-a}{2\pi}, \tag{1.19}$$

or, when viewed as a point process, $F^*(\lambda) = (\lambda/2\pi)$, $\lambda \in (-\pi, \pi]$. This is an absolutely continuous function with derivative $f(\lambda) = (1/2\pi)$, which implies that the sequence whose elements are defined by the spectral representation

$$\epsilon_t = \int_{-\pi}^{\pi} e^{it\lambda} \, dZ^*(\lambda), \tag{1.20}$$

is a white noise process with unit variance; this is so since

$$E\epsilon_{t+\tau}\overline{\epsilon}_t = \int_{-\pi}^{\pi} \int_{-\pi}^{\pi} e^{i(t+\tau)\lambda} e^{-it\theta} E dZ^*(\lambda)\overline{dZ^*(\theta)}$$

$$= \int_{-\pi}^{\pi} e^{i\tau\lambda} |\phi^*(\lambda)|^2 f_x(\lambda) d\lambda = \frac{1}{2\pi} \int_{-\pi}^{\pi} e^{i\tau\lambda} d\lambda = 1, \quad \text{if } \tau = 0,$$

and zero otherwise. Moreover $dZ_x = \phi(\lambda)dZ^*$, so that

$$X_n = \int_{-\pi}^{\pi} e^{i\lambda n} \, dZ_x(\lambda) = \int_{-\pi}^{\pi} e^{i\lambda n} \phi(\lambda) \, dZ^*(\lambda) \tag{1.21}$$

$$= \sum_{j=-\infty}^{\infty} c_j \int_{-\pi}^{\pi} e^{i(n-j)\lambda} \, dZ^*(\lambda) = \sum_{j=-\infty}^{\infty} c_j \epsilon_{n-j}.$$

q.e.d.

**Corollary 1.** If $\phi$ has unilateral Fourier series, i.e. if

$$\phi(\lambda) = \sum_{j=0}^{\infty} c_j e^{-ij\lambda},$$

$X$ has the representation

$$X_n = \sum_{j=0}^{\infty} c_j \epsilon_{t-j}.$$

Proof: Evident from Proposition 4.

**Remark 8.** When is this Corollary useful? If we wish to model real world economic processes, we would certainly reject processes that are **not** physically realizable. Thus, we would want to deal with covariance stationary processes whose spectra are positive a.e.; of course, this is not the only option. First, the reader ought to realize that the required representation need not be based mechanically on the sequence of Fourier coefficients. For example, we may take the position that the process has a **rational** spectrum, i.e.

$$f_x(\lambda) = \frac{1}{2\pi} |\phi(\lambda)|^2 = \frac{1}{2\pi} \frac{|A(\lambda)|^2}{|B(\lambda)|^2},$$

where $A(\lambda) = \sum_{j=0}^m a_j e^{-ij\lambda}$, $B(\lambda) = \sum_{s=0}^n b_s e^{-is\lambda}$, and $a_0 = b_0 = 1$. **If the roots of** $\sum_{s=0}^n b_s z^s = B(z) = 0$ **all obey** $|z_j| > 1$, by the fundamental theorem of algebra we can write (as was the case with lag operator polynomials)

$$B(z) = \prod_{j=1}^n (1 - \nu_j z), \quad \nu_j = \frac{1}{z_j}, \quad |\nu_j| < 1.$$

Consequently,

$$\frac{1}{B(z)} = \sum_{j=0}^\infty \gamma_j z^j, \quad \frac{A(z)}{B(z)} = \sum_{k=0}^\infty c_k z^k, \quad c_k = \sum_{s=0}^k \gamma_{k-s} a_s.$$

From the construction above, and Corollary 1, it follows that

$$X_n = \sum_{k=0}^\infty c_k \epsilon_{n-k}.$$

Stochastic sequences that can be represented (uniquely) as unilateral $MA(\infty)$, as above, are said tp be **causal** in the literature of statistics.
**Remark 9.** In Remark 8 we had intimated that all covariance stationary processes with a.e. positive rational spectral density, have a representation as a unilateral $MA(\infty)$. Notice that, although this is a **very large class** of functions, it has the property that the $MA(\infty)$ process in question contains only $n + m$ free parameters! Hence, it cannot represent the class of all spectral densities that are positive a.e. On the other hand, given any such spectral density, we can approximate it by a **rational spectral density** as

closely as we desire, by taking $n$ and $m$ sufficiently large. This may be done as follows: Suppose

$$f_x(\lambda) = \frac{1}{2\pi} \left| \sum_{j=0}^{\infty} c_j e^{-ij\lambda} \right|^2 = \frac{1}{2\pi} |\phi(\lambda)|^2,$$

and we wish to approximate

$$\phi(\lambda) \sim \frac{A(\lambda)}{B(\lambda)}.$$

If equality held, we would have the relation $B(\lambda)\phi(\lambda) = A(\lambda)$. Equating coefficients we obtain

$$\sum_{s=0}^{k} c_{k-s} b_s = a_k, \quad k = 0, 1, 2, 3, \ldots, n \tag{1.22}$$

$$= 0, \quad \text{otherwise.}$$

Notice that this is a recursive system that can be used to obtain the coefficient sequence $\{c_k : k \geq 0\}$, **if the polynomials** $A(\lambda), B(\lambda)$ **are given**; or, conversely, to obtain the polynomials $A(\lambda)$, $B(\lambda)$ for given sequence $\{c_k : k \geq 0\}$, although in the latter case not always uniquely so. To this end note that, since by convention $c_0 = a_0 = b_0 = 1$, we can write

$$c_k = a_k - \sum_{s=1}^{k} c_{k-s} b_s, \tag{1.23}$$

and it is evident that the coefficients $b_s$ are well behaved.[14] Consequently, we may compute as many $c_k$ as we wish; such $c_k$, however many, are functions **only** of the $n$ $b_s's$ and the $m$ $a_j's$.

Conversely, suppose the sequence $\{c_k : k \geq 0\}$ is given, and it is desired to find a rational polynomial approximation. For this case, assuming that $b_k = 0$, $k > n$, we can write the recursion above as

$$b_k = a_k - \sum_{s=1}^{k} c_{k-s} b_s, \quad 0 \leq k \leq m,$$

---

[14] This is the equivalent of requiring the roots of $|B(z)| = 0$ to be greater than unity in modulus.

$$b_k = -\sum_{s=1}^{k} c_{k-s} b_s, \quad m < k \leq n,$$

$$b_k = -\sum_{s=1}^{k} c_{k-s} b_s = 0, \quad k > n. \tag{1.24}$$

Of particular interest is the case where we impose the condition $a_k = 0$, $k \geq 1$, i.e., we wish to represent an $MA(\infty)$ by an $AR(n)$. Evidently, the restrictions imposed by the last equation above need not hold, **so that not all** $MA(\infty)$ **processes are representable** as finite autoregressions; the converse, however, is true in the sense that all **stable** finite autoregressions are representable as $MA(\infty)$. Despite these facts, we can certainly investigate whether it is possible to represent a $MA(\infty)$ as an $AR(\infty)$. To do so we consider again the equations above, but we **eliminate the last equation** having altered the second equation to read $k > m$. Obtaining a few of the coefficients we find

$$b_0 = 1, \quad b_2 = -(c_2 - c_1^2), \quad b_3 = -(c_3 - 2c_1 c_2 + c_1^2)$$

$$b_4 = -(c_4 - 2c_1 c_3 - c_2^2 + 3c_1^2 c_2 - c_1^4), \quad \text{etc.}$$

For the $AR(\infty)$ to have meaning, we require $\sum_{j=0}^{\infty} |b_j| < \infty$, which implies certain restrictions on the sequence $\{c_k : k \geq 0\}$. Thus, with suitable restrictions, an $MA(\infty)$ process is representable as an $AR(\infty)$ process. A case in point is one in which the spectral density is positive a.e., and it has the representation

$$f_x(\lambda) = \frac{1}{2\pi} |\phi(\lambda)|^2, \quad \phi(z) = \frac{A(z)}{B(z)}, \quad z = e^{-i\lambda},$$

where $A(z), B(z)$ are **finite-order polynomials** of order $n, m$ respectively, all of whose roots, $z_j$, **obey** $|z_j| > 1$. We give below a few examples to illustrate the applications of Proposition 4.

**Example 1.** Consider the $WN(\sigma^2)$ (white noise) sequence $\epsilon$. We have the representation

$$\epsilon_t = \int_{-\pi}^{\pi} e^{it\lambda} \, dZ_\epsilon,$$

where $Z_\epsilon$ is an appropriate stochastic measure for the white noise sequence. The autocovariance function evidently obeys $\kappa(\tau) = \sigma^2$, for $\tau = 0$ and zero otherwise. Thus, the spectral density function is, by Eq. (1.8),

$$f_\epsilon(\lambda) = \frac{1}{2\pi} \sum_{s=-\infty}^{\infty} \kappa(s) e^{-is\lambda} = \frac{\sigma^2}{2\pi}.$$

We see therefore that the white noise process has a **constant or uniform spectral density**. Moreover,

$$EdZ_\epsilon(\lambda)\overline{dZ_\epsilon(\theta)} = \frac{\sigma^2}{2\pi}d\lambda, \quad \text{if } \lambda = \theta \ ,$$

and zero otherwise.

**Example 2.** Consider the bilateral $MA(\infty)$ process:

$$u_t = \sum_{j=-\infty}^{\infty} a_j \epsilon_{t-j}.$$

First, we note that this **cannot represent a physically realizable phenomenon** since the values assumed by $u$ at "time" $t$ depend on values to be assumed by the white noise process in the "remote future". This suggests that for physically realizable processes we should deal only with the **unilateral** $MA(\infty)$ process. The spectral representation of such a process is easily obtained as

$$u_t = \sum_{s=0}^{\infty} a_s \int_{-\pi}^{\pi} e^{i(t-s)\lambda} dZ_\epsilon = \int_{-\pi}^{\pi} e^{it\lambda} dZ_u, \quad dZ_u = \phi_u(\lambda)dZ_\epsilon \qquad (1.25)$$

and $\phi_u(\lambda) = \sum_{s=0}^{\infty} a_s e^{-is\lambda}$. [15] It follows immediately that the spectral density of the $MA(\infty)$ process is given by

$$f_u(\lambda) = \frac{\sigma^2}{2\pi}|\phi_u(\lambda)|^2 = \frac{\sigma^2}{2\pi}\left|\sum_{s=0}^{\infty} a_s e^{-is\lambda}\right|^2 = \frac{\sigma^2}{2\pi}|A(\lambda)|^2.$$

**Example 3.** Let $u$ be an $AR(m)$, as given in Definition 10. To determine the nature of its spectral representation, let $Z_u$ be the relevant stochastic measure, and set

$$\epsilon_t = \int_{-\pi}^{\pi} e^{it\lambda} dZ_\epsilon = \int_{-\pi}^{\pi} e^{i\lambda t} \left(\sum_{j=0}^{m} b_j e^{-ij\lambda}\right) dZ_u = \sum_{j=0}^{m} b_j u_{t-j}, \quad b_0 = 1. \quad (1.26)$$

---

[15] Note that $\phi_u$ is simply **the Fourier transform of the sequence**, $\{a_s : s \in \mathcal{N}_0\}$.

Since this is an identity, we must have

$$dZ_\epsilon = \phi_u(\lambda)dZ_u, \quad \phi_u(\lambda) = \sum_{j=0}^{m} b_j e^{-ij\lambda} = B(\lambda). \qquad (1.27)$$

It follows, therefore, that

$$f_u(\lambda) = \frac{\sigma^2}{2\pi} \left| \sum_{j=0}^{m} b_j e^{-ij\lambda} \right|^2 = \frac{\sigma^2}{2\pi} |B(\lambda)|^2. \qquad (1.28)$$

**Example 4.** Let $u$ be an $ARMA(m, n)$ process and set

$$\sum_{j=0}^{m} b_j u_{t-j} = \int_{-\pi}^{\pi} e^{it\lambda} B(\lambda) \, dZ_u = \int_{-\pi}^{\pi} e^{it\lambda} A(\lambda) \, dZ_\epsilon = \sum_{s=0}^{n} a_s \epsilon_{t-s}. \qquad (1.29)$$

It follows immediately that

$$B(\lambda)dZ_u = A(\lambda)dZ_\epsilon \qquad (1.30)$$

and, moreover, that the spectral density of the ARMA process is given by

$$f_u(\lambda) = \frac{\sigma^2}{2\pi} \frac{|A(\lambda)|^2}{|B(\lambda)|^2}. \qquad (1.31)$$

**Remark 10.** Notice that, as required, $f \geq 0$ in all four examples above; however, in Examples 3 and 4 there is a possiblity that the denominator obeys $|B(\lambda)|^2 = 0$, for some $\lambda \in [-\pi, \pi]$. When this is so, the spectral density is not well defined, and the finite-order autoregression cannot have a unilateral $MA(\infty)$ representation. This is made quite clear if one employs the lag operator representation

$$b(L)u_t = \epsilon_t, \quad \text{or} \quad u_t = \frac{I}{b(L)}\epsilon_t, \quad b(L) = \sum_{j=0}^{n} b_j L^j,$$

and refers to the discussion of such matters in Dhrymes (1971), Dhrymes (1984), or Dhrymes (1994b). It is shown therein that the algebra of polynomials in the lag operator $L$ is **isomorphic** to the algebra of polynomials in a real or complex indeterminate. This means that if $L$ is replaced, say, by the complex indeterminate, $z$, the desired operations are performed and then $z$ and its powers are replaced by $L$ and its powers, the result would be

precisely what is appropriate. Thus, for example, consider $B(z) = 1 - b_1 z$.
Its only root is $z_1 = (1/b_1)$; if $|z_1| > 1$, **we must require** $|b_1| < 1$. Thus,

$$\frac{1}{1 - b_1 z} = \sum_{j=0}^{\infty} b_1^j z^j, \quad \text{hence} \quad \frac{I}{I - b_1 L} = \sum_{j=0}^{\infty} b_1^j L^j.$$

Consequently, if $n = 1$ and $z_1$ is the root of $B(z) = 0$,

$$\frac{I}{b(L)} \epsilon_t = \sum_{j=0}^{\infty} b_1^j \epsilon_{t-j}.$$

If $|z_1| > 1$ then $|b_1| < 1$, and the expansion above converges absolutely, since $\sum_{j=0}^{\infty} |b_1|^j < \infty$; moreover, it converges in quadratic mean as well, since $\sum_{j=0}^{\infty} |b_1|^{2j} < \infty$. A similar result holds, by the same reasoning, for general $n$; precisely, we require that **all** roots of the characteristic equation $B(z) = 0$ **be outside the unit circle.** When some roots lie **on the unit circle**, i.e. when for some $s$, $|z_s| = 1$, we have a particular form of **nonstationarity**, referred to as the **unit root problem**, which we examine at a later stage.

## 1.4.2 Multivariate Sequences

In most econometric applications we deal with random vectors (elements), rather than random variables. In this section we shall modify the definitions and representations above to the case of sequences of random elements, particularly multivariate sequences. All references in this section are to $q$-element random vectors.

We begin by extending the result in Proposition 4 to multivariate (vector) covariance stationary processes, as follows:

**Proposition 5.** Let $X = \{X_{n\cdot} : n \in \mathcal{N}\}$ be a vector covariance stationary sequence in $\mathcal{H}^2(\mathcal{P})$, and $f_x(\lambda)$ its spectral matrix; if $f_x > 0$ a.e.,[16] in the sense that the spectral matrix is positive definite a.e., there exists a linear time invariant filter, $C = \{C_j : j \in \mathcal{N}\}$ and a (multivariate) white noise sequence, $MWN(I_q)$, such that

$$X'_{n\cdot} = \sum_{j=-\infty}^{\infty} C_j \epsilon'_{n-j\cdot}, \tag{1.32}$$

---

[16] This notation means almost everywhere with respect to Lebesgue measure, defined on the Borel $\sigma$-algebra $\mathcal{B}(\psi)$, where $\psi = [-\pi, \pi]$.

i.e. **every covariance stationary sequence with positive definite spectral matrix a.e. has a $MMA(\infty)$ representation**.

Proof: Since $X$ is covariance stationary, there exists a (vector) stochastic measure, $Z_x$, such that the elements of the sequence can be represented as

$$X'_{n\cdot} = \int_{-\pi}^{\pi} e^{i\lambda n} dZ_x(\lambda), \quad EZ_x(I)\overline{Z}_x(I) = F_x(I),$$

where $I \in \mathcal{B}(\psi)$; $F_x(\lambda)$ is the matrix of spectral and cross spectral distribution measures, corresponding to the $X$-process, evaluated at the set $I_\lambda = (-\infty, \lambda]$. The spectral matrix, $f_x$, is the derivative of $F_x$, viewed as a point function on $\psi$, provided it ($F_x$) is absolutely continuous over $\psi$. Since $f_x > 0$ a.e. its characteristic roots are real and positive a.e., and there exists a unitary matrix $Q(\lambda)$ such that

$$f_x(\lambda) = \frac{1}{2\pi} Q(\lambda)D(\lambda)\overline{Q}(\lambda)' \tag{1.33}$$
$$D = \mathrm{diag}[r_1(\lambda), r_2(\lambda), \ldots, r_q(\lambda)], \quad r_i(\lambda) > 0, \text{ a.e.}$$

Evidently, $Q$ is the unitary matrix of the characteristic vectors of $f_x$ and $D$ is the **diagonal** matrix of the **real** and **positive** corresponding characteristic roots. Define now

$$A(\lambda) = QD^{(1/2)}, \quad \text{and note that} \quad A^{-1}(\lambda) = D^{-(1/2)}\overline{Q}'; \tag{1.34}$$

consequently,

$$A^{-1}(\lambda)f_x(\lambda)\overline{A}'^{-1} = \frac{1}{2\pi}D^{-(1/2)}\overline{Q}'QD\overline{Q}'QD^{-(1/2)} = \frac{1}{2\pi}I_q. \tag{1.35}$$

From the theory of Fourier series we have that

$$A(\lambda) = \sum_{j=-\infty}^{\infty} C_j e^{-ij\lambda}, \quad C_j = \frac{1}{2\pi}\int_{-\pi}^{\pi} e^{ij\lambda}A(\lambda)\,d\lambda, \quad j = 0, \pm1, \pm2, \pm3, \ldots$$
$$\tag{1.36}$$

in the sense of $L^2$ convergence, i.e,

$$\int_{-\pi}^{\pi} \left\| A(\lambda) - \sum_{j=-N}^{N} C_j e^{-ij\lambda} \right\|^2 d\lambda \to \infty.$$

Define now the stochastic measure $Z^*$ by

$$Z^*(I) = \int_I A^{-1}(\lambda)\,dZ_x(\lambda), \quad \text{for all } I \in \mathcal{B}(\psi), \tag{1.37}$$

and note that

$$\epsilon'_{t.} = \int_{-\pi}^{\pi} e^{i\lambda t} dZ^*(\lambda), \quad t = 0, \pm 1, \pm 2, \ldots \tag{1.38}$$

is a $MNW(I_q)$ process, owing to the fact that

$$E\epsilon'_{t+\tau.}\bar{\epsilon}_{t.} = \int\int e^{i\lambda(t+\tau)} e^{-i\theta t} E dZ^*(\lambda)\overline{dZ^*(\theta)} \tag{1.39}$$

$$= \frac{1}{2\pi} \int_{-\pi}^{\pi} e^{i\lambda\tau} I_q d\lambda = I_q, \quad \text{if } \tau = 0, \tag{1.40}$$

and zero otherwise. In view of Eq. (1.38), this becomes quite evident from the fact that $dZ^*(\lambda) = A^{-1}(\lambda) dZ_x(\lambda)$, and hence

$$E dZ^*(\lambda)\overline{dZ^*(\theta)} = A^{-1}(\lambda) f_x(\lambda)\overline{A}'^{-1} d\lambda = \frac{1}{2\pi} I_q d\lambda$$

if $\lambda = \theta$ and zero otherwise. Using Eq. (1.39) we have the representation

$$X'_{t.} = \int_{-\pi}^{\pi} e^{i\lambda t} dZ_x(\lambda) = \int_{-\pi}^{\pi} e^{i\lambda t} A(\lambda) dZ^*(\lambda)$$

$$= \sum_{j=-\infty}^{\infty} C_j \int_{-\pi}^{\pi} e^{i\lambda(t-j)} dZ^*(\lambda) = \sum_{j=-\infty}^{\infty} C_j \epsilon'_{t-j.} \tag{1.41}$$

q.e.d.

Similarly, the definitions for various classes of covariance stationary processes are given below for vectors.

**Definition 8a.** Let $\epsilon = \{\epsilon'_{t.} : t \in \mathcal{N}]\}$ be a sequence of i.i.d. random vectors, with mean zero and positive definite covariance matrix, $\Sigma$, defined on the probability space $(\Omega, \mathcal{A}, \mathcal{P})$. Such a sequence is termed **multivariate white noise** (MWN) and is denoted by $MWN(\Sigma)$.

**Definition 9a.** Let $\epsilon$ be a $MWN(\Sigma)$ sequence as in Definition 8a, and define

$$u = \{u'_t : u'_{t.} = \sum_{j=-\infty}^{\infty} A_j \epsilon'_{t-j.}, \quad t \in \mathcal{N}\}.$$

The sequence $u$ is said to be a (bilateral or two sided) multivariate moving average process of infinite extent, $MMA(\infty)$; if $A_j = 0$, $j < 0$, it is said to be a (unilateral or one sided) $MMA(\infty)$; if, in addition $A_j = 0$, $j > n$, it is said to be a MMA **of order** $n$, denoted by $MMA(n)$, where the $A_j$ are **square matrices** of order $q$.

**Remark 7a.** Notice that for the $MMA(\infty)$ process we cannot leave both the series coefficients and the covariance matrix of the $\epsilon$-sequence unrestricted. It is **common in econometrics to impose the normalization convention** $A_0 = I$. Notice further that for the series to be defined, we need to impose the restriction (at least) of the form $\sum_{j=-\infty}^{\infty} ||A_j|| < \infty$, where the norm may be taken to be the **square root of the largest characteristic root of** $A_j'A_j$ or, alternatively, the square root of its trace. An adaptation of the result of Proposition 3 to the multivariate case yields, *mutatis mutandis*, the same results so that in this case too we have convergence with probability one to an a.c. finite random vector. Note further that when second moments exist, as in the case of scalar covariance stationary processes, the result above implies that the series converges in $L^2$ mode as well.

**Definition 10a.** Let $\epsilon$ be a MWN sequence as in Definition 8a, and define

$$u = \{u_{t\cdot}' : \sum_{j=-\infty}^{\infty} B_j u_{t-j\cdot}' = \epsilon_{t\cdot}', \quad t \in \mathcal{N}\}.$$

The sequence $u$ is said to be a (bilateral) **multivariate autoregression of infinite order**, and is denoted by $MAR(\infty)$. If $B_j = 0$, for $j < 0$ the sequence is said to be a unilateral (or physically realizable) $MAR(\infty)$; if $B_j = 0$ for $|j| > N$ the sequence is said to be a (finite) multivariate autoregression of order $n$ and is denoted by $MAR(n)$.

**Definition 11a.** Let $\epsilon$ be a MWN sequence as in Definition 8a, and define

$$u = \{u_{t\cdot}' : \sum_{j=-\infty}^{\infty} B_j u_{t-j\cdot}' = \sum_{s=-\infty}^{\infty} A_s \epsilon_{t-s\cdot}', \quad t \in \mathcal{N}\};$$

$u$ is said to be a (bilateral) **multivariate autoregressive moving average process of infinite order**, denoted by $MARMA(\infty, \infty)$.

If $B_j = 0$, for $j < 0$ **and** $A_s = 0$, for $s < 0$, the sequence is said to be a unilateral (or physically realizable) $MARMA(\infty, \infty)$; finally, if $B_j = 0$, for $|j| > n$ **and** $A_s = 0$, for $|s| > m$, the sequence is said to be a (finite) autoregressive moving average process of order $(n, m)$, and denoted by $MARMA(n, m)$.[17]

---

[17] Note the notational convention: The order of the autoregression appears first; thus $MARMA(m, n)$ **does not mean** the same thing as $MARMA(n, m)$.

**Remark 11.** As in the scalar case, the comments about physically realizable processes apply in the multivariate case as well; moreover, the processes described in Definitions 8a and 9a have a representation as $MMA(\infty)$ processes,[18] if the roots of the characteristic equation of the system $|B(z)| = 0$, **lie outside the unit circle**, where

$$B(z) = I_q + B_1 z + B_2 z^2 + B_3 z^3 + \ldots + B_m z^m \tag{1.42}$$

$z$ is the complex indeterminate, and $|B(z)|$ is the **determinant of the matrix** $B(z)$. Letting $I$ be the identity operator, if the roots above lie **outside the unit circle**, i.e. if the roots $z_j$ obey $|z_j| > 1$ then

$$b(L) = |B(L)| = \prod_{j=1}^{mq} (I - \nu_j L), \quad \nu_j = (1/z_j). \tag{1.43}$$

Moreover,

$$u'_{t\cdot} = \frac{H(L)}{b(L)} \epsilon_{t\cdot} \tag{1.44}$$

for the $MAR(n)$ process, and

$$u'_{t\cdot} = \frac{H(L)A(L)}{b(L)} \epsilon'_{t\cdot} \tag{1.45}$$

in the case of the $MARMA(m, n)$ process, where $H(L)$ is the adjoint of the matrix $B(L)$, $L$ being the usual **lag operator**.

We should point out that the same argument shows that $MMA(m)$, as well as $MARMA(m, n)$, can be expressed as $AR(\infty)$. Thus, consider the $MA(m)$ process

$$u'_{t\cdot} = \sum_{j=0}^{m} A_j \epsilon'_{t-j\cdot} = A(L)\epsilon'_{t\cdot}, \quad \text{such that} \quad A(L) = \prod_{j=1}^{m} (I - \rho_j L), \quad |\rho_j| < 1.$$

The matrix operator $A(L)$ is therefore invertible and its inverse is given by

$$A(L)^{-1} = \frac{K(L)}{a(L)}, \quad \text{where } K(L) \text{ is the adjoint of } A(L) \text{ and } a(L) = |A(L)|.$$

Hence,

$$\epsilon'_{t\cdot} = \frac{K(L)}{a(L)} u'_{t\cdot} = \sum_{s=0}^{\infty} K_s u'_{t-s\cdot},$$

---

[18] The argument establishing this result is not trivial; the reader interested in details is referred to Dhrymes (1984), Ch. 5, secs. 2 and 3.

displaying the process $u$ as a $MAR(\infty)$. Similarly, in the $MARMA(n, m)$

$$\sum_{j=0}^{n} B_j u'_{t-j\cdot} = \sum_{s=0}^{m} C_s \epsilon'_{t-s\cdot}, \quad \text{or} \quad B(L) u'_{t\cdot} = C(L) \epsilon'_{t\cdot},$$

$$B(L) = \sum_{j=0}^{n} B_j L^j, \quad C(L) = \sum_{s=0}^{m} C_s L^s,$$

if the characteristic roots of $|C(z)| = 0$ are all outside the unit circle we obtain formally

$$C(L)^{-1} B(L) u'_{t\cdot} = \epsilon'_{t\cdot}, \quad \text{or} \quad \sum_{k=0}^{\infty} D_k u'_{t-k\cdot} = \epsilon'_{t\cdot},$$

which displays the $MARMA(n, m)$ process $u$ as a $MAR(\infty)$ process. This leads to the next definition.

**Definition 12.** A matrix polynomial operator, $A(L) = \sum_{j=0}^{m} A_j L^j$, **all** of whose characteristic roots, i.e. the roots of the determinantal polynomial equation $0 = a(z) = |A(z)|$, lie **outside the unit circle** is said to be **invertible**. Its inverse operator has the formal representation

$$A(L)^{-1} = \frac{K(L)}{a(L)} = \sum_{s=0}^{\infty} K_s L^s, \quad \sum_{s=0}^{\infty} \| K_s \| < \infty.$$

We have therefore established Assertion 2:

**Assertion 2.** Let $u$ be the $MARMA(m, n)$ process

$$B(L) u'_{t\cdot} = A(L) \epsilon_{t\cdot},$$

where $\epsilon$ is a MWN process and $B(L)$, $A(L)$ are (matrix) polynomials in the lag operator $L$, of degree $m$ and $n$, respectively. If the roots of $|B(z)| = 0$, $z_j$, **all obey** $|z_j| > 1$, i.e. if $B(L)$ is **invertible**, the process in question has a representation as a unilateral $MMA(\infty)$, with MWN process $\epsilon$.

An immediate corollary is Assertion 3:

**Assertion 3.** Let $u$ be a stochastic process as in Assertion 2.

i. If $A(L)$ is invertible, the process in question has a representation as a $MAR(\infty)$, i.e. we have

$$\sum_{j=0}^{\infty} C_j u'_{t-j\cdot} = \epsilon'_{t\cdot}.$$

ii. If $A(L)$ and $B(L)$ are **both** invertible, the process in question has representations as both $MMA(\infty)$ and $MAR(\infty)$.

**Remark 12.** In the literature of statistics sequences of the form $u = \{u_{t\cdot} : B(L)u'_{t\cdot} = \epsilon'_{t\cdot}\}$, where $\epsilon = \{\epsilon_{t\cdot} : t \in \mathcal{N}\}$ is $MWN(\Sigma)$, and the roots of $|B(z)| = 0$, **all** lie outside the unit circle, are said to be **causal**. When this is so the $MMA(\infty)$ representation of Definition 12 is typically said to be a **causal** representation. In many ways this representation is analogous to the reduced form in simultaneous equations models, which provides us with the canonical representation of the conditional mean of the endogenous variables and their covariance matrix. Similarly, in this context the simplest and most lucid representation of the covariance function, or covariance kernel of an AR process is found through its "causal" representation.

### 1.4.3 Identification Issues in $MARMA(m, n)$ Processes

As noted earlier a $MARMA(m, n)$ has the representation

$$B(L)u'_{t\cdot} = A(L)\epsilon'_{t\cdot}, \quad B(L) = I_q - \sum_{j=1}^{m} B_j L^j, \quad A(L) = I_q + \sum_{i=1}^{n} A_i L^i. \quad (1.46)$$

In the literature of statistics, if the roots of $|B(z)| = 0$ are outside the unit circle, the process is said to be **causal**, as in the case of AR processes. If the roots of $|A(z)| = 0$ are all outside the unit circle the process is said to be **invertible**. Thus, a causal-invertible ARMA process has, according to Assertions 1 and 2, both a $VAR(\infty)$ and a $MMA(\infty)$ representation.

There is another aspect of ARMA processes that sets them apart; they present the possibility of an identification problem. Again employing the terminology of the statistical literature the "causal" representation of the process is

$$u'_{t\cdot} = C(L)\epsilon'_{t\cdot}, \quad C(L) = \sum_{j=0}^{\infty} C_j L^j, \quad C(L) = [B(L)]^{-1}A(L). \quad (1.47)$$

The MARMA process is characterized by the two parameters $m, n$, and the white noise covariance matrix $\Sigma$. It is evident that if we define, as in simultaneous equations, $B^*(L) = H(L)B(L)$, $A^*(L) = H(L)A(L)$, where

$H(L)$ is an invertible poynomial lag operator, the "causal" representation of the process is not disturbed, since

$$C^*(L) = [B^*(L)]^{-1} A^*(L) = [B(L)]^{-1} A(L) = C(L), \qquad (1.48)$$

so that the covariance structure of the process is not altered. However, again to employ a term from the simultaneous equations identification literature, the transformation above is not admissible since it may violate the "prior information" that the process is of order $(m, n)$. In other words, the polynomial operator $H(L)$ cannot be unrestricted since then it will certainly change the orders of $B^*(L)$ and $A^*(L)$ to something other than $(m, n)$. Thus, the class of admissible operators $H(L)$ has to be restricted to the class of **unimodular operators**; such operators have the property that $|H(L)| = c_0 > 0$, where $c_0$ is a **constant and** $H(L)B(L)$, $H(L)A(L)$ remain of order $(m, n)$, respectively. [19]

Finally identification also requires a **rank condition**, viz. that with $(m, n)$ as small as possible, rank $(B_m, A_n) = q$. More detail on identification issues may be found in Chapter 2, Hannan and Deistler (1988). We may summarise the preceding discussion in

**Proposition 6.** For the $MARMA(m, n)$ process of Eq. (1.46) to be physically realizable (causal) and invertible, and for the coefficient matrices in $A(L)$ and $B(L)$ to be identifiable from the "causal" representation of the process the following conditions must hold:

i. The characteristic equations $|B(z)| = 0$, $|A(z)| = 0$ have roots that lie outside the unit circle (for causality and invertibility).

ii. The polynomial operators $A(L)$, $B(L)$ are left coprime, i.e. if a matrix operator $H(L)$ exists such that $H(L)B(L)$ and $H(L)A(L)$ are of orders $(m, n)$, respectively, then $|H(L)| = c_0 > 0$, where $c_0$ is a **constant** (for identification).

iii. Rank $(B_m, A_n) = q$ [for minimal orders $(m, n)$].

---

[19] Matrix operators that satisfy this condition are said to be **left coprime**, and the representation $C(L) = [B(L)]^{-1} A(L)$ is said to be irreducible.

## 1.5 Nonstationary and Integrated Sequences

Nonstationary sequences are defined primarily in contradistinction to stationary sequences. Thus, we have the following:

**Definition 13.** In the context of Definitions 2 and 3, if a sequence is not stationary in the strict sense or in the wide sense, it is said to be **nonstationary**.

**Remark 13.** It is evident from the definition that the class of nonstationary sequences exhibits a great deal of variability. Thus, a sequence of independent **not identically** distributed random variables **is** a nonstationary sequence. Moreover, since in classical econometrics we are dealing with random elements that possess **finite second moments** and whose distributional characteristics are left unspecified, a sequence that fails to possess second moments will be considered to be nonstationary in the context of this discussion. Consider, for example, the sequence

$$A_0 u'_{t\cdot} - \sum_{k=1}^{m} A_k u'_{t-k\cdot} = \epsilon'_{t\cdot}, \qquad (1.49)$$

where $A_0 = I_q$, and $\epsilon$ is a $MWN(\Sigma)$; its associated characteristic equation is

$$a(z) = |A(z)| = 0, \quad \text{where} \quad A(z) = I_q - A_1 z - A_2 z^2 - \ldots - A_m z^m. \quad (1.50)$$

To make the point with clarity, consider the special case $m = 1$. In this case the determinant above becomes

$$0 = |I_q - A_1 z| = |z(\lambda I_q - A_1)| = z^q |\lambda I_q - A_1|, \quad \lambda = \frac{1}{z}. \qquad (1.51)$$

Since $z = 0$ is not a root, we see that the roots of the equation $|A(z)| = 0$ are **precisely** the inverse of the characteristic roots of the matrix $A_1$. Let $C$ be the unitary matrix of characteristic vectors, and $\Lambda$ the diagonal matrix of the corresponding characteristic roots, so that we have the relation

$$A_1 = C \Lambda \overline{C}'.$$

Given the formal expansion through the lag operator $L$ we obtain

$$u'_{t\cdot} = \sum_{s=0}^{\infty} A_1^s \epsilon'_{t-s\cdot} = C \sum_{s=0}^{\infty} \Lambda^s \overline{C}' \epsilon'_{t-s\cdot}. \qquad (1.52)$$

Since $\epsilon$, in Eq. (1.49), is a $MWN(\Sigma)$, the $u$-process would be a stationary process if it obeyed $\sum_{s=0}^{\infty} \parallel A_1 \parallel^s < \infty$, and thus had well-defined first and second moments. Since the $\epsilon$-process has well-defined first (zero) and second ($\Sigma$) moments, the question of whether the $u$-process is stationary depends solely on the **roots of the characteristic equation** in Eq. (1.50). Denote the roots of $|A(z)| = 0$ by $z_j$. If all such roots lie outside the unit circle, the characteristic roots of $A_1$, namely the $\lambda_j = (1/z_j)$, are less than one in absolute value, and the expansion of Eq. (1.52) converges absolutely. This means that $u$-process is well-defined as a $MMA(\infty)$ and, thus, has well-defined finite first **and** second moments; moreover it is **stationary** in both the wide and strict sense.

On the other hand, if one or more of the roots lie **on the unit** circle, or **inside** the unit circle, one or more of the $\lambda_j$ are either unity or greater than unity. The reader might well think that even in this case the $u$-process should retain its stationarity property since its basic makeup is not disturbed by this eventuality; however, the process is now **nonstationary** since neither its first nor its second moments are well defined [20] and, more importantly, the very expansion above is not well defined. Thus, it has no **stationary representation** in either the strict or wide sense; thus, by Definition 13, **it is not stationary**.

If one or more of the roots, $z_j$, lie inside the unit circle then, in the expansion above, one or more of the diagonal elements of $\Lambda$ would be greater than unity, and the expansion will "explode". To be precise, suppose $|z_q| < 1$ implying that $|\lambda_q| > 1$. The last element of the vector $\sum_{s=0}^{\infty} \Lambda^s \epsilon_{t-s}^{*'}$ becomes

$$u_{tq}^* = \sum_{s=0}^{\infty} \lambda_q^s \epsilon_{t-s,q}^*, \quad \text{where} \quad \epsilon_{t.}^* = \epsilon_{t.}\overline{C}.$$

Since $|\lambda_q| > 1$, the behavior of the component in question is **dominated** by the remote past. This is referred to as the **explosive case**, and is not considered in the balance of this volume, except for a brief discussion in Chapter 2. It bears stressing that, in terms of relevance to economic applications, it is not the fact that $u_{tq}^*$ is not well defined that rules the explosive case out of consideration; it is rather that the "behavior" of the series at "time" $t$

---

[20] A process that does not possess finite second moments could hardly be termed **covariance stationary**!

is dominated by the remote past in the sense that for $N$ sufficiently large, $|\lambda_q^N \epsilon_{t-N,q}^*|$ a.c. (almost certainly) dwarfs $|\lambda_q \epsilon_{t-1,q}^*|$. In the **unit root** case, i.e. when $|\lambda_q| = 1$, current values of the series **are not dominated** by the remote past; rather, it is as if the "system" never forgets, and whether something "occurred a long time ago" or "only recently" it registers the same impact on "the present". The expansion above, however, is still invalid and even with unit roots the process is still nonstationary. This is the type of nonstationarity we examine below.

Return now to Eq. (1.50) and consider the case where $r$ of the $z_j$ lie **on** the unit circle, and the remaining $qm - r$ roots lie **outside** the unit circle. Moreover, to simplify exposition, suppose that the $r$ roots that lie on the unit circle are actually real and **equal to** unity. If $z_j$ are the roots of $a(z)$, let $\lambda_j = (1/z_j)$. By the fundamental theorem of algebra we may write

$$a(z) = \prod_{j=1}^{qm}(1 - \lambda_j z), \quad \text{or} \quad a(L) = \prod_{j=1}^{mq}(I - \lambda_j L). \tag{1.53}$$

Without loss of generality suppose $\lambda_j = 1$, for $j = 1, 2, \ldots, r$; moreover, let $H(L)$ be the **adjoint matrix** relative to the matrix (lag) operator $A(L)$, as defined implicitly in Eq. (1.49). Since $H(L)A(L) = a(L)$, it follows immediately that

$$a(L)u'_{t\cdot} = H(L)\epsilon'_{t\cdot}, \text{ or } (I-L)^r u'_{t\cdot} = \frac{H(L)}{a^*(L)}\epsilon'_{t\cdot}, \quad a^*(L) = \prod_{j=r+1}^{mq}(I-\lambda_j L). \tag{1.54}$$

Notice that since the roots, $\lambda_j$, contained in $a^*(L)$ are **less than one** (in absolute value) we have the representation

$$\frac{H(L)}{a^*(L)} = \sum_{s=0}^{\infty} H_s L^s, \text{ so that } (I-L)^r u'_{t\cdot} = \sum_{s=0}^{\infty} H_s \epsilon'_{t-s\cdot}, \quad \sum_{s=0}^{\infty} \| H_s \| < \infty. \tag{1.55}$$

Evidently, this implies that the right member of the last equation above is **covariance stationary**; from the left member we conclude that **even though the $u$-process is nonstationary**, if we difference it $r$ times, the resulting differenced process **is covariance stationary**. Such processes are said to be **integrated**. We conclude with another definition.

**Definition 14.** In the context of Definitions 2 and 3, let $X = \{X_{t\cdot} : t \in \mathcal{N}\}$ be a stochastic process; if, for $r \in \mathcal{N}_0$, $Y = \{Y_{t\cdot} : Y_{t\cdot} = (I - L)^r X_{t\cdot}, \ t \in \mathcal{N}\}$

is (covariance) stationary, [21] **the process** $X$ is said to be **an integrated process of order** $r$, and is denoted by $I(r)$.

**Remark 14.** By convention, a **stationary process** is denoted by $I(0)$; this usage is in the spirit of the definition because, by convention, $(I - L)^0 = I$, i.e. it is the **identity operator**. Consequently, if $X$ is **stationary** $(I - L)^0 X$ is also stationary and thus, by Definition 14, $X$ is $I(0)$!

**Remark 15.** Just as we have defined MA, AR, or ARMA processes in the stationary case, we can similarly define MA, AR, or ARMA types of integrated processes.

**Definition 15.** In the context of Definitions 2 and 3, let $X = \{X_t. : t \in \mathcal{N}\}$ be a stochastic process;

i. if $Y = \{Y_t. : Y_t. = (I - L)^r X_t., \ t \in \mathcal{N}\}$ is a $MMA(\infty)$, i.e. if there exists a $MWN(\Sigma)$, say $\epsilon$, such that

$$Y_t'. = \sum_{j=0}^{\infty} A_j \epsilon_{t-j}'. , \quad \sum_{j=0}^{\infty} \| A_j \| < \infty,$$

$X$ is said to be a multivariate integrated (of order $r$) moving average process of infinite extent (MIMA), and is denoted by $MIMA(r, \infty)$; if $A_j = 0$ for $j > m$, it is said to be a finite MIMA and is denoted by $MIMA(r, m)$;

ii. if $Y$ of part i is a $MAR(\infty)$, i.e. if there exist matrices $B_s$ and a $MWN(\Sigma)$ process $\epsilon$ such that

$$B_0 Y_t'. - \sum_{j=1}^{\infty} B_j Y_{t-j}'. = \epsilon_t'. , \quad \sum_{j=0}^{\infty} \| B_j \| < \infty,$$

$X$ is said to be a multivariate integrated (of order $r$) autoregressive process of infinite extent (MIAR) and is denoted by $MIAR(r, \infty)$; if $B_j = 0$ for $j > n$, it is said to be a finite MIAR and is denoted by $MIAR(r, n)$;

---

[21] In many instances we may also require the spectral density of $Y$ to obey $f(0) > 0$ and $f(\lambda) < \infty$. As we saw in Proposition 4, a covariance stationary sequence that obeys the conditions $0 < f(\lambda) < \infty$ a.c. admits of a representation as a bilateral moving average of infinite extent.

iii. if $Y$ of part i is a $MARMA(\infty)$, i.e. if there exist matrices $B_j$, $C_s$, and a $MWN(\Sigma)$ process $\epsilon$ such that

$$B_0 Y'_{t\cdot} - \sum_{j=1}^{\infty} B_j Y'_{t-j\cdot} = \sum_{s=0}^{\infty} C_s \epsilon'_{t-s\cdot}, \quad \sum_{j=0}^{\infty} \| B_j \| < \infty, \quad \sum_{s=0}^{\infty} \| C_s \| < \infty,$$

$X$ is said to be a multivariate integrated (of order $r$) autorgressive moving average process (MIARMA) of infinite extent, and is denoted by $MIARMA(r, \infty, \infty)$; if $B_j = 0$ for $j > m$ and $C_s = 0$ for $s > n$, $X$ is said to be a finite MIARMA process and is denoted by $MIARMA(r, m, n)$.

**Remark 16.** The special case of an MIMA process, $MIMA(1, 0)$, is said to be a **multivariate random walk** (MRW); in the scalar case, it is called a **random walk** (RW).

# Chapter 2

# Prediction and Estimation

## 2.1 Preliminaries

In this chapter we take up the problem of best linear prediction, in the context of the Hilbert space, $\mathcal{H}^2$, introduced earlier. The major motivation for our discussion is that much of contemporary macroeconometrics, especially analysis involving rational expectations, is carried out in the framework of Hilbert space.

We assume that the reader is familiar with the (standard) Hilbert space defined on $R^n$, which we shall denote by $\mathcal{H}(R^n)$; in $\mathcal{H}(R^n)$ the inner product and the norm are defined, respectively, by

$$(x, y) = \sum_{j=1}^{n} x_j y_j, \quad x, y \in R^n, \quad \| x \| = (x, x)^{1/2}.$$

In fact, this is the space over which we typically solve the least squares problem, without reference to the probabilistic context of the general linear model (GLM). Thus, if $y, x_{\cdot i} \in R^n$, $i = 0, 1, 2, \ldots k$, the least squares problem may be posed as

$$\min_{\beta} \left\| y - \sum_{j=0}^{k} \beta_j x_{\cdot j} \right\|, \quad \beta = (\beta_0, \beta_1, \ldots, \beta_k)', \quad k < n.$$

The resulting estimator is simply the projection of the vector $y \in R^n$, on the subspace of $R^n$ spanned by the vectors $\{x_{\cdot j} : j = 0, 1, 2, \ldots, k\}$. The first order conditions are

$$-2(y, x_{\cdot j}) + 2 \sum_{s=0}^{k} (x_{\cdot j}, x_{\cdot s}) \beta_s = 0.$$

If the vectors $\{x_{\cdot j} : j = 0, 1, 2, \ldots k\}$ are **an orthonormal set**, i.e. if $(x_{\cdot s}, x_{\cdot j}) = 1$, for $s = j$ and zero otherwise, the solution is given by

$$\beta_j = (y, x_{\cdot j}), \quad j = 0, 1, 2, \ldots, k.$$

If the orthonormality condition does not hold, the solution is given, in the familiar matrix form, by $\hat{\beta} = (X'X)^{-1}X'y$, where the $x_{\cdot j}$ are the $k+1$ columns of $X$, assuming that $n \geq k+1$. The projection of $y$ on the subspace spanned by the columns of the matrix $X$ is given by $\hat{y} = \sum_{j=0}^{k}(y, x_{\cdot j})x_{\cdot j}$, and $\hat{y} = X(X'X)^{-1}X'y$, respectively, when the orthonormality condition does, and does not hold.

The entity $\hat{y}$ exists, under appropriate conditions, and is the best linear predictor of $y$ given $\{x_{\cdot j} : j = 0, 1, 2, \ldots k\}$. It is often denoted by

$$\hat{y} = \sum_{j=0}^{k}(y, x_{\cdot j})x_j = E(y|x_{\cdot 0}, x_{\cdot 1}, \ldots, x_{\cdot k}),$$

and is occasionally referred to as the conditional expectation of $y$ given the vectors in the matrix $X$. This terminology, however, is not appropriate **unless** $y$ and the vectors in question are **jointly** normal.

The preceding shows that orthonormality simplifies arguments considerably. It is, thus, comforting to know that there exists a routine procedure that transforms any set of linearly independent vectors in $R^n$ into an orthonormal set. We discuss this issue, however, in the context of the more complex Hilbert space, $\mathcal{H}^2(\Omega, \mathcal{A}, \mathcal{P})$, also denoted by $\mathcal{H}^2(\mathcal{P})$. This is the space of (complex) zero mean, finite second moment random elements defined on the probability space ($\Omega$, $\mathcal{A}$, $\mathcal{P}$), with the inner product and the norm defined, respectively, by [1]

$$< X, Y > = EX\overline{Y}' = \int_{\Omega} X\overline{Y}' \, d\mathcal{P}, \quad \| X \| = < X, X >^{1/2}. \qquad (2.1)$$

---

[1] In the context of the complex Hilbert space $\mathcal{H}^2(\mathcal{P})$ the inner product is typically denoted by angle brackets $< \cdot, \cdot >$; in the (real) Hilbert space $\mathcal{H}(R^n)$, it is typically denoted by parentheses $(\cdot, \cdot)$, as in Chapter 1. Notice, further, that for notational simplicity we do not use parentheses or brackets to denote the object of the expectation operator $E$, except when necessary to make the meaning clear. Thus, we typically will write $EX$, **not** $E(X)$.

Before we proceed to this aspect, we give the following useful result.

**Proposition 1** (Parallelogram Law). Let $X, Y \in \mathcal{H}^2(\mathcal{P})$. Then,

$$\| X + Y \|^2 + \| X - Y \|^2 = 2 \| X \|^2 + 2 \| Y \|^2 .$$

Proof: By definition,

$$\psi = \| X + Y \|^2 + \| X - Y \|^2 = <X,X> +2<X,Y> + <Y,Y>$$

$$+ <X,X> -2<X,Y> + <Y,Y> = 2 \| X \|^2 + 2 \| Y \|^2 .$$

q.e.d.

Suppose we have a collection of (linearly independent) zero mean random variables, $\{X_j : X_j \in \mathcal{H}^2(\Omega, \mathcal{A}, \mathcal{P}), j = 1, 2, \ldots n\}$, i.e. such that $\sum_{j=1}^n c_j X_j = 0$ implies $c_j = 0$, for all $j$; is it always possible to transform it into a set of **orthonormal random variables**, and if so, how is it done? The answer is yes and the procedure is given in the proof of the next proposition.

**Proposition 2** (Gram-Schmidt Orthogonalization). Let $X$ be a set of $n$ zero mean, linearly independent, random variables in $\mathcal{H}^2(\mathcal{P})$, with $\mathrm{Var}(X_j) = \sigma_j^2$, $j = 1, 2, \ldots, n$. Then $X$ may be transformed into an equivalent **orthonormal set**, say $\epsilon = \{\epsilon_j : j = 1, 2 \ldots, n\}$.

Proof: Define $\epsilon_1 = \frac{X_1}{\|X_1\|}$. Moreover,

$$\mathrm{Var}(\epsilon_1) = \| \epsilon_1 \|^2 = \frac{1}{\| X_1 \|^2} \int_\Omega X_1 \overline{X_1} \, d\mathcal{P}$$

$$= \frac{\sigma_1^2}{\sigma_1^2} = 1 \tag{2.2}$$

If $k$ such $\epsilon$'s have been selected, define the $(k+1)^{st}$ element by

$$\epsilon_{k+1} = \frac{X_{k+1} - \hat{X}_{k+1}}{\| X_{k+1} - \hat{X}_{k+1} \|}, \quad \hat{X}_{k+1} = \sum_{j=1}^k <X_{k+1}, \epsilon_j> \epsilon_j, \tag{2.3}$$

and note that all $\epsilon$'s have mean zero. Moreover, for $0 < s < k \leq n$,

$$
< \epsilon_k, \epsilon_s > = \frac{1}{\parallel X_k - \hat{X}_k \parallel} \int_\Omega \left[ X_k \bar{\epsilon}_s - \sum_{j=1}^{k-1} < X_k, \epsilon_j > \epsilon_j \bar{\epsilon}_s \right] d\mathcal{P}
$$

$$
= \frac{1}{\parallel X_k - \hat{X}_k \parallel} [< X_k, \epsilon_s > - < X_k, \epsilon_s >] = 0. \qquad (2.4)
$$

That the variance of $\epsilon_{k+1}$ is unity is quite evident by construction.

q.e.d.

**Remark 1**. Gram-Schmidt orthogonalization may be represented as a matrix transformation,

$$
\epsilon = DX - A\epsilon, \quad \epsilon = \begin{pmatrix} \epsilon_1 \\ \epsilon_2 \\ \vdots \\ \epsilon_n \end{pmatrix}, \quad X = \begin{pmatrix} X_1 \\ X_2 \\ \vdots \\ X_n \end{pmatrix}, \qquad (2.5)
$$

where $A$ is a **lower triangular** matrix, and $D$ is a diagonal matrix which has the inverse of the norm of $X_k - \hat{X}_k$ for its $k^{th}$ diagonal element . More precisely,

$$
A = \begin{bmatrix} 0 & 0 & 0 & \dots & 0 \\ a_{22} < X_2, \epsilon_1 > & 0 & 0 & \dots & 0 \\ a_{33} < X_3, \epsilon_1 > & a_{33} < X_3, \epsilon_2 > & 0 & \dots & 0 \\ \vdots & \vdots & \vdots & \vdots & \vdots \\ a_{nn} < X_n, \epsilon_1 > & a_{nn} < X_n, \epsilon_2 > & a_{nn} < X_n, \epsilon_3 > & \dots & 0 \end{bmatrix},
$$

$$
D = \text{diag}\,(a_{11}, a_{22}, a_{33}, \dots, a_{nn}), \qquad (2.6)
$$

where

$$
a_{11} = \frac{1}{\parallel X_1 \parallel}, \quad a_{jj} = \frac{1}{\parallel X_j - \hat{X}_j \parallel}, \quad \hat{X}_j = \sum_{s=1}^{j-1} < X_j, \epsilon_s > \epsilon_s, \quad \text{for } j \geq 2.
$$

Two observations about the procedure above are useful. First, when we consider the set of linear combinations

$$
\{Y : Y = \sum_{j=1}^{n} c_j X_j, \quad c_j \in \mathcal{C}\},
$$

where $\mathcal{C}$ is the field of complex numbers, it is clear that it is **equivalent to the set of linear combinations generated** by the **orthonormal variables**, i.e. the $\epsilon$'s just defined. Second, if $X_0$ is to be "predicted" by the **orthonormal set**, and we write the prediction as $\hat{X}_0 = \sum_{j=1}^{n} < X_0, \epsilon_j > \epsilon_j$ then $X_0 - \hat{X}_0$ is **orthogonal to every element** $\epsilon_j$ **used in "predicting"** $X_0$. Typically, this is indicated by the notation

$$X_0 - \hat{X}_0 \perp \epsilon_j, \quad j = 1, 2, \ldots, n.$$

For the GLM, the orthogonality statement above is simply the statement that, **within the sample**, the least squares residuals, $\hat{u} = y - \hat{y}$, obey $\hat{u} \perp X$. This is so because $\hat{y} = P_x y$, $P_x = X(X'X)^{-1}X'$, $X = (x_{\cdot 0}, x_{\cdot 1}, \ldots, x_{\cdot k+1})$, and thus $X'\hat{u} = 0$.

**Remark 2**. Though the entities $X_j$ and $\epsilon_j$ above are **scalar random variables**, all of the results obtained remain valid even if we consider them to be random vectors. In the **vector** case we shall define **orthonormality** by the condition that the **vectors in question** obey $E \epsilon'_{j\cdot} \bar{\epsilon}_{k\cdot} = 0$, if $j \neq k$, and $E \epsilon'_{j\cdot} \bar{\epsilon}_{k\cdot} = I_q$, when $j = k$. This is because in the vector case, the inner product operator orthogonality condition $< \epsilon_{j\cdot}, \epsilon_{k\cdot} >= 0$ only implies

$$\text{tr}[\text{Cov}(\epsilon'_{j\cdot}, \ \epsilon'_{k\cdot})] = 0.$$

Such a condition, however, **does not imply** that the (cross) covariance matrix is the **zero matrix**, as required for orthogonality, unless the matrix in question is at least **positive semidefinite**.

Finally, we note tat Remark 2 has a bearing **only on how** we estimate the coefficients of the projection, an issue we examine in a later section.

## 2.2   Orthogonal Complements

In this section, we formalize the concepts just discussed, and establish the important projection theorem.

We begin by noting that a Hilbert space is a normed linear vector space with an inner product, and that it is **complete**; the latter means that all Cauchy sequences converge to an element in that space.

**Definition 1.** Let $X = \{X_j : j \in \mathcal{N}\}$ be a set of random elements defined on the probability space $(\Omega, \mathcal{A}, \mathcal{P})$. The collection $M_n(X) = \{Y : Y = \sum_{j=-n}^{n} c_j X_j, \ c_j \in \mathcal{C}\}$ is said to be the $n$-dimensional **linear manifold** generated by $X$.

The closure of $M_n(X)$, i.e. the collection that consists of $M_n(X)$ and the $L^2$ limits of sequences in $M_n(X)$, is said to be the $n$-dimensional **closed linear manifold** generated by $X$, and is denoted by $\mathcal{M}_n(X)$.

The collection $\mathcal{M}(X) = \bigcap_{n=-\infty}^{\infty} \mathcal{M}_n(X)$ is said to be the (closed) **linear manifold** of processes generated by $X$, i.e. of processes that, for every $n$, can be approximated by elements $\eta_n \in \mathcal{M}_n(X)$, say $\eta_n = \sum_{j=-n}^{n} c_j X_j$.

**Definition 2.** Let $\mathcal{S} \subseteq H$ be a subspace of a Hilbert space $H$. The collection

$$\mathcal{Q} = \{q : q \perp x, \quad \text{for every } x \in \mathcal{S}\},$$

is said to be the **orthogonal complement** of $\mathcal{S}$, and is often denoted by $\mathcal{S}^\perp$.

**Definition 3.** The subspace $\mathcal{S}$ (of a Hilbert space) is said to be **closed** if and only if, for $q_n \in \mathcal{S}$ and $\| q_n - q \| \longrightarrow 0$, $q \in \mathcal{S}$, i.e. if $\mathcal{S}$ contains the (mean square) limits of sequences in $\mathcal{S}$.

An immediate consequence of the definitions above is the following proposition.

**Proposition 3.** Let $\mathcal{S}$ be a (linear) subspace of a Hilbert space, $\mathcal{H}$; then, $\mathcal{S}^\perp$ is a closed subspace of $\mathcal{H}$.

Proof: We need to show that if $q_1, q_2 \in \mathcal{S}^\perp$ and $c_1, c_2 \in \mathcal{C}$ **then** $(c_1 q_1 + c_2 q_2, x) = c_1(q_1, x) + c_2(q_2, x) = 0$, for every element $x \in \mathcal{S}$. Since, evidently, $0 \in \mathcal{S}^\perp$, $\mathcal{S}^\perp$ is a linear subspace. To prove closure, let $\{q_n : q_n \in \mathcal{S}^\perp, \ n \geq 1\}$ be a sequence converging to $q$; we complete the proof by showing that $q \in \mathcal{S}^\perp$. But

$$|(q_n, x) - (q, x)| = |(q - q_n, x)| \leq \| x \| \| q - q_n \| \longrightarrow 0. \qquad (2.7)$$

Since, for every $n$, $(q_n, x) = 0$, it follows that $(q, x) = 0$, or $q \in \mathcal{S}^\perp$.

q.e.d.

**Proposition 4** (Projection Theorem).  Let $\epsilon = \{\epsilon_n : n \in \mathcal{N}\}$ be an orthonormal sequence of random elements in $\mathcal{H}^2(\Omega, \mathcal{A}, \mathcal{P})$, $\mathcal{M}$ be the closed linear manifold generated by $\epsilon$, as in Definition 1, and note that $\mathcal{M} \subseteq \mathcal{H}^2(\Omega, \mathcal{A}, \mathcal{P})$.  Given **any** element $X \in \mathcal{H}^2(\Omega, \mathcal{A}, \mathcal{P})$, the following statements are true:

   i. there exists a unique element, $Y \in \mathcal{M}$, such that

$$\| X - Y \| = \inf_{\zeta \in \mathcal{M}} \| X - \zeta \|; \tag{2.8}$$

   ii.  $Y = \sum_{j=-\infty}^{\infty} c_j \epsilon_j, \quad c_j = <X, \epsilon_j>$;

   iii.  $X - Y \perp \xi$, **for every** $\xi \in \mathcal{M}$, i.e.  $X - Y \in \mathcal{M}^{\perp}$.

   iv.  $\mathcal{H}^2(\Omega, \mathcal{A}, \mathcal{P}) = \mathcal{M} \oplus \mathcal{M}^{\perp}$, i.e.  every element $X \in \mathcal{H}^2(\Omega, \mathcal{A}, \mathcal{P})$ can be written (uniquely) $X = X^s + X^r$ such that $X^s \in \mathcal{M}$, and $X^r \in \mathcal{M}^{\perp}$.

Proof: Let $d = \inf_{\zeta \in \mathcal{M}} \| X - \zeta \| = \| X - Y \|$.  Choose a sequence $\{d_n : n \in \mathcal{N}\}$ and a corresponding sequence $\{\zeta_n : n \geq 1\}$ such that $\| X - \zeta_n \| = d_n$, and $d_n \longrightarrow d$.  From Proposition 1 we have
$\| (\zeta_n - X) + (X - \zeta_m) \|^2 + \| (\zeta_n - X) - (X - \zeta_m) \|^2 = 2 \| \zeta_n - X \|^2 + 2 \| X - \zeta_m \|^2$, which implies

$$\| \zeta_n - \zeta_m \|^2 = 2 \| \zeta_n - X \|^2 + 2 \| X - \zeta_m \|^2 - 4 \left\| \frac{\zeta_n + \zeta_m}{2} - X \right\|^2$$

$$\leq 2d_n^2 + 2d_m^2 - 4d^2. \tag{2.9}$$

Letting $n, m \longrightarrow \infty$, Eq. (2.9) implies $\| \zeta_n - \zeta_m \|^2 \longrightarrow 0$, so that $\{\zeta_n : n \geq 1\}$ is a Cauchy sequence. Since the Hilbert space is complete, the sequence converges to an element, say $Y \in \mathcal{M}$, thus establishing existence.  For uniqueness, suppose $Z \in \mathcal{M}$ is another element such that $d = \| X - Z \|$. We show that $Y = Z$, in the sense that $\| Y - Z \| = 0$.  Using the parallelogram law again, we find

$$\| Y - Z \|^2 = 2 \| X - Y \|^2 + 2 \| X - Z \|^2 - 4 \left\| \frac{Y + Z}{2} - X \right\|^2$$

$$= 4d^2 - 4 \left\| \frac{Y + Z}{2} - X \right\|^2 \leq 0, \tag{2.10}$$

which is a contradiction, and thus completes the proof of i.

To prove part ii, note that $\epsilon$ is an orthonormal basis for the Hilbert space $\mathcal{M}$, and $Y \in \mathcal{M}$. Thus, by the properties of Hilbert space, $Y = \sum_{j=-\infty}^{\infty} c_j \epsilon_j$, where $c_j = \langle Y, \epsilon_j \rangle$, which completes the proof of part ii.

To prove part iii, we again proceed by contradiction. Let $b \in \mathcal{C}$, $\zeta \in \mathcal{M}$ be arbitrary, and note that $\| X - Y - b\zeta \|^2 = \| X - Y \|^2 + | b |^2 \| \zeta \|^2 - \bar{b}(X - Y, \zeta) - b(\zeta, X - Y)$. Since $b$ is arbitrary, let $\lambda$ be **real**, such that $0 < \lambda < (2/ \| \zeta \|^2)$ and put $b = \lambda(X - Y, \zeta)$. It follows then that $\lambda^2 \| \zeta \|^2 - 2\lambda < 0$ and, moreover,

$$\| X - Y - b\zeta \|^2 = \| X - Y \|^2 + [\lambda^2 \| \zeta \|^2 - 2\lambda] \,| \,(X - Y, \zeta) \,|^2,$$

$$< \|X - Y\|^2. \tag{2.11}$$

**This is a contradiction, unless** $(X - Y, \zeta) = 0$, for all $\zeta \in \mathcal{M}$, i.e. unless $X - Y \perp \zeta$, for every $\zeta \in \mathcal{M}$.

The proof of iv follows immediately from i and iii.

q.e.d.

**Remark 3.** The proposition above has established the existence (and uniqueness) of an element, $Y$, in the closed linear manifold generated by the orthonormal sequence $\epsilon$, namely $\mathcal{M} \subseteq \mathcal{H}^2(\Omega, \mathcal{A}, \mathcal{P})$, which is **closest** to any given element, say $X \in \mathcal{H}^2(\Omega, \mathcal{A}, \mathcal{P})$. It is reasonable to think of $Y$ as being an outcome of an operator, transformation, or mapping,

$$P : \mathcal{H}^2(\Omega, \mathcal{A}, \mathcal{P}) \longrightarrow \mathcal{M},$$

and to write $P(X) = Y$.[2] The operator $P$ is termed a **projection** (or projection operator), and $Y$ is said to be the **orthogonal projection** of $X$ **onto** $\mathcal{M}$. If we introduce the **identity operator,** $I$, such that for all $X \in \mathcal{H}^2(\Omega, \mathcal{A}, \mathcal{P})$, $I(X) = X$, we may further define the (projection) operator

$$I - P = P^* : \mathcal{H}^2(\Omega, \mathcal{A}, \mathcal{P}) \longrightarrow \mathcal{M}^\perp$$

---

[2] For clarity we ought to write $P_\epsilon$ or $P(X; \epsilon)$, instead of $P$, thus indicating the entity that generates the subspace on which the projection is made.

so that $P^*(X)$ is **the orthogonal projection of** $X$ **onto** $\mathcal{M}^\perp$, and thus orthogonal to every element $\zeta \in \mathcal{M}$. In fact, by Proposition 4, we can decompose any element $X \in \mathcal{H}^2(\Omega, \mathcal{A}, \mathcal{P})$ it can be decomposed into the orthogonal components, $P(X)$ and $P^*(X)$, such that the first lies in $\mathcal{M}$, and the second lies in its orthogonal complement $\mathcal{M}^\perp$.

The next proposition sets forth a number of useful properties of projection operators.

**Proposition 5.** Let $\mathcal{H}^2(\Omega, \mathcal{A}, \mathcal{P})$ be the Hilbert space of Proposition 4 and $\epsilon = \{\epsilon_n : n \in \mathcal{N}\}$ be an orthonormal sequence of random elements. Let $\mathcal{M}$ be the closed linear manifold generated by $\epsilon$ – so that $\mathcal{M} \subseteq \mathcal{H}^2(\Omega, \mathcal{A}, \mathcal{P})$ – and define the projection operators [3]

$$P : \mathcal{H}^2(\Omega, \mathcal{A}, \mathcal{P}) \longrightarrow \mathcal{M},$$

$$P^* : \mathcal{H}^2(\Omega, \mathcal{A}, \mathcal{P}) \longrightarrow \mathcal{M}^\perp,$$

as in Proposition 4 and Remark 3. The following statements are true:

i. the projection operators are linear, i.e given $\alpha_i \in \mathcal{C}$ and $X_i \in \mathcal{H}^2(\mathcal{P})$, $i = 1, 2$,
$$P(a_1 X_1 + a_2 X_2) = \alpha_1 P(X_1) + \alpha_2 P(X_2);$$

ii. for every $X \in \mathcal{H}^2(\Omega, \mathcal{A}, \mathcal{P})$, there exists a unique decomposition $X = P(X) + P^*(X)$;

iii. for every $X \in \mathcal{H}^2(\Omega, \mathcal{A}, \mathcal{P})$, $\| X \|^2 = \| P(X) \|^2 + \| P^*(X) \|^2$;

iv. if $X_n \in \mathcal{H}^2(\Omega, \mathcal{A}, \mathcal{P})$, $n \geq 1$, and such that $\| X_n - X \| \longrightarrow 0$, then $\| P(X_n) - P(X) \| \longrightarrow 0$;

v. $X \in \mathcal{M}$ if and only if $P(X) = X$ and $X \in \mathcal{M}^\perp$ if and only if $P(X) = 0$;

---

[3] To avoid excessive elaboration, we carry out the proof **only in terms of the operator** $P$. Since $P^* = I - P$, it is clear that, *mutatis mutandis*, similar properties hold for $P^*$ as well.

vi. let $\mathcal{M}_i$, $P_i$, $i = 1, 2$, be the closures of (two) linear manifolds and their respective projection operators; then $\mathcal{M}_1 \subseteq \mathcal{M}_2$ if and only if for every $X \in \mathcal{H}^2(\Omega,\ \mathcal{A},\ \mathcal{P})$, $P_1 \circ P_2(X) = P_1(X)$, where the notation $P_1 \circ P_2(X)$ means $P_1[P_2(X)]$.

Proof: To prove i, it will suffice to show that

$$\alpha_1 X_1 + \alpha_2 X_2 - \alpha_1 P(X_1) - \alpha_2 P(X_2) \in \mathcal{M}^\perp.$$

This implies that $\alpha_1 P(X_1) + \alpha_2 P(X_2) \in \mathcal{M}^\perp$ is the projection of $\alpha_1 X_1 + \alpha_2 X_2$. We may rewrite the expression above as

$$\alpha_1 P^*(X_1) + \alpha_2 P^*(X_2) = \alpha_1[X_1 - P(X_1)] + \alpha_2[X_2 - P(X_2)]$$

$$= \alpha_1 X_1 + \alpha_2 X_2 - \alpha_1 P(X_1) - \alpha_2 P(X_2).$$

Since $P^*(X_i) \in \mathcal{M}^\perp$, $i = 1, 2$, the conclusion follows from the linearity of the space $\mathcal{M}^\perp$.

The proof of ii was given in the proof of part i of Proposition 4.

The proof of part iii is a simple consequence of the orthogonal decomposition of $X$ since, for every element $X \in \mathcal{H}^2(\Omega,\ \mathcal{A},\ \mathcal{P})$, $< P(X), P^*(X) >= 0$.

To prove part iv we observe that from parts i and ii, $P(X_n - X) = P(X_n) - P(X)$ and, moreover,

$$\| X_n - X \|^2 = \| P(X_n - X) \|^2 + \| P^*(X_n - X) \|^2$$

$$\geq \| P(X_n - X) \|^2. \tag{2.12}$$

Since $\| X_n - X \| \longrightarrow 0$, it follows from Eq. (2.12) that $\| P(X_n - X) \|^2 \longrightarrow 0$, which completes the proof of iv.

To prove part v, note that by the definition of projection operators for every $X \in \mathcal{M}$ we have

$$X = P(X) + P^*(X) = Y + Z.$$

Hence if $X \in \mathcal{M}$, $Z = 0$ so that $X = P(X)$; conversely, if $Z = 0$, $X \in \mathcal{M}$. On the other hand, if $X \in \mathcal{M}^\perp$ then $Y = 0$, so that $P(X) = 0$; conversely if $P(X) = 0$, $X = Z$ so that $X = P^*(X) \in \mathcal{M}^\perp$, which completes the proof of part v.

To prove part vi, let $\mathcal{M}_i$ be closed linear manifolds, $P_i$ their respective projection operators, $i = 1, 2$, and consider

$$X = P_2(X) + P_2^*(X), \quad P_1(X) = P_1[P_2(X)] + P_1[P_2^*(X)]. \qquad (2.13)$$

If $\mathcal{M}_1 \subseteq \mathcal{M}_2$ then, evidently, $\mathcal{M}_2^\perp \subseteq \mathcal{M}_1^\perp$; thus, since $P_2^*(X) \in \mathcal{M}_2^\perp$ and $\mathcal{M}_2^\perp \subseteq \mathcal{M}_1^\perp$, by part v, above, we conclude

$$P_1[P_2^*(X)] = 0, \quad \text{which implies} \quad P_1(X) = P_1[P_2(X)].$$

Conversely, suppose that $P_1(X) = P_1[P_2(X)]$; this means that $P_1[P_2^*(X)] = 0$, which, again by part v, above, implies that $P_2^*(X) \in \mathcal{M}_1^\perp$. By construction, $P_2^*(X) \in \mathcal{M}_2^\perp$; thus

$$\mathcal{M}_2^\perp \subseteq \mathcal{M}_1^\perp \quad \text{or} \quad \mathcal{M}_1 \subseteq \mathcal{M}_2,$$

which completes the proof of part vi.

$$\text{q.e.d.}$$

**Remark 4**. Propositions 4 and 5 have been stated in terms of subspaces (closed linear manifolds) of a Hilbert space of square integrable random elements; such subspaces are generated by orthonormal sets. A careful look at the proofs, however, shows that orthonormality of the basis of the manifolds is not required; nor is it necessary to deal with an underlying probability space, provided the inner product definition is changed *mutatis mutandis*. The relationships dealt with are essentially geometric, and, on a Euclidean space, they have to do with finding the closest distance between a point in a space and an appropriate subspace; once the point in the subspace is found, in terms of which the distance of the original point from the subspace is to be measured, we can set up a correspondence between the two. Proposition 4 deals with this issue; Proposition 5 deals with the properties of such correspondence.

## 2.3  Wold's Decomposition

Claims related to Wold's decomposition are ubiquitous in the literature of macroeconometrics. This juncture is particularly appropriate for a discussion of Wold's decomposition, because the latter is simply an application of the projection theorem which we proved in the previous section.

In Eq. (1.5) (of Chapter 1) we exhibited, **in the frequency domain**, a decomposition of a covariance stationary process, in terms of sinusoids with random amplitude. In this section we give another decomposition of covariance stationary processes due to the Swedish econometrician and statistician Herman Wold. [4]  The decomposition is **in the time domain**, and states that every covariance stationary process can be written as the sum of two uncorrelated processes. Thus, if $Y$ is covariance stationary

$$Y = Y^d + Y^{nd},$$

where $Y^d$ is said to be a **deterministic or singular** process, $Y^{nd}$ is said to be a **nondeterministic or regular** process, and the two are uncorrelated, or orthogonal in a Hilbert space context.

**Definition 4.** Let $X = \{X_j : j \in \mathcal{N}\}$ be a sequence in $\mathcal{H}^2(\Omega, \mathcal{A}, \mathcal{P})$, $M_n(X)$ **the linear manifold spanned by** $\{X_j : -\infty < j \leq n\}$, $\mathcal{M}_n(X)$ the corresponding closed linear manifold, and $\mathcal{M}(X) = \bigcap_{n=-\infty}^{\infty} \mathcal{M}_n(X)$. Let $\mathcal{R}_n(X)$ be the orthogonal complement of $\mathcal{M}_{n-1}(X)$ in $\mathcal{M}_n(X)$, so that

$$\mathcal{M}_n(X) = \mathcal{M}_{n-1}(X) \oplus \mathcal{R}_n(X).$$

If $X_{n+1} \in \mathcal{M}_n(X)$, for every $n \in \mathcal{N}$ or, more precisely, if $X_n \in \mathcal{M}(X)$ for every $n$, the process is said to be **purely deterministic or singular;** if $\mathcal{M}(X) = \{0\}$, i.e. the only process that may be predicted from the past of $X_n$ is the zero process, $X$ is said to be **purely nondeterministic or regular.** [5]

---

[4] Our discussion will not be very extensive since Wold's decomposition is, basically, a consequence of the projection theorem and the characterization of **nondeterministic** or **regular** processes. Propositions 4 and 5 of Chapter 1 provide the same or more latitude for the uses to which Wold's decomposition is put in macroeconometrics.

[5] Note that if a vector process $X_{n+1} \in \mathcal{M}(X)$ for every $n \in \mathcal{N}$, there exists a sequence

More extensively, given a scalar sequence $X$ consider the projection of $X_{n+1}$ on $\mathcal{M}_n(X)$, say $P_n$, and the entity

$$\phi = \parallel X_{n+1} - P_n(X_{n+1}) \parallel^2 .$$

If $\phi = 0$, $\{X_n : n \in \mathcal{N}\}$ is said to be strictly, or purely, deterministic [6] and there exists a sequence, say $\{b_j : j \in \mathcal{N}_0\}$, such that

$$X_{n+1} = \sum_{j=0}^{\infty} b_j X_{n-j};$$

on other hand, if $\phi > 0$ we know from the projection theorem that

$$(X_{n+1} - \hat{X}_{n+1}) \perp \hat{X}_{n+1},$$

and from this relationship we may ultimately construct the purely determin- istic and purely nondeterministic processes, respectively $X_n^d$, and $X_n^{nd}$, such that

$$X_n = X_n^d + X_n^{nd}, \quad \text{with} \quad X_n^d \perp X_n^d.$$

**If $X_n^d = 0$, the sequence is seen to be purely nondeterministic, or regular; if $X_n^{nd} = 0$ the sequence is seen to be purely deterministic, or singular.** One would expect that a regular, or purely nondeterministic, process has a representation as

$$X_n = \sum_{j=0}^{\infty} c_j \epsilon_{n-j},$$

where $\epsilon$ is an orthonormal $WN(1)$ process. We state this as a proposition without proof.

---

of matrices, say $\{B_j : j \in \mathcal{N}_0\}$, such that for every $n$

$$X_{n+1} = \sum_{s=0}^{\infty} B_s X_{n-s} = \hat{X}_{n+1}, \quad \text{in the sense that} \quad \|X_{n+1} - \hat{X}_{n+1}\|^2 = 0.$$

When $X_n$ is a scalar process the entities $B_j$ are simply scalars.

[6] For a vector process to be purely deterministic the equivalent requirement is that $\| X_{n+1}' - P_n(X_{n+1}') \|^2 = 0$. Note that this implies that the covariance matrix of the difference is null. In the following discussion it is implicitly assumed that there are no singularities, and that if the process is not purely deterministic, the covariance matrix of the difference is **nonsingular**.

**Proposition 6.** Let $X$ be a sequence in $\mathcal{H}^2(\Omega, \mathcal{A}, \mathcal{P})$; the sequence $X$ is nondeterministic, or regular, if and only if there exists a $WN(1)$ (orthonormal) sequence $\epsilon \in \mathcal{H}^2(\Omega, \mathcal{A}, \mathcal{P})$, and a sequence $c = \{c_j : c_j \in \mathcal{C}, j \in \mathcal{N}_0\}$ such that

$$X_n = \sum_{j=0}^{\infty} c_j \epsilon_{n-j}.$$

The Wold decomposition theorem can now be stated.

**Proposition 7** (Wold's Decomposition). Let $X$ be a real, zero mean, covariance stationary (vector) sequence in $\mathcal{H}^2(\Omega, \mathcal{A}, \mathcal{P})$. There exists a unique purely deterministic or singular process, say $X^d$; a unique purely nondeterministic or regular process, say $X^{nd}$; an orthonormal set ($MWN(I_q)$ process) $\epsilon$; and matrices $B = \{B_j : B_j \in \mathcal{C}, j \in \mathcal{N}_0\}$, $C = \{C_j : C_j \in \mathcal{C}, j \in \mathcal{N}_0\}$ such that

$$X_{n\cdot}' = X_{n\cdot}^{d'} + X_{n\cdot}^{nd'}, \quad X_{n\cdot}^{nd'} = \sum_{j=0}^{\infty} C_j \epsilon_{n-j\cdot}',$$

$$X_n^{d'} = \sum_{j=0}^{\infty} B_j X_{n-1-j\cdot}', \quad \sum_{j=0}^{\infty} \| B_j \|^2 < \infty, \quad \sum_{j=0}^{\infty} \| C_j \|^2 < \infty.$$

Proof: As before, let $\mathcal{M}_t$ be the closed linear manifold generated by $\{X_{j\cdot} : -\infty < j \leq t\}$, let $P_{t-1}$ be the projection operator that takes $X_{t\cdot}$ into $\mathcal{M}_{t-1}$, and define

$$\epsilon_{t\cdot}^{*'} = X_{t\cdot}' - P_{t-1}X_{t\cdot}'. \tag{2.14}$$

Two properties are immediately obvious from Eq. (2.14); first, $\epsilon_{t\cdot}^* \in \mathcal{M}_t$ and second, $\epsilon_{t\cdot}^* \in \mathcal{M}_{t-1}^{\perp}$. We next show that it is an **orthogonal** sequence and derive from it an **orthonormal** sequence $\{\epsilon_{t\cdot} : t \in \mathcal{N}\}$, which is $MWN(I_q)$. To this end consider $< \epsilon_{t\cdot}^*, \epsilon_{t+\tau}^* >$, $\tau > 0$. Evidently,

$$\epsilon_{t+\tau}^* \in \mathcal{M}_{t+\tau} \supset \mathcal{M}_{t-1}. \tag{2.15}$$

Because $\epsilon_{t\cdot}^* \in \mathcal{M}_{t-1}^{\perp}$ and Eq. (2.15) implies $\mathcal{M}_{t+\tau}^{\perp} \subset \mathcal{M}_{t-1}^{\perp}$, we conclude that

$$< \epsilon_{t\cdot}^*, \epsilon_{t+\tau}^* > = 0 \tag{2.16}$$

in view of the fact that $\epsilon_{t+\tau}^* \in \mathcal{M}_{t+\tau}$. Since $X$ is a (real) zero mean, weakly stationary process we conclude that $E\epsilon_{t\cdot}^* = 0$. Its covariance matrix is given by

$$E\epsilon_{t\cdot}^{*\prime}\epsilon_{t\cdot}^{*} = \sum_{j=0}^{\infty}\sum_{j'=0}^{\infty} B_j^* K(j'-j) B_{j'}^{*\prime} = \Phi, \qquad (2.17)$$

which evidently does not depend on $t$, where $B_0^* = I_q$ and $B_j^* = -B_j$. By the conventions adopted $\Phi > 0$ so that it has a triangular decomposition, [see Dhrymes (1984), pp. 68-69], say $\Phi = T'T$. Define now the sequence

$$\epsilon_{t\cdot} = \epsilon_{t\cdot}^* T^{-1}, \quad t \in \mathcal{N}, \qquad (2.18)$$

and note that $E\epsilon_{t\cdot}'\epsilon_{t\cdot} = I_q$. Hence

$$\epsilon = \{\epsilon_{t\cdot} : t \in \mathcal{N}\} \qquad (2.19)$$

is an $MWN(I_q)$ process as claimed. Let $\mathcal{G}_n$ be the closed linear manifold generated (spanned) by $\{\epsilon_{s\cdot} : -\infty < s \le n\}$ and define by $Q_t$ the projection operator that takes $X_{t\cdot}$ into $\mathcal{G}_t$, so that

$$Q_t X_{t\cdot}' = \sum_{s=0}^{\infty} C_s \epsilon_{t-s\cdot}', \quad C_s = EX_{t\cdot}'\epsilon_{t-s\cdot}. \qquad (2.20)$$

and define

$$S_{t\cdot}' = X_{t\cdot}' - Q_t X_{t\cdot}'. \qquad (2.21)$$

We now show that the closed linear manifold, $\mathcal{F}_t$, generated by $\{S_{j\cdot} : -\infty < j \le t\}$ obeys

$$\mathcal{F}_t = \bigcap_{j=0}^{\infty} \mathcal{M}_{t-j}(X), \quad \text{and moreover,} \quad \mathcal{F}(S) = \bigcap_{t=-\infty}^{\infty} \mathcal{F}_t = \mathcal{M}(X).$$

It is evident that for any $j \in \mathcal{N}_0$,

$$ES_{t\cdot}'\epsilon_{t-j\cdot} = EX_{t\cdot}'\epsilon_{t-j\cdot} - \sum_{s=0}^{\infty} C_s E\epsilon_{t-s\cdot}'\epsilon_{t-j\cdot} = C_j - C_j = 0.$$

Since

$$X_{t-j\cdot}' = P_{t-1-j}X_{t-j\cdot}' + \epsilon_{t-j\cdot}^*,$$

$P_{t-1-j}X_{t-j\cdot}' \in \mathcal{M}_{t-1-j} \subset \mathcal{M}_{t-1}$, and $\epsilon_{t\cdot} \in \mathcal{M}_{t-1}^{\perp}$, it follows that

$$ES_{t-j\cdot}'\epsilon_{t\cdot} = 0, \quad \text{for all } j \in \mathcal{N}_+,$$

so that the two sequences $S = \{S_{t\cdot} : t \in \mathcal{N}\}$ and $\epsilon = \{\epsilon_{t\cdot} : t \in \mathcal{N}\}$ are mutually uncorrelated. To complete the proof, we show that for all $t$

$S_{t.} \in \mathcal{M}(X)$, thereby showing that it is purely deterministic. To do so we note that $S_{t.} \perp \mathcal{G}_{t-1}$, for all $t$; this implies that $S_{t.} \in \mathcal{M}_{t-2}$. Proceeding in this fashion we may thus show that, for all $t$,

$$S_{t.} \in \bigcap_{j=0}^{\infty} \mathcal{M}_{t-j}, \text{ and consequently } \mathcal{F}(S) \subseteq \mathcal{M}(X). \tag{2.22}$$

Conversely, consider a process $Z \in \bigcap_{j=0}^{\infty} \mathcal{M}_{t-j}(X)$. This means that $Z \in \mathcal{M}_{s-1}$ for every $s$. Consequently, $EZ'\epsilon_{s.} = 0$ for every $s$. In turn, this means that $Z \in \mathcal{F}_s$ for every $s$, or in other words $\mathcal{F}_t \supseteq \bigcap_{j=0}^{\infty} \mathcal{M}_{t-j}$. We conclude that $\mathcal{F}_t = \bigcap_{j=0}^{\infty} \mathcal{M}_{t-j}$, and moreover that $\mathcal{F}(S) = \mathcal{M}(X)$. Since $S_{t.} \in \mathcal{F}_t$ for every $t$, $S$ is a purely deterministic process. Putting $X_{t.}^d = S_{t.}$ and $X_{t.}^{nd} = \sum_{s=0}^{\infty} C_s \epsilon_{t-s.}$, where the latter is purely nondeterministic, we have proved that every (real) zero mean covariance stationary process may be written as

$$X_{t.} = X_{t.}^d + X_{t.}^{nd},$$

such that the two processes in the right member are mutually uncorrelated.

The uniqueness aspect is not pursued here since it would add nothing to the reader's understanding of the nature and implications of this result. If one requires greater rigor or detail, one is referred to Brockwell and Davis (1991), pp. 187-191.

# 2.4 Best Linear Predictors (BLP)

## 2.4.1 Preliminaries

In this section we examine a number of the stochastic sequences introduced in Chapter 1 and deal with problems of prediction and estimation associated with them. We begin by recalling from the previous chapter that $X_{t.} \in \mathcal{H}^2(\mathcal{P})$ is said to be **stationary**, and **causal**, or **physically realizable**, if and only if there exists a $MWN(0, \Sigma)$, say $\epsilon_{t.}$ and nonrandom matrices $\{B_j : j \in \mathcal{N}_0\}$ such that

$$X_{t.}' = \sum_{j=0}^{\infty} B_j \epsilon_{t-j.}', \tag{2.23}$$

the convergence of the right member being either $L^2$, or a.c.

**Remark 5**. The term "physically realizable" has an obvious interpretation. The term "stationary" was defined in Chapter 1; the term "causal" is common in the literature of statistics and means precisely physically realizable, i.e. that the behavior of $X$ "today" depends only on events occurring today ($\epsilon_t$.) and earlier ($\epsilon_{t-i}$, $i \geq 1$), but not on what will happen in the future ($\epsilon_{t+j}$, $j \geq 1$). As we saw in Chapter 1 invertible AR or MA sequences have stationary and "causal" representation. The point of the example below is to show that this may also be the case for certain nonstationary sequences.

**Example 1**. Consider the AR(1) (scalar) process of the form

$$y_t = \gamma y_{t-1} + \epsilon_t.$$

If $|\gamma| < 1$, it is an **invertible process** by the arguments of Chapter 1; using the operator notation developed therein we find

$$y_t = \frac{I}{I - \gamma L} \epsilon_t = \sum_{j=0}^{\infty} \gamma^j \epsilon_{t-j},$$

so that $y$ is an invertible AR(1) process and thus "causal" or physically realizable.

Next consider the case where $|\gamma| > 1$; now the expansion above is not well defined, and evidently the sequence is not invertible. The question is whether the AR(1) viewed as a **difference equation** has a stationary solution. We may rewrite the AR(1) as

$$y_{t-1} = \lambda y_t - \lambda \epsilon_t, \quad \lambda = \frac{1}{\gamma}, \quad |\lambda| < 1.$$

Again using operator methods, we may solve the equation above using the **forward operator**, say $L^{-1}$, such that $L^{-1} y_t = y_{t+1}$. Doing so yields

$$y_t = -\frac{\lambda I}{I - \lambda L^{-1}} \epsilon_{t+1} = -\lambda \sum_{j=0}^{\infty} \lambda^j \epsilon_{t+1+j},$$

which is evidently a **stationary solution**; it is not, however, physically realizable or "causal". This shows that "causality" and stationarity need not be synonymous.

It is interesting nonetheless that we may represent the stationary solution above in standard form by reformulating the white noise process. To this end, premultiply by the **invertible operator** $(I - \lambda L)$ to obtain

$$(I - \lambda L)y_t = \eta_t, \quad \eta_t = -\frac{\lambda I - \lambda^2 L}{I - \lambda L^{-1}}\epsilon_{t+1} = \lambda^2\epsilon_t - (1 - \lambda^2)\sum_{j=1}^{\infty}\lambda^j\epsilon_{t+j}.$$

The result above is obtained as follows: Dividing $\lambda I - \lambda^2 L$ by $I - \lambda L^{-1}$ yields the quotient $-\lambda^2 L$ and the remainder $\lambda(1 - \lambda^2)I$, which, when expanded, results in the last member of the equation above.

We now show that $\eta$ is a sequence of uncorrelated random variables with mean zero and variance $\lambda^2\sigma^2$, i.e. that $\{\eta_t : t \in \mathcal{N}\}$ is a $WN(\lambda^2\sigma^2)$ process, as required. To this end, it is evident that the process has mean zero; moreover, for $|s| \geq 0$,

$$E\eta_t\eta_{t+s} = \lambda^4 E\epsilon_{t+s}\epsilon_t - \lambda^2(1 - \lambda^2)\sum_{j=1}^{\infty}\lambda^j E\epsilon_{t+s+j}\epsilon_t - \lambda(1 - \lambda^2)$$

$$\times \sum_{r=1}^{\infty}\lambda^r E\epsilon_{t+s}\epsilon_{t+r} + (1 - \lambda^2)^2\sum_{j=1}^{\infty}\sum_{r=1}^{\infty}\lambda^{j+r}E\epsilon_{t+s+j}\epsilon_{t+r}$$

$$= 0 - 0 - \lambda^2(1 - \lambda^2)\lambda^{|s|} + \lambda^{|s|}(1 - \lambda^2)^2\sum_{j=1}^{\infty}\lambda^{2j} = 0, \quad \text{for } s \neq 0,$$

$$= \text{Var}(\eta_t) = \lambda^2\sigma^2, \quad \text{for } s = 0,$$

thus establishing that $\eta$ is a white noise process.

Unfortunately, however, the same argument cannot be employed in the case where $|\gamma| = \pm 1$.

In this volume we deal with invertible and integrated processes only; we **do not deal** with explosive processes. As the example suggests there is very little loss of relevance in ignoring explosive processes, ( $|\gamma| > 1$ ), since they generally possess stationary solutions, which are equivalent to invertible processes ( $|\lambda| < 1$ ) with somewhat altered white noise. In subsequent chapters we deal with processes for which $\gamma = 1$, but **not** with processes with **complex or negative** unit roots.

## 2.4.2 Derivation of BLP

The content of this section is best understood as the solution of this problem: Suppose we are given the autocovariance function $\kappa(\cdot)$, for scalar processes, or the autocovariance kernel, $K(\cdot)$, for multivariate or vector processes. How can we produce the best linear predictor of the process realization at time $T+1$, given realizations (observations) of the process for $t = 1, 2, \ldots, T$?[7]

To this end, consider the process $\{X_t. : t \in \mathcal{N}\}$ defined on $\mathcal{H}^2(\mathcal{P})$ and let it be desired to "predict" (linearly) $X_{T+1}.$ given $X_{(T)} = \{X_t. : t = 1, 2, \ldots T\}$. We begin with a definition.

**Definition 5.** Let $\{X_t. : t \in \mathcal{N}_+\}$ be defined on $\mathcal{H}^2(\mathcal{P})$. The best linear predictor of $X_{T+h}.$, $h \geq 1$, is the linear combination $\sum_{j=1}^{T} X_{T-j}. A_j'$, that minimizes (with respect of the matrices $A_j$, subject to $A_0 = I_q$) the function

$$\phi_{T+h}^2 = \left\| X_{T+h}. - \sum_{j=1}^{T} X_{T-j}. A_j' \right\|^2 \tag{2.24}$$

or, alternatively, the projection of $X_{T+h}.$ on the linear manifold $M_T$, spanned by $\{X_t. : t = 1, 2, \ldots, T\}$. This projection is denoted by

$$\hat{X}_{T+h}. = P_T(X_{T+h}.|X_1., X_2., \ldots X_T.), \quad \text{or more simply}, \quad \hat{X}_{T+h}. = P_T(X_{T+h}.).$$

When $h = 1$, this is termed the "one-step-ahead predictor"; when $h > 1$, it is termed the "many-steps-ahead predictor".

Utilizing the apparatus developed in earlier sections, let $M_T$ be the linear manifold generated by $X_{(T)}$, and let $P_T$ be the corresponding projection operator. The desired "one-step-ahead prediction" is given by

$$\hat{X}_{T+1}.' = P_T(X_{T+1}.'|X_1., X_2., \ldots, X_T.)$$

and, from Definition 5, the BLP is obtained by minimizing

$$\left\| X_{T+1}. - \hat{X}_{T+1}. \right\|^2 = \phi_{T+1}^2, \tag{2.25}$$

which is the mean squared prediction error (actually the trace of the mean squared error matrix). Since the projection operator **minimizes** this error and the projection takes $X_{T+1}.$ into the **linear manifold**, the term "best linear predictor" is completely justified. We now derive an expression for the

---

[7] In all subsequent discussions in this chapter we deal with **real processes**.

$h$-steps ahead BLP, of which the one-step-ahead BLP is a special case with $h = 1$. Since

$$X_{T+h\cdot} = \sum_{j=1}^{T} X_{T+1-j\cdot} B_{T,j}, \qquad (2.26)$$

we now determine the form of the matrices $B_{T,j+1}$. If we define

$$x_{(T)} = (X_{1\cdot}, X_{2\cdot}, \dots, X_{T\cdot}), \quad B_T = (B'_{T,1}, B'_{T,2}, \dots, B'_{T,T})'), \qquad (2.27)$$

we can write Eq. (2.26) as

$$X_{T+h\cdot} = x_{(T)} B_T; \qquad (2.28)$$

utilizing the result in Proposition 90, pp. 105-107, Dhrymes (1984), we may rewrite Eq. (2.25) as

$$\phi^2_{T+h} = < X'_{T+h\cdot} - \text{vec}(x_{(T)} B_T), \ X'_{T+h\cdot} - \text{vec}(x_{(T)} B_T) >, \quad \text{or}$$

$$\phi^2_{T+h} = \text{tr} E \left( X'_{T+h\cdot} X_{T+h\cdot} - X'_{T+h\cdot} b'_T (I_q \otimes x'_{(T)}) - (I_q \otimes x_{(T)}) b_T X_{T+h\cdot} \right)$$

$$+ \text{tr} E b_T b'_T (I_q \otimes x'_{(T)} x_{(T)}), \quad b_T = \text{vec}(B_T). \qquad (2.29)$$

The first and last terms in the rightmost member above are easily dealt with. The second term may be evaluated as follows:

$$E\text{tr}(X'_{T+h\cdot} b'_T (I_q \otimes x'_{(T)}) = E b'_T (I_q \otimes x'_{(T)}) X'_{T+h\cdot}$$

$$= E\text{vec}(A'_1)'(I_q \otimes A_2)\text{vec}(A_3), \quad \text{with}$$

$$A_1 = B'_T, \quad A_2 = x'_{(T)}, \quad A_3 = X_{T+h\cdot},$$

so that what we are dealing with is

$$E b'_T (I_q \otimes x'_{(T)}) X'_{T+h\cdot} = \text{tr} B'_T E x'_{(T)} X_{T+h\cdot} = \text{vec}(\kappa'_{(T,h)})' b_T,$$

where

$$\kappa_{(T,h)} = E X'_{T+h\cdot} x_{(T)} = [K(T+h-1), K(T+h-2), \dots, K(h)]'. \qquad (2.30)$$

It follows, therefore, that

$$\phi_{T+h} = \text{tr} K(0) - 2\text{vec}(\kappa'_{(T,h)})' b_T + b'_T K_{(T)} b_T, \quad K_{(T)} = [K(i-j)], \qquad (2.31)$$

i.e.  $K_{(T)}$ is a **block** matrix whose (i, j) block is defined by $K(i - j) = EX'_{i.}X_{j.}$ .

To determine the coefficient matrices of the BLP, we simply minimize the expression above with respect to the vector $b_T$ , thus obtaining

$$b_T = (I_q \otimes K_{(T)})^{-1}\text{vec}(\kappa'_{(T,h)}), \quad \text{or} \quad B_T = K_{(T)}^{-1}\kappa'_{(T,h)}. \qquad (2.32)$$

Evidently, the **one-step-ahead BLP**, based on $T$ observations is given by

$$\hat{X}_{T+1.} = x_{(T)}B_T, \quad \text{with} \quad B_T = K_{(T)}^{-1}\kappa'_{(T,1)}. \qquad (2.33)$$

Often it is convenient or necessary to produce the BLP repeatedly, and in such a case the successive inversion of matrices of the form $K_{(T)}$ becomes quite onerous as $T$ increases. The burden may be eased by using the partitioned inverse form [see Dhrymes (1984), pp. 37-39] and the identity

$$(C + XDY)^{-1} = C^{-1}X(D^{-1} + YC^{-1}X)^{-1}YC^{-1}.$$

However, there are more efficient algorithms that rely on econometric theory motivation, and we pursue those instead.

### Durbin-Levinson-Whittle (DLW) Algorithm

In the DLW algorithm, given in this form by Whittle (1963), we employ not only a "forward" but also a "backward" predictor. For this reason we distinguish between the two by denoting the matrices defining the forward predictor by $F_{T,j}$ and those that define the backward predictor by $B_{T,T-j}$ . The basic idea of this algorithm is to take the linear manifold $M_T$ and partition it into two orthogonal subspaces. By writing the BLP in the corresponding two forms, we obtain an identity that connects the matrices $F_{T,j}$ defining the operator, say $P_T$, that takes $X_{T+1.}$ into the manifold $M_T$ , and the matrices $F_{T-1,j}$ . The price we pay is that we need a backward predictor as well.

Let $P_T$ be the projection operator based on $T$ observations and define

$$\hat{X}'_{T+1.} = P_T(X_{T+1.}|X_{T.}, X_{T-1.}, \ldots, X_{1.}). \qquad (2.34)$$

This is the usual one-step-ahead BLP derived earlier and which in this discussion can be represented, according to our new convention, by

$$\hat{X}'_{T+1.} = \sum_{j=1}^{T} F_{T,j} X'_{T+1-j.} \qquad (2.35)$$

Consider next the subspace generated by $\{X_{j.} : j = 2, 3, \ldots, T\}$ and the projection operator based on $T-1$ observations, i.e. $P_{T-1}$. We may certainly use this operator to "backward predict" [8]

$$\hat{X}'_{1.} = P_{T-1}(X_{1.}|X_{2.}, X_{3.}, \ldots, X_{T.}) = \sum_{j=1}^{T-1} B_{T-1,T-j} X'_{T+1-j.} \qquad (2.36)$$

Evidently, the residual vector, say

$$S_{1.} = X_{1.} - \hat{X}_{1.}, \qquad (2.37)$$

**is orthogonal** to the space generated by the set $\{X_{j.} : j = 2, 3, \ldots, T\}$. Consequently, if we append to that set the residual vector $S_{1.}$ we have an alternative basis for the manifold $M_T$ and are able to write the alternative representation of its projection as

$$\hat{X}'_{T+1.} = P_{T-1}(X_{T+1.}|X_{2.}, X_{3.}, \ldots, X_{T.}) + D S'_{1.}, \qquad (2.38)$$

where $D$ is chosen so that $X_{T+1} - S_1 D' \perp S_{1.}$. This amounts therefore to writing the projection operator $P_T$ in terms of $P_{T-1}$, plus another operator which is orthogonal to it. Evidently, we can rewrite Eq. (2.38) as

$$\hat{X}'_{T+1.} = \sum_{j=1}^{T-1} F_{T-1,j} X'_{T+1-j.} + D X'_{1.} - D \sum_{j=1}^{T-1} B_{T-1,T-1-j} X'_{T+1-j.}$$

$$= \sum_{j=1}^{T-1} \left( F_{T-1,j} - D B_{T-1,T-j} \right) X'_{T+1-j.} + D X'_{1.} \qquad (2.39)$$

Comparing with Eq. (2.35) we conclude

$$F_{T,j} = F_{T-1,j} - D B_{T-1,T-j}, \quad j = 1, 2, \ldots, T-1, \quad F_{T,T} = D. \qquad (2.40)$$

---

[8] This is certainly a verbal monstrosity; the alternative is to use the term "backcast", by analogy with "forecast", which is not any more agreeable. By the same standard the term "postdict" in contradistinction to predict, at least preserves an all Latin root word!

To determine the matrix $D$ we note from its definition that

$$DES'_1.S_1. = EX'_{T+1}.S_1. = E[X'_{T+1}. - P_{T-1}(X_{T+1\cdot|X_2.,\ldots,X_T.})]X_1.$$

$$= K(T) - \sum_{j=1}^{T-1} F_{T-1,j}K(T-j) = Q_{1,T-1},$$

$$D = Q_{1,T-1}V_{0,T-1}^{-1}, \quad V_{0,T-1} = ES'_1.S_1.. \tag{2.41}$$

The entity $V_{0,T-1}$ is recognized as the **mean squared error matrix** of the one-step-**backward** BLP based on the subsequent $T-1$ observations.

We now produce a similar recursion for the coefficient matrices of the backward predictor. By definition, the backward predictor based on $T$ observations is given by

$$\hat{X}'_0. = P_T(X_0.|X_1., X_2., \ldots, X_T.) = \sum_{j=1}^{T} B_{T,T+1-j}X'_{T+1-j}., \tag{2.42}$$

and thus carries $X_0.$ into the linear manifold $M_T$; we may define a subspace of the latter spanned by $\{X_j. : j = 1, 2, \ldots, T-1\}$ and let $P_{T-1}$ be the operator that projects entities into that subspace. In particular, define

$$S'_T. = X'_T. - P_{T-1}(X_T.|X_1., X_2., \ldots, X_{T-1}.) = X'_T. - \sum_{j=1}^{T-1} F_{T-1,j}X'_{T-j}. \tag{2.43}$$

By construction $S_T.$ is orthogonal to every vector in the linear manifold spanned by $\{X_j. : j = 1, 2, \ldots, T-1\}$. Hence, we may write

$$\hat{X}'_0. = \sum_{j=1}^{T-1} B_{T-1,T-j}X'_{T-j}. + HS'_T.,$$

$$= \sum_{j=1}^{T-1}(B_{T-1,T-j} - HF_{T-1,j})X'_{T-j}. + HX'_T., \tag{2.44}$$

and $H$ is chosen by the condition that $X_0. - S_T.H' \perp S_T.$, i.e.

$$HES'_T.S_T. = EX'_0.S_T. = E(X'_0. - \sum_{j=1}^{T-1} B_{T-1,T-j}X'_{T-j}.)X_T.$$

$$= K(-T) - \sum_{j=1}^{T-1} B_{T-1,T-j}K(-j) = Q_{0,T-1};$$

$$ES'_{T.}S_{T.} = ES'_{T.}X_{T.} = K(0) - \sum_{j=1}^{T-1} F_{T-1,j}K(-j) = V_{1,T-1},$$

$$H = Q_{0,T-1}V_{1,T-1}^{-1}. \tag{2.45}$$

The entity $V_{1,T-1}$ is recognized as the mean squared error matrix of the one-step-**forward** predictor based on the previous $T-1$ observations. Comparing Eqs. (2.44) and (2.42) we have the result

$$B_{T,T-j} = B_{T-1,T-j} - B_{T,T}F_{T-1,j}, \quad j = 1, 2, \ldots, T-1, \quad B_{T,T} = H, \tag{2.46}$$

which completes the recursion. We have therefore proved the following proposition.

**Proposition 8** (Durbin-Levinson-Whittle Algorithm). Let $\{X_{t.} : t \in \mathcal{N}\}$ be a covariance stationary process. The one-step-forward BLP produced by the DLW algorithm and based on the preceding $T$ observations is

$$\hat{X}'_{T+1.} = \sum_{j=1}^{T} F_{T,j}X'_{T+1-j.}, \quad \text{for } T \geq 1$$

$$= 0, \quad \text{for } T = 0.$$

Moreover, the coefficient matrices of the BLP of Eq. (2.35) are given recursively by

$$F_{T,T} = Q_{1,T-1}V_{0,T-1}^{-1}, \quad B_{T,T} = Q_{0,T-1}V_{1,T-1}^{-1}$$

$$B_{T,T-j} = B_{T-1,T-j} - B_{T,T}F_{T-1,j},$$

$$F_{T,j} = F_{T-1,j} - F_{T,T}B_{T-1,T-j}, \quad j = 1, 2, \ldots, T-1$$

$$Q_{0,T-1} = K(-T) - \sum_{j=1}^{T-1} B_{T-1,T-j}K(-j),$$

$$Q_{1,T-1} = K(T) - \sum_{j=1}^{T-1} F_{T-1,j}K(T-j),$$

$$V_{0,T-1} = K(0) - \sum_{j=1}^{T-1} B_{T-1,T-j}K(T-j),$$

$$V_{1,T-1} = K(0) - \sum_{j=1}^{T-1} F_{T-1,j} K(-j),$$

$$V_{0,0} = V_{10} = K(0), \quad Q'_{0,0} = Q_{1,0} = K(1), \tag{2.47}$$

where the last equation set simply states the initial conditions.

**Remark 6.** Notice that beginning with the initial conditions we are able to determine $F_{1,1}$, $B_{1,1}$, from the first equation in Eq. (2.47).

Given the latter we may compute

$$V_{0,1} = K(0) - B_{1,1} K(1), \quad V_{1,1} = K(0) - F_{1,1} K(-1),$$

$$Q_{0,1} = K(-2) - B_{1,1} K(-1), \quad Q_{1,1} = K(2) - F_{1,1} K(1), \quad \text{and thus}$$

$$F_{2,2} = Q_{1,1} V_{0,1}^{-1}, \quad B_{2,2} = Q_{0,1} V_{1,1}^{-1}$$

$$F_{2,1} = F_{1,1} - F_{2,2} B_{1,1}, \quad B_{2,1} = B_{1,1} - B_{2,2} F_{1,1},$$

and so on.

**The Innovations Algorithm**

Let $X$ be a zero mean (at least) covariance stationary process; suppose we have observations $\{X_{j\cdot} : j = 1, 2, \ldots, T\}$ and want to obtain the BLP of $X_{T+1\cdot}$. As before we define the BLP

$$\hat{X}'_{t\cdot} = P_{t-1}(X_{t\cdot}|X_{t-1\cdot}, X_{t-2\cdot}, \ldots, X_{1\cdot}) = \sum_{j=1}^{t-1} F_{t-1,j} X'_{t-j\cdot}, \tag{2.48}$$

and the associated entities

$$S_{t\cdot} = X_{t\cdot} - \hat{X}_{t\cdot}, \quad t = 1, 2, \ldots, T. \tag{2.49}$$

The entities in Eq. (2.49) are termed **innovations**.[9] Two aspects need be noted: (1) They are very nearly the product of a Gram-Schmidt-like

---

[9] If the underlying vectors are jointly normal, and we place our discussion in the context of a probability space, we may define the stochastic basis $\{\mathcal{G}_t : \mathcal{G}_t = \sigma(X_{s\cdot}, 1 \leq s \leq t, t \in \mathcal{N}_+\}$, and note that under joint normality $\hat{X}_{t\cdot} = E[X_{t\cdot}|\mathcal{G}_{t-1}]$; therefore, $E[S_{t\cdot}|\mathcal{G}_{t-1}] = 0$. Since the $S_{t\cdot}$ are certainly integrable, and $\mathcal{G}_t$-measurable, it follows that they are a sequence of **martingale differences**. If the joint normality assumption is dropped, they would not necessarily be martingale differences unless we redefined $\hat{X}_{t\cdot}$ to be $E[X_{t\cdot}|\mathcal{G}_{t-1}]$.

orthogonalization procedure, except that their variances are not normalized; (2) they are mutually **orthogonal** and, moreover, they are a **basis** for the linear manifold $M_T$.

Hence any vector in $M_T$ may be written as a linear combination of the innovations, and we may define the BLP by

$$\hat{X}'_{T+1\cdot} = \sum_{j=1}^{T} C_{T,j} S'_{T+1-j\cdot} \tag{2.50}$$

We now derive a recursion for computing the coefficient matrices $C_{T,j}$ and the covariance matrix of the prediction error $S_{T+1\cdot}$. To this end, post multiply by $S_{r+1\cdot}$ to obtain

$$E\hat{X}'_{T+1\cdot} S_{r+1\cdot} = \sum_{j=1}^{T} C_{T,j} E S'_{T+1-j\cdot} S_{r+1\cdot} = C_{T,T-r} V_r, \quad V_r = E S'_{r+1\cdot} S_{r+1\cdot} \tag{2.51}$$

Since

$$S_{T+1\cdot} \perp S_{r+1\cdot}, \text{ we find } E\hat{X}'_{T+1\cdot} S_{r+1\cdot} = E X'_{T+1\cdot} S_{r+1\cdot},$$

so that

$$C_{T,T-r} V_r = E X'_{T+1\cdot} S_{r+1\cdot} = K(T-r) - \sum_{j=1}^{r} E X'_{T+1\cdot} S_{r+1-j\cdot} C'_{r,j}$$

$$= K(T-r) - \sum_{j=1}^{r} C_{T,T-(r-j)} V_{r-j} C'_{r,j}$$

$$= K(T-r) - \sum_{i=0}^{r-1} C_{T,T-i} V_i C'_{r,r-i}, \tag{2.52}$$

which provides the desired recursion for $r = 0, 1, 2, \ldots T - 1$. Finally, since $X'_{T+1\cdot} = S'_{T+1\cdot} + \sum_{j=1}^{T} C_{T,j} S'_{T+1-j\cdot}$, and the $S_{j\cdot}$ are **mutually orthogonal**, we find, taking the covariance matrix of both sides,

$$K(0) = V_T + \sum_{j=0}^{T-1} C_{T,T-j} V_j C'_{T,T-j}. \tag{2.53}$$

The third equality in the equation set Eq. (2.52) is valid because of Eq. (2.51) and the third results from the change in index $i = r - j$.

We have therefore proved the following proposition.

**Proposition 9.**  Under the conditions of Proposition 8, the innovations algorithm yields the one-step-ahead BLP

$$\hat{X}'_{T+1\cdot} = \sum_{j=1}^{T} C_{T,j} S'_{T+1-j\cdot}, \quad \text{for } T \geq 1,$$

$$= 0 \quad \text{for } T = 0,$$

where $S_{j\cdot}$ are as defined in Eqs. (2.48) and (2.49), with prediction error covariance matrix [10]

$$V_T = E S'_{T+1\cdot} S_{T+1\cdot} = K(0) - \sum_{j=0}^{T-1} C_{T,T-j} V_j C'_{T,T-j}.$$

The coefficient matrices are given by the recursions

$$C_{s,s-r} = \left( K(s-r) - \sum_{i=0}^{r-1} C_{s,s-i} V_i C'_{r,r-i} \right) V_r^{-1},$$

for $r = 0, 1, 2, \ldots, s-1$ and $s = 1, 2, 3, \ldots, T$, with **initial condition** $V_0 = K(0)$.

**Remark 7.**  Given the initial condition, we obtain for $s = 1$

$$C_{1,1} = \left( K(1) - C_{1,1} K(0) C'_{0,0} \right) = K(1)[K(0)]^{-1},$$

since by convention matrices of the form $C_{0,s}$ or $C_{s,0}$ are null for any $s$. Next we obtain $V_1 = K(0) - C_{1,1} K(0) C'_{1,1}$.

For $s = 2$ we find

$$C_{2,2} = K(2)[K(0)]^{-1} \quad \text{(for } r = 0),$$

$$C_{2,1} = \left( K(1) - C_{2,2}[K(0)]^{-1} C'_{1,1} \right) V_1^{-1}, \text{(for } r = 1),$$

$$V_2 = K(0) - \sum_{j=0}^{1} C_{2,2-j} V_j^{-1} C'_{2,2-j},$$

---

[10] More generally, we have

$$V_s = E S'_{s+1\cdot} S_{s+1\cdot} = K(0) - \sum_{j=0}^{s-1} C_{s,s-j} V_j C'_{s,s-j}.$$

and so on. Thus, the only matrix inversion involved in this algorithm is the inversion of the residual covariance matrices $V_j$.

**Remark 8.** If we begin with the supposition that we are dealing with stationary processes that are representable as $VAR(\infty)$ or $MMA(\infty)$, or both, the intuition behind the DLW algorithm is that it produces successively the coefficient matrices in

$$X'_{t\cdot} = \sum_{j=1}^{\infty} F_j X'_{t-j\cdot} + \epsilon'_{t\cdot},$$

except that it is based only on observations for $t \in \mathcal{N}_+$. The intuition behind the innovations algorithm, under the same proviso, is that it produces successively the coefficient matrices in

$$X'_{t\cdot} = \sum_{i=1}^{\infty} A_i \epsilon'_{t-i\cdot} + \epsilon'_{t\cdot}.$$

This leads us to the **conjecture** that the entities $V_{1,T}$ and $V_T$ of the DLW and innovation algorithms, respectively, converge to the covariance matrix of the white noise process that defines the two representations. We give a demonstration (although not a rigorous proof) of this fact in the case of the DLW algorithm. To this end, suppose the true model is

$$X'_{t\cdot} = \sum_{j=1}^{n} B_j X'_{t-j\cdot} + \epsilon'_{t\cdot},$$

and its characteristic equation $|I_q - \sum_{j=1}^{n} B_j z^j| = 0$ has all roots outside the unit circle so that the sequence is stationary and invertible. However, the order, $n$, of the VAR is **unknown** to the investigator. The process of obtaining the BLP, based on $T$ observations, yields

$$S'_{T+1\cdot} = X'_{T+1\cdot} - \hat{X}'_{T+1\cdot} = \epsilon'_{T+1\cdot} - \sum_{j=1}^{T}(F_{T,j} - B_j)X'_{T+1-j\cdot},$$

it being understood that $B_j = 0$, $j > n$. The BLP procedure obtains the coefficient matrices $F_{T,j}$ by finding the global minimum of

$$\mathrm{tr} E S'_{T+1\cdot} S_{T+1\cdot} = \mathrm{tr}\left[\Sigma + \sum_{i=1}^{T}\sum_{j=1}^{T}(F_{T,j} - B_j)K_x(i-j)(F_{T,i} - B_i)'\right].$$

From Proposition 2 in Chapter 1 the autocovariance kernel is a positive semidefinite sequence; if we also assume that the matrix $K_{(T)}$ of Eq. (2.31) is positive definite, then in the limit the expression above is minimized only if

$$\lim_{T \to \infty} F_{T,j} = B_j, \quad j = 1, 2, \ldots, n$$

and otherwise converges to zero. This shows, however, that

$$\lim_{T \to \infty} V_{1,T} = \lim_{T \to \infty} E S'_{T+1.} S_{T+1.} = \Sigma.$$

A more roundabout way would be to use the stationarity of the roots of the system to argue that $\lim_{\tau \to \infty} K(\tau) = 0$; since the coefficient matrices depend on the autocovariance kernel, $K(\cdot)$, the result would imply that $\| F_{T,j} - B_j \| \to 0$, where here the notation $\| \cdot \|$ indicates the usual Euclidean norm in $R^k$.

An similar argument could be made regarding the $MMA(\infty)$ representation and the innovations algorithm.

**Remark 9.** An interesting by-product of the discussion in Remark 8 is that the residual vectors, i.e. the innovations $S_{T+1.}$, behave, for large $T$, like the white noise realizations $\epsilon_{T+1.}$. To see this, utilize the representation of the preceding remark to obtain

$$\psi^{*'}_{T+1.} = S'_{T+1.} - \epsilon'_{T+1.} = - \sum_{j=1}^{T} (F_{T,j} - B_j) X'_{T+1-j.},$$

$$\| \psi^{*}_{T+1.} \|^2 = \, < \psi^{*}_{T+1.}, \, \psi^{*}_{T+1.} > \, = \operatorname{tr} E (S_{T+1.} - \epsilon_{T+1.})' (S_{T+1.} - \epsilon_{T+1.})$$

$$= \operatorname{tr} \left[ \Sigma + \sum_{i=1}^{T} \sum_{j=1}^{T} (F_{T,j} - B_j) K_x (i-j) (F_{T,i} - B_i)' - \Sigma - \Sigma + \Sigma \right],$$

which **converges to zero with** $T$. Thus, we conclude

$$S_{T+1.} \xrightarrow{\mathrm{L}^2} \epsilon_{T+1.}.$$

**Remark 10** (Partial Autocorrelations). The partial autocorrelation coefficient between two random variables, say $z_1, z_2$ is defined as the correlation coefficient between them **in the context of their conditional distribution**, given an appropriate $\sigma$-field.

If the sequence $\{X_{t.} : t \in \mathcal{N}_+\}$ is Gaussian, we show that the parameter $D = F_{T,T}$ is the partial autocorrelation coefficient at lag $T$. If it is not Gaussian, we may still define it to be so by convention. We demonstrate this in the case of **scalar** sequences for simplicity of exposition.

To this end return to Eqs. (2.35) and (2.36), which give the BLP of $X_{T+1}$ and $X_1$, respectively, given the observations $X_{(2)} = \{X_j : j = 2, 3, \ldots T\}$. Under joint normality

$$P_{T-1}(X_{T+1}) = E(X_{T+1}|X_{(2)}), \quad P_{T-1}(X_1) = E(X_1|X_{(2)}).$$

Similarly, the conditional variances of $X_{T+1}$ and $X_1$, given $X_{(2)}$ are obtained, respectively, as

$$ES^2_{T+1}, \quad ES^2_1, \quad \text{with} \quad ES^2_1 = ES^2_{T+1},$$

where $S_{T+1} = X_{T+1} - P_{T-1}(X_{T+1})$ and $S_1 = X_1 - P_{T-1}(X_1)$. The equality of the two conditional variances may be argued in terms of the stationarity of the sequence or may be readily verified directly. Thus, by definition the conditional correlation coefficient in question, say $\rho_{T+1,1:2,3,\ldots,T}$, is

$$\rho_{T+1,1:2,3,\ldots,T} = \frac{< S_{T+1}, \; S_1 >}{\| S_1 \|^2}.$$

But the numerator of the fraction above is simply $Q_{1,T-1}$, and the denominator is $V_{0,T-1}$, thus leading to the conclusion that

$$\rho_{T+1,1:2,3,\ldots,T} = D = F_{T,T}.$$

Notice that $F_{T,T}$ is the coefficient of the most distant lag in the representation of Eq. (2.35). Thus, if it is "nearly zero", this will give an indication that the **order** of the autoregression is not too different from $T$.

## 2.4.3   BLP in $MARMA(m,n)$ Sequences

In this section we add to the requirements of the preceding discussion the condition that we predict $X_{T+1.}$ given $X_{(T)}$ as above **through**, or **utilizing**, a $MARMA(m,n)$ specification. Evidently, we may use the procedures developed above, but a very considerable simplification occurs if we utilize the following device, introduced into the literature by Ansley (1979). Thus, write

$$B(L)X'_{t\cdot} = A(L)\epsilon'_{t\cdot}, \quad B(L) = I_q - \sum_{j=1}^{m} B_j L^j, \quad A(L) = \sum_{i=0}^{n} A_i L^i, \quad A_0 = I_q.$$

$$(2.54)$$

The device in question consists of defining a new process

$$W'_{t\cdot} = X'_{t\cdot}, \quad \text{for } t = 1, 2, \ldots, p, \quad p = \max(m, n)$$

$$= B(L)X'_{t\cdot}, \quad t > p. \tag{2.55}$$

Note that for $t > p$, $W'_{t\cdot} = A(L)\epsilon'_{t\cdot}$. Thus, the autocovariance kernel of the $W$-process, say $K_w(\cdot, \cdot)$, is completely determined by the autocovariance kernel of the $X$-process, say $K_x(\cdot, \cdot)$, and the coefficient matrices of the AR and MA part, respectively, $B_j$, $A_i$, $j = 1, 2, \ldots, m$, $i = 1, 2, \ldots n$. Since we are dealing with real processes, $K(i-j) = K(j-i)'$. Thus, we need only determine such entities for $i \leq j$; the others may be found by transposition. In fact, we have

$$K_w(i,j) = EW'_{i\cdot}W_{j\cdot} = EX'_{i\cdot}X_{j\cdot} = K_x(i-j), \quad \text{for } 1 \leq i \leq j \leq p$$

$$= K_x(i-j) - \sum_{r=1}^{m} K_x(i-j+r)B'_r, \quad \text{for } 1 \leq i \leq p \leq j \leq 2p$$

$$= \sum_{i=0}^{n} A_i \Sigma A'_i, \quad \text{for } p < i \leq j \leq i + p$$

$$= 0, \quad \text{for } p < i \leq j, \text{ and } j > i + p. \tag{2.56}$$

Using the innovations algorithm we may produce the BLP of $W$ as follows:

$$\hat{W}'_{T+1\cdot} = \sum_{j=1}^{T} C_{T,j}(W'_{T+1-j\cdot} - \hat{W}'_{T+1-j\cdot}), \quad \text{for } 1 \leq T < p$$

$$= \sum_{j=1}^{n} C_{T,j}(W'_{T+1-j\cdot} - \hat{W}'_{T+1-j\cdot}), \quad \text{for } T \geq p. \tag{2.57}$$

To retrieve from Eq. (2.57) the BLP for $X$ we note that

$$\hat{W}'_{t\cdot} = \hat{X}'_{t\cdot}, \quad \text{for } t \leq p$$

$$= \hat{X}'_{t.} - \sum_{j=1}^{m} B_j X'_{t-j.}, \quad \text{for } t > p, \quad \text{so that}$$

$$W'_{t.} - \hat{W}'_{t.} = \sum_{j=1}^{m} B_j X'_{t-j.} + X'_{t.} - \hat{X}'_{t.}, \quad \text{for all } t. \tag{2.58}$$

Inserting this in Eq. (2.57), we determine the BLP for $X$ to be

$$\hat{X}'_{T+1.} = \sum_{j=1}^{T} C_{T,j}(X'_{T+1-j.} - \hat{X}'_{T+1-j.}), \quad \text{for } 1 \leq T < p \tag{2.59}$$

$$= \sum_{j=1}^{n} C_{T,j}(X'_{T+1-j.} - \hat{X}'_{T+1-j.}), \quad \text{for } T \geq p,$$

which completely solves the BLP problem, and at the same time results in substantial computational saving. Note, for example, that if we proceeded to apply the innovations algorithm directly to the $X$-process we would have obtained similar results, **except** that the upper limit of summation in the last equation would have been $T$, not $n$!

The remaining parameter to be determined is $\Sigma$, and this may be determined through the residual covariance matrix

$$V_T = ES'_{T+1.}S_{T+1.} = K_x(0) - \sum_{j=1}^{T} C_{T,T-j}\Sigma C'_{T,T-j}. \tag{2.60}$$

## 2.5 Estimation in Stationary Sequences

The stationary sequences we deal with in this section are $VAR$, $MMA$, and $MARMA$, all of finite order. The problem of estimating the parameters of such sequences has two major parts: (1) how to obtain estimators, given that the order of the process is **known**, and (2) how to estimate the parameters and the orders involved simultaneously, when the orders are **not known**.

### 2.5.1 Estimation of Sequences of Known Order

Before we proceed we note that if $\{X_{t.} : t \in \mathcal{N}_+\}$ is a covariance stationary process with nonzero mean $\mu$ then evidently $\{Y_{t.} : Y_{t.} = X_{t.} - \mu, \; t \in \mathcal{N}_+\}$ is a covariance stationary process with zero mean. Moreover, since the sample mean, say $\bar{X}_T$ based on $T$ observations, converges at least in probability to $\mu$ little loss of relevance is occasioned by considering **only** zero mean covariance stationary processes. We begin this discussion with sequences which are $AR(m)$ and for which $m$ is known.

### Yule-Walker Equations and Estimators

The simplest estimator of the parameters of the autoregression

$$X'_{t\cdot} = \sum_{j=1}^{m} B_j X'_{t-j\cdot} + \epsilon'_{t\cdot} \tag{2.61}$$

is based on the method of moments. This entails multiplying the equation on the right by $X_{t-i\cdot}$, $i = 0, 1, 2, \ldots, m$, and taking expectations to obtain for $t > m$

$$K(0) = \sum_{j=1}^{m} B_j K(-j) + \Sigma, \tag{2.62}$$

$$K(i) = \sum_{j=1}^{m} B_j K(i-j), \quad i = 1, 2, \ldots, m.$$

The relations in Eq. (2.62) are referred to as the Yule-Walker equations. Evidently, we may base an estimator of the AR coefficient matrices $B_j$, and the covariance of the white noise process $\Sigma$, on these relations by simply replacing the (lag) autocovariance matrices by their sample estimators, say

$$\hat{K}(\tau) = \frac{1}{T} \sum_{t=1}^{T-\tau} X'_{t+\tau\cdot} X_{t\cdot}, \quad \tau = 0, 1, 2, \ldots, m. \tag{2.63}$$

Since we are dealing with **real** processes, $K(-\tau) = K'(\tau)$. The Yule-Walker equations yield the solution

$$B_{(m)} = \kappa_{(m)} K_{(m)}^{-1}, \quad B_{(m)} = (B_1, B_2, \ldots, B_m),$$

$$K_{(m)} = \begin{bmatrix} K(0) & K(-1) & \cdots & K(-m+1) \\ K(1) & K(0) & \cdots & K(-m+2) \\ \vdots & \vdots & \vdots & \vdots \\ K(m) & K(m-1) & \cdots & K(0) \end{bmatrix},$$

$$\kappa_{(m)} = [K(1), K(2), \cdots, K(m)], \tag{2.64}$$

**provided** $K_{(m)}$ is invertible, which is assumed. Evidently, if we substitute the sample autocovariances $\hat{K}(\cdot)$, we obtain the Yule-Walker **estimators**

$$\hat{B}_{(m)} = \hat{\kappa}_{(m)} \hat{K}_{(m)}^{-1}. \tag{2.65}$$

By Proposition 34 in Dhrymes (1989), pp. 364-365, the sample autocovariance matrix of lag $\tau$, say $\hat{K}(\tau)$, converges at least $L^2$ (in mean square) to the corresponding parameter, provided the underlying process obeys suitable fourth moment conditions. Thus, the Yule-Walker estimators are at least consistent. Consequently, when dealing with least squares or maximum likelihood equations (first order conditions) which are nonlinear in the parameters we seek to estimate, they may serve as preliminary, or initial estimators.

**Remark 11.** In the general linear model, systems of seemingly unrelated regressions (SUR), and in simultaneous equations the data matrices we deal with are of the form $Y = (y_{t\cdot})$, where $y_{t\cdot}$ is a **row** vector containing the relevant observations in the entire system at time $t$. For example in the SUR model we deal with

$$Y = XB + U,$$

where $X$ is the data matrix of the "explanatory" variables $B$ is the matrix of regression coefficients, and $U = (u_{t\cdot})$. The least squares estimator of $B$ is given by $(X'X)^{-1}X'Y$. If we insisted in writing the vectors of SUR model as **column** vectors, as we did for the covariance stationary process above, out of deference to the literature, we would have to write the SUR model as

$$Y^* = B^* X^* + U^*,$$

where the starred matrices are simply the transposes of the original matrices, and thus write the least squares estimator of $B^*$ as $Y^* X^{*\prime} (X^* X^{*\prime})^{-1}$. Of course, we get precisely the same estimator as above, but the notation robs the reader of the sense of comfortable familiarity with material already well known.

### Likelihood Functions for Stationary Processes

In this section we take advantage of the development of the recursive methods developed in connection with BLP, in order to simplify the likelihood or least squares functions involved in estimating the parameters of stationary sequences. In order for the reader the appreciate the essential simplicity of the resulting entities we begin with a scalar process. Thus, suppose $\{X_t : t \in \mathcal{N}_+\}$ is at least a covariance stationary process; further suppose that it is Gaussian and we have $T$ observations from which to estimate its underlying parameters. Let $X_{(T)} = (X_1, X_2, \ldots, X_T)'$ be the **vector of observations** and set

$$K_{(T)} = [\kappa(i - j)], \quad i, j = 1, 2, \ldots T \tag{2.66}$$

be the covariance matrix of $X_{(T)}$. The log likelihood function (LF) of the observations is given by

$$L(K_{(T)}; X_{(T)}) = -\frac{T}{2}\ln(2\pi) - \frac{1}{2}\ln|K_{(T)}| - \frac{1}{2}X'_{(T)}K_{(T)}^{-1}X_{(T)}. \tag{2.67}$$

Reverting now to the innovations algorithm of a previous section, we may write

$$P_{t-1}(X_t) = E(X_t|X_{t-1}, X_{t-2}, \ldots, X_1), \tag{2.68}$$

with initial condition $\hat{X}_1 = 0$. The conditional expectation representation is justified in view of the joint normality of the observations. Define

$$S_t = X_t - P_{t-1}(X_t), \quad t = 1, 2, \ldots T, \quad S_1 = X_1, \tag{2.69}$$

and note that $\{S_t : t = 1, 2, \ldots, T\}$ is an orthogonal system that serves as a basis of the linear manifold generated by $X_{(T)}$. Thus, using the innovation algorithm we obtain

$$\hat{X}_{t+1} = \sum_{j=1}^{t} c_{t,j} S_{t+1-j}, \quad t = 1, 2, \ldots, T - 1. \tag{2.70}$$

Define now the (lower triangular) matrix

$$C = \begin{bmatrix} c_{0,0} & 0 & 0 & 0 & \cdots & 0 \\ c_{1,1} & c_{1,0} & 0 & 0 & \cdots & 0 \\ c_{2,2} & c_{2,1} & c_{2,0} & 0 & \cdots & 0 \\ c_{3,3} & c_{3,2} & c_{3,1} & c_{3,0} & \cdots & 0 \\ \vdots & \vdots & \vdots & \vdots & \vdots & \vdots \\ c_{T-1,T-1} & c_{T-1,T-2} & c_{T-1,T-3} & c_{T-1,T-4} & \cdots & c_{T-1,0} \end{bmatrix}, \tag{2.71}$$

where it is understood that $c_{t,0} = 1$, for $t = 0, 1, \ldots, T - 1$, and note that

$$\hat{X}_{(T)} = CS_{(T)} - S_{(T)} = (C - I_T)S_{(T)}, \quad X_{(T)} = CS_{(T)}. \tag{2.72}$$

Viewing Eq. (2.72) as a transformation from $X_{(T)}$ to $S_{(T)}$, we note that the Jacobian of the transformation is **unity** and, moreover, that

$$\text{Cov}(X_{(T)}) = CV_{(T)}C', \quad V_{(T)} = \text{Cov}(S_{(T)}) = \text{diag}(v_0, v_1, \ldots, v_{T-1}), \quad (2.73)$$

where

$$v_t = \kappa(t+1, t+1) - \sum_{j=1}^{t} c_{t,j}^2 v_j, \quad t = 0, 1, \ldots, T-1, \quad (2.74)$$

with initial condition $v_0 = \kappa(1,1) = \kappa(0) = \text{Var}(X_t)$. Hence, the LF may be rewritten as

$$L(K_{(T)}; X) = -\frac{T}{2}\ln(2\pi) - \frac{1}{2}\sum_{t=1}^{T}\ln v_{t-1} - \frac{1}{2}\sum_{t=1}^{T}\left(\frac{S_t^2}{v_{t-1}}\right). \quad (2.75)$$

Given an estimator for the covariance function, say $\hat{\kappa}(\tau)$, we obtain $S_{(T)}$, employing the innovations algorithm, so that the LF in Eq. (2.75) depends **only** on the entities $v_t$ as represented in Eq. (2.74).

If the nature of the process $X$ is specified, the LF is a function of the parameters defining the process. Let this be denoted by the vector $\theta$. The maximum likelihood estimator of the underlying parameters satisfies the equations, for $j = 0, 1, 2, \ldots T-1$,

$$\frac{\partial L}{\partial v_j}\frac{\partial v_j}{\partial \theta} = 0. \quad (2.76)$$

Evidently,

$$\frac{\partial L}{\partial v_j} = -\frac{1}{2v_j} + \frac{1}{2}\frac{S_{j+1}^2}{v_j^2}, \quad (2.77)$$

and it follows immediately that the equations defining the ML estimator of the parameters of the process constitute a **nonlinear** system that may be solved **only** by iteration. Since in iterations it is important that we begin with an **initial consistent** estimator, an important aspect of estimation is the ability of a procedure to produce such an initial estimator.

### Initial Consistent Estimators

An initial consistent estimator for VAR models is easily obtained through the Yule-Walker estimator. We produce below an initial consistent estimator for the $MMA(n)$ and $MARMA(m,n)$ models. In previous discussions we established that the innovations algorithm produces the orthogonal entities, $S_{t\cdot}$, such that

$$S_{T\cdot} - \epsilon_{T\cdot} \xrightarrow{\text{L}^2} 0.$$

If the model under consideration is

$$X_{t\cdot}' = A(L)\epsilon_{t\cdot}' = \sum_{j=0}^{n} A_j \epsilon_{t-j\cdot}', \qquad (2.78)$$

with $n$ **known**, we may use Eq. (2.52) to obtain recursively

$$C_{n,n-r}V_r = K(n-r) - \sum_{j=1}^{r-1} C_{n,n-j}V_j C_{r,r-j}'; \qquad (2.79)$$

if in the recursions above we substitute for the autocovariance matrices $K(\tau)$ their sample analog $\hat{K}(\tau)$, we may conjecture that this procedure will produce consistent estimators of the coefficient matrices $A_{n-r}$, $r = 0, 1, \ldots, n-1$. Unfortunately, as shown in Brockwell and Davis (1988) this will not be so, unless we choose [11] $k = o(T^{1/3})$ and obtain instead the entities

$$C_{k,k-r}V_r = \hat{K}(k-r) - \sum_{j=1}^{r-1} C_{k,k-j}V_j C_{r,r-j}', \qquad (2.80)$$

for $r = 0, 1, \ldots, k-1$. It will then follow that

$$\operatorname*{plim}_{T\to\infty} C_{k,s} = A_s, \quad s = 1, 2, \ldots, n, \quad k > n, \qquad (2.81)$$

which thus disposes of the problem of obtaining initial consistent estimators for the underlying parameters of the $MMA(n)$ model.

In dealing with the $MARMA(m,n)$ model's initial consistent estimators, let the model be given by

$$B(L)X_{t\cdot}' = A(L)\epsilon_{t\cdot}', \quad B(L) = I_q - \sum_{i=1}^{m} B_i L^i, \quad A_0 = I_q.$$

Assuming that it is stable (or "causal" in the literature of statistics), we have the representation

$$X_{t\cdot}' = D(L)\epsilon_{t\cdot}', \quad D(L) = \sum_{i=0}^{\infty} D_i L^i.$$

---

[11] The notation $k = o(T^{1/3})$ means that $k/T^{1/3} \to 0$ and $k \to \infty$ as $T \to \infty$.

Since by definition $D(L) = [B(L)]^{-1}A(L)$ or, equivalently, $B(L)D(L) = A(L)$, equating like powers of $L$ we find

$$A_s = D_s - \sum_{i=1}^{m \wedge s} B_i D_{s-i}, \quad s = 1, 2, \ldots, n$$

$$D_s = \sum_{i=1}^{m} B_i D_{s-i}, \quad s > p = \max(m, n). \tag{2.82}$$

If we fit the $MMA(\infty)$ representation of the model above by the innovations algorithm, we shall estimate consistently the matrices $D_s$ by $C_{k,s}$ as in Eq. (2.80), with $k = o(T^{1/3})$. We observe that in Eq. (2.82) we are dealing with a decomposable system, so that if we look at the **last** $m$ equations we have

$$d_{(m)} = b_{(m)}D_{(m)}, \quad d_{(m)} = [D_{p+1}, D_{p+2}, \ldots, D_{p+m}],$$

$$b_{(m)} = [B_1, B_2, \ldots, B_m]$$

$$D_{(m)} = \begin{bmatrix} D_p & D_{p+1} & D_{p+2} & \cdots & D_{p+m-1} \\ D_{p-1} & D_p & D_{p+1} & \cdots & D_{p+m-2} \\ \vdots & \vdots & \vdots & \vdots & \vdots \\ D_{p-m+1} & D_{p-m+2} & D_{p-m+3} & \cdots & D_p \end{bmatrix}. \tag{2.83}$$

If for the matrices $D_j$ we substitute their consistent estimator $C_{k,j}$, as noted in the preceding discussion, we obtain a consistent estimator of $b_{(m)}$, viz.

$$\hat{b}_{(m)} = \hat{d}_{(m)}\hat{D}_{(m)}^{-1}. \tag{2.84}$$

Substituting that in the first equation set of Eq. (2.82) we shall find consistent estimators for the matrices $A_s$, viz.

$$\hat{A}_s = \hat{D}_s - \sum_{i=1}^{m \wedge s} \hat{B}_i \hat{D}_{s-i}, \quad s = 1, 2, \ldots, n, \tag{2.85}$$

thus completing the problem of finding initial consistent estimators for the parameters of the $MARMA(m, n)$ model.

## 2.5.2 Estimation of the $MARMA(m, n)$ Model

Consider the "causal" $ARMA(m, n)$ model

$$B(L)X'_{t.} = A(L)\epsilon'_{t.}, \quad B(L) = I_q - \sum_{j=1}^{m} B_j L^j, \quad A(L) = I_q + \sum_{i=1}^{n} A_i L^i,$$

such that $|B(z)| = 0$ and $|A(z)| = 0$ both have roots **outside** the unit circle. In this context, given the stochastic basis $\{\mathcal{G}_t : \mathcal{G}_t = \sigma(\epsilon_{s\cdot},\ s \leq t)\}$, we may allow the $\epsilon$-sequence to be a martingale difference (MD) process, with $E(\epsilon_{t\cdot}|\mathcal{G}_{t-1}) = 0$ and $\mathrm{Cov}(\epsilon_{t\cdot}|\mathcal{G}_{t-1}) = \Sigma > 0$ – or more generally an MD process obeying $\sup_t E|\epsilon_{t\cdot}|^{2+\delta} < \infty$, $\delta > 0$. The discussion in this section is based on Reinsel, Basu, and Yap (1992).

In dealing with estimation issues raised by the MARMA model above, we are seeking, in part, to estimate parameters, the matrices $A_i$, which correspond to variables that **are not observed**. This complicates matters considerably and leads to an elaborate framework and a highly nonlinear estimation problem. We first illustrate how the nonobservability of the $\epsilon_{t-i\cdot}$ impacts on estimation issues. To simplify our discussion, we rewrite the equation above as

$$X'_{t\cdot} = \sum_{j=1}^{m} B_j X'_{t-j\cdot} + \sum_{i=1}^{n} A_i \epsilon'_{t-i\cdot} + \epsilon^{*'}_{t\cdot},$$

and attempt to estimate its parameters **subject to certain initial conditions**. Putting $p = \max(m,n)$, we note that the first observation we can fully utilize is the $(p+1)^{st}$ observation. Thus, the usable observations on the model, for **estimation purposes**, are best written as

$$X'_{p+t\cdot} = \sum_{j=1}^{m} B_j X'_{p+t-j\cdot} + \sum_{i=1}^{n} A_i \epsilon^{*'}_{p+t-i\cdot} + \epsilon^{*'}_{p+t\cdot},\quad t = 1, 2, \ldots, T.$$

Consequently, if we have a sample of size $T + p$ on the vector $X_{s\cdot}$, the effective sample for estimation purposes is of size $T$!

The initial conditions under which we operate are that the entities $X_{s\cdot}$, $s = 1, 2, \ldots, p$, are fixed numbers, and $\epsilon^{*}_{s\cdot} = 0$, for $s \leq p$.

To take advantage of the preceding we introduce the more convenient notation

$$y_{t\cdot} = X_{p+t\cdot}, \quad x_{t\cdot} = (X_{p+t-1\cdot}, X_{p+t-2\cdot}, \ldots, X_{p+t-m\cdot}),$$

$$z_{t\cdot} = (\epsilon_{t-1\cdot}, \epsilon_{t-2\cdot}, \ldots, \epsilon_{t-n\cdot}), \quad C_1 = (B_1,\ B_2, \ldots, B_m)',$$

$$C_2 = (A_1, A_2, \ldots, A_n)', \quad \epsilon_{t\cdot} = \epsilon^{*}_{p+t\cdot},\ t = 1, 2, \ldots, T, \tag{2.86}$$

so that

$$y_{t\cdot} = x_{t\cdot}C_1 + z_{t\cdot}C_2 + \epsilon_{t\cdot}, \quad t = 1, 2, \ldots, T, \quad \text{or more compactly,}$$

$$Y = XC_1 + ZC_2 + U, \tag{2.87}$$

which exhibits the model in the familiar form of SUR. To motivate the estimation procedure we assume [12] that the error process is one of i.i.d. random vectors with mean zero and covariance matrix $\Sigma > 0$. We may thus write the log-likelihood function as

$$L(C_1, C_2, \Sigma) = -\frac{qT}{2}\ln(2\pi) - \frac{T}{2}\ln|\Sigma| - \frac{1}{2}\sum_{t=1}^{T}\epsilon_{t\cdot}\Sigma^{-1}\epsilon_{t\cdot}'$$

$$= -\frac{qT}{2}\ln(2\pi) - \frac{T}{2}\ln|\Sigma| - \frac{1}{2}\text{tr}\Sigma^{-1}U'U.$$

Two things are evident from the preceding: (1) The LF contains no observables; (2) if we were to "estimate" $U$, say by $\tilde{U}$, we can easily obtain the ML estimator $\hat{\Sigma} = (\tilde{U}'\tilde{U})/T$. Thus, in the discussion of the estimation procedure we shall consider only the estimation of the matrices $C_1$, $C_2$.

If the elements of matrix $Z$ were observable, we would simply restate the LF in terms of the transformation from $U$ to $Y$. Maximizing the resulting expression with respect to $\gamma_1 = \text{vec}(C_1)$, $\gamma_2 = \text{vec}(C_2)$, we would obtain the first-order conditions

$$\frac{\partial L}{\partial \gamma_1} = [y - (I_m \otimes X)\gamma_1 - (I_n \otimes Z)\gamma_2]'(\Sigma^{-1} \otimes I_T)(I_m \otimes X) = 0,$$

$$\frac{\partial L}{\partial \gamma_2} = [y - (I_m \otimes X)\gamma_1 - (I_n \otimes Z)\gamma_2]'(\Sigma^{-1} \otimes I_T)(I_m \otimes Z) = 0,$$

$$\frac{\partial L}{\partial \Sigma^{-1}} = \frac{T}{2}\Sigma - \frac{1}{2}(Y - XC_1 - ZC_2)'(Y - XC_1 - XC_2) = 0.$$

The solution of these equations would be a perfectly routine problem, were it not for the fact that **the elements of the matrix $Z$ are not observable!** We are thus forced to estimate the coefficients of variables which **are**

---

[12] As we shall see, when we deal with the limiting distribution aspects of this problem, the normality assumption is superfluous; it only provides the motive for the function to be extremized in order to obtain parameter estimators. The limiting distribution results continue to hold under the assumption that the error process $\{\epsilon_{t\cdot} : t \in \mathcal{N}_+\}$ is an MD sequence.

**not directly** observed, but whose effect is exercised through $X_t.$, and thus observed only indirectly. This converts the problem to a highly nonlinear one, which may only be solved by **iteration given an initial consistent estimator** (ICE) of the elements of the matrices $C_1$ and $C_2$.

The special circumstances of this problem require a somewhat different orientation than is implicit in the SUR context above. In the latter, we operate with entities of the form $\mathrm{vec}(U)$, which **display the columns** of $U$ in one long column. Here we need to **display the rows** of $U$ in one long **row**, and then transpose it so that it becomes a column vector. To this end, introduce the operator $\mathrm{rvec}(\cdot)$, which is defined by

$$\mathrm{rvec}(U) = (\epsilon_{1.}, \epsilon_{t.}, \ldots, \epsilon_{T.}),$$

and note that, for any matrix $M$

$$\mathrm{rvec}(M)' = \mathrm{vec}(M'). \tag{2.88}$$

To adapt our discussion to this requirement, return to Eq. (2.87) and note the following: [13]

$$U' = Y' - C_1'X' - C_2'Z' = Y' - \sum_{i=1}^{m} B_i Y_{-i}' - \sum_{j=1}^{n} A_j U_{-j}', \quad Y_{-i} = L^i Y,$$

$$U_{-j} = L^j U, \quad \epsilon = \mathrm{vec}(U') = y - \sum_{i=1}^{m}(L^i Y \otimes I_q)b_i - \sum_{j=1}^{n}(L^j U \otimes I_q)a_j$$

$$\epsilon \sim N(0, I_T \otimes \Sigma), \quad b_i = \mathrm{vec}(B_i), \quad a_j = \mathrm{vec}(A_j). \tag{2.89}$$

We may thus write the LF, in the new notational framework, as

$$L(B, A, \Sigma) = -\frac{T}{2}\ln(2\pi) - \frac{T}{2}\ln|\Sigma| - \frac{1}{2}\epsilon'(I_T \otimes \Sigma^{-1})\epsilon. \tag{2.90}$$

Rearranging the third set of equations in Eq. (2.89), we obtain

$$w = y - \sum_{i=1}^{m}(L^i Y \otimes I_q)b_i = \epsilon + \sum_{j=1}^{n}(L^j U \otimes I_q)a_j = \mathrm{vec}(U' + C_2 Z'). \tag{2.91}$$

---

[13] Note the unusual form of the covariance matrix of the error process; this is a consequence of **defining** $\epsilon$ to be $\mathrm{vec}\,(U')$, instead of $\mathrm{vec}\,(U)$, as would be the case in the standard SUR context.

It turns out that the rightmost member of the equation above plays a very important role in our discussion, and for this reason certain alternative representations are extremely useful. We list them below with a brief explanation of their derivation.

$$U + ZC_2 = U + \sum_{j=1}^{n} L^j U A_j';$$

$$\text{vec}[U' + C_2'Z'] = (I_{Tq} + A^*)\epsilon, \quad A^* = \sum_{j=1}^{n}(N^j \otimes A_j)$$

$$\text{vec}[U' + C_2'Z'] = \epsilon + \sum_{j=1}^{n}(L^j U \otimes I_q)a_j, \tag{2.92}$$

where $N = (n_{ij})$ is $T \times T$, such that $n_{ij} = 1$ **if** $i + 1 = j$, and zero otherwise. [14] The first representation in Eq. (2.92) follows from the definition of $Z$ and $C_2$. The second representation follows immediately if we note that the **rows** of $U + ZC_2$ are given by

$$U + ZC_2 = (\epsilon_{t\cdot} + \sum_{j=1}^{t-1} \epsilon_{t-i\cdot} A_j'), \quad \text{if } t < n$$

$$= (\epsilon_{t\cdot} + \sum_{j=1}^{n} \epsilon_{t-i\cdot} A_j'), \quad t \geq n. \tag{2.93}$$

If one writes these in one long row, one obtains $\epsilon' + \epsilon' A^{*'}$, where $A^*$ is a $Tq \times Tq$ block matrix, whose $i, j$ blocks obey

$$A_{i,j}^* = A_s, \quad \text{for } s + i = j, \quad s = 1, 2, \ldots, n, \tag{2.94}$$

and zero otherwise. It is then evident that $A^* = \sum_{j=1}^{n}(N^j \otimes A_j)$, since the latter indicates a (lower triangular) block matrix whose first sub-diagonal blocks (all) consist of $A_1$, and more generally the $s^{th}$ sub-diagonal blocks (all) consist of $A_s$, $s = 1, 2, \ldots n$, and all other blocks are null. Note further that the last equation set in Eq. (2.92) exhibits $\text{vec}(C_2'Z')$ in terms of the

---

[14] Notice that powers of the matrix $N$ obey $N^s = (n_{ij}^{(s)})$, with $n_{ij}^{(s)} = 1$, if $i + s = j$, and zero otherwise. In particular, $N^T = 0$. A matrix, $N$, is said to be **nilpotent**, if some power of it is **null**; the smallest integer $\nu$ such that $N^\nu = 0$, is said to be the **index of nilpotency**. See Gantmacher (1959), Vol. I, p. 226.

vectors $a_j$ , while the penultimate equation exhibits the latter in terms of the
vector $\epsilon$. We shall find both representations quite useful in our discussion.

Return now to Eq. (2.91) and note that we may write it is as

$$w = y - \sum_{i=1}^{m}(L^i Y \otimes I_q)b_i = A\epsilon, \quad A = (I_{Tq} + A^*). \tag{2.95}$$

Since $A$ is evidently nonsingular we have

$$\epsilon = A^{-1}w, \quad \text{or} \quad \epsilon = w - A^*\epsilon. \tag{2.96}$$

Treating the first representation as a transformation from $\epsilon$, which is not observable, to $w$ which is, and noting that the Jacobian of this transformation
is unity we may rewrite the LF of Eq. (2.90) as

$$L(B, A, \Sigma) = -\frac{Tq}{2}\ln(2\pi) - \frac{T}{2}\ln|\Sigma| - \frac{1}{2}w' A'^{-1}(I_T \otimes \Sigma^{-1})A^{-1}w,$$

which is now displayed in a form that consists solely of observables and
unknown parameters. In maximizing the LF in order to obtain parameter
estimators, it is more convenient to use the form in Eq. (2.90). Maximizing
that form with respect to the parameters $b_i$, $i = 1, 2, \ldots, m$, and $a_j$, $j = 1, 2, \ldots, n$, yields the normal equations [15]

$$\frac{\partial L}{\partial b_s} = -\epsilon'(I_T \otimes \Sigma^{-1})\left(\frac{\partial \epsilon}{\partial b_s}\right) = 0, \quad \frac{\partial L}{\partial a_r} = -\epsilon'(I_T \otimes \Sigma^{-1})\left(\frac{\partial \epsilon}{\partial a_s}\right) = 0,$$

for $s = 1, 2, \ldots, m$, $r = 1, 2, \ldots, n$. To evaluate the first derivative above,
we note from Eqs. (2.95) and (2.96) that $(\partial \epsilon/\partial b_s) = -A^{-1}(L^s Y \otimes I_q)$; to
evaluate the second derivative, we note that from a computational (estima-
tion) point of view $\epsilon$ is a **nonlinear** function of the $a_r$ . For example, in the
second representation of Eq. (2.96), $\epsilon$ depends on the $a_r$ **directly** through
the matrix $A^*$, and **indirectly** through the lagged values $\epsilon_{t-i}$. in the right
member of that equation. Thus, the derivative $\partial \epsilon/\partial a_r$, has two components:
one in which the **direct** dependence on $A^*$ is held "constant", and one in

---

[15] We remind the reader of our convention: If $y$ is $m \times 1$, $x$ is $n \times 1$, $A$ is $m \times n$,
and $y = Ax$, then

$$\frac{\partial y}{\partial x} = \left(\frac{\partial y_i}{\partial x_j}\right) = A.$$

Thus, if $y$ is a scalar and $x$ is a (column) vector the derivative $\partial y/\partial x$ is a **row vector**.

which the (relevant) lagged values, the $\epsilon_{t-i.}$, are held "constant". The first component may be obtained from Eq. (2.96), while the second may be obtained from the third equation set of Eq. (2.89). Combining the two we have

$$\frac{\partial \epsilon}{\partial a_r} = -A^* \frac{\partial \epsilon}{\partial a_r} - (L^r U \otimes I_q).$$

Thus, we have established

$$\frac{\partial \epsilon}{\partial b_s} = A^{-1} \frac{\partial w}{\partial b_s} = -A^{-1}(L^s Y \otimes I_q),$$

$$\frac{\partial \epsilon}{\partial a_r} = -A^{-1}(L^r U \otimes I_q), \tag{2.97}$$

and consequently, the first-order conditions for the ML estimators of the $MARMA(m,n)$ model are given by

$$\frac{\partial L}{\partial b_s} = w' A'^{-1} (I_T \otimes \Sigma^{-1}) A^{-1}(L^s Y \otimes I_q) = 0, \quad s = 1, 2, \ldots, m,$$

$$\frac{\partial L}{\partial a_s} = \epsilon'(I_T \otimes \Sigma^{-1}) A^{-1}(L^r U \otimes I_q) = 0, \quad r = 1, 2, \ldots, n. \tag{2.98}$$

Notice that the first equation above is expressed in terms of observables and unknown parameters only; the second, however, **is not entirely** expressed in terms of observables. Moreover, the equations are highly nonlinear in the parameters of interest, and may only be solved by iteration. The nonobservable entities, viz. the elements of $L^r U$, may be "estimated" given an initial consistent estimator (ICE) of the matrices $A_j$, $B_i$, as $\tilde{\epsilon} = \tilde{A}^{-1}(y - \sum_{i=1}^{m}(L^i Y \otimes I_q)\tilde{b}_i)$. The solution of the system then proceeds by iteration.

**Solution by Iteration**

To set up the iteration procedure, we introduce some additional notation. Put

$$V = (V_1, V_2), \quad V_1 = (LY, L^2 Y, \ldots, L^m Y), \quad V_2 = (LU, L^2 U, \ldots, L^n U),$$

$$\gamma = (b_1', b_2', \ldots, b_m', a_1', a_2', \ldots, a_n')', \quad Q = (V \otimes I_q), \quad V = (v_t.), \tag{2.99}$$

so that

$$\left(\frac{\partial L}{\partial \gamma}\right)^{'} = -\left(\frac{\partial \epsilon}{\partial \gamma}\right)^{'} (I_T \otimes \Sigma^{-1})\epsilon = Q^{'}A^{'-1}(I_T \otimes \Sigma^{-1})(y - Q\gamma),$$

$$\epsilon = y - Q\gamma, \quad \left(\frac{\partial \epsilon}{\partial \gamma}\right) = -A^{-1}Q, \tag{2.100}$$

Next, we require an expression for the Hessian of the LF. From the first equation set of Eq. (2.100) we have, by definition,

$$\frac{\partial^2 L}{\partial \gamma \partial \gamma} = -\left(\frac{\partial \epsilon}{\partial \gamma}\right)^{'} (I_T \otimes \Sigma^{-1})\left(\frac{\partial \epsilon}{\partial \gamma}\right) - \left\{\left[\epsilon^{'}(I_T \otimes \Sigma^{-1})\right] \otimes I_\nu\right\} \frac{\partial \mathrm{vec}[(\partial \epsilon / \partial \gamma)^{'}]}{\partial \gamma}, \tag{2.101}$$

where $\nu = (n+m)q^2$. Since it may be shown that terms of the form

$$\left(\left[\epsilon^{'}(I_T \otimes \Sigma^{-1})\right] \otimes I_{(n+m)q^2}\right)\left(\frac{\partial \mathrm{vec}\left[(\partial \epsilon / \partial \gamma)^{'}\right]}{\partial \gamma}\right),$$

upon division by $T$ converge to a **null matrix** at least in probability, we may neglect them in the iteration process. Thus, we may use the approximation

$$\frac{\partial^2 L}{\partial \gamma \partial \gamma} \approx -\left(\frac{\partial \epsilon}{\partial \gamma}\right)^{'} (I_T \otimes \Sigma^{-1})\left(\frac{\partial L}{\partial \gamma}\right) = -Q^{'}A^{'-1}(I_T \otimes \Sigma^{-1})A^{-1}Q. \tag{2.102}$$

The iteration used to solve the estimation problem employs the Newton-Raphson algorithm, which is basically a procedure that relies on the mean value expansion

$$\left(\frac{\partial L_T}{\partial \gamma}(\hat{\gamma})\right)^{'} = \left(\frac{\partial L_T}{\partial \gamma}(\tilde{\gamma})\right)^{'} + \left(\frac{\partial^2}{\partial \gamma \partial \gamma}(\gamma^*)\right)(\hat{\gamma} - \tilde{\gamma}), \quad L_T = \frac{1}{T}L(A, B, \Sigma), \tag{2.103}$$

where $\tilde{\gamma}$ is an ICE of $\gamma$ and $|\gamma^* - \tilde{\gamma}| < |\hat{\gamma} - \tilde{\gamma}|$. Using the representations and approximations above, the iteration procedure, at the $k^{th}$ step, consists of the equations

$$\tilde{\gamma}_{(k)} = \tilde{\gamma}_{(k-1)} - \left(\frac{\partial^2 L_T}{\partial \gamma \partial \gamma}(\tilde{\gamma}_{(k-1)})\right)^{-1}\left(\frac{\partial L_T}{\partial \gamma}(\tilde{\gamma}_{(k-1)})\right)^{'}$$

$$= \tilde{\gamma}_{(k-1)} + \left(Q^{'}A^{'-1}(I_T \otimes \Sigma^{-1})A^{-1}Q\right)^{-1}_{(k-1)}[Q^{'}A^{'-1}(I_T \otimes \Sigma^{-1})\epsilon]_{(k-1)}$$

$$\tilde{\Sigma}_{(k)} = \frac{1}{T}\tilde{U}^{'}_{(k)}\tilde{U}_{(k)}. \tag{2.104}$$

Noting that the iteration scheme **requires** the computation of the elements of $Q$, **some of which are not observable**, return to Eqs. (2.87), (2.93), (2.96) and write

$$\epsilon_{t\cdot} = y_{t\cdot} - x_{t\cdot}C_1 - \sum_{i=1}^{n} \epsilon_{t-i\cdot}A_i', \quad t = 1, 2, \ldots, T.$$

Given an ICE of $B_i$ and $A_j$, say $\tilde{B}_i$, $\tilde{A}_j$, $i = 1, 2, \ldots, m$, $j = 1, 2, \ldots, n$, we may "predict" or estimate the $\epsilon's$ in view of the initial conditions $\epsilon_{s\cdot} = 0$, for $s \leq p$. Computing a few recursively, we find, with $\tilde{w}_t = y_{t\cdot} - x_{t\cdot}\tilde{C}_1$,

$$\tilde{\epsilon}_{1\cdot} = \tilde{w}_{1\cdot}, \quad \tilde{\epsilon}_{2\cdot} = \tilde{w}_{2\cdot} - \tilde{\epsilon}_{1\cdot}\tilde{A}_1',$$

or, more generally, for $t = 1, 2, \ldots T$,

$$\tilde{\epsilon}_{t\cdot} = y_{t\cdot} - x_{t\cdot}\tilde{C}_1 - \sum_{i=1}^{t-1} \tilde{\epsilon}_{t-i\cdot}\tilde{A}_i', \quad t \leq n$$

$$= y_{t\cdot} - x_{t\cdot}\tilde{C}_1 - \sum_{i=1}^{n} \tilde{\epsilon}_{t-i\cdot}\tilde{A}_i', \quad t > n. \tag{2.105}$$

The ML estimator of $\gamma$ and $\Sigma$ is simply the converging iterate as given by Eq. (2.104). An ICE is easily obtainable by the methods of the preceding section, or by the method given in An, Chen, and Hannan (1983), as well as many other methods.

### Limiting Distribution

To derive the limiting distribution of the ML estimator $\hat{\gamma}$, which is the converging iterate of the procedure examined above, we employ again the mean value representation

$$\frac{\partial L_T}{\partial \gamma}(\hat{\gamma}) = 0 = \frac{\partial L_T}{\partial \gamma}(\gamma^\circ) + \left(\frac{\partial^2 L_T}{\partial \gamma \partial \gamma}(\gamma^*)\right)(\hat{\gamma} - \gamma^\circ), \quad \text{or equivalently}$$

$$\sqrt{T}(\hat{\gamma} - \gamma^\circ) = -\left[\left(\frac{1}{T}\frac{\partial^2 L}{\partial \gamma \partial \gamma}(\gamma^*)\right)\right]^{-1} \frac{1}{\sqrt{T}} Q' A'^{-1} (I_T \otimes \Sigma^{-1})\epsilon, \tag{2.106}$$

where $|\gamma^* - \gamma^\circ| < |\hat{\gamma} - \gamma^\circ|$, and the last vector of the second equation set above is evaluated at the true parameter matrices $A_j$, $B_i$. Since $\hat{\gamma}$ is consistent, it may be shown that

$$\frac{1}{T} \frac{\partial^2 L}{\partial \gamma \partial \gamma}(\gamma^*) \sim -\frac{1}{T} \left( Q' A'^{-1} (I_T \otimes \Sigma^{-1}) A^{-1} Q \right) \xrightarrow{\text{P}} \Phi > 0. \qquad (2.107)$$

Thus, we need only establish the limiting distribution of the rightmost vector in the second equation set of Eq. (2.106). To this end, note from Eq. (2.95) that

$$A'^{-1} = I_{Tq} + A_*, \quad A_* = \sum_{j=1}^{n} (N'^j \otimes F_j),$$

$$A'^{-1}(I_T \otimes \Sigma^{-1}) = (I_T \otimes \Sigma^{-1}) + \sum_{j=1}^{n} (N'^j \otimes F_j \Sigma^{-1}),$$

where $F_j$ is the (common) block on the $j^{th}$ super-diagonal of $A'^{-1}$. Hence

$$D = Q' A'^{-1}(I_T \otimes \Sigma^{-1}) = \sum_{j=0}^{n} (V' N'^j \otimes F_j \Sigma^{-1}) = (D_1, D_2, \ldots, D_T)$$

$$D_t = \sum_{j=0}^{n} (v'_{t-j\cdot} \otimes F_j \Sigma^{-1}), \quad N^0 = I_T, \quad F_0 = I_q, \qquad (2.108)$$

and the $D_t$, $t = 1, 2, \ldots, T$, are each $(n+m)q \times q$ matrices. Thus

$$\xi_{(T)} = \frac{1}{\sqrt{T}} Q' A'^{-1}(I_T \otimes \Sigma^{-1})\epsilon = \sum_{t=1}^{T} D_t \epsilon'_{t\cdot}. \qquad (2.109)$$

Recalling the definition of the stochastic basis at the beginning of the section, we note that the sequence $\{ D_t \epsilon'_{t\cdot} : t \in \mathcal{N}_+ \}$ is $\mathcal{G}_t$-measurable, $E|D_t \epsilon_{t\cdot}| < \infty$, and

$$E \left( D_t \epsilon'_{t\cdot} | \mathcal{G}_{t-1} \right) = 0, \quad \text{Cov} \left( D_t \epsilon'_{t\cdot} | \mathcal{G}_{t-1} \right) = D_t \Sigma D'_t > 0,$$

and thus a square integrable $MD$ sequence. Moreover, if

$$\frac{1}{T} \sum_{t=1}^{T} D_t \Sigma D'_t \xrightarrow{\text{P}} \Psi, \qquad (2.110)$$

by Proposition 21 in Dhrymes (1989), p. 337,

$$\xi_{(T)} \xrightarrow{\text{d}} N(0, \Psi), \qquad (2.111)$$

provided the $\epsilon$-sequence obeys a Lindeberg condition. We further note that

$$\frac{1}{T} \sum_{t=1}^{T} D_t \Sigma D'_t = \frac{1}{T} D(I_T \otimes \Sigma)D' = \frac{1}{T} Q' A'^{-1}(I_T \otimes \Sigma^{-1}) A^{-1} Q \xrightarrow{\text{P}} \Psi.$$

Thus we conclude that

$$\sqrt{T}(\hat{\gamma} - \gamma) \xrightarrow{\text{d}} N(0, \Psi^{-1}). \tag{2.112}$$

We summarize the discussion of this entire section in the following proposition.

**Proposition 10.** Consider the standard ("causal") $MARMA(m,n)$ model, $B(L)X'_{t.} = A(L)\epsilon'_{t.}$, such that $|B(z)| = 0$ and $|A(z)| = 0$ both have roots **outside** the unit circle. If the error process is also normal, the likelihood function is given by

$$L(B, A, \Sigma) = -\frac{T}{2}\ln(2\pi) - \frac{T}{2}\ln|\Sigma| - \frac{1}{2}w'A'^{-1}(I_T \otimes \Sigma^{-1})A^{-1}w,$$

where $w$ is as defined in terms of observables in Eq. (2.95). The following statements are true:

i. given an ICE, the ML estimator is found by iteration, as given in Eq. (2.104); the converging iterate is the ML estimator and it converges, at least in probability, to the true parameter $\gamma^\circ$, as the latter is defined in Eq. (2.99);

ii. its limiting distribution is given by

$$\sqrt{T}(\hat{\gamma} - \gamma^\circ) \xrightarrow{\text{d}} N(0, \Psi^{-1}), \quad \Psi = \operatorname*{plim}_{T \to \infty} \frac{1}{T}Q'A'^{-1}(I_T \otimes \Sigma^{-1})A^{-1}Q,$$

where $Q$ and $A$ are as defined in Eqs. (2.99), and (2.95), respectively;

iii. all previous results remain valid even if the error process is a MD obeying

$$\sup_t E|\epsilon_{t.}|^{2+\delta} < \infty, \quad \text{for} \quad \delta > 0,$$

$\operatorname{Cov}(\epsilon'_{t.}) = \Sigma > 0$ and a Lindeberg condtion; the estimator $\hat{\gamma}$, minimizing the "LF" function above, is then a "pseudo" ML estimator.

## 2.5.3 Estimation of Sequences of Unknown Order

One possibility of dealing with this problem is simply to take advantage of our discussion of BLP. Thus, suppose $\{X_{t.} : t \in \mathcal{N}_+\}$ is a covariance stationary process. The (one-step) BLP discussed earlier expresses the prediction of $X_{T+1 \cdot}$, in terms of certain underlying parameters of the **autocovariance** function or kernel. Evidently, a simple way in which we may turn that result into an estimation procedure is simply to **replace** the autocovariance kernel or function by its **estimator**. By Proposition 34 in Dhrymes (1989), pp. 364-365, the sample autocovariance matrix of lag $\tau$, say $\hat{K}(\tau)$, converges at least $L^2$ (in mean square) to the corresponding parameter, provided the underlying process obeys suitable fourth moment conditions. Hence, if the $VAR(n)$ is stated as

$$X_{t \cdot}^{'} = \sum_{j=1}^{n} B_j X_{t-j \cdot}^{'} + \epsilon_{t \cdot}^{'}, \qquad (2.113)$$

then the coefficient matrices of the estimated forward predictor obey

$$\plim_{T \to \infty} \hat{F}_{Tj} = B_j, \quad j = 1, 2, \ldots, n \text{ and zero otherwise.} \qquad (2.114)$$

Moreover, the estimated mean squared (prediction) error matrix, $\hat{V}_{1,T}$ is obtained as follows: Set

$$X_{T \cdot}^{'} - \hat{X}_{T \cdot}^{'} = \epsilon_{T \cdot}^{'} - \sum_{j=1}^{T-1} (\hat{F}_{T,j} - B_j) X_{T-j \cdot}^{'}, \qquad (2.115)$$

and note that, for sufficiently large $T$, we have

$$\text{Cov}(X_{T \cdot}^{'} - \hat{X}_{T \cdot}^{'}) \approx \Sigma. \qquad (2.116)$$

If a distribution theory is obtained, we may use the apparatus of Remark 9 to determine the order by testing the "partial correlation" parameter $F_{n+1,n+1}$. If it is deemed to be zero, we may conclude that the order of the sequence is $n$.

A similar procedure may be applied to MMA and MARMA models.

A more satisfactory approach is to deal with the problem of estimation directly through least squares, or maximum likelihood, methods if the distribution of the process is specified. In this context, the order of the VAR or MARMA is estimated through the Akaike information criterion or variants of it, such as the bias corrected Akaike information criterion, or the Bayesian information criterion, or other similar devices. Basically, what all these procedures do is penalize for excessive parametrization.

**Why Overfitting Is Undesirable**

Overfitting is the term applied to a situation in which the true model is MARMA with AR and MA orders, respectively $(m_1, n_1)$, but we fit a model with parameters $m \geq m_1$ and/or $n \geq n_1$, at least one of the inequalities being strict.

For ease of exposition, we illustrate the problems that may be created by overfitting in the case of a VAR model. Suppose we estimate the parameters of this model, say $X'_{t\cdot} = \sum_{i=1}^m B_i X'_{t-i\cdot} + \epsilon'_{t\cdot}$, on the basis of a realization (sample) $\{X_{t\cdot} : t = 1, 2, 3, \ldots, T\}$. Let the estimators be given by $\hat{b} = (\hat{b}'_1, \ldots, \hat{b}'_m)'$, $\hat{\Sigma}$. Let $\{Y_{t\cdot} : t = 1, 2, \ldots, T\}$ be another **independent** realization of the **same process** and suppose we use the estimator obtained from the other realization to obtain an estimate of the BLP of $Y_{T+1\cdot}$. The mean square prediction error matrix is, by definition,

$$\psi_{T+1|T} = E(Y_{T+1\cdot} - \hat{Y}_{T+1\cdot})'(Y_{T+1\cdot} - \hat{Y}_{T+1\cdot})$$

$$= E[\epsilon'_{T+1\cdot} - (V_{T\cdot} \otimes I_q)(\hat{b} - b)][\epsilon'_{T+1\cdot} - (V_{T\cdot} \otimes I_q)(\hat{b} - b)]'$$

$$= E_Y E_{X|Y} [\epsilon_{T+1\cdot} - (V_{T\cdot} \otimes I_q)(\hat{b} - b)]'[\epsilon_{T+1\cdot} - (V_{T\cdot} \otimes I_q)(\hat{b} - b)]$$

$$= E_Y \left( \epsilon'_{T+1\cdot}\epsilon_{T+1\cdot} + (V_{T\cdot} \otimes I_q)(K^{-1}_{(m)} \otimes \Sigma)(V_{T\cdot} \otimes I_q) \right)$$

$$\approx E_Y \left( \epsilon'_{T+1\cdot}\epsilon_{T+1\cdot} + [\frac{1}{T}V_{T\cdot}K^{-1}_{(m)}V'_{T\cdot}) \otimes \Sigma)] \right)$$

$$\approx \left( 1 + \frac{mq}{T} \right) \Sigma, \quad V_{t\cdot} = (X_{t-1\cdot}, X_{t-2\cdot}, \ldots, X_{t-m\cdot}). \tag{2.117}$$

For $T$ sufficiently large, and $\hat{\Sigma}$ the ML estimator of $\Sigma$, $(T/T - m)E\hat{\Sigma} \approx \Sigma$. Thus, replacing $\Sigma$ in the expression above, we find that the prediction error [16]

$$FPEM(m) \approx \frac{T+m}{T-m} \hat{\Sigma}. \tag{2.118}$$

Notice that irrespective of what the true order is, increasing $m$ initially reduces the "size" of the estimate of the covariance matrix, but **always**

---

[16] In this literature the mean square prediction error is said to be the **final prediction error(FPE)** and in the multivariate case the **final prediction error matrix (FPEM)**, terminology we employ as well.

**increases the fraction** $(T + m/T - m)$. Once $m$ exceeds the true order, however, such reductions will become minimal. Hence, overfitting implies that when the order of the fitted autoregression is higher than is warranted by the true order, the FPEM will begin to increase; consequently, predictions generated by an overfitted model have larger than necessary mean square (prediction) errors.

**Information-Based Criteria for Order Selection**

As we have pointed out, the order selection problem is routinely solved by incorporating in the ML or least squares procedure a penalty for overfitting. Just what the penalty is or should be and how it is justified are the topics discussed in this section. In fact, minimization of the FPEM was also proposed by Akaike (1969) as a criterion for order selection. In this section we discuss more general selection criteria such as the Akaike information criterion (AIC), its bias-corrected alternative (AICC), suggested by Akaike (1973) and Hurvich and Tsai (1989), respectively, as well as the Bayesian information criterion (BIC), also suggested by Akaike (1978). An alternative BIC has been proposed by Schwarz (1978). Of these the AIC has been shown in Monte Carlo studies to lead to overfitting, which the AICC attempts to correct. The FPE, AIC, and AICC are known to be efficient, in the sense that for an infinite order AR they lead to an optimal rate of convergence to the true FPE. These procedures, however, are not **consistent**, in the sense that the fitted order, say $\hat{m}$, does not necessarily converge (in probability) to the true order, say $m$. By contrast, the BIC is a **consistent procedure**.

The AIC and AICC criteria employ the apparatus of Kullback information, [see e.g. Dhrymes (1994b), pp. 279-281]. We recall that if $P, Q$, are two probability measures defined on the probability space ($\Omega$, $\mathcal{A}$, $\mathcal{P}$), and $\mu$ is a **dominant measure**, there exist non-negative integrable functions, $f, g$ such that

$$P(A) = \int_A f \, d\mu, \quad Q(A) = \int_A g \, d\mu, \quad \text{for any } A \in \mathcal{A}.$$

The Kullback information (KI) of $P$ on $Q$ is defined by

$$K(P, Q) = \int_\Omega f \ln\left(\frac{f}{g}\right) d\mu = \int_\Omega k(s) g \, d\mu, \quad k = s \ln s + 1 - s, \quad s = \frac{f}{g}.$$

Since $k(s) \geq 0$ and is equal to zero if and only if $s = 1$, we conclude that

$$K(P, Q) \geq 0, \quad K(P, Q) = 0 \quad \text{if and only if} \quad P = Q \ a.e., \ \text{or} \ f = g, \ a.e.$$

Now, if $X = \{X_{t.} : t \in \mathcal{N}_+\}$ is a stochastic process, in our discussion an AR, MA, or ARMA process, defined on $(\Omega, \ \mathcal{A}, \ \mathcal{P})$, i.e. if for every $t$, $X_{t.} : \Omega \to R^q$, where $R^q$ is the $q$-dimensional Euclidean space, let $(R^q, \mathcal{B}(R^q), P)$ be the probability space of the **range**, where $\mathcal{B}(R^q)$ is the Borel $\sigma$-algebra and $P$ a probability measure. We may consider $X$ to be defined on this (range) probability space by appropriate assignments. Thus, consider $A = \{\omega : X_{t.}(\omega) \in B, \ B \in \mathcal{B}(R^q)\}$; since $\mathcal{P}(A)$ is the **probability** that $X_{t.} \in B$, we may **define** $P(B) = \mathcal{P}(A)$ and thus take $X$ to be defined in terms of the probability space $(R^q, \mathcal{B}(R^q), P)$. In this context, let $L^*$ be the likelihood function (**not** the log-likelihood function) of a realization $(X_1., X_2., \ldots, X_{T.})$; by analogy with the preceding discussion, we may define the KI in terms of the parameters characterizing the stochastic sequence $X$. To this end, let $B$ be any set that belongs to the Borel $\sigma$-algebra of the **product space** that consists of $T$ copies of $(R^q, \mathcal{B}(R^q), P)$, let $\theta$ and $\psi$ be two (admissible) parameter points, and put

$$P_\theta(B) = \int_B L^*(x; \theta) \, dx, \quad P_\psi(B) = \int_B L^*(x; \psi) \, dx,$$

$$K(\theta, \psi) = \int_{R^{qT}} L^*(x; \theta) \ln\left(\frac{L^*(x; \psi)}{L^*(x; \theta)}\right) dx.$$

If we interpret $\theta$ to be the true parameter, and $\psi$ another point, the Kullback information (of $\theta$ on $\psi$) gives a measure of how close $\psi$ is to $\theta$. It is shown in Dhrymes (1994b) that if $L(X; \theta)$ is the log-likelihood function, for a sample $X$, the joint density function of which depends on an underlying parameter vector $\theta$, whose true value is $\theta^\circ$, and its admissible space is $\Theta$, the ML estimator of $\theta^\circ$, say $\hat{\theta}_T$, may be found by the operation

$$\inf_{\theta \in \Theta} H_T(\theta), \quad H_T(\theta) = [L_T(X; \theta^\circ) - L_T(X; \theta)]; \quad L_T = \frac{1}{T}L.$$

Tt is further shown therein that $H_T(\hat{\theta}_T)$ converges at least in probability to the KI. More precisely

$$H_T(\hat{\theta}_T) \xrightarrow{P} K(\bar{\theta}, \theta^\circ) \quad \text{where} \quad \inf_{\theta \in \Theta} K(\theta, \theta^\circ) = K(\bar{\theta}, \theta^\circ). \tag{2.119}$$

Evidently, if the model is **correctly specified** and $K$ has a unique global minimum (identification), this limit is **zero**, and thus the ML estimator converges to the true parameter at least in probability.

**The AIC and AICC criteria examined in this section, seek to determine the order of a $MARMA(m, n)$ model by minimizing an estimated KI measure for finite samples.** To mimic the situation above we resort to the same technique employed earlier. To this end consider the $MARMA(m, n)$ model of Section 2.5.2, let $(X_1., X_2., \ldots, X_T.)$ and $(Y_1., Y_2., \ldots, Y_T.)$ be two **independent** realizations and let $\hat{\gamma}$, $\hat{\Sigma}$ be the ML estimators obtained through the $X$ realization. We may mimic the small sample analog of the KI function by the approximate expectation of $-L(Y; \hat{\gamma}, \hat{\Sigma}) + L(X; \hat{\gamma}, \hat{\Sigma})$. From Eqs. (2.95), (2.96), and (2.101), we find

$$L(Y; \hat{\gamma}, \hat{\Sigma}) = -\frac{Tq}{2}\ln(2\pi) - \frac{T}{2}\ln|\hat{\Sigma}| - \frac{1}{2}\hat{w}_y' \hat{A}'^{-1}(I_T \otimes \hat{\Sigma}^{-1})\hat{A}^{-1}\hat{w}_y$$

$$= -\frac{Tq}{2}\ln(2\pi) - \frac{T}{2}\ln|\hat{\Sigma}| - \frac{1}{2}(y - Q_y\hat{\gamma})(I_T \otimes \hat{\Sigma}^{-1})(y - Q_y\hat{\gamma}).$$

Adding and substracting $S(X; \hat{\gamma}, \hat{\Sigma}) = \frac{1}{2}(x - Q_x\hat{\gamma})(I_T \otimes \hat{\Sigma}^{-1})(x - Q_x\hat{\gamma})$ the relation above becomes

$$L(Y; \hat{\gamma}, \hat{\Sigma}) = -\frac{Tq}{2}\ln(2\pi) - \frac{T}{2}\ln|\hat{\Sigma}| - S(X; \hat{\gamma}, \hat{\Sigma}) + S(X; \hat{\gamma}, \hat{\Sigma}) - S(Y; \hat{\gamma}, \hat{\Sigma})$$

$$= L(X; \hat{\gamma}, \hat{\Sigma}) + \frac{Tq}{2} - S(Y; \hat{\gamma}, \hat{\Sigma}).$$

Hence, the small sample KI approximation is given by

$$K_T(\hat{\gamma}, \gamma^\circ) = -2[L(Y; \hat{\gamma}, \hat{\Sigma}) - L(X; \hat{\gamma}, \hat{\Sigma})] = 2S(Y; \hat{\gamma}, \hat{\Sigma}) - Tq.$$

To find a small sample approximation for the expected value of the rightmost member above note that $\epsilon = y - Q_y\gamma^\circ$, so that $\epsilon_t.$ "depends" on the elements of $\gamma$ not only directly, as it multiplies elements of $Q_y$, but also indirectly through the dependence of $\epsilon_t.$ on the lags $\epsilon_{t-j}.$, which serve to define it. In fact we had shown in Eq. (2.101) that $(\partial\epsilon/\partial\gamma)' = -A^{-1}Q$. Expand $S(Y; \cdot, \cdot)$ by Taylor's series (in $\gamma$), using up to second derivative terms to obtain

$$S(Y; \hat{\gamma}, \hat{\Sigma}) = S(Y; \gamma^\circ, \hat{\Sigma}) + 2\left(\frac{\partial\epsilon}{\partial\gamma}(\gamma^\circ)\right)(I_T \otimes \hat{\Sigma}^{-1})(\hat{\gamma} - \gamma^\circ) +$$

$$(\hat{\gamma} - \gamma^{\circ})' \left(\frac{\partial \epsilon}{\partial \gamma}(\gamma^{\circ})\right)' (I_T \otimes \hat{\Sigma}^{-1}) \left(\frac{\partial \epsilon}{\partial \gamma}(\gamma^{\circ})\right) (\hat{\gamma} - \gamma^{\circ}) + R,$$

where $R$ contains terms that converge to zero in probability. We observe that the expectation of the derivative $(\partial\epsilon/\partial\gamma)(\gamma^{\circ})$ is zero, and moreover by construction it is independent of the terms that multiply it. Thus, ignoring it, and terms that converge to zero in probability, we find

$$S(Y; \hat{\gamma}, \hat{\Sigma}) \approx S_1 + S_2 = S(Y; \gamma^{\circ}, \hat{\Sigma}) \tag{2.120}$$

$$+ (\hat{\gamma} - \gamma^{\circ})' \left[\left(\frac{\partial \epsilon}{\partial \gamma}(\gamma^{\circ})\right)' (I_T \otimes \hat{\Sigma}^{-1}) \left(\frac{\partial \epsilon}{\partial \gamma}(\gamma^{\circ})\right)\right] (\hat{\gamma} - \gamma^{\circ})$$

$$= \epsilon'(I_T \otimes \hat{\Sigma}^{-1})\epsilon + (\hat{\gamma} - \gamma^{\circ})' Q_y A'^{-1}(I_T \otimes \hat{\Sigma}^{-1}) A^{-1} Q_y (\hat{\gamma} - \gamma^{\circ}),$$

whose (approximate) expectation we need to determine. We note that because of approximate normality $\hat{\Sigma}$ and $\hat{\gamma} - \gamma^{\circ}$ are (approximately) mutually independent; moreover, by construction these entities are independent of all other terms, since the latter depend only on the $Y$-realization. We first take the expectation of $\hat{\Sigma}^{-1}$. This is proportional to a random matrix that has the **inverse Wishart** distribution [see Johnson and Kotz (1972), pp. 162-163]. Therefore we find [17]

$$E\hat{\Sigma}^{-1} \approx \frac{Tq}{Tq - (n + m)q^2 - 2} \Sigma^{-1} \tag{2.121}$$

Hence the first term in Eq. (2.120) yields

$$ES_1 = E[\operatorname{tr}(I_T \otimes \hat{\Sigma}^{-1})\epsilon\epsilon' = \frac{Tq}{Tq - (n + m)q^2 - 2}\operatorname{tr}(I_T \otimes \Sigma^{-1})(I_T \otimes \Sigma)$$

$$= \frac{T^2 q^2}{Tq - (n + m)q^2 - 2}.$$

For the second term, first taking the expectation of $\hat{\Sigma}^{-1}$ and $\hat{\gamma} - \gamma^{\circ}$, and then the expectation with respect to $Y$, we find

$$ES_2 = E_Y E_{X|Y} \left(\operatorname{tr} Q_y' A'^{-1}(I_T \otimes \hat{\Sigma}^{-1}) A^{-1} Q_y\right) (\hat{\gamma} - \gamma^{\circ})(\hat{\gamma} - \gamma^{\circ})'$$

---

[17] The term $Tq$ is "due" to the fact that it is $\hat{\Sigma}^{-1}/T$ that has the inverse Wishart distribution; the term $-(n + m)q^2$ is the number of parameters estimated; the term minus 2 is "due" to the Jacobian of the transformation from the Wishart to the inverse Wishart.

$$= \frac{Tq}{Tq - (m+n)q^2 - 2} \text{tr} E_Y \left( Q'_y A'^{-1} (I_T \otimes \Sigma^{-1}) A^{-1} Q_y \right) (T\Psi)^{-1}$$

$$= \frac{Tq}{Tq - (m+n)q^2 - 2} \text{tr} \left( (T\Psi)(T\Psi)^{-1} \right) = \frac{Tq(m+n)q^2}{Tq - (m+n)q^2 - 2}.$$

Combining the results above, we find

$$K_T(\hat{\gamma}, \gamma^\circ) \approx ES_1 + ES_2 - Tq = \frac{2Tq[(m+n)q^2 + 1]}{Tq - (m+n)q^2 - 2}. \tag{2.122}$$

To summarize the discussion above we note that the **AICC criterion obtains estimators of the parameters and order of the** $MARMA(m,n)$ **model by minimizing**

$$-2L(B, A, \Sigma) + \frac{2Tq[(m+n)q^2 + 1]}{Tq - (m+n)q^2 - 2}.$$

The AIC criterion is essentially the same, except that it does not penalize to the same degree for the number of parameters being estimated. Specifically: The **AIC criterion obtains estimators of the parameters and order of the** $MARMA(m,n)$ **model by minimizing**

$$-2L(B, A, \Sigma) + 2(m+n)q^2 + 1.$$

For the scalar case we have, setting $q = 1$:

The AICC criterion is: $-2L(B, A, \sigma^2) + \frac{2T[(m+n)+1]}{T-(m+n)-2}$.

The AIC criterion is $-2L(B, A, \sigma^2) + 2(m+n) + 1$.

Finally, the Akaike BIC criterion is given by:

$$\text{BIC}(A, B, \sigma^2) = (T - n - m)\ln\left(\frac{T\hat{\sigma}^2}{T - n - m}\right) + T(1 + \ln\sqrt{2\pi})$$

$$+(n+m)\ln\left(\frac{\sum_{t=1}^{T} X_t^2 - T\hat{\sigma}^2}{n+m}\right).$$

The Schwarz BIC criterion is obtained by formulating the issue as one of finding the Bayesian solution to the problem of choosing one of a number of competing models, under a particular set of priors; these priors are not absolutely continuous in that they put positive probability on certain subspaces of the general parameter space, viz. those corresponding to the competing models. The criterion is obtained from the first two terms of the asymptotic expansion of the Bayes solution. Since the first term is simply the ML estimator, and the second term depends only on the singularities of the priors, but not on their shapes, this criterion is of general applicability. It minimizes the expression stated below:

$$\text{Schwarz BIC} = -2L(A, B, \sigma^2) + (m + n)\ln T,$$

where as in all the preceding the notation $L(A, B, \sigma^2)$ indicates the maximum of the LF, and $(n + m)$ indicates the number of parameters employed in the scalar case. [18] Evidently, the results above give us, as a special case, the various criteria for AR and MA models.

## 2.5.4 Estimation of Structural VAR Models

The VAR or MARMA models examined earlier have been extensively applied in empirical macro studies. It is fair to say, however, that VAR models are devoid of any information regarding the contemporaneous interaction of the economic variables examined. In point of fact, they shed very little light on the workings of the economic systems to which they are applied, even though they may offer an excellent representation of how the past shapes the present and future evolution of the system, *ceteris paribus*.

To illuminate aspects of the economic structure examined, it is necessary to return to the methods of simultaneous equations, in the form of the **structural VAR model**. The latter is simply a set of equations of the form

$$X_t.\Pi = \sum_{i=1}^{n} X_{t-i}.\Pi_i + \epsilon_t.. \tag{2.123}$$

We note that Eq. (2.123) represents a **simultaneous equations model with no exogenous variables**. A model of this variety was examined by Mann and Wald (1943), who gave conditions for identification of its parameters, and produced the ML estimator of such parameters, as well as its

---

[18] In the vector case $n + m$ is to be replaced by $(n + m)q^2$, since each of the matrices $B_i$, $A_j$ contains $q^2$ parameters.

limiting distribution. In the prevailing intellectual climate of the time, identification was attained by putting "prior" restrictions on the elements of the
matrices $\Pi$, $\Pi_i$, $i = 1, 2, \ldots, n$. In its contemporary resurrection, as represented, for example, in Blanchard (1989), the structural VAR model above
is estimated under the following set of conditions:

 i. $\{\epsilon_{t\cdot} : t \in \mathcal{N}\}$ is a sequence of i.i.d. random vectors, each distributed
    as $N(0, I_q)$;

 ii. in order to produce identification, there are prior restrictions on the
    matrix $\Pi$, **but not on the** $\Pi_i$, $i = 1, 2, \ldots, n$.

To render this model in the usual GLM or SUR mode, put

$$y_{t\cdot} = X_{t\cdot}, \quad x_{t\cdot} = (X_{t-1\cdot}, X_{t-2\cdot}, \ldots, X_{t-n\cdot}), \quad Y = (y_{t\cdot}), \quad X = (x_{t\cdot}),$$

$$\Pi_* = (\Pi_1', \Pi_2', \ldots, \Pi_n')', \tag{2.124}$$

and write the $T$ observations in the form $Y\Pi = X\Pi_* + U$, where $U = (\epsilon_{t\cdot})$.
The LF is given by [19]

$$L(\Pi, \Pi_*) = -\frac{qT}{2}\ln(2\pi) + \frac{T}{2}\ln|\Pi\Pi'| - \frac{1}{2}\mathrm{tr}(Y\Pi - X\Pi_*)(Y\Pi - X\Pi_*)', \tag{2.125}$$

which is to be maximized **subject to the prior restrictions**

$$R\mathrm{vec}(\Pi) = r, \quad R = \mathrm{diagonal}(R_1, R_2, \ldots, R_q), \quad R_i \text{ is } k_i \times q, \text{ of rank } r_i. \tag{2.126}$$

Partially maximizing with respect to $\Pi_*$, we find

$$\hat{\Pi}_* = (X'X)^{-1}X'Y\Pi, \quad \frac{1}{2}\mathrm{tr}(Y\Pi - X\hat{\Pi}_*)(Y\Pi - X\hat{\Pi}_*)' = \frac{T}{2}\mathrm{tr}\Pi'\hat{\Omega}\Pi,$$

$$\hat{U} = [I_T - X(X'X)^{-1}X']Y, \quad \hat{\Omega} = \frac{1}{T}\hat{U}'\hat{U}. \tag{2.127}$$

We further note that by Proposition 90 in Dhrymes (1984), p. 106,

$$\mathrm{tr}\Pi'\hat{\Omega}\Pi = \mathrm{vec}(\Pi)'(I_q \otimes \hat{\Omega})\mathrm{vec}(\Pi).$$

---

[19] This form of the LF is obtained by first writing the LF of the observations on $\epsilon_{t\cdot}$
and then **treating the equation** $y_{t\cdot}\Pi - x_{t\cdot}\Pi_* = \epsilon_{t\cdot}$ **as a transformation from** $\epsilon_{t\cdot}$
to $y_{t\cdot}$. The term $\frac{1}{2}\ln|\Pi\Pi'|$ **represents the logarithm of the Jacobian of this
transformation**.

The concentrated LF is now to be maximized subject to the constraint in Eq. (2.126). We note that this is a special case of a nonlinear estimation problem, under constraint, treated extensively in Chapter 5 of Dhrymes (1994b). Let

$$S_T(\theta) = L_T(\Pi) + \lambda'(r - R\text{vec}(\Pi)) \qquad (2.128)$$

$$= -\frac{q}{2}\ln(2\pi) + \frac{1}{2}\ln|\Pi\Pi'| - \frac{1}{2}\beta'(I_q \otimes \hat{\Omega})\beta,$$

$$+\lambda'(r - R\beta), \quad \theta = (\beta', \lambda')', \quad \beta = \text{vec}(\Pi).$$

The first order derivatives are given by

$$\frac{\partial S_T}{\partial \beta} = \frac{1}{2}\frac{\partial \ln|\Pi\Pi'|}{\partial \text{vec}(\Pi\Pi')}\frac{\partial \text{vec}(\Pi\Pi')}{\partial \beta} - \beta'(I_q \otimes \hat{\Omega}) - \lambda'R,$$

$$\frac{\partial S_T}{\partial \lambda} = (r - R\beta)'. \qquad (2.129)$$

From Proposition 106 in Dhrymes (1984), pp. 124-125, we find

$$\frac{\partial \ln|\Pi\Pi'|}{\partial \text{vec}(\Pi\Pi')} = \text{vec}[(\Pi\Pi')^{-1}]', \quad \frac{\partial \text{vec}(\Pi\Pi')}{\partial \beta} = [I_{q^2} + I_{(q,q)}](\Pi \otimes I_q),$$

$$I_{(q,q)} = (I_q \otimes e_{\cdot 1}, I_q \otimes e_{\cdot 2}, \ldots, I_q \otimes e_{\cdot q})',$$

where $e_{\cdot i}$ is a $q$-element (column) vector of zeros, except for the $i^{th}$ element, which is unity.

The representation of the second derivative above follows from the properties of the permutation matrix $I_{(q,q)}$ relevant to our discussion, which are:

i. $I_{(q,q)}$ is symmetric and $I'_{(q,q)}I_{(q,q)} = I_{q^2}$ ;

ii. for any square matrix of dimension $q$, $\text{vec}(F') = I_{(q,q)}\text{vec}(F)$ ;

iii. for any square matrices $A, B$ of dimension $q$, $(B \otimes A) = I_{(q,q)}(A \otimes B)I_{(q,q)}$, or $(B \otimes A) = I'_{(q,q)}(A \otimes B)I'_{(q,q)}$ .

A proof may be found in Lemmata 9 and 10 in Dhrymes (1994b), pp. 215-217, or in Henderson and Searle (1981).

From the preceding we note that $I'_{(q,q)}\mathrm{vec}(\Pi'^{-1}\Pi^{-1}) = \mathrm{vec}(\Pi'^{-1}\Pi^{-1})$ because of the symmetry of $\Pi'^{-1}\Pi^{-1}$; thus we have

$$\left(\frac{\partial S_T}{\partial \beta}\right)' = \mathrm{vec}(\Pi'^{-1}) - \mathrm{vec}(\hat{\Omega}\Pi) - R'\lambda, \quad \left(\frac{\partial S_T}{\partial \lambda}\right)' = r - R\beta. \qquad (2.130)$$

Evidently, this is a nonlinear system whose solution may only be found by iteration. To this end, we employ the Newton-Raphson algorithm, which requires us to obtain the Hessian of the objective function. This method relies on a Taylor series expansion about an initial point, say $\tilde{\theta}$, and ignores third order terms which may be shown to converge to zero, at least in probability. More precisely, we have

$$0 = \left(\frac{\partial S_T}{\partial \theta}(\hat{\theta})\right)' \approx \left(\frac{\partial S_T}{\partial \theta}(\tilde{\theta})\right)' + \left(\frac{\partial^2 S_T}{\partial \theta \partial \theta}(\tilde{\theta})\right)'(\hat{\theta} - \tilde{\theta}).$$

If we interpret $\tilde{\theta}$ as the estimator of $\theta$ at the $(s-1)^{st}$ iteration, $\hat{\theta}$ as the estimator at the $s^{th}$ iteration, respectively $\theta_{(s-1)}$, $\theta_{(s)}$, and note that the Hessian is

$$\frac{\partial^2 S_T}{\partial \theta \partial \theta} = -\begin{bmatrix} (\tilde{\Pi}^{-1} \otimes \tilde{\Pi}'^{-1})I'_{(q,q)} + (I_q \otimes \hat{\Omega}) & R' \\ R & 0 \end{bmatrix} = -\begin{bmatrix} \tilde{A}_{11} & A_{12} \\ A_{21} & A_{22} \end{bmatrix} = -\tilde{A}, \qquad (2.131)$$

the iteration algorithm becomes

$$\theta_{(s)} = \theta_{(s-1)} + \begin{bmatrix} (\Pi^{-1}_{(s-1)} \otimes \Pi'^{-1}_{(s-1)})I'_{(q,q)} + (I_q \otimes \hat{\Omega}) & R' \\ \\ R & 0 \end{bmatrix}^{-1}$$

$$\times \begin{pmatrix} \mathrm{vec}(\Pi'^{-1}_{(s-1)}) - (I_q \otimes \hat{\Omega})\beta_{(s-1)} - R'\lambda_{(s-1)} \\ \\ r - R\beta_{(s-1)} \end{pmatrix}, \text{ or}$$

$$\beta_{(s)} = B_{11}[\mathrm{vec}(\Pi'^{-1}_{(s-1)}) - (I_q \otimes \hat{\Omega})\beta_{(s-1)} - R'\lambda_{(s-1)}] + B_{12}(r - R\beta_{(s-1)})$$

$$\lambda_{(s)} = B_{21}[\mathrm{vec}(\Pi'^{-1}_{(s-1)}) - (I_q \otimes \hat{\Omega})\beta_{(s-1)} - R'\lambda_{(s-1)}] + B_{22}(r - R\beta_{(s-1)}),$$

$$B_{11} = V_{11} - V_{11}R'(RV_{11}R')^{-1}RV_{11}, \quad V_{11}^{-1} = A_{11} + R'R,$$

$$B_{12} = V_{11}R'(RV_{11}R')^{-1}, \quad B_{21} = B'_{12},$$

$$B_{22} = I - (RV_{11}R')^{-1}, \tag{2.132}$$

where the terms $B_{ij}$ in the right member of the equations for $\beta_{(s)}$ and $\lambda_{(s)}$ are evaluated at $\theta_{(s-1)}$. Finally, the converging iterate, say $\hat{\theta} = (\hat{\beta}', \hat{\lambda}')'$ is the ML estimator of $\theta$.

## 2.5.5  Identification Issues

The existence of the inverse in the iteration algorithm above requires the non-singularity of the limit (parameter version) of the matrix $\tilde{A}$ in Eq. (2.131). In turn, from Proposition A.1 in Dhrymes (1984), pp. 142-145, this requires the further conditions that $V_{11}^{-1} = A_{11} + R'R$ and $R'V_{11}R$ be of full rank. Since by one of the fundamental assumptions of such models $R$ is of full (column) rank, the essential requirement is that $V_{11}$ be **nonsingular**, i.e. positive definite.

The matrix $(\Pi^{-1} \otimes \Pi'^{-1})I'_{(q,q)}$ is symmetric, as demonstrated in

$$[(\Pi^{-1} \otimes \Pi'^{-1})I'_{(q,q)}]' = \{I_{(q,q)}[I'_{(q,q)}(\Pi^{-1} \otimes \Pi'^{-1})I'_{(q,q)}]\}' = (\Pi^{-1} \otimes \Pi'^{-1})I'_{(q,q)}.$$

Moreover, by assumption, $\Omega = (\Pi'^{-1}\Pi^{-1}) > 0$, so that $A_{11}$ is symmetric and at least positive semidefinite. However, it **cannot be positive definite**; if $A_{11}$ were **nonsingular**, it would mean that we could identify the parameters of $\Pi$, uniquely, without imposing **any** restrictions. However, it is evident from Eq. (2.123) that this cannot possibly be so since **post multiplying** that equation by an **orthogonal matrix** yields a model that satisfies all conditions placed upon the original model in Eq. (2.123), and is thus **observationally equivalent to that model**. This is an aspect of the **identification problem** in structural VAR models. We now show that $A_{11}$ is **singular**, so that the restrictions we have imposed, or at least some of them, are **strictly required** for identification.

Since $R'R$ is of rank $k$, the **requirement that** $V_{11} > 0$ implies that $\text{rank}(A_{11}) \geq q^2 - k$. This is the **rank condition** for identification in structural VAR models.

To establish the singularity of $A_{11}$, we may rewrite it as

$$A_{11} = (I_q \otimes \Pi'^{-1})[(\Pi^{-1} \otimes I_q)I'_{(q,q)} + (I_q \otimes \Pi^{-1})]$$

$$= (I_q \otimes \Pi'^{-1})\{I'_{(q,q)}[I'_{(q,q)}(\Pi^{-1} \otimes I_q)I'_{(q,q)}] + (I_q \otimes \Pi^{-1})\}$$

$$= (I_q \otimes \Pi'^{-1})[I_{(q,q)}(I_q \otimes \Pi^{-1}) + (I_q \otimes \Pi^{-1})]$$

$$= (I_q \otimes \Pi'^{-1})(I_{(q,q)} + I_{q^2})(I_q \otimes \Pi^{-1}). \tag{2.133}$$

Since $I_q \otimes \Pi$ is **nonsingular**, $\mathrm{rank}\,(A_{11}) = \mathrm{rank}(I_*$, where $I_* = I_{q^2} + I_{(q,q)}$. We note that

$$I_* = I_{(q,q)} + I_{q^2} = (I_q \otimes e_{.1}, I_q \otimes e_{.2}, \ldots, I_q \otimes e_{.q})' + (e_{.1} \otimes I_q, e_{.2} \otimes I_q, \ldots, e_{.q} \otimes I_q)'.$$

Let $x = (x'_{.1}, x'_{.2}, \ldots, x'_{.q})'$ be a $q^2$ element (column) vector such that each $x_{.i}$ has $q$ elements. The solution of the equation

$$I_* x = 0,$$

is easily seen to satisfy the conditions

    i. $x_{ii} = 0$, for all $i$;

    ii. $x_{ji} = -x_{ij}$, for $i \neq j$.

Thus $I_*$ is singular. The number of linearly independent solutions are $q(q - 1)/2$, and consequently

$$\mathrm{rank}(I_*) = \mathrm{rank}(A_{11}) = q^2 - \frac{q(q - 1)}{2} = \frac{q(q + 1)}{2}. \tag{2.134}$$

The rank condition $\mathrm{rank}\,(V_{11}) = q^2 - k$ requires $\mathrm{rank}(A_{11}) \geq q^2 - k$, which in conjunction with Eq. (2.134 ) implies

$$\frac{q(q + 1)}{2} \geq q^2 - k, \quad \text{or} \quad k \geq \frac{q(q - 1)}{2},$$

which is the **order condition** for identification.

**Limiting Distribution**

To determine the limiting distribution we use the mean value theorem to obtain

$$0 = \left(\frac{\partial S_T}{\partial \theta}(\hat{\theta})\right)' = \left(\frac{\partial S_T}{\partial \theta}(\theta^\circ)\right)' + \frac{\partial^2 S_T}{\partial \theta \partial \theta}(\theta^*)\,(\hat{\theta} - \theta^\circ),$$

where $\theta^\circ$ is the true parameter vector and $|\theta^{*'} - \theta^\circ| \leq |\hat{\theta} - \theta^\circ|$. Since the ML estimator is consistent, this condition implies that the Hessian above converges to the Hessian evaluated at the true parameter point. We also note that under the null that the restrictions are valid the "true" Lagrange multiplier is zero, so that

$$\sqrt{T}(\hat{\theta} - \theta^\circ) = \sqrt{T}\begin{pmatrix} \hat{\beta} - \beta^\circ \\ \hat{\lambda} \end{pmatrix} = \begin{pmatrix} B_{11} \\ B_{21} \end{pmatrix} [\sqrt{T}\frac{\partial L_T}{\partial\partial\beta}(\beta^\circ)],$$

where $B_{11}$ and $B_{21}$ are as given in Eq. (2.132), and are defined in terms of the limit of $\tilde{A}_{11}$, i.e. in terms of

$$A_{11} = (\Pi^{-1} \otimes \Pi'^{-1})I_{(q,q)} + (I_q \otimes \Omega). \tag{2.135}$$

We further note that

$$\sqrt{T}\left(\frac{\partial L_T}{\partial\partial\beta}(\beta^\circ)\right)' = -(\Pi' \otimes I_q)\frac{1}{\sqrt{T}}\sum_{t=1}^{T} \zeta'_{t\cdot}, \quad \zeta_{t\cdot} = u_{t\cdot} \otimes u_{t\cdot} - \omega', \quad \omega = \text{vec}(\Omega). \tag{2.136}$$

The sequence $\{\zeta_{t\cdot} : t \in \mathcal{N}\}$ is i.i.d. with mean zero and covariance matrix [20]

$$\text{Cov}(\zeta_{t\cdot}) = \Omega \otimes \Omega + (\Omega \otimes \omega_{\cdot 1}, \Omega \otimes \omega_{\cdot 2}, \ldots, \Omega \otimes \omega_{\cdot q})' = [I_{q^2} + I_{(q,q)}](\Omega \otimes \Omega) = \Phi. \tag{2.137}$$

Consequently,

$$\sqrt{T}\left(\frac{\partial L_T}{\partial\partial\beta}(\beta^\circ)\right)' \xrightarrow{d} N[0, (\Pi' \otimes I_q)\Phi(\Pi \otimes I_q)], \tag{2.138}$$

and thus we conclude

$$\sqrt{T}(\hat{\theta} - \theta^\circ) \xrightarrow{d} N(0, \Psi), \quad \Psi = \begin{bmatrix} \Psi_{11} & \Psi_{12} \\ \Psi_{21} & \Pi_{22} \end{bmatrix}, \tag{2.139}$$

$$\sqrt{T}[\text{vec}(\hat{\Pi}) - \text{vec}(\Pi^\circ)] \xrightarrow{d} N(0, \Psi_{11}), \quad \Psi_{11} = B_{11}(\Pi' \otimes I_q)\Phi(\Pi \otimes I_q)B'_{11},$$

$$\sqrt{T}\hat{\lambda} \xrightarrow{d} N(0, \Psi_{22}), \quad \Psi_{22} = B_{21}(\Pi' \otimes I_q)\Phi(\Pi \otimes I_q)B'_{21},$$

We now simplify the expressions involved in the matrix $\Psi$. It is evident, by direct verification, that

$$(\Omega \otimes \omega_{\cdot 1}, \Omega \otimes \omega_{\cdot 2}, \ldots, \Omega \otimes \omega_{\cdot q})' = I_{(q,q)}(\Omega \otimes \Omega),$$

---

[20] For a similar derivation, see Dhrymes (1994b), pp. 270-271.

a fact we have used above; moreover, we have that $\Phi = [I_{q^2} + I_{(q,q)}](\Omega \otimes \Omega)$, and as noted above $I_{(q,q)}(A \otimes B)I_{(q,q)} = (B \otimes A)$. Consequently,

$$(\Pi' \otimes I_q)\Phi(\Pi \otimes I_q) = (I_q \otimes \Omega) + (\Pi' \otimes I_q)I_{(q,q)}(\Omega \otimes \Omega)(\Pi \otimes I_q)$$

$$= (I_q \otimes \Omega) + (\Pi' \otimes I_q)[I_{(q,q)}(\Omega \otimes \Omega)I_{(q,q)}][I'_{(q,q)}(\Pi \otimes I_q)I'_{(q,q)}]I_{(q,q)}$$

$$= (I_q \otimes \Omega) + (\Pi^{-1} \otimes \Pi'^{-1})I_{(q,q)} = A_{11}.$$

Thus,

$$\Psi_{11} = B_{11}A_{11}B'_{11}, \quad \Psi_{22} = B_{21}A_{11}B'_{21}, \quad \Psi_{12} = B_{11}A_{11}B'_{21}, \qquad (2.140)$$

$$\Psi_{12} = 0, \quad \text{since} \quad V_{11}R' = 0, \quad \text{and} \quad A_{11}B'_{21} = R'[(RV_{11}R')^{-1} - I_{q^2}].$$

We have therefore proved Proposition 11.

**Proposition 11.** Consider the model in Eq. (2.123) together with:

i. the sequence $\{\epsilon_{t\cdot} : t \in \mathcal{N}\}$ is one of i.i.d. $N(0, I_q)$ random vectors;

ii. there are no prior restrictions on the coefficient matrices $\Pi_i$, $i = 1, 2, \ldots, q$, but the elements of the matrix $\Pi$ obey the prior restrictions $R\beta = r$, where $\beta = \text{vec}(\Pi)$, $R = \text{diag}(R_1, R_2, \ldots, R_q)$, and $R_i$ is $k_i \otimes q$, of rank $k_i$, $i = 1, 2, \ldots, q$.

Then the following statements are true:

1. The parameters of the model are identified if

$$A_{11} + R'R = (\Pi^{-1} \otimes \Pi'^{-1})I_{(q,q)} + (I_q \otimes \Omega) + R'R$$

is of full rank (positive definite);

2. the matrix $A_{11}$ is of rank $q(q+1)/2$ and since the rank of $R'R$ is $k = \sum_{i=1}^{q} k_i$, the order condition for identification is $k \geq q(q-1)/2$;

3. The ML estimator of the parameters of the model is found from Eq. (2.127) and the iteration algorithm of Eq. (2.132), and it is simply the converging iterate; the iteration may commence with $\hat{\Pi}$ such that $\hat{\Pi}\hat{\Pi}' = \hat{\Omega}^{-1}$.[21]

---

[21] Evidently, the estimated parameters obey $R\hat{\beta} = r$, and thus the covariance matrix of their limiting distribution is **singular**.

4. The limiting distribution of the feedback parameters and the Lagrange
   multipliers, under the null $R\beta = r$, is given by

$$\sqrt{T}\begin{pmatrix} \hat{\beta} - \beta^{\circ} \\ \hat{\lambda} \end{pmatrix} \overset{\mathrm{d}}{\to} N(0, \Psi), \quad \Psi = \begin{bmatrix} \Psi_{11} & \Psi_{12} \\ \Psi_{21} & \Psi_{22} \end{bmatrix},$$

   where the $\Psi_{ij}$, $i, j = 1, 2$, are as in Eq. (2.140).

5. The Lagrange multiplier estimators ($\hat{\lambda}$) are asymptotically **independent** of the feedback parameter estimators ($\hat{\beta}$).

**Remark 12.** In several empirical applications the identification issue is neglected, and the authors operate with the entity $\hat{\Omega}$, or more precisely $\hat{\Omega}^{-1}$,
which they proceed to decompose as $\hat{\Omega}^{-1} = \hat{\Pi}\hat{\Pi}'$. The problem with this
approach is that the matrix of the decomposition is **not unique**. If $Q$ is
**an arbitrary orthogonal** matrix, $\hat{\Pi}^* = \hat{\Pi}Q$ also decomposes the matrix
in question, i.e. $\hat{\Omega}^{-1} = \hat{\Pi}^*\hat{\Pi}^{*'}$. Thus, the estimator $\hat{\Pi}$ as above conveys no
**information on the economics of the problem** over and above that con-
tained in the reduced form estimator, which is very little. In fact, the same
reduced form evidence may be consistent with many diametrically opposed
economic interpretations of the system, depending on which orthogonal matrix we pair with some basic solution, say $\tilde{\Pi}$. To overcome this identification
problem, some empirical papers impose the minimal number of conditions
for uniqueness of the decomposition. By Proposition 11, this minimal number is $q(q-1)/2$, which has the effect if $\Pi^{\circ}$ is the true parameter matrix,
and if $\Pi^{(1)}$ is an admissible matrix [22] then $\Pi^{(1)} = \Pi^{\circ}Q$ is satisfied **only
with** $Q = I_q$. [23] When only the minimal number of restrictions is imposed
we have the case of **just identification**. Again from the theory of simultaneous equations [see in particular Dhrymes (1994b), pp. 249-256, and the
appendix to Dhrymes (1994a)], we note that a just identified structure is

---

[22] In the terminology of simultaneous equations theory an admissible structure, or in
this case matrix, is one that obeys all known a priori restrictions.

[23] Notice that this minimal number of restrictions is **in addition** to the restrictions
imposed by the requirement that $\epsilon_t \sim N(0, I_q)$, instead of $N(0, \Sigma)$, $\Sigma > 0$. An equivalent
procedure would be to take $\Sigma$ to be a **diagonal matrix**, in which case we may impose
the requirement that the **diagonal elements of** $\Pi$ are all unities. Nothing of substance
will change if this convention is imposed.

simply a reparameterization of the reduced form, and as such conveys no more information than the reduced form. In point of fact, there are many ways in which the minimal ( $q(q-1)/2$ ) restrictions may be chosen; each may well lead to a very distinctive interpretation of the nature of the economic phenomenon modeled and **each may claim empirical validity with the same cogency**; the empirical authority **for all is** $\hat{\Omega}^{-1}$.

To summarize, structural VAR's obtained by the **"just identification decomposition"** of the estimated reduced form covariance matrix are **useless** in providing information regarding the economic structure underlying the phenomenon being modeled by the "structural VAR". To obtain such information, one must in effect employ simultaneous equations methods, i.e. the quantity $k$ of Proposition 11 should obey $k > q(q-1)/2$. The validity of the $k - q(q-1)/2$ overidentifying restrictions may be tested through the Lagrange multiplier test given in Dhrymes (1994b), pp. 122-140, or Dhrymes (1994a). If their validity is accepted, the economic implications of the structural VAR will have a cogency beyond that conferred by the reduced form covariance matrix, $\hat{\Omega}$. If, on the other hand, it is **rejected** the investigator will have to conclude that the empirical evidence provides little if any information about the structure of the phenomenon being studied, at least as formulated in the estimated structural VAR model.

## 2.6    State Space Models

### 2.6.1    Fundamental Definitions and Properties

In this section we present the fundamental ideas of state space modeling and some examples. This subject has a vast literature which we cannot cover in any degree of detail. What we shall attempt to do is provide an exposition of the salient aspects of this topic. We begin with the basic concepts.

**Definition 6**. The set of equations

$$Y_{t\cdot}^{'} = J_t Z_{t\cdot}^{'} + v_{t\cdot}^{'}, \quad Z_{t+1\cdot}^{'} = K_t Z_{t\cdot}^{'} + w_{t\cdot}^{'}, \quad t = 1, 2, \ldots, \qquad (2.141)$$

is said to be a state space model, or more appropriately a model in **state space form** (SSF). The first set of equations is termed the **observation**, or **measurement, equation**, while the second is termed the **transitional equation**, or the **state space equation**. The entity $Z_t$ is said to be the **state vector**. The entities $(J_t, K_t)$ are known and nonstochastic and are said to be the **system matrices**. Let the dimension of the observation vector be $q$, and the dimension of the state vector $p$, so that the system matrices are of dimension, respectively, $q \times p$ and $p \times p$.

The SSF is typically a representation of a more basic model, and is assumed to obey the following conditions:

i. the sequence $\{\xi_t = (v_t, w_t) : t = 1, 2, \ldots\}$ is a zero mean square integrable orthogonal sequence with $\mathrm{Cov}\,(\xi_t) = \Sigma_t$;

ii. the initial condition (or state) $Z_1$, obeys $EZ'_1 = \bar{Z}_1$, $\mathrm{Cov}\,(Z_1) = C_z$ and is **orthogonal** to the $\xi$-sequence, i.e. $EZ'_1 \xi_t = 0$ for all $t$;

iii. the state vector $Z_t$ need not be observable, but the vector $Y_t$ is observable.

**Remark 13.** The transitional equation is also written as $Z'_{t+1} = K_t Z'_t + A_t u'_t + w'_t$, or $Z'_{t+1} = K_t Z'_t + c'_t + w'_t$. In the first variant, $u_t$ denotes an observable entity under the control of the investigator, and is termed the **control vector**; in the second variant, the vector $c_t$ generally denotes seasonal "dummies", or variables with similar interpretation that modify the equation from period to period.

**Remark 14.** The convenience of the SSF of a model lies in the simplicity of the second equation, since it is generally rather easy to solve first order difference equations.

The origins of this formulation lie in the systems control literature. The model may describe, for example, the problem of determining the coordinates of a space craft, which was launched under a specified set of conditions that determines the initial state $Z_1$. The matrices $K_t$ are known from the engineering specification of the craft and the laws of physics. The space coordinates $Z_t$ are not directly observed, but are measured through telemetry that yields the "noisy" observations in the vector $Y_t$. The matrices $J_t$ are known from the properties of the measuring instruments.

**Remark 15**. Note that even if the $\xi$-sequence is also assumed to be time homogeneous, i.e. $\Sigma_t = \Sigma$, the state-space and observation vectors are typically not, owing to the time varying specification of the system matrices. In typical applications in econometrics, the system matrices are **time homogeneous**, i.e. $J_t = J$ and $K_t = K$. If, **in addition**, $K$ is a stable matrix, i.e. its characteristic roots are less than unity in absolute value, both the state and observation vectors are covariance stationary and the state-space equation have a unique stationary solution, as demonstrated below.

**Stationary Representations**

In the general case, solving the transitional equation recursively we find

$$Z'_{t\cdot} = F_{t-1}Z'_{1\cdot} + \sum_{j=0}^{t-2} F_j w'_{t-1-j\cdot}$$

$$F_j = \sum_{i=0}^{j-1} K_{t-i}, \quad j = 1, 2, \ldots, t-1, \quad F_0 = I_p, \tag{2.142}$$

When the system matrices are time homogeneous, $F_t = K^t$ and extending the system backward we conclude that

$$Z'_{t\cdot} = \sum_{j=0}^{\infty} K^j w'_{t-1-j\cdot}, \quad Y'_{t\cdot} = JZ'_{t\cdot} + v'_{t\cdot}. \tag{2.143}$$

Since the $w$-process is an orthogonal sequence with constant covariance matrix and $K^j$ converges to zero exponentially, [24] we find that the state-space vector is a covariance stationary process; since the observation vector is a linear transformation of the state-space vector, and $\xi_{t\cdot}$ is an orthogonal sequence with a time homogeneous covariance matrix, it follows that the former is also covariance stationary. Moreover, the observation and state-space vectors are **jointly** covariance stationary.

A major inference from the preceding is:

i. $Z_{t\cdot} \perp \xi_{s\cdot}$, for $t \leq s$;

---

[24] This is due to the fact that the characteristic roots of $K$ are less than one in absolute value.

ii. when the system matrices are time homogeneous, the matrix $K$ is stable and $\xi$ is an orthogonal process with time homogeneous covariance matrix

$$\text{Cov}(\xi_{t\cdot}) = \Sigma = \begin{bmatrix} \Sigma_{11} & \Sigma_{12} \\ \Sigma_{21} & \Sigma_{22} \end{bmatrix},$$

the process $\{\Upsilon_{t\cdot} = (Y_{t\cdot}, \ Z_{t\cdot}) : t \in \mathcal{N}_+\}$ is covariance stationary, and its second moment properties are given by

$$C_z(0) = \sum_{j=0}^{\infty} K^j \Sigma_{22} (K')^j \tag{2.144}$$

$$\text{Cov}(\Upsilon'_{t\cdot}, \Upsilon_{t\cdot}) = (J', I_p)' C_z(0)(J, I_p) + \begin{bmatrix} \Sigma_{11} & 0 \\ 0 & 0 \end{bmatrix},$$

$$\text{Cov}(\Upsilon_{t+\tau\cdot}, \Upsilon_{t\cdot}) = (J', I_p)'[K^\tau C_z(0)](J', I_p).$$

**Illustrations of SSF**

As we noted in the definition, the state space representation is more accurately conceived of as the SSF of some other basic or underlying model.

We discuss below the state space form representation of several basic models.

**Example 2**. [Stationary $VAR(m)$] Consider the $VAR(m)$

$$B(L)X'_{t\cdot} = \epsilon'_{t\cdot}, \quad B(L) = I_q - \sum_{j=1}^{m} B_j L^j. \tag{2.145}$$

Its SSF may be rendered as the so-called companion matrix representation. To this end, define

$$Z_{t\cdot} = (X_{t\cdot}, X_{t-1\cdot}, \dots, X_{t-m+1\cdot}), \quad w_{t\cdot} = (\epsilon_{t\cdot}, 0, 0, \dots, 0),$$

$$K = \begin{bmatrix} B_1 & B_2 & B_3 & \cdots & B_{m-1} & B_m \\ 0 & I_q & 0 & \cdots & 0 & 0 \\ 0 & 0 & I_q & \cdots & 0 & 0 \\ 0 & 0 & 0 & \cdots & 0 & \\ \vdots & \vdots & \vdots & \vdots & \vdots & \vdots \\ 0 & 0 & 0 & \cdots & I_q & 0 \end{bmatrix},$$

$$Z'_{t+1\cdot} = KZ'_{t\cdot} + w'_{t\cdot}, \quad Y'_{t\cdot} = JZ'_{t\cdot}, \quad J = (I_q, 0, 0, \dots, 0). \tag{2.146}$$

Note that here the state vector is observable for $t \geq m + 1$ and, moreover, that the characteristic roots of $K$ are less than one in absolute value, if the underlying $VAR(m)$ is **stable**; in this case the state vector has the stationary solution $Z'_{t.} = \sum_{j=0}^{\infty} K^j w'_{t-1-j.}$.

Notice, further, that this representation is **not unique**. For example, if $Q$ is **any** nonsingular matrix, pre-multiply the relations above by $Q$ to obtain,

$$QZ'_{t+1.} = (QKQ^{-1})QZ'_{t.} + Qv'_{t.},$$

$$QY'_{t.} = (QJQ^{-1})QZ'_{t.} + Qv'_{t.}. \qquad (2.147)$$

In the SSF above the state vector is $QZ'_{t.}$, the new observation vector is $QY'_{t.}$ and the new transition matrix is $QKQ^{-1}$, which has **the same characteristic roots** as $K$. Thus, none of the essential properties of the SSF representation are disturbed by this operation. The objective in this aspect of the problem is to choose from among the many possible state vectors the one with **minimal dimension**.

**Example 3** [SSF of a $MMA(n)$]. Consider the multivariate moving average model

$$X'_{t.} = A(L)\epsilon'_{t.}, \quad A(L) = I_q + \sum_{i=1}^{n} A_i L^i. \qquad (2.148)$$

Define

$$Z_{t.} = (X_{t.}, \epsilon_{t.}, \epsilon_{t-1.}, \ldots, \epsilon_{t-n+1.}), \quad K = \begin{bmatrix} 0 & A_1 & A_2 & A_3 & \cdots & A_n \\ 0 & 0 & 0 & \cdots & 0 & 0 \\ 0 & I_q & 0 & \cdots & 0 & 0 \\ 0 & 0 & I_q & \cdots & 0 & 0 \\ \vdots & \vdots & \vdots & \vdots & \vdots & \vdots \\ 0 & 0 & 0 & \cdots & I_q & 0 \end{bmatrix},$$

$$v'_{t.} = (I_q, I_q, 0, 0, \ldots, 0)' \epsilon'_{t+1.}, \quad Z'_{t+1.} = KZ'_{t.} + v'_{t.},$$

$$Y'_{t.} = (I_q, 0, 0, \ldots, 0)Z'_{t.}. \qquad (2.149)$$

Notice that $Z_{s.} \perp v_{t.}$, $s \leq t$, that the state vector is not completely observable and that $\{\xi_{t.} : \xi_{t.} = (v_{t.}, 0), t \in \mathcal{N}_+\}$ is an orthogonal sequence, as required for models in SSF. The matrix $K$, however, is evidently singular, and its roots are not necessarily less than one in absolute value, even if the $MMA(n)$ is invertible.

**Example 4** (Vector Random Walk with Drift and Noise). Consider the vector random walk with drift

$$X'_{t,1\cdot} = X'_{t-1,1\cdot} + \beta + \epsilon'_{t1\cdot},$$

which is not directly observable; what is observable is

$$X'_{t2\cdot} = X'_{t1\cdot} + \epsilon'_{t2\cdot}.$$

The SSF of this model may be constructed as follows: set

$$Z_{t\cdot} = (X_{t1\cdot}, \beta'), \quad Z'_{t+1\cdot} = KZ_{t\cdot} + I_* \epsilon'_{t1\cdot}, \quad I_* = (I_q, 0)', \quad K = \begin{bmatrix} I_q & I_q \\ 0 & I_q \end{bmatrix} \quad (2.150)$$

and note that the observation equation is

$$Y'_{t\cdot} = (I_q, 0)Z'_{t\cdot} + \epsilon'_{t2\cdot}. \quad (2.151)$$

**Example 5** (Seasonal Process with Error). The seasonal process with period $p$ obeys the conditions $EX_{t+p\cdot} = EX_{t\cdot} = s_t$ and $\sum_{j=1}^{p} EX_{j+\tau\cdot} = 0$, for any $\tau > 0$; thus the seasonal process with error is given by

$$X_{t\cdot} = -\sum_{j=1}^{p} X_{t-j\cdot} + \epsilon_{t\cdot}.$$

The state vector is given by $Z_{t\cdot} = (X_{t\cdot}, X_{t-1\cdot}, \ldots, X_{t-p+1\cdot})$; the transition and measurement equations are given, respectively, by

$$Z'_{t+1\cdot} = KZ_{t\cdot} + (I_q, 0, \ldots, 0)' \epsilon'_{t\cdot}, \quad Y'_{t\cdot} = JZ'_{t\cdot},$$

$$K = \begin{bmatrix} -I_q & -I_q & -I_q & \cdots & -I_q & -I_q \\ 0 & I_q & 0 & \cdots & 0 & 0 \\ 0 & 0 & I_q & \cdots & 0 & 0 \\ \vdots & \vdots & \vdots & \vdots & \vdots & \vdots \\ 0 & 0 & \cdots & \cdots & I_q & 0 \end{bmatrix}, \quad J = (I_q, 0, \ldots, 0).$$

If $p = 4$ so that we are dealing with quarterly seasonals, and we begin our observations with the first quarter of the first year and number all observations by quarter only, [25] we find, for a scalar process,

$$X_1 = (1-2.5)+\epsilon_1, \quad X_2 = (2-2.5)+\epsilon_2, \quad X_3 = (3-2.5)+\epsilon_3, \quad X_4 = (4-2.5)+\epsilon_4,$$

and thereafter $X_{np+i} = s_i + \epsilon_{np+i}$, for $n \geq 0$, where $s_i = (i - 2.5)$ and $i = 1,2,3,4$.

**Example 6** (Random Walk with Trend and Seasonal). To obtain the SSF representation of this process we simply combine the results from Examples 4 and 5 as follows: Let

$$Z_{t1.} = (X_{t1.}, \beta'), \quad Z_{t2.} = (X_{t,2.}, X_{t-1,2.}, \ldots, X_{t-p+1,2.}), \quad Z_{t.} = (Z_{t1.}, Z_{t2.});$$

$$K = \begin{bmatrix} K_1 & 0 \\ 0 & K_2 \end{bmatrix}, \quad K_1 = \begin{bmatrix} I_q & I_q \\ 0 & I_q \end{bmatrix}, \quad K_2 = - \begin{bmatrix} I_q & I_q & I_q & \cdots & I_q & I_q \\ 0 & I_q & 0 & \cdots & 0 & 0 \\ 0 & 0 & I_q & \cdots & 0 & 0 \\ \vdots & \vdots & \vdots & \vdots & \vdots & \vdots \\ 0 & 0 & \cdots & \cdots & I_q & 0 \end{bmatrix},$$

$$Z'_{t+1.} = K Z'_{t.}, \quad Y'_{t.} = (I_q, 0, I_q, 0, \ldots, 0) Z'_{t.} + \epsilon'_{t.}.$$

It is clear that the measurement equation yields $Y_{t.} = X_{t1.} + X_{t2.} + \epsilon_{t.}$. In this context,

$$X_{t1.} = X_{t-1,1.} + \beta' + \epsilon_{t1.}, \quad \text{representing the random walk with drift, while}$$

$$X_{t2.} = -\sum_{j=1}^{p-1} X_{t-j.} + \epsilon_{t2.}; \quad \text{thus,}$$

$$Y'_{t.} = X'_{t1.} + X'_{t2.} + \epsilon'_{t.}.$$

Thus, the measurement equation gives us a representation of a (noisy) random walk with drift, plus a seasonal process with error, as desired.

## 2.6.2 Kalman Procedures

As we pointed out in previous discussion on the BLP, the various algorithms used therein to obtain the coefficient matrices of the predictor are **not new**

---

[25] Otherwise we need to use a double subscript, one for years and another for quarters.

or distinct estimation techniques, but rather convenient computational procedures for obtaining the relevant estimators, recursively, fully taking into account computations made (earlier) with fewer observations. The same is true with Kalman procedures, which we illustrate, for Kalman prediction, in a very simple context utilizing the GLM.

Thus, suppose we have $T$ observations, contained in the vector and matrix, respectively,

$$y_{(T)} = (y_1, y_2, \ldots, y_T)', \quad X_{(T)} = (x_{t\cdot}), \quad t = 1, 2, \ldots, T.$$

The model is represented, in the usual notation, as $y_{(T)} = X_{(T)}\beta + u_{(T)}$, and the problem is to obtain the BLP of $y_{T+1}$, given $T$ observations and given $T+1$ observations. [26] Examining this problem in the least squares context, the BLP of $y_{T+1}$ given $T$ observations is

$$\hat{y}_{T+1|T} = x_{T+1\cdot}\hat{\beta}_{(T)}, \quad \hat{\beta}_{(T)} = [X'_{(T)}X_{(T)}]^{-1}X'_{(T)}y_{(T)}, \tag{2.152}$$

while the BLP of the same, given $T+1$ observations is

$$\hat{y}_{T+1|T+1} = x_{T+1\cdot}\hat{\beta}_{(T+1)}, \quad \hat{\beta}_{(T+1)} = [X'_{(T+1)}X_{(T+1)}]^{-1}X'_{(T+1)}y_{(T+1)}. \tag{2.153}$$

We now produce two recursions, one linking $\hat{\beta}_{(T)}$ to $\hat{\beta}_{(T+1)}$, and another linking $\hat{y}_{T+1|T}$ to $\hat{y}_{T+1|T+1}$. To this end, note that

$$y_{(T+1)} = \begin{pmatrix} y_{(T)} \\ y_{T+1} \end{pmatrix}, \quad X_{(T+1)} = \begin{bmatrix} X_{(T)} \\ x_{T+1\cdot} \end{bmatrix}. \tag{2.154}$$

Consequently,

$$\hat{\beta}_{(T+1)} = [X'_{(T)}X_{(T)} + x'_{T+1\cdot}x_{T+1\cdot}]^{-1}[X'_{(T)}y_{(T)} + x'_{T+1\cdot}y_{T+1}]. \tag{2.155}$$

By Proposition 32, Corollary 5, in Dhrymes (1984), pp. 38-39, and the identity given in Section 2.4.2, with $C = X'_{(T)}X_{(T)}$, $X = x_{T+1\cdot}$, $D = 1$, $Y = x'_{T+1\cdot}$, we find

$$\hat{\beta}_{(T+1)} = \hat{\beta}_{(T)} + \frac{1}{\alpha}[X'_{(T)}X_{(T)}]^{-1}x'_{T+1\cdot}(y_{T+1} - x_{T+1\cdot}\hat{\beta}_{(T)}),$$

$$\alpha = 1 + x_{T+1\cdot}[X'_{(T)}X_{(T)}]^{-1}x'_{T+1\cdot}. \tag{2.156}$$

---

[26] It may seem odd to the reader that we wish to obtain the BLP of $y_{T+1}$, given $T+1$ observations, since in that case, evidently, $y_{T+1}$ is known. However, in state-space models, where this procedure originated, the state vector is generally not observable.

This shows that if we have estimated the regression coefficients based on a sample of size $T$, and another observation becomes available, it is **not necessary** to invert the matrix $X'_{(T+1)}X_{(T+1)}$ in order to obtain the regression coefficient vector based on $T+1$ observations! This implies, in turn, that if the matrix $X_{(T)}$ has $k+1$ columns, we need only invert $X'_{(k+1)}X_{(k+1)}$, to obtain $\hat{\beta}_{(k+1)}$, and thereafter we simply **update** the estimator by means of Eq. (2.155). This is the great convenience offered by this procedure.

Consider now the prediction of $y_{T+1}$ given $T+1$ observations. Using again Eq. (2.155) we find

$$\hat{y}_{T+1|T+1} \;=\; x_{T+1\cdot}\hat{\beta}_{(T+1)} = \hat{y}_{T+1|T} + G_{T+1}(y_{T+1} - x_{T+1\cdot}\hat{\beta}_{(T)}),$$

$$G_{T+1} \;=\; \frac{x_{T+1\cdot}[X'_{(T)}X_{(T)}]^{-1}x'_{T+1\cdot}}{1 + x_{T+1\cdot}[X'_{(T)}X_{(T)}]^{-1}x'_{T+1\cdot}}. \tag{2.157}$$

The entity $G_{T+1}$ is termed the **Kalman gain**, while $y_{T+1} - x_{T+1\cdot}\hat{\beta}_{(T)}$ is said to be the **innovation** at time $T+1$.

**Remark 16.** Notice that, in the context of the GLM, the Kalman gain is simply the ratio of two variances, i.e. the variance of the prediction $\hat{y}_{T+1|T} = x_{T+1\cdot}\hat{\beta}_{(T)}$, to the variance of the prediction error

$$y_{T+1} - x_{T+1\cdot}\hat{\beta}_{(T)} = u_{T+1} - x_{T+1\cdot}(\hat{\beta}_{(T)} - \beta).$$

The Kalman procedures for models in SSF involve three distinct aspects, although the algorithms involved are essentially the same; they are

   i. estimation or prediction of the state vector $Z_{t\cdot}$, given the observations $Y_{1\cdot}, Y_{2\cdot}, \ldots, Y_{t-1\cdot}$, with an initial vector $Y_{0\cdot}$, i.e. the operation

$$\hat{Z}'_{t\cdot} = P_{t-1}(Z_{t\cdot}|Y_{0\cdot}, Y_{1\cdot}, \ldots, Y_{t-1\cdot}).$$

    This is the **prediction problem.** [27]

---

[27] The initial vector $Y_{0\cdot}$, which is **not** generated by the observation equation $Y'_{t\cdot} = J_t Z'_{t\cdot} + v'_{t\cdot}$, is introduced in order to provide for genuine stochastic initial conditions, or to allow us the use of constant terms in the implicit "regressions" that define the BLP. In the latter case $Y_{0\cdot} = e' = (1, 1, \ldots, 1)$, or in the scalar case $Y_0 = 1$.

ii. Estimation or prediction of $Z_{t\cdot}$ given $Y_{j\cdot}$, $j = 0, 1, \ldots t$, i.e. the operation

$$\hat{Z}'_{t\cdot} = P_t(Z_{t\cdot}|Y_{0\cdot}, Y_{1\cdot}, \ldots, Y_{t\cdot}),$$

which is the **filtering problem**.

iii. Estimation or prediction of $Z_{t\cdot}$ given $Y_{j\cdot}$, $j = 0, 1, \ldots T$, $T > t$ i.e. the operation

$$\hat{Z}'_{t\cdot} = P_t(Z_{t\cdot}|Y_{0\cdot}, Y_{1\cdot}, \ldots, Y_{T\cdot}),$$

which is the **smoothing problem**.

## Prediction

In obtaining the recursions, the Kalman prediction algorithm employs the same device we had employed in Section 2.4.2 in deriving the DLW algorithm. The problem here is formulated as the prediction of $Z_{t+1\cdot}$, given a certain information set, and the operation is defined by

$$\hat{Z}'_{t+1\cdot} = P_t(Z_{t+1\cdot}|Y_{j\cdot}, j = 0, 1, \ldots t). \tag{2.158}$$

We remind the reader that the notation $P_t(\cdot|\cdot)$ refers to the projection of the vector to the left of the vertical line onto the linear manifold generated by the variables to the right of the vertical line. [28] The device employed in connection with the derivation of the DLW algorithm was to express the projection operator $P_t$ as the sum of two orthogonal operators. The same device will be employed here, by obtaining the "innovation"

$$S_{t\cdot} = Y_{t\cdot} - \hat{Y}_{t\cdot}, \quad \hat{Y}'_{t\cdot} = P_{t-1}(Y_{t\cdot}|Y_{j\cdot}, j = 0, 1, \ldots, t-1). \tag{2.159}$$

Since $S_{t\cdot} \perp Y_{j\cdot}$, for $j = 0, 1, 2, \ldots, t-1$, we may rewrite Eq. (2.158) as [29]

$$\hat{Z}'_{t+1\cdot} = P_{t-1}(Z_{t+1\cdot}|Y_{0\cdot}, Y_{1\cdot}, \ldots Y_{t-1\cdot}) + D_t S'_{t\cdot}, \tag{2.160}$$

---

[28] In the literature of econometrics the operation in the right member of Eq. (2.158) is rendered as a **conditional expectation** and is denoted by $E(\cdot|\cdot)$. This is, however, **incorrect**, unless it is the case that all entities on the right as well as the left of the vertical line **are jointly normal**.

[29] Notice that in the prediction aspect, **all predictions are conditioned on observations up to the one immediately preceding**. Thus, we have no notation for writing the prediction of $Z_{t+1\cdot}$ conditioned on $Y_{j\cdot}, j = 0, 1, 2, \ldots, t-1$. On the other hand from

where $D_t$ is defined by the condition $(Z_{t+1.} - S_t.D_t') \perp S_t.$. To determine the recursions let us establish the notation

$$\text{Cov}(\phi_{t.}) = \begin{bmatrix} \Sigma_{11(t)} & \Sigma_{12(t)} \\ \Sigma_{21(t)} & \Sigma_{22(t)} \end{bmatrix}, \quad \phi_{t.} = (v_{t.}, w_{t.}),$$

$$\text{MSE}(Z_{t.} - \hat{Z}_{t.}) = \Phi_t = \Psi_t - \Pi_t,$$

$$\Psi_t = EZ_{t.}'Z_{t.}, \quad \Pi_t = E\hat{Z}_{t.}'\hat{Z}_{t.}. \tag{2.161}$$

The last set of results is due to the fact that $(Z_{t.} - \hat{Z}_{t.}) \perp \hat{Z}_{t.}$, so that $E\hat{Z}_{t.}'Z_{t.} = E\hat{Z}_{t.}'\hat{Z}_{t.}$. A recursion for $\Psi_{t+1}$ is easily established through

$$\Psi_{t+1} = EZ_{t+1.}'Z_{t+1.} = E[K_t Z_{t.}' + w_{t.}'][K_t Z_{t.}' + w_{t.}']' = K_t \Phi_t K_t' + \Sigma_{22(t)}. \tag{2.162}$$

To determine $D_t$, from Eq. (2.160), we need to establish expressions for $F_t = EZ_{t+1.}'S_{t.}$ and $H_t = ES_{t.}'S_{t.}$, since the former is defined as the solution of $D_t H_t = F_t$. Thus,

$$F_t = EZ_{t+1.}'S_{t.} = E[K_t Z_{t.}' + w_{t.}'][(Z_{t.} - \hat{Z}_{t.})J_t' + v_{t.}] = K_t \Phi_t J_t' + \Sigma_{21(t)},$$

$$H_t = E(Y_{t.} - \hat{Y}_{t.})'(Y_{t.} - \hat{Y}_{t.}) = E[(Z_{t.} - \hat{Z}_{t.})J_t' + v_{t.}]'[(Z_{t.} - \hat{Z}_{t.})J_t' + v_{t.}]$$

$$= J_t \Phi_t J_t' + \Sigma_{11(t)}, \quad D_t = F_t H_{t(g)},$$

$$\hat{Z}_{t+1.}' = K_t \hat{Z}_{t.}' + F_t H_{t(g)} S_{t.}', \tag{2.163}$$

where the last equation is simply Eq. (2.160), taking into account the linearity of the projection operator, and the discussion in footnote 28. $H_{t(g)}$ denotes the **generalized inverse** of $H_t$ [see Dhrymes (1984), pp. 82 ff.] [30]

_____

the transition equation, and the linearity of the projection operator, we have

$$P_{t-1}(Z_{t+1.}|Y_j., j = 0, 1, \ldots, t-1) = K_t P_{t-1}(Z_{t.}') + P_{t-1}(w_{t.}') = K_t \hat{Z}_{t.}',$$

because $w_{t.} \perp Y_j.$, for $j = 0, 1, \ldots, t-1$. Thus, Eq. (2.160) may be written as

$$\hat{Z}_{t+1.}' = P_t(Z_{t+1.}) = K_t P_{t-1}(Z_{t.}) + D_t S_{t.}' = K_t \hat{Z}_{t.}' + D_t S_{t.}'.$$

[30] If $A$ is $m \times n$, its (unique) generalized inverse is defined in the context to be the $n \times m$ matrix $A_g$ such that (a) $A_g A A_g = A_g$, (b) $A A_g A = A$, (c) $A A_g$ is symmetric and (d) $A_g A$ is symmetric.

In executing the Kalman prediction procedure it is necessary to specify the initial conditions $Z_1.$ and $Y_0.$ .

The preceding has established Proposition 12.

**Proposition 12.** Consider the state space model

$$Y_t'. = J_t Z_t'. + v_t'., \quad Z_{t+1}' = K_t Z_t'. + w_t'., \quad t = 1, 2, \ldots,$$

under the conditions given in Definition 6, and let it be desired to predict the state vector $Z_{t+1}.$ given the information $\{Y_j. : j = 0, 1, \ldots t\}$. The BLP is given by

$$\hat{Z}_{t+1}' = K_t \hat{Z}_t'. + D_t S_t'., \quad S_t'. = Y_t'. - P_{t-1}(Y_t.|Y_j., j = 0, 1, 2, \ldots t - 1),$$

and is **uniquely determined** by the initial conditions, $Y_0.$ , $Z_1.$ , $\hat{Z}_1'. = P(Z_1.|Y_0.)$, $\Pi_1 = E\hat{Z}_1'. \hat{Z}_1.$ , $\Psi_1 = EZ_1'. Z_1.$ , $\Phi_1 = \Psi_1 - \Pi_1$ , through the recursions

$$F_t = K_t \Phi_t J_t' + \Sigma_{21(t)}, \quad H_t = J_t \Phi_t J_t' + \Sigma_{11(t)}, \quad D_t = F_t H_{t(g)},$$

$$\Psi_{t+1} = K_t \Psi_t K_t' + \Sigma_{22(t)}, \quad \Pi_{t+1} = K_t \Pi_t K_t' + F_t H_{t(g)} F_t',$$

$$\Phi_{t+1} = \Psi_{t+1} - \Pi_{t+1}. \tag{2.164}$$

**Remark 17.** When the initial condition obeys $Y_0. = e' = (1, 1, \ldots, 1)$, it may be shown that $\hat{Z}_1. = \mu_1. = EZ_1.$ , as follows: Since $Y_0. = e'$, the BLP is given by $\hat{Z}_1'. = De$. Then, conditionally on $Y_0.$ ,

$$MSE(Z_1. - \hat{Z}_1.) = E(Z_1. - \mu_1.)'(Z_1. - \mu_1.) + (\hat{Z}_1. - \mu_1.)'(\hat{Z}_1. - \mu_1.), \tag{2.165}$$

which is minimized only when we take $De = \hat{Z}_1'. = \mu_1'.$ .

### Filtering

In (Kalman) filtering the problem is to estimate or predict the state vector $Z_t.$ , given all observations through time $t$. Since earlier we dealt with the problem of estimating and/or predicting the state vector $Z_t.$ given observations only through time $t - 1$, and later we shall be dealing with the same problem given observations through time $T > t$, we require some additional notation, to make clear the set of the conditioning variables. To this end, let

$$\hat{Z}'_{t\cdot|s} = P_s(Z_{t\cdot}|Y_{j\cdot} : j = 0, 1, 2, \ldots, s), \text{ or simply, } \hat{Z}'_{t\cdot|s} = P_s(Z_{t\cdot}) \quad (2.166)$$

be the projection of $Z_{t\cdot}$ on the linear manifold generated by $\{Y_{j\cdot} : j = 0, 1, 2, \ldots, s\}$. What we examined in the preceding discussion is the case with $s = t - 1$, so that the entities $\hat{Z}_{t\cdot}$, or $\hat{Y}_{t\cdot}$ will be rendered in the new notation as $\hat{Z}_{t\cdot|t-1}$ or $\hat{Y}_{t\cdot|t-1}$. The entities we shall deal with in this section are $\hat{Z}_{t\cdot|t}$ or $\hat{Y}_{t\cdot|t}$, while the entities we shall deal with in the next section are $\hat{Z}_{t\cdot|r}$ or $\hat{Y}_{t\cdot|r}$ for $r > t$. Similarly, we are in need of additional notation for the mean squared error matrices; specifically, we define $\Phi_{t|s} = MSE(Z_{t\cdot} - \hat{Z}_{t\cdot|s})$, so that the entity $\Phi_t$ of the preceding discussion should be rendered as $\Phi_{t|t-1}$ in the notation introduced above.

The purpose of Kalman filtering, as opposed to prediction, is to make use of the additional observation in order to make the estimation of the state vector more accurate, by reducing the mean squared error matrix. The same purpose is served by (Kalman) smoothing, which uses **all of the sample information** in estimating or "predicting" the state vector at **each** "time" $t$.

We note that

$$P_t(Z_{t\cdot}) = P_{t-1}(Z_{t\cdot}) + DS'_{t\cdot}, \quad S'_{t\cdot} = Y'_{t\cdot} - \hat{Y}'_{t\cdot|t-1}, \quad (2.167)$$

where the matrix $D$ is obtained by the condition $(Z_{t\cdot} - \hat{Z}_{t\cdot|t-1}) \perp S_{t\cdot}$. Since $Y_{t\cdot} - \hat{Y}_{t\cdot|t-1} = (Z_{t\cdot} - \hat{Z}_{t\cdot|t-1})J'_t + v_{t\cdot}$,

$$EZ'_{t\cdot}S_{t\cdot} = E(Z_{t\cdot} - \hat{Z}_{t\cdot|t-1})'[(Z_{t\cdot} - \hat{Z}_{t\cdot|t-1})J'_t + v_{t\cdot}]$$

$$= \Phi_t J'_{t\cdot}, \text{ so that } D = \Phi_t J'_t H_{t(g)}. \quad (2.168)$$

We may thus write the Kalman filtering estimator as

$$\hat{Z}'_{t\cdot|t} = \hat{Z}'_{t\cdot|t-1} + \Phi_t J'_t H_{t(g)} S'_{t\cdot}. \quad (2.169)$$

Since

$$(Z_{t\cdot} - \hat{Z}_{t\cdot|t})' = (Z_{t\cdot} - \hat{Z}_{t\cdot|t-1})' - \Phi_t J'_t H_{t(g)} S'_{t\cdot}, \text{ it follows}$$

$$\Phi_{t|t} = \Phi_t - \Phi_t J'_t H_{t(g)} J_t \Phi_t. \quad (2.170)$$

This is due to the fact that

$$E(Z_{t\cdot} - \hat{Z}_{t\cdot|t-1})'S_{t\cdot}H_{t(g)}J_t\Phi_t = EZ'_{t\cdot}S_{t\cdot}H_{t(g)}J_t\Phi_t = \Phi_t J'_t H_{t(g)}J_t\Phi_t$$

We have therefore proved

**Proposition 13**. (Kalman Filtering). Under the conditions of Proposition 12, the filtering estimator and its mean squared error are

$$\hat{Z}'_{t\cdot|t} = \hat{Z}'_{t\cdot|t-1} + \Phi_t J'_t H_{t(g)}S'_{t\cdot}, \quad \Phi_{t|t} = \Phi_t - \Phi_t J'_t H_{t(g)}J_t\Phi_t,$$

together with the recursions and initial conditions given in Proposition 12.

**Smoothing**

This aspect of Kalman procedures is not extensively employed in econometrics, but is included in this exposition for the sake of completeness. Because the number of conditioning observations varies considerably we need to introduce the additional notation, for $r \geq t$

$$\Phi_{t,r} = E(Z_{t\cdot} - \hat{Z}_{t\cdot|r-1})'(Z_{r\cdot} - \hat{Z}_{r\cdot|r-1}). \tag{2.171}$$

As before, we have

$$P_r(Z_{t\cdot}|Y_{j\cdot}) = P_{r-1}(Z_{t\cdot}|Y_{j\cdot}) + DS'_{r\cdot}, \tag{2.172}$$

where $D$ is determined by the condition

$$(Z_{t\cdot} - \hat{Z}_{t\cdot|r-1}) \perp S_{r\cdot}, \quad S_{r\cdot} = Y_{r\cdot} - \hat{Y}_{r\cdot|r-1} = (Z_{r\cdot} - \hat{Z}_{r\cdot|r-1})J'_r + v_{r\cdot}.$$

By an argument we have employed many times in previous discussion this yields $D = \Phi_{t,r}J'_r H_{r(g)}$, so that the fixed point smoothed predictor is

$$\hat{Z}'_{t\cdot|r} = \hat{Z}'_{t\cdot|r-1} + \Phi_{t,r}J'_r H_{r(g)}S'_{r\cdot}, \tag{2.173}$$

and its conditional mean squared error matrix is easily obtained from Eq. (2.173) as the recursion

$$\Phi_{t|r} = \Phi_{t|r-1} - \Phi_{t,r}J'_r H_{r(g)}J_r\Phi_{t,r}'. \tag{2.174}$$

This is so because $E(\Phi_{t,r}J_r'H_{r(g)}S_r')\hat{Z}_{t\cdot|r-1} = \Phi_{t,r}J_r'H_{r(g)}J_r\Phi_{t,r}'$.

Finally, we need to obtain a recursion for the "cross" mean squared error matrices $\Phi_{t,r}$. Thus,

$$
\begin{aligned}
\Phi_{t,r+1} &= E(Z_{t\cdot} - \hat{Z}_{t\cdot|r})'(Z_{r+1\cdot} - \hat{Z}_{r+1\cdot|r}) = E(Z_{t\cdot} - \hat{Z}_{t\cdot|r-1})'(Z_{r+1\cdot} - \hat{Z}_{r+1\cdot|r}) \\[2mm]
&= E(Z_{t\cdot} - \hat{Z}_{t\cdot|r-1})'[(Z_{r\cdot} - \hat{Z}_{r\cdot|r-1})(K_r' - J_r'H_{r(g)}F_r') + w_{r\cdot} - v_{r\cdot}H_{r(g)}F_r'] \\[2mm]
&= \Phi_{t,r}[K_r' - J_r'H_{r(g)}F_r'], \quad F_r = K_r\Phi_{rr}J_r'.
\end{aligned}
\tag{2.175}
$$

The second equality above follows from the fact that $\hat{Z}_{t\cdot|r}$ and $\hat{Z}_{t\cdot|r-1}$ are orthogonal to $(Z_{r\cdot} - \hat{Z}_{r\cdot|r-1})$ for $t < r$; the third equality follows from Eqs. (2.141), (2.160), the discussion in footnote 30, and the representation of $S_{r\cdot}$ above; the last equality follows from the fact that $w_{r\cdot}, v_{r\cdot}$ are orthogonal to $(Z_{t\cdot} - \hat{Z}_{t\cdot|r-1})$, for $t < r$.

We have therefore proved Proposition 14.

**Proposition 14.** (Fixed Point Kalman Smoothing). Under the conditions of Proposition 12, and taking into account the results of Proposition 13, the fixed point smoothing procedure involves the following recursions,

$$
\hat{Z}_{t\cdot|r}' = \hat{Z}_{t\cdot|r-1}' + \Phi_{t,r}J_r'H_{r(g)}J_rS_{r\cdot}', \text{(fixed point smoothed predictor)}
$$

$$
\Phi_{t|r} = \Phi_{t|r-1} - \Phi_{t,r}J_r'H_{r(g)}J_r\Phi_{t,r}' \quad \text{(its mean squared error matrix)}
$$

$$
\Phi_{t,r+1} = \Phi_{t,r}[K_r' - J_r'H_{r(g)}F_r'], \quad r = t, t+1, \ldots, T-1.
\tag{2.176}
$$

Although the preceding discussion has divided the problem thereby easing of its exposition, it has also resulted in dual notation. This may make applications of these procedures more difficult than they are intrinsically. For this reason we restate all the results above in uniform notation.

## KALMAN PROCEDURES

### Fundamental Definitions

$$
\Phi_{t,r} = E(Z_{t\cdot} - Z_{t\cdot|r-1})'(Z_{r\cdot} - Z_{r\cdot|r-1}), \quad r = t, t+1, \ldots, T;
$$

$$
\Phi_{t|r} = E(Z_{t\cdot} - Z_{t-1\cdot|r})'(Z_{t\cdot} - Z_{t\cdot|r}), \quad r = t, t+1, \ldots, T;
$$

$$\Psi_t = EZ'_{t\cdot}Z_{t\cdot}, \quad t = 1, 2, \dots, T;$$

$$\Pi_{t|r} = E\hat{Z}'_{t\cdot|r}\hat{Z}_{t\cdot|r}, \quad r = t - 1, t, \dots, T,$$

$$F_t = EZ'_{t+1\cdot}S_{t\cdot}, \quad S_{t\cdot} = Y_{t\cdot} - \hat{Y}_{t\cdot|t-1}, \quad t = 2, 3, \dots T,$$

$$H_t = ES'_{t\cdot}S_{t\cdot}, \quad H_{t(g)} \text{ is its generalized inverse.} \tag{2.177}$$

Notice that $\Phi_{t,t} = \Phi_{t|t-1} = \Phi_t$, where the single subscript notation was employed in the Kalman prediction phase of our discussion.

### Initial Conditions

The following entities are specified as initial conditions: $Y_{0\cdot}$, $Z_{1\cdot}$, $\hat{Z}_{1\cdot} = P(Z_{1\cdot}|Y_{0\cdot})$, $\Psi_1 = EZ'_{1\cdot}Z_{1\cdot}$, $\Pi_{1|0} = E\hat{Z}'_{1\cdot}\hat{Z}_{1\cdot}$, $\Phi_{1|0} = \Psi_1 - \Pi_{1|0}$.

### Recursions

$$F_t = K_t\Phi_{t,t}J'_t + \Sigma_{22(t)}, \quad t = 1, 2, \dots, T; \tag{2.178}$$

$$H_t = J_t\Phi_{t,t}J'_t + \Sigma_{11(t)}, \quad t = 1, 2, \dots T;$$

$$\Psi_{t+1} = K_t\Psi_tK'_t + \Sigma_{22(t)}, \quad t = 1, 2, \dots, T-1;$$

$$\Pi_{t+1|t} = K_t\Pi_{t|t-1}K'_t + F_tH_{t(g)}F'_t, \quad t = 1, 2, \dots, T-1;$$

$$\Phi_{t+1,t+1} = \Phi_{t+1|t} = \Psi_{t+1} - \Pi_{t+1|t}, \quad t = 1, 2, \dots, T-1;$$

$$\Phi_{t,r+1} = \Phi_{t,r}(K'_r - J'_rH_{r(g)}F'_r), \quad r = t, t+1, \dots, T;$$

$$\Phi_{t|r} = \Phi_{t|r-1} - \Phi_{t,r}J'_rH_{t(g)}J_r\Phi'_{t,r}, \quad r = t, t+1, \dots, T;$$

$$\hat{Z}'_{t+1\cdot|t} = K_t\hat{Z}'_{t\cdot|t-1} + F_tH_{t(g)}S'_{t\cdot}, \quad t = 1, 2, \dots T \text{ (prediction)};$$

$$\hat{Z}'_{t\cdot|t} = \hat{Z}'_{t\cdot|t-1} + \Phi_{t,t}J'_tH_{t(g)}S'_{t\cdot}, \quad t = 2, 3, \dots T \text{ (filtering)};$$

$$\hat{Z}'_{t\cdot|r} = \hat{Z}'_{t\cdot|r-1} + \Phi_{t,r}J'_rH_{r(g)}S'_{r\cdot}, \quad r = t+1, \dots, T \text{ (f.p.s.,}$$

where f.p.s. stands for fixed point smoothing.

# Chapter 3

# Unit Roots; I(1) Regressors

## 3.1 Introduction

Consider the general linear model

$$y_t = x_t.\beta + u_t, \ t = 1, 2, \ldots T.$$

In the standard case we take the explanatory variables (regressors) represented by the $n$-element **row** vector $x_t.$ to obey

$$\frac{X'X}{T} \overset{\text{OL, P, or a.c.}}{\rightarrow} M_{xx} > 0, \quad X = (x_t.),$$

where $OL$ stands for ordinary limit, when the regressors are nonstochastic; when they are stochastic we either take them to be independent of the error process $\{u_t : t \in \mathcal{N}_+\}$ or, what amounts to the same thing more or less, we make the analysis of the problem **conditional** on them. In either case, the error process in the simplest case is one of i.i.d. random variables with mean zero and variance $\sigma^2 < \infty$. When these conditions hold, the least squares, or other appropriate estimator of the parameter vector $\beta$, say $\hat{\beta}$, is **consistent** in the sense that it converges a.c. to the parameter $\beta$ and, moreover, $\sqrt{T}(\hat{\beta} - \beta)$ converges in distribution to $N(0, \sigma^2 M_{xx}^{-1})$. What the reader may not fully appreciate is the fact that the "reason" $\sqrt{T}(\hat{\beta} - \beta)$ converges in distribution to $N(0, \sigma^2 M_{xx}^{-1})$ is intimately related to the convergence of $X'X/T$. If $X'X$ does not converge when normalized by $T$, it does not necessarily follow that the distributional conclusions above will continue to hold.

The problem we analyze in this chapter arises when the regressors are $I(1)$, in which case $X'X/T$ does not converge but, as we shall see, $X'X/T^2$ converges. An important special case is one in which the model is $y_t = \lambda y_{t-1} + u_t$, where it is asserted that in truth $\lambda = 1$, but this is ignored, or is not known to the analyst, or is treated **as a testable, rather than as a maintained** hypothesis.

The context we have created above has extensive applications in view of the fact that many macro economists hold that a number of macro variables are $I(1)$ processes. When that is the case, much of the standard inference theory available for the GLM is inapplicable and we need to find a solution to the inference problems arising therein.

The special case we have noted also has important motivational bases in terms of a number of developments in economic theory. For example, the efficient markets hypothesis, Fama (1965), (1970); the behavior of futures prices, Samuelson (1965), Mandelbrot (1966); aggregate consumption, Hall (1978); and macro time series, Nelson and Plosser (1982), all imply the special case noted above. The estimation and inference problem in this context is frequently referred to as the unit root (UR) problem. Another motivation for studying such a problem is somewhat "diagnostic" in nature and involves the following consideration. Suppose one is interested in the behavior of a certain variable, and one entertains the view that it is basically a time trend with an i.i.d. random component, i.e.

$$y_t = \alpha + \beta t + u_t, \quad t = 1, 2, \ldots, T.$$

Upon estimating the parameters of this model by least squares, the investigator finds a "high" Durbin-Watson statistic, and decides to examine the same model on the assumption that the random component is a first order autoregression. Thus, he considers the model

$$y_t = \alpha + \beta t + \frac{I}{I - \rho L} \epsilon_t, \quad t = 1, 2, \ldots, T, \tag{3.1}$$

where again the sequence $\{\epsilon_t : t \in \mathcal{N}_+\}$ is one of i.i.d. random variables with mean zero and variance $\sigma^2 < \infty$, $L$ is the **lag operator**, and $I$ is the **identity operator**, i.e. for any function $x(t)$, $Lx(t) = x(t-1)$, and $Ix(t) = x(t)$. This model may be estimated by nonlinear least squares involving a search on $\rho$, by minimizing the usual Aitken "objective" function. Alternatively, one might proceed by reducing the model to

$$y_t = a + \beta(1 - \rho)t + \rho y_{t-1} + \epsilon_t, \quad a = \alpha + (\beta - \alpha)\rho,$$

and estimate the parameters $a = \alpha + (\beta - \alpha)\rho$, $b = \beta(1 - \rho)$, and $\rho$ by least squares; thereafter one may obtain estimates for $\alpha$ and $\beta$ if desired. In relatively small samples the investigator may not detect the UR property of this model, if indeed this is the case, and may erroneously conclude from the insignificance of $b$ that $y_t$ is a sequence of nonzero mean random variables. Indeed, **if** $\rho = 1$ the model, as exhibited above, reduces to

$$y_t = y_{t-1} + \beta + \epsilon_t, \quad \text{or} \quad \Delta y_t = \beta + \epsilon_t.$$

Or, alternatively, if the model is in fact as exhibited above, the test statistic employed (earlier) in testing for significance **does not have even asymptotically** the $t$-distribution, or the normal distribution on which the investigator bases his conclusions! If, in fact $a \neq 0$ then the sequence $y_t$ is simply a random walk with a drift, and if $a = 0$, the sequence is simply a random walk and **not** one of i.i.d. random variables.

## 3.2    UR Model With i.i.d. Errors

Consider the model

$$y_t = \beta + \lambda y_{t-1} + u_t, \quad t = 1, 2, \ldots T, \tag{3.2}$$

on the assumption that $\{u_t : t \in \mathcal{N}_+\}$ is a $WN(\sigma^2)$ sequence, i.e. one of i.i.d. random variables with mean zero and variance $\sigma^2 < \infty$. In the case $|\lambda| < 1$ this is simply a distributed lag model, with a trivial exogenous variable, and has been studied extensively in the literature of econometrics [see, e.g., Dhrymes (1971)]. The consistency of the least squares estimator of $\lambda$, in the special case $\beta = 0$, has been demonstrated by Rubin (1950), for $\lambda \in R$. The limiting distribution for this estimator was obtained by White (1957) by the method of moment generating functions. The modern econometric literature of this problem, however, begins with Dickey and Fuller (1979)

(DF) and, particularly, Phillips and Perron (1987) (PP), whose formulation we shall follow in our discussion.[1]

The motivating question we ask (in the simplest case with $\beta = 0$) is: What is the (limiting) distribution of the least squares estimator of $\lambda$, its associated $t$-ratio, and the estimate of $\sigma^2$? The entities in question are given, respectively, by

$$\hat{\lambda} = \frac{\sum_{t=1}^{T} y_t y_{t-1}}{\sum_{t=1}^{T} y_{t-1}^2}, \quad \tau_{\hat{\lambda}} = \frac{(\hat{\lambda} - 1)\sqrt{\sum_{t=1}^{T} y_{t-1}^2}}{s}, \quad s^2 = \frac{1}{T}\sum_{t=1}^{T}(y_t - \hat{\lambda} y_{t-1})^2. \quad (3.3)$$

In deriving these limiting distributions, we generally operate under the hypothesis

$$H_0 : \lambda = 1, \quad \beta = 0.$$

As in Chapter 9, we place our discussion in the usual context; i.e. we assume that the (structural) error sequence above is defined over the probability space $(\Omega, \mathcal{A}, \mathcal{P})$; we further define the sequence of (sub) $\sigma$-algebras $\mathcal{G}_t = \sigma(u_s, s \leq t)$ for $t \in \mathcal{N}_+$ together with $\mathcal{G}_0 = (\emptyset, \Omega)$,[2] such that

$$\mathcal{G}_0 \subset \mathcal{G}_1 \subset \cdots \mathcal{G}_T \subset \mathcal{A}.$$

On the assumption that $y_0 = 0$, it is easy to see that the sequence $\{(y_t, \mathcal{G}_t) : t \in \mathcal{N}_+\}$ is a **martingale**, since

$$E|y_t| < \infty, \quad \text{and} \quad E(y_t | \mathcal{G}_{t-1}) = y_{t-1}, \quad \text{for all } t \in \mathcal{N}_+.$$

One could perhaps gain an insight into the problems that are generated by the UR hypothesis if one notes that solving the difference equation recursively one finds

$$y_t = \beta \sum_{j=0}^{t-1} \lambda^j (t - j) + \lambda^t y_0 + \sum_{j=0}^{t-1} \lambda^j u_{t-j}, \quad (3.4)$$

---

[1] The latter is, indeed, an important paper in that it gave the problem investigated by DF the proper formulation and, in so doing, not only revealed the precise nature of its complexity but also afforded the means of further developments. Indeed, many of the developments discussed in this chapter are variously due to Phillips and Durlauf (1986), Phillips and Hansen (1990), Park and Phillips (1988), (1989), and so on.

[2] Occasionally, this is referred to as the $\sigma$-algebra of constants.

so that in the case $\beta = 0$, if $|\lambda| < 1$, the "behavior" of the series is progres-
sively less dependent on initial conditions. If $\lambda = 1$ it is as much dependent
on the remote past as it is on the most recent past, and if $|\lambda| > 1$ the behav-
ior of the series is most importantly governed by initial conditions. In our
analysis, it will turn out that the initial condition, $y_0$, does not present any
special problem beyond the notational nuisance of having to carry it along;
thus we shall retain it, rather than assume that $y_0 = 0$.

When $\beta \neq 0$ the behavior of the series is dominated by the time trend
when $|\lambda| \leq 1$, and by its stochastic component when $|\lambda| > 1$. When $\lambda = 1$,
we rewrite Eq. (3.4) as

$$y_t = \beta t + y_0 + S_t, \quad S_t = \sum_{j=1}^{t} u_j, \tag{3.5}$$

and note that

$$E(y_t|y_0) = \beta t, \quad \text{Var}(y_t|y_0) = t\sigma^2, \tag{3.6}$$

so that the mean and the variance of the "explanatory" variable, $y_{t-1}$, are
**unbounded**.

To determine the limiting behavior of the entities in the motivating question, we write the $(T)$ observations in the vector form

$$y = X\gamma + u, \quad X = (e, \; y_{-1}), \quad \gamma = (\beta, \lambda)', \quad y = (y_1, y_2, \ldots y_T)', \tag{3.7}$$

where $e$ is a column vector of unities, $y_{-1} = (y_0, y_2, \ldots y_{T-1})'$, and obtain
the least squares estimator as

$$\hat{\gamma} - \gamma = (X'X)^{-1}X'u, \quad \text{so that} \tag{3.8}$$

$$\hat{\beta} - \beta = \frac{1}{\delta}\left(\frac{e'u}{T}(y_{-1}'y_{-1}) - \frac{e'y_{-1}}{T}(y_{-1}'u)\right), \quad \delta = y_{-1}'\left(I - \frac{ee'}{T}\right)y_{-1}$$

$$\hat{\lambda} - \lambda = \frac{1}{\delta}\left[y_{-1}'\left(I - \frac{ee'}{T}\right)\right]u.$$

## 3.2.1    The case $\beta = 0$

When it is assumed *a priori* that $\beta = 0$ all terms containing the vector of
unities $e$ disappear from the equations above. Under the hypothesis $\lambda = 1$,

the entities in the motivating question become

$$\hat{\lambda} - 1 = \frac{\sum_{t=1}^{T} u_t y_{t-1}}{\sum_{t=1}^{T} y_{t-1}^2},$$

$$\tau_{\hat{\lambda}}^2 = \frac{(\hat{\lambda} - 1)^2 \sum_{t=1}^{T} y_{t-1}^2}{s^2}, \tag{3.9}$$

$$\hat{u}_t = y_t - \hat{\lambda} y_{t-1} = u_t - (\hat{\lambda} - 1) y_{t-1}, \quad \text{so that}$$

$$s^2 = \frac{1}{T} \sum_{t=1}^{T} [u_t^2 - 2(\hat{\lambda} - 1) u_t y_{t-1} + (\hat{\lambda} - 1)^2 y_{t-1}^2].$$

The behavior of $\hat{\lambda} - 1$ is determined by the behavior of $\sum_{t=1}^{T} u_t y_{t-1}$ and $\sum_{t=1}^{T} y_{t-1}^2$; the behavior of $\tau_{\hat{\lambda}}$ is determined additionally by the behavior of $s^2$, and the behavior of the latter is determined, additionally, by the behavior of $\sum_{t=1}^{T} u_t^2 / T$. If their asymptotic behavior is established then, by the considerations given in Remark 5 of Chapter 9, or by Proposition 28, Corollary 5, Chapter 4, in Dhrymes (1989), pp. 242-244, we shall be able to determine the limiting distribution of all entities in the motivating question. To this end note

$$y_t = y_{t-1} + u_t, \quad y_t^2 = y_{t-1}^2 + u_t^2 + 2u_t y_{t-1}, \quad \text{so that}$$

$$u_t y_{t-1} = \frac{1}{2} \left( y_t^2 - y_{t-1}^2 - u_t^2 \right); \quad \text{consequently,}$$

$$\sum_{t=1}^{T} u_t y_{t-1} = \frac{1}{2} \left( y_T^2 - y_0^2 - \sum_{t=1}^{T} u_t^2 \right). \tag{3.10}$$

By Propositions 6 and 23 in Dhrymes (1989), pp. 148-149 and pp. 188-190, respectively, we have

$$\frac{y_0^2}{T} \overset{\text{a.c.}}{\to} 0, \quad \frac{\sum_{t=1}^{T} u_t^2}{T} \overset{\text{a.c.}}{\to} \sigma^2, \text{so that, asymptotically,}$$

$$\frac{\sum_{t=1}^{T} u_t y_{t-1}}{T} \sim \frac{1}{2} \left( \frac{y_T^2}{T} - \sigma^2 \right). \tag{3.11}$$

Since

$$\frac{y_T^2}{T} = \frac{1}{T} \left( y_0^2 + 2y_0 S_T + S_T^2 \right) \sim \frac{1}{T} \left( 2y_0 S_T + S_T^2 \right),$$

we need only examine the behavior of the entities $S_T$ and $S_T^2$. Both terms are amenable to a Donsker-like argument. To this end define the stochastic processes

$$X^T(r,\omega) = \frac{1}{\sqrt{T}} S_{[rT]}(\omega) = \frac{1}{\sqrt{T}} S_{t-1}(\omega), \quad \text{for } r \in \left[\frac{t-1}{T}, \frac{t}{T}\right),$$

$$= \frac{1}{\sqrt{T}} S_{[rT]}(\omega) = \frac{1}{\sqrt{T}} S_t(\omega), \quad \text{for } r \in \left[\frac{t}{T}, \frac{t+1}{T}\right), \qquad (3.12)$$

and so on. Notice from the equation set in Eq. (3.12) that the paths of the stochastic processes defined above are **discontinuous** and we are dealing with paths in the space $D(\mathcal{T}, R)$, with $\mathcal{T} = [0, 1]$. Thus, by Proposition 14 of Chapter 9, we conclude that the sequence of the distribution functions of the stochastic processes defined in Eq. (3.12) converge weakly to the Wiener measure on $\mathbf{W}$, or in our convenient notation,

$$X^T(r) \overset{d}{\to} \sqrt{\sigma^2} B(r), \quad r \in [0, 1], \qquad (3.13)$$

where $B$ is the standard BM. Moreover, since $X^T(1) = (S_T/\sqrt{T})$, and $B(1)$ is evidently an a.c. finite random variable, we have from Proposition 40, Chapter 4, in Dhrymes (1989), p. 262,

$$\frac{S_T}{T} = \frac{1}{\sqrt{T}} \frac{S_T}{\sqrt{T}} \overset{P}{\to} 0. \qquad (3.14)$$

Moreover, since

$$\frac{S_T^2}{T} = \left(\frac{S_T}{\sqrt{T}}\right)^2 \overset{d}{\to} \sigma^2 [B(1)]^2,$$

we conclude, in view of Eq. (3.11),

$$\frac{1}{T} \sum_{t=1}^{T} u_t y_{t-1} \overset{d}{\to} \frac{\sigma^2}{2} \left[B(1)^2 - 1\right]. \qquad (3.15)$$

We next consider

$$\frac{1}{T} \sum_{t=1}^{T} y_{t-1}^2 = \frac{1}{T} \sum_{t=1}^{T} (y_0^2 + 2y_0 S_{t-1} + S_{t-1}^2) = y_0^2 + 2y_0 \frac{1}{T} \sum_{t=1}^{T} S_{t-1} + \frac{1}{T} \sum_{t=1}^{T} S_{t-1}^2.$$

Unfortunately, the sum

$$\frac{1}{T} \sum_{t=1}^{T} S_{t-1} = \sum_{j=0}^{T-1} \left(1 - \frac{j}{T}\right) u_j \quad \text{does not converge.}$$

Upon renormalizing $\sum_{t=1}^{T} y_{t-1}^2$ by a higher power, say two, we find

$$\frac{1}{T^2} \sum_{t=1}^{T} y_{t-1}^2 = \frac{y_0^2}{T} + 2y_0 \frac{1}{T^2} \sum_{t=1}^{T} S_{t-1} + \frac{1}{T^2} \sum_{t=1}^{T} S_{t-1}^2,$$

and it is evident that the first term converges a.c. to zero. To deal with the second and third terms define, as before, the sequence of stochastic processes

$$X^T(r, \omega) = \left( \frac{S_{[rT]}(\omega)}{\sqrt{T}} \right), \quad \text{for} \ \ r \in \left[ \frac{t-1}{T}, \frac{t}{T} \right), \quad t = 1, 2, \ldots, T. \quad (3.16)$$

By Proposition 14, Chapter 9 of this volume, Proposition 28, Corollary 5, and Proposition 40, Chapter 4, in Dhrymes (1989), pp. 242-243 and p. 262, respectively,

$$\frac{y_0}{\sqrt{T}} \frac{1}{T^{3/2}} \sum_{t=1}^{T} S_{t-1} = \frac{y_0}{\sqrt{T}} \sum_{t=1}^{T} \int_{(t-1/T)}^{t/T} X^T(r, \omega) \ dr \xrightarrow{\text{P}} 0,$$

because

$$\frac{1}{T^{3/2}} \sum_{t=1}^{T} S_{t-1} = \sum_{t=1}^{T} \int_{(t-1/T)}^{t/T} X^T(r, \omega) \ dr \xrightarrow{\text{d}} \sqrt{\sigma^2} \int_0^1 B(r) \ dr. \quad (3.17)$$

It follows from the preceding that

$$\frac{1}{T^2} \sum_{t=1}^{T} y_{t-1}^2 \sim \frac{1}{T^2} \sum_{t=1}^{T} S_{t-1}^2 = \sum_{t=1}^{T} \int_{(t-1/T)}^{t/T} [X^T(r, \omega)]^2 \ dr \xrightarrow{\text{d}} \sigma^2 \int_0^1 B(r)^2 \ dr.$$
$$(3.18)$$

A more extensive explanation for the results above is as follows: The stochastic processes defined in Eqs. (3.12) and (3.17) are not continuous, but they are rcll (right continuous with left limits) and are elements in $D(\mathcal{T}, R)$ for $\mathcal{T} = [0, 1]$. In Proposition 14 of Chapter 9 we give a Donsker-like result for this space when it is endowed with the Skorohod topology; in that context, the error sequence $\{u_t : t \in \mathcal{N}_+\}$, in terms of which the entities $S_t$ are defined, is one of i.i.d. random variables with zero mean and finite variance. The case we are considering satisfies all the conditions of Proposition 14, Chapter 9 of this volume, which implies that

$$X^T(r) \xrightarrow{\text{d}} \sqrt{\sigma^2} B(r), \quad \text{for} \ \ r \in [0, 1].$$

Our conclusions are, thus, justified by Donsker's theorem, and Proposition 28 and Corollary 5 in Dhrymes (1989), pp. 242-243. We have therefore established that

$$T(\hat{\lambda} - 1) \xrightarrow{d} \frac{1}{2} \frac{B(1)^2 - 1}{\int_0^1 B(r)^2 \, dr}. \tag{3.19}$$

Turning now to the estimator of $\sigma^2$, we have

$$s^2 = \left( \frac{1}{T} \sum_{t=1}^{T} u_t^2 \right) - 2T(\hat{\lambda} - 1) \frac{1}{T^2} \left( \sum_{t=1}^{T} u_t y_{t-1} \right) + \left( T(\hat{\lambda} - 1) \right)^2 \frac{1}{T^3} \sum_{t=1}^{T} y_{t-1}^2.$$

Since by Proposition 23, Chapter 3, in Dhrymes (1989), pp. 188-190,

$$\frac{1}{T} \sum_{t=1}^{T} u_t^2 \xrightarrow{a.c.} \sigma^2,$$

and in view of the discussion above, we conclude that

$$s^2 \xrightarrow{P} \sigma^2. \tag{3.20}$$

Combining these results and using Eq. (3.3) we obtain

$$\tau_{\hat{\lambda}}^2 \xrightarrow{d} \frac{(B(1)^2 - 1)^2}{4(\int_0^1 B(r)^2 dr)} \quad \text{or} \quad \tau_{\hat{\lambda}} \xrightarrow{d} \frac{B(1)^2 - 1}{2 \left( \int_0^1 B(r)^2 dr \right)^{1/2}}, \tag{3.21}$$

which completes the discussion and resolution of the motivating question posed at the outset of this section.

We have therefore proved the following Proposition.

**Proposition 1** (i.i.d. errors). Consider the model in Eq. (3.2) and suppose $\beta = 0$, and $\{u_t : t \in \mathcal{N}_0\}$ is a sequence of square integrable, zero mean i.i.d. random variables whose variance is denoted by $\sigma^2 < \infty$. The least squares estimator of $\lambda$

$$\hat{\lambda} = \frac{\sum_{t=1}^{T} y_{t-1} u_t}{\sum_{t=1}^{T} y_{t-1}^2}$$

has the following properties under the hypothesis $H_0 : \lambda = 1$:

    i.   $T^\alpha(\hat{\lambda} - 1) \xrightarrow{P} 0$, for any $\alpha \in [0, 1)$;

    ii.   $T(\hat{\lambda} - 1) \xrightarrow{d} \frac{1}{2} \frac{B(1)^2 - 1}{\int_0^1 B(r)^2 \, dr}$;

    iii.   $\tau_{\hat{\lambda}} \xrightarrow{d} \frac{B(1)^2 - 1}{2 \left( \int_0^1 B(r)^2 \, dr \right)^{1/2}}$, where $\tau_{\hat{\lambda}}$ is the usual $t$-ratio associated with the least squares estimator of $\lambda$.

**Remark 1.** This is a convenient place to emphasize the differences between the inference procedures when we are dealing with the standard case, and those appropriate in the context of unit roots and/or integrated regressors (explanatory variables). The simple context in which we have operated above will highlight these differences without the distraction of complexities that will characterize our later discussion. There are two important differences to be discussed now, and one that will emerge in our next discussion, when more complex error processes are considered.

First: The limiting distribution of the entity $\hat{\lambda} - 1$ can be obtained **only if we normalize it** by $T$; by contrast, when it is asserted that $|\lambda| < 1$ the normalization that yields convergence in distribution is $\sqrt{T}$. As we noted above, this is due to the fact that, in the unit root context, $\sum_{t=1}^{T} y_{t-1}^2 / T$ **diverges** while $\sum_{t=1}^{T} y_{t-1}^2 / T^2$ converges, in distribution, to the integral of the square of a BM path. Second: An equally important facet of this comparison is that $\sum_{t=1}^{T} u_t y_{t-1} / T$ converges in distribution to a multiple of a centered chi-square variable with one degree of freedom, while in the standard case it converges a.c. to zero. Taking some license with the use of certain terms, we may put it that, asymptotically, $T(\hat{\lambda}-1)$ behaves like the ratio of a **centered** chi-square with one degree of freedom, and a "mixture" of multiples of chi-square variables.

In the standard case, the entity $\sqrt{T}(\hat{\lambda}-1)$ is asympotically $N[0, (1-\lambda^2)]$. This is so because the sequence $\zeta = \{\zeta_t : \zeta_t = y_{t-1} u_t\}$, with the stochastic basis $\{\mathcal{G}_t : \mathcal{G}_t = \sigma(u_s, \ s \leq t), \ t \in \mathcal{N}_0\}$, is a martingale difference sequence. Putting $\zeta_{tT} = \zeta_t / \sqrt{T}$ and noting that

$$\sum_{t=1}^{T} E(\zeta_{tT}^2 | \mathcal{G}_{tT}) = \sigma^2 \frac{1}{T} \sum_{t=1}^{T} y_{t-1}^2 \xrightarrow{\text{a.c.}} \frac{\sigma^4}{1-\lambda^2} = \sigma^4 \phi,$$

we conclude from Proposition 21 in Dhrymes (1989), p. 337, that

$$\frac{1}{\sqrt{T}} \sum_{t=1}^{T} u_t y_{t-1} \xrightarrow{\text{d}} N(0, \sigma^4 \phi).$$

Since

$$\frac{1}{T} \sum_{t=1}^{T} y_{t-1}^2 \xrightarrow{\text{a.c.}} \frac{\sigma^2}{1-\lambda^2},$$

we have that $\sqrt{T}(\hat{\lambda} - 1)$ converges to $N[0, (1 - \lambda^2)]$. In the stable case, the variance (of the asymptotic distribution) of the estimator of $\lambda$ declines quadratically as $\lambda \to 1$, and converges to zero. This means that $\sqrt{T}(\hat{\lambda} - 1)$ converges, in distribution, to the **degenerate** random variable (zero) as $\lambda \to 1$ and this is in accord with the result established in Eq. (3.19).

Since $s^2 \xrightarrow{\text{P}} \sigma^2$ there is no difference here.

Turning now to the $t$-ratio we see that it behaves like the ratio of a **centered** chi-square with one degree of freedom, and the **square root** of a "mixture" of (multiples of) chi-square variables. The justification for this claim is that we may approximate the integral

$$\int_0^1 B(r)^2 \, dr \approx \sum_{i=1}^{\infty} r_i \left( \frac{B_{r_i}^2}{r_i} \right),$$

where the $\{r_i : i \geq 1\}$ are **rationals** in $[0, 1]$ and, evidently, $B_{r_i}^2 / r_i$ are chi-square variables with one degree of freedom. Thus the denominator of the $t$-ratio may be approximated by a weighted sum of chi-square variables with one degree of freedom. These chi-square variables, however, are not independent since the paths of BM have independent **increments**, over the interval $[0, 1]$, but **not** independent ordinates. Specifically, if $t_1 \leq t_2$ are in $[0, 1]$, $B(t_1), B(t_2)$ are **not** independent, but $B(t_1)$ and $B(t_2) - B(t_1)$ are. By contrast, in the standard case the $t$-ratio behaves asymptotically like a unit normal.

Evidently, these are very significant differences and it would be very unsatisfactory to rely for our inference on the tabulations given for the standard case. Tabulations for the appropriate distribution in this case have been given in Dickey and Fuller (1979), and tests based on these tabulations are referred to as **Dickey-Fuller tests.**

**Remark 2.** Occasionally, in the literature of econometrics, the result stated in Eq. (3.19) is referred to as "superconsistency". This is occasioned by the fact that since the left member of the equation above converges in distribution to an **a.c. finite random variable**, it follows therefore that for **any** $\alpha > 0$

$$T^{1-\alpha}(\hat{\lambda} - 1) \xrightarrow{\text{P}} 0;$$

i.e. **even if one multiplies** $\hat{\lambda} - 1$ by a power of $T$ (less than one), the resulting entity **still converges (in probability) to zero.** However tempting the term may be, it should be avoided since the term "superconsistency" has a very well-defined technical meaning in the literature of statistical inference. In addition, this term implies that the "normal" definition of consistency is the condition $T^\gamma(\hat{\lambda}-1) \overset{P}{\to} 0$, for $\gamma < (1/2)$ ! A better term for this property is consistency of order $\beta < 1$, in the sense that $T^\beta(\hat{\lambda}-1)$ converges to zero. In the classical context, with uniformly bounded variances, we usually deal with estimators which are consistent of order $\beta < (1/2)$.

Before we leave this topic let us derive the limiting behavior (distribution) of two other entities which are useful in the context of the unit roots literature. These entities are

$$\frac{1}{T^{5/2}} \sum_{t=1}^{T} (t-1)S_{t-1} \text{ and } \frac{1}{T^{3/2}} \sum_{t=1}^{T} (t-1)u_{t-1}.$$

The limiting distribution of the first is given by

$$\frac{1}{T^{5/2}} \sum_{t=1}^{T} (t-1)S_{t-1} = \sum_{t=1}^{T} \int_{\frac{t-1}{T}}^{\frac{t}{T}} rX^T(r)\, dr \overset{d}{\to} \sqrt{\sigma^2} \int_{0}^{1} rB(r)\, dr.$$

As for the second, note that

$$\frac{1}{T^{3/2}} \sum_{t=1}^{T} S_{t-1} = \frac{1}{T^{3/2}} \sum_{t=1}^{T} \sum_{j=1}^{t} u_{j-1} = \frac{1}{T^{1/2}} \sum_{t=1}^{T} \left(1 - \frac{t-1}{T}\right) u_{t-1}.$$

Consequently,

$$\frac{1}{T^{3/2}} \sum_{t=1}^{T} (t-1)u_{t-1} = \frac{1}{T^{1/2}} \sum_{t=1}^{T} u_{t-1} - \frac{1}{T^{3/2}} \sum_{t=1}^{T} S_{t-1},$$

and using the results above, particularly Eq. (3.17), we obtain

$$\frac{1}{T^{3/2}} \sum_{t=1}^{T} (t-1)u_{t-1} \overset{d}{\to} \sqrt{\sigma^2} \left[B(1) - \int_{0}^{1} B(r)\, dr\right].$$

We now summarize the findings of this section.

$$\frac{1}{T^{3/2}} \sum_{t=1}^{T} S_{t-1} \overset{d}{\to} \sqrt{\sigma^2} \int_{0}^{1} B(r)\, dr;$$

$$\frac{1}{T^2} \sum_{t=1}^{T} S_{t-1}^2 \overset{d}{\to} \sigma^2 \int_{0}^{1} B(r)^2\, dr;$$

$$\frac{1}{T^{5/2}} \sum_{t=1}^{T} (t-1)S_{t-1} \overset{\mathrm{d}}{\to} \sqrt{\sigma^2} \int_0^1 rB(r)\,dr;$$

$$\frac{1}{T^{3/2}} \sum_{t=1}^{T} (t-1)u_{t-1} \overset{\mathrm{d}}{\to} \sqrt{\sigma^2} \left[ B(1) - \int_0^1 B(r)\,dr \right]. \tag{3.22}$$

## 3.2.2   The case $\beta \neq 0$, *a priori*

Here we investigate the case when the analyst runs a "regression with a constant term" and obtains the results of Eq. (3.8).

To examine the limiting behavior of $\hat{\lambda}$, $\tau_{\hat{\lambda}}$, $s^2$, in this case, we first examine the behavior of

$$\delta = y_{-1}' \left( I - \frac{ee'}{T} \right) y_{-1} = \sum_{t=1}^{T} (y_{t-1} - \bar{y}_{-1})^2.$$

From Eq. (3.4), and setting $\lambda = 1$, we find

$$y_{t-1} = y_0 + \beta(t-1) + S_{t-1}, \quad \bar{y}_{-1} = \frac{1}{T}\sum_{s=1}^{T} y_{s-1}, \quad \bar{S}_{-1} = \frac{1}{T}\sum_{t=1}^{T} S_{t-1};$$

$$\bar{y}_{-1} = y_0 + \beta \left( \frac{1}{T}\sum_{t=1}^{T}(t-1) \right) + \bar{S}_{-1}, \quad \left( \frac{1}{T}\sum_{t=1}^{T}(t-1) \right) = \frac{T-1}{2};$$

$$y_{t-1}^2 = y_0^2 + \beta^2(t-1)^2 + S_{t-1}^2 + 2y_0\beta(t-1) + 2y_0 S_{t-1} + 2\beta(t-1)S_{t-1};$$

$$\bar{y}_{-1}^2 = y_0^2 + \beta^2 \left( \frac{1}{T}\sum_{t=1}^{T}(t-1) \right)^2 + \bar{S}_{-1}^2 + 2y_0\beta \left( \frac{1}{T}\sum_{t=1}^{T}(t-1) \right)$$

$$+ 2y_0\bar{S}_{-1} + 2\beta \left( \frac{1}{T}\sum_{t=1}^{T}(t-1) \right)\bar{S}_{-1};$$

$$\delta = \sum_{t=1}^{T} y_{t-1}^2 - T\bar{y}_{-1}^2 = \beta^2 \sum_{t=1}^{T} \left[ t^2 - \left( \frac{T+1}{2} \right)^2 \right] + \sum_{t=1}^{T} S_{t-1}^2 - T\bar{S}_{-1}^2$$

$$+ 2\beta \sum_{t=1}^{T}(t-1)S_{t-1} - 2\beta \left( \frac{T(T-1)}{2} \right)\bar{S}_{-1}$$

$$+ 2y_0[\sum_{t=1}^{T} S_{t-1} - T\bar{S}_{-1}]. \tag{3.23}$$

Since

$$\sum_{t=1}^{T} \left[ t^2 - \left( \frac{T+1}{2} \right)^2 \right] = \sum_{t=1}^{T} t^2 - \frac{1}{4} T(T+1)^2$$

$$= \frac{1}{6} T(T+1)(2T+1) - \frac{1}{4} T(T+1)^2 = \frac{T^3 - T}{12},$$

it follows that for $\delta$ to converge it must be normalized (divided) by $T^3$. When we do so, by the results in the equation set Eq. (3.22), all other terms in $\delta$ converge in probability to zero and, consequently, we obtain

$$\frac{\delta}{T^3} \overset{P}{\to} \frac{\beta^2}{12}. \tag{3.24}$$

From Eqs. (3.24) and (3.8) we obtain

$$\hat{\lambda} - 1 \sim \frac{12}{\beta^2} \left( \frac{y'_{-1} u}{T^3} - \frac{\bar{y}_{-1} \bar{u}}{T^3} \right), \quad \hat{\beta} - \beta \sim \frac{12}{\beta^2} \left( \bar{u} \frac{y'_{-1} y_{-1}}{T^3} - \bar{y}_{-1} \frac{y'_{-1} u}{T^3} \right).$$

It follows immediately from the equation set Eq. (3.23) that

$$\frac{y'_{-1} y_{-1}}{T^3} \overset{P}{\to} \frac{\beta^2}{3}, \quad \frac{\bar{y}_{-1}}{T} \to \frac{\beta^2}{2},$$

$$\frac{y'_{-1} u}{T^{3/2}} \overset{d}{\to} \beta \sqrt{\sigma^2} [B(1) - \int_0^1 B(r)\, dr], \quad \sqrt{T} \bar{u} \overset{d}{\to} \sqrt{\sigma^2} B(1); \quad \text{thus,}$$

$$\sqrt{T}(\hat{\beta} - \beta) \overset{d}{\to} \sqrt{\sigma^2} B(1),$$

$$T^{3/2}(\hat{\lambda} - 1) \overset{d}{\to} \frac{12}{\beta} \sqrt{\sigma^2} \left( B(1) - \int_0^1 B(r)\, dr \right).$$

The result in Eq. (3.24) illustrates the fact that polynomial time trends cannot be seriously entertained as part of the specification of economic time series; to do so, we must be prepared to assert that eventually they (the time trends) dominate the behavior of the time series, in the sense that the second moments of the latter, when suitably normalized, **reflect only** the behavior of the deterministic trends. Notice further that, in the presence of polyno- mial trends, the parameter estimate $\hat{\lambda} - 1$ converges to zero (in probability) at the rate of $T^{-s}$, for $s \in [0, 3/2)$, and that its limiting distribution is **normal**. This is so since $B(1) \sim N(0,1)$ and, from Example 4 in Chapter 8, $\int_0^1 B(r)\, dr \sim N(0, 1/3)$. In addition, the estimator of the coefficient of the polynomial trend has a normal limiting distribution as well. This, of course, is due in large measure to the differences in the "rate of growth" of the two components of $y_t$, viz. $t$ and $S_t$. The second moment of the first diverges at the rate of $T^3$, while that of the second component diverges only at the rate of $T^2$ as Eqs. (3.24) and (3.26) make clear.

The situation is quite different if we obtain the limiting distribution of the entities above under the hypothesis

$$H_0: \ \beta = 0, \ \lambda = 1 .$$

This will allow us to determine whether the phenomenon we investigate is best described as a simple random walk, or a random walk with drift, according to whether the hypothesis $\beta = 0$ is accepted or rejected. Under $H_0$, we have

$$\frac{\delta}{T^2} = \frac{1}{T^2}\left( T\bar{S}_{-1}^2 + \sum_{t=1}^{T} S_{t-1}^2 - 2\bar{S}_{-1}\sum_{t=1}^{T} S_{t-1} \right) = \frac{1}{T^2}\sum_{t=1}^{T} S_{t-1}^2 - \frac{1}{T}\bar{S}_{-1}^2. \quad (3.25)$$

From Eqs. (3.17) and (3.18) we obtain

$$\frac{1}{T}\bar{S}_{-1}^2 = \left( \frac{1}{T^{3/2}} \sum_{t=1}^{T} S_{t-1} \right)^2 \xrightarrow{d} \sigma^2 \left( \int_0^1 B(r)\,dr \right)^2 ,$$

$$\frac{1}{T^2}\sum_{t=1}^{T} S_{t-1}^2 \xrightarrow{d} \sigma^2 \int_0^1 B(r)^2 \, dr; \quad \text{then}$$

$$\frac{\delta}{T^2} \xrightarrow{d} \sigma^2 \phi_0, \quad \text{where}$$

$$\phi_0 = \int_0^1 B(r)^2 \, dr - \left( \int_0^1 B(r)\,dr \right)^2 . \quad (3.26)$$

Under $H_0$, the limiting distribution of $\sqrt{T}(\hat{\beta} - \beta)$ is easily obtained, from Eqs. (3.15), (3.17), (3.18), and a standard CLT, as

$$\sqrt{T}(\hat{\beta} - \beta) \xrightarrow{d} \frac{\sqrt{\sigma^2}(\phi_1 - \phi_2)}{\phi_0}, \quad \text{where}$$

$$\phi_1 = B(1)\left( \int_0^1 B(r)^2 \, dr \right) , \quad (3.27)$$

$$\phi_2 = \left(\int_0^1 B(r)\,dr\right)\left[\frac{1}{2}\left(B(1)^2 - 1\right)\right], \quad \text{since}$$

$$\left(\frac{1}{\sqrt{T}}\sum_{t=1}^T u_t\right)\left(\frac{y_{-1}'y_{-1}}{T^2}\right) \overset{d}{\to} \left(\sqrt{\sigma^2}B(1)\right)\left(\sigma^2\int_0^1 B(r)^2\,dr\right),$$

$$\left(\frac{e'y_{-1}}{T^{3/2}}\right)\left(\frac{y_{-1}'u}{T}\right) \overset{d}{\to} \left(\sqrt{\sigma^2}\int_0^1 B(r)\,dr\right)\left[\frac{\sigma^2}{2}\left(B(1)^2 - 1\right)\right].$$

Under the hypothesis $\lambda = 1$, we have from Eq. (3.8)

$$T(\hat{\lambda} - 1) = \frac{T^2}{\delta}\left(\frac{y_{-1}'u}{T} - \bar{y}_{-1}\bar{u}\right).$$

We have already determined the limiting distribution of all terms in the right member above, except for

$$\bar{y}_{-1}\bar{u} = y_0\bar{u} + \left(\frac{1}{T^{3/2}}\sum_{t=1}^T S_{t-1}\right)\left(\frac{1}{\sqrt{T}}\sum_{t=1}^T u_t\right) \overset{d}{\to} \sigma^2 B(1)\int_0^1 B(r)\,dr,$$

by Eq. (3.18) and a standard CLT. Thus,

$$T(\hat{\lambda} - 1) \overset{d}{\to} \frac{(1/2)\int_0^1 B(r)^2\,dr - B(1)\int_0^1 B(r)\,dr}{\phi_0}. \tag{3.28}$$

Finally, let

$$\hat{u} = y - \hat{\beta}e - \hat{\lambda}y_{-1} = u - (\hat{\beta} - \beta) - (\hat{\lambda} - 1)y_{-1}, \quad s^2 = \frac{1}{T}\hat{u}'\hat{u},$$

and note that, again,

$$s^2 \overset{P}{\to} \sigma^2. \tag{3.29}$$

The $t$-ratios in the present case are given by

$$\tau_{\hat{\beta}}^2 = \frac{(\hat{\beta} - \beta)^2\,\delta}{s^2(y_{-1}'y_{-1}/T)} = \frac{T(\hat{\beta} - \beta)^2(\delta/T^2)}{s^2(y_{-1}'y_{-1}/T^2)} \overset{d}{\to} \frac{(\phi_1 - \phi_2)^2}{\phi_0\left(\int_0^1 B(r)^2\,dr\right)}$$

$$\tau_{\hat{\lambda}}^2 = \frac{(\hat{\lambda} - 1)^2\,\delta}{s^2} = \frac{[T(\hat{\lambda} - 1)]^2(\delta/T^2)}{s^2}$$

$$\overset{d}{\to} \frac{\left[(1/2)(B(1)^2 - 1) - B(1)\int_0^1 B(r)\,dr\right]^2}{\phi_0}. \tag{3.30}$$

The results above follow directly from the definition of $t$-ratios and Eqs. (3.22), (3.26), and (3.27).

**Remark 3**. Before we consider generalizations of the model examined above, it is useful to consider the nature and behavior of nonstochastic variables that are compatible with the framework of our discussion. As the result of Eq. (3.24) and the comments ancillary thereto make clear, not every specification of the nonstochastic variables may be convenient.

Since, in our framework, the stochastic regressors are assumed to be $I(1)$, as is the case, e.g., with $y_{t-1}$ in the discussion above, we have the result that

$$\frac{1}{T^2} \sum_{t=1}^{T} y_{t-1}^2 \xrightarrow{d} \int_0^1 B^2(r) dr;$$

thus, it would be desirable if the nonstochastic variables exhibit, asymptotically, a similar behavior. Let us illustrate what we have in mind, first, in terms of time trends. Note that

$$\frac{1}{T^2} \sum_{t=1}^{T} (t-1) = \frac{1}{T} \sum_{t=1}^{T} \left( \frac{t-1}{T} \right) = \sum_{t=1}^{T} \int_{\frac{t-1}{T}}^{\frac{t}{T}} \frac{[Tr]}{T} \, dr, \quad r \in \left[ \frac{t-1}{T}, \frac{t}{T} \right).$$

Consequently, we obtain

$$\lim_{T \to \infty} \frac{1}{T^2} \sum_{t=1}^{T} (t-1) = \int_0^1 r \, dr = \frac{1}{2}.$$

In general, employing the same reasoning we determine, for **any integer** $k < \infty$, that

$$\lim_{T \to \infty} \frac{1}{T^{k+1}} \sum_{t=1}^{T} (t-1)^k = \int_0^1 r^k \, dr = \frac{1}{k+1}.$$

The reader will be able to verify this directly when, in the appendix to Chapter 4, we give a procedure for determining the sum of (the first) $T$ powers of integers.

We conclude, then, that it is appropriate in this context that any nonstochastic variables admitted to the analysis, say $x_t$, should have the property that for some (strictly positive) increasing sequence $b_T \uparrow \infty$

$$\frac{1}{b_T} \sum_{t=1}^{T} x_{t-1} = \frac{1}{T} \sum_{t=1}^{T} \frac{T x_{[Tr]}}{b_T} = \int_{\frac{t-1}{T}}^{\frac{t}{T}} k_T(r) \, dr, \quad r \in \left[ \frac{t-1}{T}, \frac{t}{T} \right),$$

$$\lim_{T \to \infty} \frac{1}{T} \sum_{t=1}^{T} \frac{T x_{[Tr]}}{b_T} = \int_0^1 k(r) \, dr \le \int_0^1 |k(r)| \, dr < \infty.$$

## 3.3 The UR Model without i.i.d. Errors

### 3.3.1 Stationary and Ergodic Errors

In the discussions above, the i.i.d. assumption on the error processes plays an important role in that it allows us to apply Donsker's theorem (Propositions 7 and 14, of Chapter 9) in order to determine the limiting distribution of the stochastic sequences defined in Eqs. (3.12) and (3.16). However, as we also show in Chapter 9, Donsker-like theorems (FCLT) may be obtained **in the absence of the** i.i.d. assumption, provided the error processes in question are stationary, and ergodic, or mixing, or are nonstationary but obey certain mixing and moment conditions. This is done in Propositions 16, 22, 23, 24, 25, 27, 28, and 29 of Chapter 9. Thus, in the context of Proposition 16, when the sequence $\{u_t : t \in \mathcal{N}_0\}$ is square integrable, stationary (and ergodic) and obeys certain moment conditions, the stochastic processes defined in Eq. (3.16) obey

$$X(r)^T \overset{\mathrm{d}}{\to} \sqrt{\sigma_0^2} B(r), \quad r \in [0,1], \quad \text{where}$$

$$\sigma_0^2 = E(u_1^2) + 2 \sum_{k=1}^{\infty} E(u_1 u_{1+k}). \tag{3.31}$$

Under the conditions stated in that proposition $\sigma_0^2 \in (0, \infty)$.

Another important implication of the i.i.d. assumption pertains to the result in the first equation of the set in Eq. (3.11), that in Eq. (3.20), and in the one immediately preceding. All of these results depend crucially on the i.i.d. assumption. It turns out that the conditions of many of the (relevant) propositions in Chapter 9 also ensure that the conclusions above continue to hold. To give an example of why this is so, in the context of Proposition 16, note that **if** the sequence $\{u_t : t \in \mathcal{N}_+\}$ **is** ergodic then the sequence $u_t^2$ is **also stationary and ergodic.** [3] Since $Eu_t^2 < \infty$, it follows from the mean

---

[3] As in Chapter 9, let $\xi$ be a random variable defined on the probability space ( $\Omega$ , $\mathcal{A}$ , $\mathcal{P}$ ), and suppose $\xi$ is square integrable. Then there exists an ergodic measure preserving transformation, say $T$, such that $u_k = \xi \circ T^k$ ; define $\zeta_k = u_k^2 = [\xi \circ T^k]^2$ and note that for any $B \in \mathcal{B}(R)$, $\zeta_k(\omega) \in B$ if and only if $\xi \circ T^k(\omega) \in B_1$, for some appropriate set $B_1 \in \mathcal{B}(R)$. That $\{\zeta_k : k \in \mathcal{N}_+\}$ is stationary is obvious from its definition; to see that it is also ergodic, let $A_k = \zeta_k^{-1}(B)$ be an invariant set. This means that $A_k =$

ergodic theorem of Proposition 31 in Dhrymes (1989), pp. 357-359, applied
to the sequence $\{u_t^2 : t \in \mathcal{N}_+\}$, that

$$\frac{1}{T} \sum_{t=1}^{T} u_t^2 \overset{\text{a.c.}}{\to} \sigma_1^2, \quad \sigma_1^2 = E(u_1^2), \tag{3.32}$$

Propositions 19 and 20 of Chapter 9 provide conditions under which the result
above regarding a.c. convergence holds; Propositions 22, 23, 24, 25, 27, 28
of Chapter 9 provide conditions for the FCLT to apply under a variety of
circumstances for stationary mixing and simply stationary processes. When
the error process obeys the conditions of Proposition 29, we have no simple
way to establish analogous results. In point of fact, similar results hold, but
we have to argue them a bit more arduously than in the previous case. The
basis for such argument is given in Remark 13 of Chapter 9.

To apply the framework of Remark 13, and the surrounding discussion,
let

$$\xi_t = u_t^2 - \sigma_t^2, \quad c_t^2 = 4 \parallel u_t^2 - \sigma_t^2 \parallel_2^2, \quad b_T = T.$$

If $\{u_t : t \in \mathcal{N}_+\}$ is mixing, so is $\{\xi_t : \xi_t = u_t^2 - \sigma_t^2, \, t \in \mathcal{N}_+\}$; in addition,
note that the strong law of large numbers (SLLN), given in that discussion,
requires the conditions

$$4 \sum_{t=1}^{\infty} \frac{\parallel u_t^2 - \sigma_t^2 \parallel^2}{t^2} < \infty, \quad \psi_T \sim \frac{1}{T^{1/2}(\ln T)^2},$$

which will be satisfied if we work with Corollary 5 of Chapter 9, and $\beta \in
[4, \infty)$.

Thus, in all cases, the findings of the preceding section will continue to
hold in the present case as well, with only minor modifications.

**Case $\beta = 0$**

When the "true" and operational model is that of Eq. (3.2) **with $\beta = 0$**,
the modifications referred to above are as follows:

$$\frac{\sum_{t=1}^{T} u_t y_{t-1}}{T} \overset{\text{d}}{\to} \frac{\sigma_0^2}{2} \left( B(1)^2 - \nu^2 \right), \quad \nu^2 = \frac{\sigma_1^2}{\sigma_0^2},$$

---

$T^{-k} \circ \xi^{-1}(B_1) = T^{-k}(A_0) = A_0$. Since the original process is ergodic it follows therefore
that the set $A_k = A_0$, for all $k \geq 1$ is $\mathcal{P}$-trivial, i.e. its measure is either zero or one,
which establishes the ergodicity of $\{u_t^2 : t \in \mathcal{N}_+\}$.

$$\frac{y_{-1}'y_{-1}}{T^2} \xrightarrow{\text{d}} \sigma_0^2 \int_0^1 B(r)^2 \, dr, \quad s^2 \xrightarrow{\text{P}} \sigma_1^2$$

$$\frac{1}{T^{3/2}} \sum_{t=1}^T S_{t-1} = \sum_{t=1}^T \int_{(t-1/T)}^{t/T} X^T(r,\omega) \, dr \xrightarrow{\text{d}} \sqrt{\sigma_0^2} \int_0^1 B(r) \, dr,$$

$$T(\hat{\lambda} - 1) = \frac{(y_{-1}'u)T}{(y_{-1}'y_{-1}/T^2)} \xrightarrow{\text{d}} \frac{B(1)^2 - \nu^2}{2 \int_0^1 B(r)^2 \, dr},$$

$$\tau_{\hat{\lambda}}^2 = \frac{[T(\hat{\lambda} - 1)]^2 (y_{-1}'y_{-1}/T^2)}{s^2}$$

$$\xrightarrow{\text{d}} \frac{(B(1)^2 - \nu^2)^2}{4(\nu^2 \int_0^1 B(r)^2 dr)}, \quad \text{or}$$

$$\tau_{\hat{\lambda}} \xrightarrow{\text{d}} \frac{B(1)^2 - \nu^2}{2\nu \left( \int_0^1 B(r)^2 dr \right)^{1/2}}. \tag{3.33}$$

We have therefore proved Proposition 2.

**Proposition 2** (Stationary and Ergodic or Mixing Errors). Consider the model in Proposition 1, but suppose that the error sequence $\{u_t : t \in \mathcal{N}_0\}$ satisfies the conditions of (any of) Propositions 16, 22 through 25, or 27 or 28 of Chapter 9. Then, under the hypothesis $\beta = 0$, $\lambda = 1$ the following is true:

$$T(\hat{\lambda} - 1) \xrightarrow{\text{d}} \frac{B(1)^2 - \nu^2}{2 \int_0^1 B(1)^2 \, dr}, \quad \nu^2 = \frac{\sigma_1^2}{\sigma_0^2},$$

$$\tau_{\hat{\lambda}} \xrightarrow{\text{d}} \frac{B(1)^2 - \nu^2}{2\nu \left( \int_0^1 B(r)^2 \, dr \right)^{1/2}}, \quad \text{where}$$

$$\sigma_1^2 = Eu_1^2, \quad \sigma_0^2 = \sigma_1^2 + 2\sigma_1^*, \quad \sigma_1^* = \sum_{j=1}^{\infty} Eu_1 u_{1+j}.$$

**Remark 4.** The reader should note that the distributions above contain the **nuisance** parameters $\sigma_0^2$ and $\sigma_1^2$, through the ratio $\nu^2$, which appears nearly everywhere in Eq. (3.33). This makes it impossible to utilize the tabulations given by Dickey and Fuller (1979), for the Dickey-Fuller tests. The difficulty, however, is more apparent than real. In the expression

for the limiting distribution of $T(\hat{\lambda} - 1)$ the nuisance parameters appear
in the term $-(\nu^2/2) \int_0^1 B(r)^2 \, dr$. What we need to do is "replace" it by
$-(1/2) \int_0^1 B(r)^2 \, dr$, which is the term appearing in the simpler case of i.i.d.
errors, which is the basis for the Dicky-Fuller tabulations. This is easily accomplished by the addition of a term, say $h_{1T}$, such that $T(\hat{\lambda}-1) + h_{1T}$ has
the "Dickey-Fuller" distribution given in Eq. (3.19). We wish to introduce a
term whose limiting distribution "neutralizes" the term involving $\nu^2$ (in the
numerator of the limit) and introduces instead the term $-1$. If $\hat{\sigma}_i^2$ are at
least consistent estimators of $\sigma_i^2$, $i = 0, 1$, respectively, the desired entity is

$$h_{1T} = -\frac{2\hat{\sigma}_1^*}{2y_{-1}'y_{-1}/T^2} \xrightarrow{\text{d}} \frac{\nu^2 - 1}{2 \int_0^1 B(r)^2 \, dr},$$

$$-2\hat{\sigma}_1^* = \hat{\sigma}_1^2 - \hat{\sigma}_0^2 \quad \text{and} \quad T(\hat{\lambda} - 1) + h_{1T} \xrightarrow{\text{d}} \frac{B(1)^2 - 1}{2 \int_0^1 B(r)^2 \, dr},$$

which thus enables us to use the tabulations given by Dickey and Fuller.
Evidently, we have the consistent estimator $\hat{\sigma}_1^2 = s^2$, which was examined
earlier. As we pointed out then, a consistent estimator of $\sigma_0^2$ in the stationary ergodic case may be obtained as $\hat{\sigma}_0^2 = 2\pi \hat{f}(0)$, where $f$ is the **spectral
density function** of the sequence, and $\hat{f}$ is a consistent estimator of the
spectral density. For a discussion of estimators of the spectral densities of stationary processes see, for example, Brockwell and Davis (1991), or Dhrymes
(1970). In the nonstationary case, to be discussed in the following section,
we have again, *mutatis mutandis*, $\hat{\sigma}_1^2 = s^2$. In that context the **existence** of
$\sigma_0^2$ is **assumed** and one may provide an argument to establish that, in view
of Corollary 5 in Chapter 9, a consistent estimator for $\sigma_0^2$ exists as well.

   A similar statement may be made with respect to the $t$-ratio. Execu-
tion of this transformation is somewhat more delicate; nonetheless we may
transform the test statistic $\tau_{\hat{\lambda}}$ so that it has the same limiting distribution
as the one tabulated by Dickey and Fuller and given in the discussion of the
model with i.i.d. errors, Eq. (3.21). It is easily verified that the desired
transformation is

$$\hat{\nu}\tau_{\hat{\lambda}} + h_{2T}, \quad h_{2T} = \frac{\hat{\sigma}_1^2 - \hat{\sigma}_0^2}{2\sqrt{\hat{\sigma}_0^2(y_{-1}'y_{-1}/T^2)}}$$

This is so since

$$\hat{\nu}\tau_{\hat{\lambda}} \xrightarrow{\text{d}} \frac{B(1)^2 - \nu^2}{2\left(\int_0^1 B(r)^2 dr\right)^{1/2}},$$

$$h_{2T} \xrightarrow{\text{d}} \frac{\sigma_1^2 - \sigma_0^2}{2\sigma_0^2\left(\int_0^1 B(r)^2\, dr\right)^{1/2}}, \quad \text{so that}$$

$$\hat{\nu}\tau_{\hat{\lambda}} + h_{2T} \xrightarrow{\text{d}} \frac{B(1)^2 - 1}{2\left(\int_0^1 B(r)^2 dr\right)^{1/2}},$$

as required. We have therefore proved the following proposition.

**Proposition 3.** In the context of Proposition 2, the following is true:

$$T(\hat{\lambda} - 1) + h_{1T} \xrightarrow{\text{d}} \frac{B(1)^2 - 1}{2\int_0^1 B(1)^2\, dr},$$

$$\hat{\nu}\tau_{\hat{\lambda}} + h_{2T} \xrightarrow{\text{d}} \frac{B(1)^2 - 1}{2\left(\int_0^1 B(r)^2 dr\right)^{1/2}},$$

$$h_{1T} = -\frac{\hat{\sigma}_1^*}{y'_{-1}y_{-1}/T^2}, \quad h_{2T} = \frac{\hat{\sigma}_1^2 - \hat{\sigma}_0^2}{2\sqrt{\hat{\sigma}_0^2(y'_{-1}y_{-1}/T^2)}},$$

where $\hat{\sigma}_0^2$, $\hat{\sigma}_1$ are consistent estimators of the corresponding parameters, $2\hat{\sigma}_1^* = \hat{\sigma}_0^2 - \hat{\sigma}_1^2$ and $\hat{\nu}^2 = (\hat{\sigma}_0^2/\hat{\sigma}_1^2)$.

Despite the complexity of the operations above, the transformations employed are, in principle, of the same type as the transformations one employs in the following simple context: Suppose we have a "statistic" $\bar{x}$ which is asserted to be (asymptotically) normal with mean $\mu$ and **unknown** variance $\phi$; we want to test the null hypothesis $\mu = 1$ and thus, under the null, $\bar{x} - 1$ is asymptotically normal with mean **zero** and unknown variance, $\phi$. Unfortunately, tabulations are not available for **this** distribution; what are available are tabulations of the distribution $N(0,1)$. If we employ the transformation $(\bar{x} - 1)/\hat{\phi}^{1/2}$, where $\hat{\phi}$ is a **consistent** estimator of $\phi$ then the transformed entity is asymptotically $N(0,1)$ and tabulations **are available for this distribution**. In PP (1987) the transformations above are termed the $Z$-transforms.

Although slightly more complicated, the situation is analogous in the following sections and this argument will not necessarily be repeated in every instance.

**Case** $\beta \neq 0$ **,** *a priori*

When the "true" model is that of Eq. (3.2) **with** $\beta = 0$ [4] but in the operational model ( *a priori* ) $\beta \neq 0$, the modifications referred to above are as follows: In all the equations in the set Eq. (3.26) replace $\sigma^2$ by $\sigma_0^2$ and, in addition,

$$\left( \frac{1}{\sqrt{T}} \sum_{t=1}^{T} u_t \right) \left( \frac{y'_{-1} y_{-1}}{T^2} \right) \xrightarrow{\text{d}} \left( \sqrt{\sigma_0^2} B(1) \right) \left( \sigma_0^2 \int_0^1 B(r)^2 \, dr \right),$$

$$\left( \frac{e' y_{-1}}{T^{3/2}} \right) \left( \frac{y'_{-1} u}{T} \right) \xrightarrow{\text{d}} \left( \sqrt{\sigma_0^2} \int_0^1 B(r) \, dr \right) \left[ \frac{\sigma_0^2}{2} \left( B(1)^2 - \nu^2 \right) \right],$$

$$T(\hat{\lambda} - 1) \xrightarrow{\text{d}} \frac{(1/2)(B(1)^2 - \nu^2) - B(1) \int_0^1 B(r) \, dr}{\phi_0},$$

$$\sqrt{T}(\hat{\beta} - \beta) \xrightarrow{\text{d}} \frac{\sqrt{\sigma_0^2}(\phi_1 - \phi_2)}{\phi_0} \quad \text{where now} \qquad (3.34)$$

$$\phi_1 = B(1) \left( \int_0^1 B(r)^2 \, dr \right),$$

$$\phi_2 = \left( \int_0^1 B(r) \, dr \right) \left[ \frac{1}{2} \left( B(1)^2 - \nu^2 \right) \right],$$

$$s^2 \xrightarrow{\text{P}} \sigma_1^2 \qquad (3.35)$$

and $\phi_0$ is as defined in Eq. (3.26). The $t$-ratios are given by

$$\tau_{\hat{\lambda}}^2 = \frac{(\hat{\lambda} - 1)^2 \, \delta}{s^2} = \frac{[T(\hat{\lambda} - 1)]^2 (\delta/T^2)}{s^2}$$

$$\xrightarrow{\text{d}} \frac{\left[ (1/2)(B(1)^2 - \nu^2) - B(1) \int_0^1 B(r) \, dr \right]^2}{\nu^2 \phi_0}, \qquad (3.36)$$

---

[4] Or, at least, we obtain the limiting distribution of the relevant entities on the assumption that $\beta = 0$.

$$\tau_{\hat{\beta}}^2 = \frac{(\hat{\beta} - \beta)^2 \, \delta}{s^2(y_{-1}'y_{-1}/T)} = \frac{T(\hat{\beta} - \beta)^2(\delta/T^2)}{s^2(y_{-1}'y_{-1}/T^2)} \xrightarrow{d} \frac{(\phi_1 - \phi_2)^2}{\nu^2 \phi_0 \left( \int_0^1 B(r)^2 \, dr \right)}.$$

We have therefore proved Proposition 4.

**Proposition 4.** Consider the model of Eq. (3.2) and suppose the error process stasifies the requirements of (any of) Propositions 16, 22-25, or 27, 28 of Chapter 9; the following statements are true:

$$T(\hat{\lambda} - 1) \xrightarrow{d} \frac{(1/2)(B(1)^2 - \nu^2) - B(1) \int_0^1 B(r) \, dr}{\phi_0},$$

$$\sqrt{T}(\hat{\beta} - \beta) \xrightarrow{d} \frac{\sqrt{\sigma_0^2}(\phi_1 - \phi_2)}{\phi_0},$$

$$\tau_{\hat{\lambda}}^2 = \frac{(\hat{\lambda} - 1)^2 \, \delta}{s^2} \xrightarrow{d} \frac{\left[ (1/2)(B(1)^2 - \nu^2) - B(1) \int_0^1 B(r) \, dr \right]^2}{\nu^2 \phi_0},$$

$$\tau_{\hat{\beta}}^2 = \frac{(\hat{\beta} - \beta)^2 \, \delta}{s^2(y_{-1}'y_{-1}/T)} \xrightarrow{d} \frac{(\phi_1 - \phi_2)^2}{\nu^2 \phi_0 \left( \int_0^1 B(r)^2 \, dr \right)}.$$

The modified or transformed entities are given by

$$T(\hat{\lambda} - 1) + h_{1T} \xrightarrow{d} \frac{(1/2)(B(1)^2 - 1) - B(1) \int_0^1 B(r) \, dr}{\phi_0},$$

$$\hat{\nu}\tau_{\hat{\lambda}} + h_{2T} \xrightarrow{d} \frac{(1/2)(B(1)^2 - 1) - B(1) \int_0^1 B(r) \, dr}{\phi_0^{1/2}},$$

$$h_{1T} = -\frac{\hat{\sigma}_1^*}{(\delta/T^2)}, \quad h_{2T} = -\frac{\hat{\sigma}_1^*}{[\hat{\sigma}_0^2(\delta/T^2)]^{1/2}},$$

$$\frac{1}{\hat{\sigma}_0} \left( \sqrt{T}(\hat{\beta} - \beta) + h_{3T} \right) \to \frac{\phi_1 - \phi_2^*}{\phi_0}, \quad h_{3T} = \frac{\hat{\sigma}_1^*(\bar{y}_{-1}/\sqrt{T})}{(\delta/T^2)},$$

$$\hat{\nu}\tau_{\hat{\beta}} + h_{4T} \xrightarrow{d} \frac{\phi_1 - \phi_2^*}{[\phi_0(\int_0^1 B(r)^2 \, dr)]^{1/2}},$$

$$\phi_2^* = \frac{1}{2} \left( \int_0^1 B(r) \, dr \right) \left( B(1)^2 - 1 \right),$$

$$h_{4T} = \frac{(\hat{\sigma}_1^*/\hat{\sigma}_0)(\bar{y}_{-1}/\sqrt{T})}{[(\delta/T^2)(y_{-1}'y_{-1}/T^2)]^{1/2}}.$$

### 3.3.2   Nonstationary (Heterogeneous) Errors

When the error process $\{u_t : t \in \mathcal{N}_+\}$ is heterogeneous, i.e. nonstationary in the sense that elements of the sequence may have different moments, the problem is solved by Proposition 29 of Chapter 9 (or its corollaries), which substitutes moment and mixing conditions for stationarity and/or ergodicity. In the context of Proposition 29, the existence of $\sigma_0^2$ is **assumed** [see the condition in Eqs. (9.47) and (9.48) therein] and the existence of $\sigma_1^2$ is implied by the mixing condition, which also implies that

$$\frac{1}{T}\sum_{t=1}^{T} u_t^2 \xrightarrow{\mathrm{P}} \sigma_1^2. \tag{3.37}$$

**Thus, all the results in the preceding discussion apply in this case as well.**

## 3.4   Vector Random Walks

### 3.4.1   Symmetric Parameter Matrices

To deal with the multivariate (vector) version of the problem above, nothing of great substance need be altered in the discussion of the previous sections; nonetheless we shall examine the problem below. We do so, first to establish convenient notation for the multivariate case and, second, to prepare the argument for the somewhat more general case of integrated regressors, instrumental variables, and the like. Consider the generalization of the simple model in Eq. (3.2), on the assumption that $\beta = 0$, viz. [5]

$$y_{t\cdot} = y_{t-1\cdot}A + u_{t\cdot}, \quad t = 1, 2, \ldots, T, \quad Y = Y_{-1}A + U, \tag{3.38}$$

---

[5] We shall attempt to preserve the notational practice of the original papers, except for the requirements of maintaining notational continuity and simplicity. We shall also conform to the notational practice in the GLM, of which the present discussion is a generalization. For example, to preserve the convention of writing the ($T$ observations on a) GLM as $y = X\beta + u$, a single observation must be written as $y_t = x_{t\cdot}\beta + u_t$, where $x_{t\cdot}$ is the **row vector** containing the explanatory variables. For the so-called SUR models (seemingly unrelated regressions) we write typically $Y = XB + U$, which dictates that the observation at "time" $t$ be written as $y_{t\cdot} = x_{t\cdot}B + u_{t\cdot}$, where $y_{t\cdot}$ is the vector of ($m$) dependent variables and $x_{t\cdot}$ is the vector of ($G$) independent variables, or regressors.

where $y_{t.}$ and $u_{t.}$ are $q$-element vectors, while $A$ is a $q \times q$ matrix of parameters. If the characteristic roots of $A$ are all less than one in absolute value we have

$$y_{t.}' = \sum_{j=0}^{\infty} A'^j u_{t-j.}', \qquad (3.39)$$

which is a simple distributed lag model; if some of the characteristic roots are unity, and others are not, we have the case of cointegration [6] in which some linear combination(s) of the elements of the vector $y$ may be stationary, and others may be integrated of order one. Finally, if all characteristic roots are unity, the model may be rendered as $y_{t.} = y_{t-1.} + u_{t.}$. It is the distribution of the least squares estimator $\hat{A}$ under this hypothesis that we shall examine below. In this section we shall deal with the case where $A$ is a symmetric matrix; strictly speaking its least squares estimator is given by

$$\tilde{A} = (Y_{-1}'Y_{-1})^{-1}Y_{-1}'Y = A + (Y_{-1}'Y_{-1})^{-1}Y_{-1}'U. \qquad (3.40)$$

which of course need not be symmetric. To preserve symmetry we shall treat the least squares estimator as if it were $\hat{A} = (\tilde{A} + \tilde{A}')/2$, and examine its limiting distribution under the null hypothesis

$$H_0 : A = I_q,$$

which is, of course, a symmetric matrix. Note that under $H_0$ we have

$$y_{t.} = y_{0.} + S_{t.}, \quad S_{t.} = \sum_{j=1}^{t} u_{j.}, \quad t = 1, 2, \ldots, \qquad (3.41)$$

and consequently,

$$\frac{Y_{-1}'Y_{-1}}{T^2} = \frac{y_{0.}'y_{0.}}{T} + \frac{y_{0.}'}{\sqrt{T}} \frac{1}{T^{3/2}} \sum_{t=1}^{T} S_{t-1.} + \left( \frac{1}{T^{3/2}} \sum_{t=1}^{T} S_{t-1.}' \right) \frac{y_{0.}}{\sqrt{T}}$$

$$+ \frac{1}{T^2} \sum_{t=1}^{T} S_{t-1.}'S_{t-1.}, \quad \frac{Y_{-1}'U}{T} = y_{0.}'\bar{u} + \frac{1}{T} \sum_{t=1}^{T} S_{t-1.}'u_{t.}. \quad (3.42)$$

Once again, define

$$X^T(r, \omega) = \frac{1}{\sqrt{T}} S_{t-1.}(\omega) = \frac{1}{\sqrt{T}} S_{[rT].}(\omega), \quad r \in \left[ \frac{t-1}{T}, \frac{t}{T} \right). \qquad (3.43)$$

---

[6] Cointegration will be examined extensively in the following chapter. What is termed *cointegration* in the literature of econometrics is referred to as **partial nonstationarity** in the statistical literature.

Using the obvious multivariate analog of our previous discussion, and invoking either (any of) Proposition 16, 22 through 25, 27, 28, or Proposition 29 and (any of) its corollaries, of Chapter 9, we conclude

$$X^T(r, \omega) \xrightarrow{d} B(r)P_0, \quad \Sigma_0 = P_0'P_0, \quad \text{so that}$$

$$\frac{1}{T^2} \sum_{t=1}^{T} S_{t-1.}' S_{t-1.} = \sum_{t=1}^{T} \int_{t-1/T}^{t/T} X^{T'}(r, \omega) X^T(r, \omega) dr,$$

$$\xrightarrow{d} P_0' \left[ \int_0^1 B'(r)B(r) \, dr \right] P_0,$$

$$\frac{1}{T^{3/2}} \sum_{t=1}^{T} S_{t-1.}' \xrightarrow{d} P_0' \int_0^1 B(r)' \, dr, \quad \text{and consequently}$$

$$\frac{Y_{-1}'Y_{-1}}{T^2} \xrightarrow{d} \Phi = P_0' \left[ \int_0^1 B(r)' B(r) \, dr \right] P_0, \tag{3.44}$$

where $B$ is the SMBM (standard multivariate Brownian motion) of Chapter 7, written in **row form**, i.e. $B(r) = (B_1(r), B_2(r), \ldots, B_q(r))$. The meaning of the "nuisance" parameter $\Sigma_0$, appearing in the equation set Eq. (3.44), depends on the assumptions made regarding the error process, and thus on the proposition we invoke for determining the limiting distributions. **If** we assume that the error process is i.i.d. then Proposition 7 is appropriate and consequently

$$\Sigma_0 = \Sigma_1 = E(u_{1.}'u_{1.}), \quad \text{when Proposition 7 is invoked.} \tag{3.45}$$

**If** the error process is assumed to be stationary and ergodic and and to obey the conditions of Proposition 16 (or any of the several other FCLT for stationary and mixing processes), we have

$$\Sigma_0 = E(u_{1.}'u_{1.}) + \sum_{t=2}^{\infty} E(u_{1.}'u_{t.}) + \sum_{t=2}^{\infty} E(u_{t.}'u_{1.}) = \Sigma_1 + \Sigma_1^* + \Sigma_1^{*'}. \tag{3.46}$$

**If** the error process is assumed to be nonstationary, but obeys certain mixing and moment conditions given in Proposition 29, or its corollaries, we find

$$\Sigma_0 = \lim_{T \to \infty} \frac{1}{T} E(S_{T.}'S_{T.}) = \Sigma_1 + \Sigma_1^* + \Sigma_1^{*'}, \quad \text{where}$$

$$\Sigma_1 = \lim_{T \to \infty} \frac{1}{T} \sum_{t=1}^{T} E u_{t.}'u_{t.}, \quad \Sigma_1^* = \lim_{T \to \infty} \frac{1}{T} \sum_{t<s} E u_{t.}'u_{s.}. \tag{3.47}$$

Finally note that the matrix $P_0$ of the decomposition of $\Sigma_0$ occasionally termed its square root [7] may be chosen conveniently as an **upper triangular** matrix, a result given in Proposition 58 of Dhrymes (1984), pp. 68-69.

To examine the behavior of the term $Y'_{-1}U/T$ we note that

$$\frac{1}{T}Y'_{-1}U \;=\; \frac{1}{T}\left(\sum_{t=1}^{T} y'_{t-1.}u_{t.}\right) = \frac{1}{T}y'_{0.}S_{T.} + \frac{1}{T}\sum_{t=1}^{T} S'_{t-1.}u_{t.},$$

$$S_{t.} \;=\; S_{t-1.} + u_{t.}, \quad S'_{t.}S_{t.} = S'_{t-1.}S_{t-1.} + u'_{t.}u_{t.} + S'_{t-1.}u_{t.} + u'_{t.}S_{t-1.}.$$

Thus,

$$\frac{1}{T}\sum_{t=1}^{T}\left(u'_{t.}S_{t-1.} + S'_{t-1.}u_{t.}\right) \;=\; \frac{1}{T}\sum_{t=1}^{T}\left(S'_{t.}S_{t.} - S'_{t-1.}S_{t-1.} - u'_{t.}u_{t.}\right)$$

$$=\; \frac{1}{T}\left(S'_{T.}S_{T.} - u'_{0.}u_{0.} - \sum_{t=1}^{T} u'_{t.}u_{t.}\right).$$

This last sum converges to a term which has different meaning depending on the assumptions made. Thus,

$$\frac{1}{T}\sum_{t=1}^{T}\left(u'_{t.}S_{t-1.} + S'_{t-1.}u_{t.}\right) \;\xrightarrow{\text{d}}\; P'_0[B(1)'B(1) - P'^{-1}_0\Sigma_1 P^{-1}_0]P_0 = \Psi^*, \quad (3.48)$$

where

$$\Sigma_0 \;=\; \Sigma_1 \;=\; Eu'_{1.}u_{1.}, \quad \text{when Proposition 7 is invoked}$$

$$=\; Eu'_{1.}u_{1.}, \quad \text{when Proposition 16, 22, 23, 24, 25, 27, or 28 is invoked}$$

$$=\; \lim_{T\to\infty}\frac{1}{T}\sum_{t=1}^{T} Eu'_{t.}u_{t.}, \quad \text{when Proposition 29 is invoked}, \quad (3.49)$$

and $\Sigma_1$ is defined analogously. To complete this derivation, we need to determine the limiting distribution of $(1/T)\sum_{t=1}^{T} S'_{t-1.}u_{t.}$ which, in the scalar case, obeys $\xrightarrow{\text{d}} \frac{1}{2}\Psi^*$.

---

[7] Other definitions of the square root of a positive definite matrix are possible; for example, if $Q$ is the (orthogonal) matrix of its characteristic vectors, and $\Lambda$ the diagonal matrix containing the corresponding (ordered) characteristic roots, we have $\Sigma_0 = Q\Lambda Q'$ and another square root is $P_1 = Q\Lambda^{1/2}Q'$, which also obeys $\Sigma_0 = P'_1 P_1$.

Since $A$ is a symmetric matrix, we are able to use the result of Eq. (3.48) **directly**, by simply following the **convention** noted above and, therefore, conclude that

$$T(\hat{A} - A) \xrightarrow{\text{d}} \frac{1}{2} \Phi^{-1} \Psi^* \tag{3.50}$$

$$= \frac{1}{2} P_0^{-1} \left( \int_0^1 B(r)' B(r) \, dr \right)^{-1} \left( B(1)' B(1) - P_0^{-1} \Sigma_1 P_0^{-1} \right) P_0.$$

This is a particularly attractive choice since under the null hypothesis of (all) unit roots we have $A = I_q$, which is evidently symmetric, and the task of obtaining the limiting distribution is simplified.

To complete the solution of the multivariate problem, we need to obtain the limiting distribution of

$$\hat{\Sigma}_1 = \frac{1}{T} \hat{U}' \hat{U}, \quad \text{where} \quad \hat{U} = Y - Y_{-1} \hat{A} = U - Y_{-1}(\hat{A} - A), \tag{3.51}$$

under the hypothesis $H_0 : A = I_q$. Since

$$\frac{1}{T} \hat{U}' \hat{U} = \frac{1}{T} \sum_{t=1}^T u'_{t\cdot} u_{t\cdot} - T(\hat{A} - A)' \frac{Y'_{-1} U}{T^2} - \frac{U' Y_{-1}}{T^2} T(\hat{A} - A)$$

$$+ T(\hat{A} - A)' \frac{Y'_{-1} Y_{-1}}{T^3} T(\hat{A} - A),$$

it follows that

$$\hat{\Sigma}_1 \xrightarrow{\text{P}} \Sigma_1,$$

and hence in distribution as well, where $\Sigma_1$ is as defined in Eq. (3.49), according to whether Proposition 7, 16, 22 through 25, 27, 28, or Proposition 29 (Corollary 5) is invoked.

We conclude by giving a more convenient **vector** representation to the estimators, and an appropriate scalar test statistic for testing the hypothesis $H_0 : A = I_q$.

Consider again the second equation set in Eq. (3.38). Using the results in Dhrymes (1984), pp. 102-106, **vectorize** that set of equations to read

$$y = (I_q \otimes Y_{-1}) a + u, \quad y = \text{vec}(Y), \; a = \text{vec}(A), \; u = \text{vec}(U). \tag{3.52}$$

It follows immediately from the discussion above that the least squares estimator of $a$, say

$$\hat{a} = (I_q \otimes Y'_{-1}Y_{-1})^{-1}(I_q \otimes Y'_{-1})y, \quad \text{obeys}$$

$$T(\hat{a} - a) \xrightarrow{d} \frac{1}{2}(I_q \otimes \Phi)^{-1}\psi^*, \quad \psi^* = \text{vec}(\Psi^*). \tag{3.53}$$

The OLS estimated covariance matrix of that vector is given by

$$\hat{S}_* = \hat{\Sigma}_1 \otimes \left(\frac{Y'_{-1}Y_{-1}}{T^2}\right)^{-1}.$$

Thus, the usual "chi-squared" or "$F$-statistic" is given by

$$\eta^{*2} = T^2(\hat{a} - a)'\left(\hat{\Sigma}_1^{-1} \otimes \frac{Y'_{-1}Y_{-1}}{T^2}\right)(\hat{a} - a)$$

$$\xrightarrow{d} \frac{1}{4}\psi^{*'}(\Sigma_1^{-1} \otimes \Phi^{-1})\psi^*. \tag{3.54}$$

For the test to be carried out one takes $a = \text{vec}(I_q)$, which is the null hypothesis, and carries out the necessary modifications so that the "nuisance" parameters $P'_0P_0 = \Sigma_0$ and $\Sigma_1$ are eliminated.

We now examine the modifications required in the present case. For notational convenience put

$$K = \int_0^1 B(r)'B(r)\,dr, \quad \Phi = P'_0KP_0, \quad \Psi^* = P'_0B(1)'B(1)P_0 - \Sigma_1.$$

It is easily verified that if

$$H_{1T} = \frac{1}{2}\left(\frac{Y'_{-1}Y_{-1}}{T^2}\right)^{-1}\hat{\Sigma}_1 \tag{3.55}$$

then

$$\hat{P}_0[T(\hat{A} - A) + H_{1T}]\hat{P}_0^{-1} \xrightarrow{d} \frac{1}{2}K^{-1}B(1)'B(1), \tag{3.56}$$

which has a distribution that contains no nuisance parameters, and thus may be usefully tabulated. In the preceding $\hat{P}_0$, $\hat{\Sigma}_1$ are **consistent** estimators of the corresponding parameter matrices.

Alternatively, one may wish to deal with a scalar statistic like the chi-square statistic of Eq. (3.54); that statistic, however, is grossly inconvenient

to deal with in our context, and no particular virtue attaches to it other than
the fact that this is what one would employ if one were, in fact, dealing with
a standard regression problem. Instead we consider the following: define the
matrix

$$S^* = \hat{\Sigma}_0^{-1} \otimes \frac{Y_{-1}'Y_{-1}}{T^2}$$

and note that

$$\eta^2 = [T(\hat{a} - a) + h_{1T}]' \left( \hat{\Sigma}_0^{-1} \otimes \frac{Y_{-1}'Y_{-1}}{T^2} \right) [T(\hat{a} - a) + h_{1T}]$$

$$\xrightarrow{d} \frac{1}{4} B(1)[I_q \otimes B(1)K^{-1}](P_0 \otimes P_0'^{-1})(\Sigma_0^{-1} \otimes P_0'KP_0)(P_0' \otimes P_0^{-1})$$

$$\cdot [I_q \otimes K^{-1}B(1)']B(1)'$$

$$= \frac{1}{4}[B(1)B(1)'][B(1)K^{-1}B(1)'],$$

which is the product of a chi-square variable with $q$ degrees of freedom, and
a quadratic form in unit normals involving a **random** matrix which is a.c.
positive definite. Thus, it contains **no unknown** parameters, and hence it
may be usefully tabulated.

We have therefore proved Proposition 5.

**Proposition 5.** (Symmetric $A$). Consider the model in Eq. (3.38) and
suppose $A$ is taken to be symmetric; under the hypothesis $H_0 : A = I_q$,
and considering its estimator to be $\hat{A} = (\tilde{A} + \tilde{A}')/2$, where $\tilde{A}$ is the least
squares estimator, the following is true:

$$T(\hat{A} - A) \xrightarrow{d} \frac{1}{2}\Phi^{-1}\Psi^*$$

$$= \frac{1}{2}P_0^{-1} \left( \int_0^1 B(r)'B(r)\,dr \right)^{-1} \left( B(1)'B(1) - P_0^{-1}\Sigma_1 P_0^{-1} \right) P_0,$$

where $P_0$ is, say, the matrix of the **triangular** decomposition of $\Sigma_0$; the matrices $\Sigma_0$, $\Sigma_1$ are as defined in Eqs. (3.46), (3.47), and (3.49). To eliminate
the dependence of the distribution on these nuisance parameters, define

$$H_{1T} = \frac{1}{2} \left( \frac{Y_{-1}'Y_{-1}}{T^2} \right)^{-1} \hat{\Sigma}_1$$

in which case

$$\hat{P}_0[T(\hat{A} - A) + H_{1T}]\hat{P}_0^{-1} \overset{d}{\to} \frac{1}{2}K^{-1}B(1)'B(1),$$

where $\hat{\Sigma}_0$, $\hat{\Sigma}_1$ are consistent estimators of the corresponding parameter matrices.

A scalar statistic for testing the null hypothesis $H_0 : A = I_q$ may be based on

$$\eta^2 = [T(\hat{a} - a) + h_{1T}]' \left(\hat{\Sigma}_0^{-1} \otimes \frac{Y_{-1}'Y_{-1}}{T^2}\right)[T(\hat{a} - a) + h_{1T}]$$

$$\overset{d}{\to} \frac{1}{4}[B(1)B(1)'][B(1)K^{-1}B(1)'],$$

which does not depend on nuisance parameters.

## 3.4.2 Nonsymmetric Parameter Matrices

The convenient (symmetry) assumption leading to the results in Eq. (3.50) is completely inappropriate and cannot be maintained in the general case, especially in the following sections, which discuss the GLM with $I(1)$ regressors or instrumental variables estimators. Thus, we need actually to derive the limiting distribution of the entity $(1/T)\sum_{t=1}^{T} S_{t-1}' . u_t$. In so doing we are, of course, aided and guided by the result of Eq. (3.48) which serves as a point of departure. To this end, recall that from the multivariate Ito formula (see Proposition 5 in Chapter 8), we have for a twice differentiable function $h$

$$h(X_t) = h(X_0) + \sum_{j=1}^{q} \int_0^t h_i(X_s)\, dX_s^{(i)} + \frac{1}{2}\sum_{i,j=1}^{q} \int_0^t h_{ij}(X_s)\, d[X^{(i)}, X^{(j)}]_s, \quad (3.57)$$

where $X^{(i)}$ indicates the $i^{th}$ component of the vector $X$ and the notation $[X^{(i)}, X^{(j)}]_t$ indicates quadratic (co)variation of the two components, on the interval $[0, t]$. If we take $X$ to correspond to the $q$-element SMBM, $B$, and $h(B) = B_i B_j$ where $B_i$ is the $i^{th}$ component of the (row) vector $B$ we find by Eq. (3.57)

$$B_i(t)B_j(t) = \int_0^t B_i(s)\, dB_j(s) + \int_0^t B_j(s)\, dB_i(s) + \delta_{ij}t,$$

where $\delta_{ij}$ is the Kronecker delta; this is so since the components of the SMBM are independent and thus their quadratic (co)variation vanishes. It follows, therefore, that

$$B(t)'B(t) = \int_0^t B(s)\,dB(s)' + \int_0^t B(s)'\,dB(s) + I_q t,$$

and for $t = 1$ we have

$$B(1)'B(1) = \int_0^1 B(s)\,dB(s)' + \int_0^1 B(s)'\,dB(s) + I_q.$$

Therefore,

$$P_0'B(1)'B(1)P_0 - \Sigma_1 = P_0'\left(\int_0^1 B(s)\,dB(s)' + \int_0^1 B(s)'\,dB(s)\right)P_0 + \Sigma_0 - \Sigma_1.$$
$$(3.58)$$

To explore the matter further return to Eqs. (3.46) and (3.49), which constitute the definition of $\Sigma_0$ in the context of Proposition 16 (and other results relating to stationary and/or mixing cases) and Proposition 29 and its corollaries, respectively. The equation set in Eq. (3.49) defines $\Sigma_1$ in the three cases; moreover,

$$\Sigma_1^* = 0, \quad \text{for Proposition 7,} \tag{3.59}$$

$$\Sigma_1^* = \lim_{T\to\infty}\frac{1}{T}\sum_{t<s}Eu_t'.u_s. = \sum_{j=1}^{\infty}Eu_1'.u_{1+j}., \quad \text{for Proposition 16, etc.}$$

$$= \lim_{T\to\infty}\frac{1}{T}\sum_{t<s}Eu_t'.u_s., \quad \text{for Proposition 29 and its corollaries,}$$

which completes the description of the covariance relations required for further development of the topic. Since in Eq. (3.57) the left member is the (equivalence class) limit of the matrix in Eq. (3.48), and since $\Sigma_0 - \Sigma_1 = \Sigma_1^* + \Sigma_1^{*'}$, we have therefore established that

$$\frac{1}{T}\sum_{t=1}^{T}\left(u_t'.S_{t-1}. + S_{t-1}'.u_t.\right) \xrightarrow{d} \Psi + \Psi', \quad \Psi = P_0'\Upsilon P_0 + \Sigma_1^*,$$

$$\Upsilon = \int_0^1 B(s)'\,dB(s). \tag{3.60}$$

Consequently, we may write

$$\frac{1}{T}\sum_{t=1}^{T}S_{t-1}'.u_t. \xrightarrow{d} P_0'\Upsilon P_0 + \Sigma_1^* = \Psi. \tag{3.61}$$

We have threfore proved the following proposition.

**Proposition 6** (Nonsymmetric $A$). Consider the model in Eq. (3.38), where $A$ is a matrix not specified to be symmetric; under the hypothesis $H_0 : A = I_q$ the least squares estimator $\hat{A} = (Y'_{-1}Y_{-1})^{-1}Y'_{-1}Y$ has the limiting distribution

$$T(\hat{A} - A) \overset{d}{\to} \Phi^{-1}(P'_0 \Upsilon P_0 + \Sigma_1^*)$$

$$= P_0^{-1}K^{-1}P_0'^{-1}\left[P'_0\left(\int_0^1 B(r)'\,dB(s)\right)P_0 + \Sigma_1^*\right].$$

The modified estimator, which does not depend on the nuisance parameters $\Sigma_0$ and $\Sigma_1^*$, is given by

$$\hat{P}_0\hat{P}_0\left(T(\hat{A} - A) + H_{1T}\right)\hat{P}_0^{-1}, \quad H_{1T} = -\left(\frac{Y'_{-1}Y_{-1}}{T^2}\right)^{-1}\hat{\Sigma}_1^*$$

and has the limiting distribution

$$\hat{P}_0\left(T(\hat{A} - A) + H_{1T}\right)\hat{P}_0^{-1} \overset{d}{\to} K^{-1}\Upsilon,$$

which is free of nuisance parameters. Alternatively, we may define the scalar "chi-squared" statistic

$$\eta^2 = [T(\hat{a} - a) + h_{1T}]'\left(\hat{\Sigma}_0^{-1} \otimes \frac{Y'_{-1}Y_{-1}}{T^2}\right)[T(\hat{a} - a) + h_{1T}], \quad h_{1T} = \text{vec}(H_{1T}),$$

which has the limiting distribution

$$\eta^2 \overset{d}{\to} \text{vec}(\Upsilon)'(I_q \otimes K^{-1})\text{vec}(\Upsilon),$$

which, similarly, does not depend on nuisance parameters. In the preceding, $\hat{P}_0$ and $\hat{\Sigma}_1^*$ are consistent estimators of $\Sigma_0 = P'_0P_0$ and $\Sigma_1^*$, respectively.

**Remark 5**. The distribution of the test statistic given above may also be represented in the form

$$\eta^2 \overset{d}{\to} \text{tr}\,\Upsilon'K^{-1}\Upsilon,$$

which, *mutatis mutandis*, is the distribution of the "trace test" for testing a hypothesis on cointegration rank, as given in Proposition 7 of Chapter 5, and tabulated by Osterwald-Lenum (1992), as well as the Tables of Appendix II to Chapter 6 for the special case $r_0 = q$.

## 3.5   Cointegration

The subject of cointegration is studied more systematically in the next chapter. Here we seize on the development of the previous section to give an illustration of how cointegration may arise in the context of autoregressive models[8] that have extensive applications. For the purposes of this discussion we define cointegration to mean a situation in which we have a vector, say $y_{t\cdot}$, **all of whose elements are individually** $I(1)$, but there exists at least one linear combination, say $z_{ti} = y_{t\cdot}\gamma_{\cdot i}$, which is $I(0)$, i.e. it is (strictly) stationary. Thus, recall the model of Eq. (3.38) and suppose the error process is (strictly) stationary. Let

$$\Lambda = \text{diag}(\lambda_1, \lambda_2, \cdots, \lambda_q), \quad A = C\Lambda C^{-1},$$

assuming that $A$ is diagonalizable [see Dhrymes (1984) pp. 44-47 and pp. 53-55].[9] In the preceding the $\lambda_i$ are the characteristic roots and $C$ is the matrix of the characteristic vectors of $A$. Further suppose one of the roots, say the first, obeys $\lambda_1 = 1$ and all the others obey $|\lambda_i| < 1$, for $i = 2, 3, \ldots, q$. In view of the relations above we may write the model of Eq. (3.38) as

$$y_{t\cdot}C = y_{t-1\cdot}C\Lambda + u_{t\cdot}C, \quad \text{or} \quad z_{t\cdot} = z_{t-1\cdot}\Lambda + v_{t\cdot}, \quad z_{t\cdot} = y_{t\cdot}C, \quad v_{t\cdot} = u_{t\cdot}C. \quad (3.62)$$

Since $C$ is nonsingular, the process $\{v_{t\cdot} : t \in \mathcal{N}_+\}$ is strictly stationary. Letting $\gamma_{\cdot 1}$ denote the first column of $C$, we may write the first element of the vector as

$$z_{t1} = y_{t\cdot}\gamma_{\cdot 1} = z_{t-1,1} + v_{t1}. \quad (3.63)$$

Since $(I - L)z_{t1} = v_{t1}$ and the latter is stationary, it follows that $z_{t1}$ is $I(1)$. By contrast, let $\gamma_{\cdot i}$ be **any other** column of $C$ and note that

$$z_{ti} = y_{t\cdot}\gamma_{\cdot i} = \lambda_i z_{t-1,i} + v_{ti}, \quad i = 2, 3, \ldots, q, \quad (3.64)$$

---

[8] The existence of cointegration is frequently asserted in the literature of econometrics and its implications are derived for estimation and other ends. It is, however, never constructively demonstrated. The development of this section relies on Tsay and Tiao (1990) and Reinsel (1993), to which the reader is referred for additional material.

[9] We examine a more general case in the discussion below.

which means that the $z_{ti}$ are **strictly stationary** for $i = 2, 3, \ldots, q$. Moreover, since

$$y_{t\cdot} = z_{t\cdot}C^{-1} = \sum_{j=1}^{q} z_{tj}c^{j\cdot}, \qquad (3.65)$$

where $c^{j\cdot}$ is the $j^{th}$ **row** of $C^{-1}$, it follows that $y_{t\cdot}$ is $I(1)$ since **every element** of the vector contains an $I(1)$ component, viz. $z_{t1}$.

It is **a simple implication of the preceding** that even though every element of the vector $y_{t\cdot}$ is $I(1)$, there exist $q - 1$ linear combinations of its elements which are **strictly stationary**. Thus, the elements of the vector $y_{t\cdot}$ may be said to be cointegrated of rank $q - 1$, which is simply the number of linearly independent vectors leading to **stationary linear combinations**. If more than one of the roots is **unity**, the cointegration rank is correspondingly reduced, i.e. the number of linearly independent vectors leading to stationary linear combinations is correspondingly reduced.

### 3.5.1   Cointegration in General VAR Models

In the context of a general VAR model, i.e. a multivariate autoregressive model of order $n$, cointegration presents a far more complicated problem than that of the preceding section. Thus, consider

$$y_{t\cdot} = y_{t-1}.A_1 + \ldots + y_{t-k}.A_k + u_{t\cdot}, \qquad (3.66)$$

where $\{u_{t\cdot} : t \in \mathcal{N}_+\}$ is strictly stationary. Introducing the lag operator $L$, we may rewrite the model above as

$$y_{t\cdot} = y_{t-1}.A(L) + u_{t\cdot}, \quad A(L) = \sum_{j=1}^{k} A_j L^j. \qquad (3.67)$$

The approach employed in the context of the simple model in Eq. (3.57) is not available now, since it is not **generally** possible to diagonalize $k$ matrices simultaneously. What is available in this context is a general result, [see Gantmacher (1955) Vol. I Chapter 8], which states that polynomial matrices, i.e. matrices whose elements are polynomials in a real or complex indeterminate, are similar to a diagonal matrix whose nonnull elements are the invariant polynomials. Specifically, in this case it states that

$$A(L) = P(L)D(L)Q(L), \quad |P(L)| > 0, \quad |Q(L)| > 0,$$

the determinants being independent of $L$. Thus, we may only obtain

$$y_t.Q(L)^{-1} = y_{t-1}.P(L)D(L) + u_t.Q(L)^{-1},$$

which is not particularly helpful in determining what connection, if any, we can establish between the properties of the matrix operator $A(L)$ and the notion of cointegration. Thus, we must employ an alternative, and rather indirect, approach. To this end, transform the system to [10]

$$\zeta_t. = \zeta_{t-1}.A + v_t., \quad v_t. = e_1. \otimes u_t.,$$

$$\zeta_t. = (y_t., y_{t-1}., \ldots, y_{t-k+1}.), \qquad (3.68)$$

$$A = \begin{bmatrix} A_1 & I & 0 & \cdots & 0 \\ A_2 & 0 & I & \cdots & 0 \\ \vdots & \vdots & \vdots & \vdots & \vdots \\ A_{k-1} & 0 & 0 & \cdots & I \\ A_k & 0 & 0 & \cdots & 0 \end{bmatrix},$$

where $e_1.$ is a $k$-element row vector all of whose elements are zero except the first, which is unity. With this transformation we are reduced to the original simple case, since we are dealing with the system in the first equation of the equation set Eq. (3.68)

$$\zeta_t. = \zeta_{t-1}.A + v_t..$$

In the context above we must make the connection between unit roots of the characteristic equation of the stochastic difference equation in Eq. (3.67) and the presence of cointegration. The characteristic equation, of the stochastic difference equation in Eq. (3.67), is given by

$$|I_q - rA_1' - r^2 A_2' - \ldots - r^k A_k'| = 0, \qquad (3.69)$$

where, in this discussion, $r$ is a complex indeterminate. The characteristic equation **of the matrix** $A'$ in Eq. (3.68), on the other hand, is given by

$$|\lambda I_{kq} - A'| = 0. \qquad (3.70)$$

---

[10] For a discussion of this device see Dhrymes (1984), pp. 133-136. It should also be noted that the matrix $A$ is occasionally referred to in the literature as the **companion matrix**.

The connection between the characteristic equations of Eqs. (3.69) and (3.70) is given in Dhrymes (1984), pp. 136-141, but it is repeated here in abbreviated form. Define

$$F_{11} = \lambda I_q - A_1', \quad F_{12} = (-A_2', -A_3', \ldots, -A_k'), \tag{3.71}$$

$$F_{21} = (-I_q, 0, \ldots, 0)', \quad F_{22} = \Lambda_{k-1} \otimes I_q, \quad F = \begin{bmatrix} F_{11} & F_{12} \\ F_{21} & F_{22} \end{bmatrix},$$

$$\Lambda_{k-1} = \begin{bmatrix} \lambda & 0 & 0 & \ldots 0 \\ -1 & \lambda & 0 & \ldots \\ \vdots & \vdots & \vdots & \vdots \\ 0 & \ldots & -1 & \lambda \end{bmatrix}, \quad \lambda I_{kq} - A' = \begin{bmatrix} F_{11} & F_{12} \\ F_{21} & F_{22} \end{bmatrix},$$

and note that

$$|F| = |F_{22}| \, |F_{11} - F_{12} F_{22}^{-1} F_{21}|$$

$$= \lambda^{q(k-1)} \, |\lambda I_q - A_1' - \lambda^{-1} A_2' - \cdots - \lambda^{-(k-1)} A_k'|$$

$$= |\lambda^k I_q - \lambda^{k-1} A_1' - \lambda^{k-2} A_2' - \cdots - A_k'|$$

$$= \lambda^{qk} |I_q - A_1' r - A_2' r^2 - \ldots - A_k' r^k|, \quad r = \lambda^{-1}. \tag{3.72}$$

The penultimate member of the relation above gives the characteristic equation $|\lambda I_q - A'| = 0$; the determinantal part of the last member gives the characteristic equation of the operator $I_q - A'(L)$, obtained by replacing the operator $L$ and its powers by the (real or complex) indeterminate $r$. Hence, **the roots of the latter** are the **inverses of the roots of the former**. What is the consequence of one (or more) of these roots being unity, say $\lambda_i$, $i = 1, 2, \ldots, r_0$, while all other roots obey $|\lambda_j| < 1$, for $j = r_0 + 1, r_0 + 2, \ldots, qk$? In this connection we have Assertion 1.

**Assertion 1.** Consider the general VAR of Eq. (3.67), and suppose the matrix $A'$ of the equation set Eq. (3.68) has $r_0$ unit roots. The following statements are true:

i. the characteristic equation of the operator $I - A'(L)$ has $r_0$ unit roots, and the operator is **not invertible,** i.e. the VAR does not have a representation as a unilateral $MMA(\infty)$;

ii. the **matrix** $I_q - A^{*'}$ has $r_0$ **zero** roots, where $A^* = \sum_{j=1}^{k} A_j$ ;

iii. the matrix $A^*$ has $r_0$ **unit** roots.

Proof: To prove i we proceed as in Remark 11 of Chapter 1; thus let $H(L)$ be the matrix adjoint to the matrix operator[11] $I - A'(L)$, so that it is a matrix whose elements are polynomials of degree at most $(q-1)k$ in the lag operator $L$; by the definition of an inverse we have

$$H(L)\,[I - A'(L)] = a(L)I_q, \quad a(L) \;=\; |I - A'(L)|.$$

Since, by the discussion immediately preceding, the characteristic roots of the matrix $A'$ are the **inverses** of the corresponding roots of the characteristic equation of the polynomial operator $I - A'(L)$, we have

$$a(L) = (I - L)^{r_0} a^*(L),$$

where $a^*(L)$ is invertible. Thus, the inverse $[I - A(L)]^{-1}$ does not exist since it requires the division of the adjoint by $a(L)$, which obeys $a(1) = 0$. This concludes the proof of part i.

   To prove ii, note that the characteristic equation of $I_q - A^{*'}$ is given by

$$|\mu I_q - (I_q - A^{*'})| = 0,$$

and for $\mu = 0$ we obtain $|I_q - A^{*'}| = 0$; the latter, however, is simply the characteristic equation of the polynomial operator, i.e. $|I - A'(r)| = 0$ evaluated at $r = 1$. Since the latter has $r_0$ **unit roots** we conlcude that the **matrix** $I_q - A^{*'}$ has $r_0$ **zero roots** and is thus of rank $q - r_0$, which conlcudes the proof of part ii.

   To prove part iii, we note that the characteristic equation of the **matrix** $A^{*'}$ is given by

$$|\nu I_q - A^{*'}| = 0,$$

---

[11] Strictly speaking this operator should be written as $I_q I - A'(L)$, since $I$ is the identity **operator** while $I_q$ is the identity **matrix** of order $q$. In the interest of notational simplicity we shall consistently omit the matrix $I_q$, it being understood that when the identity operator stands alone its coefficient is the unit element appropriate to the context, i.e. in a scalar context its coefficient is the number one, and in a multivariate context its coefficient is the appropriately dimensioned identity matrix.

which, when evaluated at $\nu = 1$, yields $|I_q - A^{*'}| = 0$. Since the latter has $r_0$ zero roots, it follows that $A^{*'}$ has $r_0$ unit roots.

q.e.d.

We now show that the incidence of unit roots in the characteristic equation of the operator $I - A'(L)$ leads to the existence of cointegration.

**Remark 6.** Before we proceed it is important to clear up certain ambiguities regarding the use of the terminology "unit roots"; in the literature of econometrics there does not exist a clearcut differentiation between the case where unit roots may **imply** cointegration, and where they may not. Consider, for example, the case of the general VAR

$$B'(L)y'_{t\cdot} = \epsilon'_{t\cdot},$$

and suppose that it is possible to write

$$B'(L) = (I - L)^d B^{*'}(L)$$

such that $B^{*'}(L)$ is an **invertible** polynomial operator of degree $k-d$. This condition is referred to as **unit root factorization**. Note that this means that the characteristic equation of the operator has a unit root of multiplicity $dq$, where $d$ is an **integer**; in such a case **cointegration is not possible**. In the absence of unit root factorization, cointegration **is present only** when the **determinant** of the operator above has a factorization of the form

$$|B'(L)| = b(L) = (I - L)^{r_0} b^*(L), \quad r_0 < dq,$$

and the roots of $b^*(L) = 0$ are **greater** than unity. In the literature of statistics what we have called unit root factorization is referred to as the **nonstationary** case, while the determinantal factorization above is referred to as the **partially nonstationary** case. As we shall see below this terminology is not sufficiently flexible and we do not maintain this distinction in the next chapter where we study cointegration more fully.

Another aspect that is not made clear in the literature of econometrics is that the "problem" of unit roots arises whenever some root of an appropriate characteristic equation, say $\lambda$, obeys $|\lambda| = 1$. Note that this can occur in

**four different forms**, viz. $\lambda = 1$, $\lambda = -1$, $\lambda = e^{i\theta}$, or $\lambda = e^{-i\theta}$, where $i^2 = -1$, and $\theta \in [-\pi, \pi]$. The last two always occur as a conjugate pair. Typically, the econometrics literature concentrates only on the first case, which leads to great simplications through differencing.

We conclude this section by showing constructively that in the case of partial nonstationarity there is cointegration, i.e. there exist linear combinations of elements of $y_t$, which are stationary and others which are nonstationary; we do so in the context of the model in Eq. (3.67), which is a general VAR model with $k > 1$ lags. To this end, suppose that the polynomial operator $I - A'(L)$ of Eq. (3.67) has $r_0$ unit roots, say $\lambda_1 = 1$ repeated $r_0$ times; equivalently, let $\lambda_1$ be a root of multiplicity $r_0$ of the matrix $A'$. If the assumption of diagonalizability is not made, as it was in the earlier discussion, we cannot be assured that the matrix of characteristic vectors (of the matrix $A'$) is nonsingular. This is so since there exist matrices with repeated roots which do not have a nonsingular matrix of characteristic vectors. For an example of this see Bellman (1960), p. 190. Hence, to develop our argument solely in terms of matrix properties, we must employ the Jordan canonical form, [see Gantmacher (1955), pp. 147-158, where it is termed the Jordan normal form]. The Jordan canonical form of the matrix $A'$ is

$$A' = PJP^{-1}, \quad J = \text{diag}(J_1, J_2, \ldots, J_m),$$

$$J_i = \begin{bmatrix} \lambda_i & 1 & 0 & 0 & \ldots & 0 \\ 0 & \lambda_i & 1 & 0 & \ldots & 0 \\ \vdots & \vdots & \vdots & \vdots & \vdots & \vdots \\ 0 & 0 & 0 & \ldots & \lambda_i & 1 \\ 0 & 0 & 0 & 0 & 0 & \lambda_i \end{bmatrix}, \tag{3.73}$$

where $P$ is the matrix of characteristic vectors, $J_i$ is a matrix of dimension $m_i$, and $m_i$ is the multiplicity of the root $\lambda_i$, such that $\sum_{i=1}^{m} m_i = m = qk$. Thus, the Jordan matrix, $J$, contains the $m$ distinct roots of $A'$, one each in the upper triangular blocks, $J_i$. Since we assert that $A'$ has $r_0$ unit roots we may, without loss of generality, assume that $\lambda_1 = 1$ and that $m_1 = r_0$. Hence the system in the first equation of the set Eq. (3.68) may be written as $P^{-1}\zeta'_{t.} = J(P^{-1}\zeta'_{t-1.}) + P^{-1}v'_{t.}$, or alternatively,

$$z'_{t.} = Jz'_{t-1.} + w'_{t.}, \quad z'_{t.} = P^{-1}\zeta'_{t.}, \quad w'_{t.} = P^{-1}v'_{t.}. \tag{3.74}$$

In view of the **block diagonal** nature of $J$ we may write the (first) subvector of $z_t.$, consisting of (its first) $r_0$ elements, as

$$z_{t.}^{(1)'} = J_1 z_{t-1}^{(1)'} + w_{t.}^{(1)'}, \qquad (3.75)$$

so that the nonstationary component will be a **vector of order** $r_0$, rather than the scalar component we encountered in the earlier discussion with a **single unit root**. Note, however, that in the previous context $r_0 = 1$, so that the result therein is simply a special case of Eq. (3.75). Let $Q = P^{-1}$, $Q_1$ be the submatrix consisting of the first $r_0$ **rows**, and use the definition of $v_t.$ in Eq. (3.68) to obtain

$$w_{t.}^{(1)'} = Q_1(e_{.1} \otimes u_{t.}'). \qquad (3.76)$$

The virtue of the Jordan canonical form representation in Eq. (3.73) is that it leads to Eqs. (3.75) and (3.77), so that the behavior of each subvector $z_{t.}^{(i)'}$ may be handled **independently of the behavior of the other subvectors**. Since $\lambda_1 = 1$, while $|\lambda_i| < 1$, for $i = 2, 3, \ldots, m$, it follows that, indeed, $z_{t.}^{(1)}$ is the only nonstationary component of $z_t.$, all others being stationary. To verify this claim, note that for $i = 1, 2, \ldots, m$,

$$z_{t.}^{(i)'} = J_i z_{t-1}^{(i)'} + w_{t.}^{(i)'}, \quad \text{and thus,} \quad z_{t.}^{(i)'} = \sum_{s=0}^{t-1} J_i^s w_{t-s.}^{(i)'}. \qquad (3.77)$$

Moreover, the matrix $J_1$ of Eq. (3.75) is of the form

$$J_1 = I_{r_0} + D_{r_0}, \qquad (3.78)$$

where

$$D_{r_0} = \begin{bmatrix} 0 & 1 & 0 & 0 & \cdots & 0 \\ 0 & 0 & 1 & 0 & \cdots & 0 \\ 0 & 0 & 0 & 1 & \cdots & 0 \\ \vdots & \vdots & \vdots & \vdots & \vdots & \vdots \\ 0 & 0 & 0 & 0 & \cdots & 1 \\ 0 & 0 & 0 & 0 & \cdots & 0 \end{bmatrix}.$$

We note that $D_{r_0}$ is a **nilpotent matrix of order** $r_0$, meaning that $D_{r_0}^{r_0} = 0$. Thus, for any integer $s \geq 1$,[12]

$$J_1^s = \sum_{j=0}^{s \wedge r_0} \binom{s}{j} D^j.$$

---

[12] The subscript on $I_{r_0}$ and $D_{r_0}$ will be dropped below for notational simplicity.

Indeed, a close examination of $J_1$ and its powers reveals that the nonzero elements of $J_1^t$ are obtainable, for $t = 1, 2, 3, \ldots$, from the rows of the Pascal triangle [which is given below from the zeroth to the sixth power of the binomial expansion of $(x + y)^k$].

$$
\begin{array}{ccccccc}
1 \\
1 & 1 \\
1 & 2 & 1 \\
1 & 3 & 3 & 1 \\
1 & 4 & 6 & 4 & 1 \\
1 & 5 & 10 & 10 & 5 & 1 \\
1 & 6 & 15 & 20 & 15 & 6 & 1
\end{array}
$$

For example, to evaluate the nonzero elements of $J_1^s$ [13] for $s \leq r_0$ we proceed as follows, using the example $r_0 = 3$, $s = 3$. The first row is obtained, in order, from the $(s+1)^{th}$ row of the Pascal triangle; since we need only three elements, we have $(1, 3, 3)$. The second row is similarly obtained from the $(s + 1)^{st}$ row of the Pascal triangle; since we need only two elements, we have $(1, 3)$. The third row is similarly obtained, and since we need only one element we have $(1)$. The same procedure applies more generally; using the example $r_0 = 3, s = 4$, we find that the first row of $J_1^4$ consists of the fifth row of the Pascal triangle. Since we need three elements, we find $(1, 4, 6)$; the second row is again obtained from the fifth row of the Pascal triange; since we need only two elements, it consists of $(1, 4)$, and so on. For $s > r_0$, the sum (and thus the expansion) terminates at $r_0$, since the matrix $D^s$ is **null** for $s$. The Pascal triangle is obtained from the expansion of $(x + y)^s$; thus, the sum of the weights applied to the vectors $w_{t-s}^{(1)'}$ in Eq. (3.77) is of the order $2^s$, although not uniformly. More precisely, using the representation of Eq. (3.77), we find

$$
z_t^{(1)'} = \sum_{s=0}^{t-1} \left[ \left( \sum_{i=0}^{s \wedge r_0} \binom{s}{i} D^i \right) \right] w_{t-s}^{(1)'}.
$$

To show that the covariance of $z_t^{(1)}$, is unbounded and thus the latter is

---

[13] Note that $J_1^s$ is an upper triangular matrix.

nonstationary, return to Eq. (3.76) and rewrite it as

$$w_{t.}^{(1)'} = Q_1 \begin{pmatrix} u_{t.}' \\ 0 \\ \vdots \\ 0 \end{pmatrix} = Q_{11} u_{t.}', \quad Q_1 = (Q_{11}, Q_{12}),$$

such that $Q_{11}$ is $r_0 \times q$, and $Q_{12}$ is $r_0 \times (k-1)q$. Hence,

$$\text{Cov}(z_{t.}^{(1)'}) = \sum_{s=0}^{t-1} \left( \sum_{j=0}^{s \wedge r_0} \binom{s}{j} D_{r_0}^j \right) \Phi_{11} \left( \sum_{j=0}^{s \wedge r_0} \binom{s}{j} D_{r_0}^j \right)', \quad \Phi_{11} = Q_{11} \Sigma Q_{11}',$$

is **unbounded**, in view of the fact that the last term in parentheses above is, for $t \geq r_0$,

$$\sum_{j=0}^{t-1 \wedge r_0} \binom{t-1}{j} = t + \frac{(t-2)(t-1)}{2} + \ldots + \frac{\prod_{i=1}^{t-1}(t-i)}{r_0!}.$$

Evidently, this diverges with $t$ and, thus, demonstrates that the component $z_{t.}^{(1)}$ is **nonstationary**.

We next show that the other components are **stationary**. Return to Eq. (3.77) and note that for the **stationary roots**, $\lambda_i$, $i > 1$,

$$J_i = \lambda_i I_{m_i} + D_{m_i},$$

where, again, $D_{m_i}$ is **nilpotent** of order $m_i$, which is the multiplicity of the root $\lambda_i$.[14] Operating in precisely the same manner as above we find, again dropping the subscript from $I_{m_i}$ and $D_{m_i}$,

$$J_i^s = \sum_{j=0}^{s \wedge m_i} \binom{s}{j} \lambda_i^{s-j} D^j.$$

We now verify that the component $z_{t.}^{(i)}$, for $i > 1$, is stationary. Since the $w$-sequence is **stationary**, we need only verify that the covariance of $z_{t.}^{(i)}$ is **bounded**. Partition

$$Q = \begin{bmatrix} Q_1 \\ Q_2 \\ \vdots \\ Q_m \end{bmatrix},$$

---

[14] Notice that if $m_i = 1$, $D_{m_i}$ is the **null matrix** and $J_i$ is the **scalar** $\lambda_i$.

such that $Q_i$ is $m_i \times qk$, corresponding to the multiplicity of the $i^{th}$ **distinct root**. [15] It follows, then, that

$$w_{t\cdot}^{(i)'} = Q_{i1} u_{t\cdot}', \quad \text{where} \quad Q_i = (Q_{i1}, Q_{i2}), \quad i \geq 2.$$

Consequently, for $i \geq 2$

$$\text{Cov}(z_{t\cdot}^{(i)'}) = \sum_{s=0}^{t-1} \left( \sum_{j=0}^{s \wedge m_i} \binom{s}{j} \lambda_i^{s-j} D_{m_i}^j \right) \Phi_{ii}' \left( \sum_{j=0}^{s \wedge m_i} \binom{s}{j} \lambda_i^{s-j} D_{m_i}^j \right), \quad \Phi_{ii} = Q_{i1} \Sigma Q_{i1}'.$$

The terms in parentheses behave, essentially, like

$$\nu_t = \sum_{s=m_i}^{t-1} \left( \lambda_i^s + s \lambda_i^{s-1} + \frac{s(s-1)}{2!} \lambda_i^{s-2} + \ldots + \frac{\prod_{j=0}^{m_i}(s-j)}{m_i!} \right).$$

It is quite apparent that $\lim_{t \to \infty} \nu_t < \infty$, which establishes the stationarity of the components $z_{t\cdot}^{(i)}$, for $i \geq 2$.

Return now to Eq. (3.74) and reverse the transformation to obtain

$$\zeta_{t\cdot}' = P z_{t\cdot}', \quad \text{so that} \quad y_{t\cdot}' = P_1 z_{t\cdot}'.$$

Evidently, $P_1$ is $q \times kq$ of rank $q$; partition

$$P_1 = (P_{11}, P_{12}, \cdots, P_{1m}), \quad P_{11} \text{ is } q \times r_0, \quad P_{1i} \text{ is } q \times m_i, \quad i \geq 2, \qquad (3.79)$$

and note that

$$y_{t\cdot}' = P_{11} z_{t\cdot}^{(1)'} + \sum_{i=2}^{m} P_{1i} z_{t\cdot}^{(i)'}, \qquad (3.80)$$

where $z_{t\cdot}^{(1)}$ is the **only nonstationary**, and $z_{t\cdot}^{(i)}$, $i \geq 2$, are the **stationary** components of the vector $z_{t\cdot}$. Eq. (3.79) shows clearly that every element of $y_{t\cdot}$ is **nonstationary** owing to its dependence on the nonstationary component $z_{t\cdot}^{(1)}$.

Since $P_1$ is of rank $q$, the row null spaces of $P_{11}$ and $P_{1i}$, $i \geq 2$ have **no elements in common**, save the null vector. This is so since if they had a vector in common, say $\gamma$, then

$$\gamma' P_1 = (\gamma' P_{11}, \gamma' P_{12}, \gamma' P_{13}, \ldots, \gamma' P_{1m}) = 0,$$

---

[15] Since we had earlier put $m_1 = r_0$, corresponding to the **unit root**, this notation is consistent with the notation of the preceding discussion.

which would contradict the rank condition on $P_1$. Consequently there exists a matrix $B_1$ of dimension $r_0 \times q$, and a matrix $B_2$ of dimension $q - r_0 \times q$, such that

$$B_1 P_{11} \neq 0, \quad B_1 P_{1i} = 0, \quad B_2 P_{11} = 0, \quad B_2 P_{1i} \neq 0, \quad i \geq 2,$$

and moreover,

$$B_1 y_{t\cdot}' = B_1 P_{11} z_{t\cdot}^{(1)'}, \qquad B_2 y_{t\cdot}' = B_2 \sum_{i=2}^{m} P_{1i} z_{t\cdot}^{(i)'}. \tag{3.81}$$

It is clear that the first set of linear combinations in Eq. (3.80) is **purely nonstationary**, while the second set is **stationary**, thus establishing cointegration.

## 3.6 Regressors Which Are $I(1)$

In this section we examine the GLM

$$y_{t\cdot} = x_{t\cdot} B + u_{t1\cdot}, \quad t = 1, 2, \ldots, T, \tag{3.82}$$

where $y_{t\cdot}$ is the $q$-element vector of dependent variables, while $x_{t\cdot}$ is the $k$-element vector of the $I(1)$ explanatory variables, or regressors. There is very little difference, in terms of the complexity of the argument, between the scalar case, $q = 1$, and the multivariate (vector) case, $q > 1$. We choose to deal with the multivariate case and define the explanatory variables by

$$x_{t\cdot} = x_{t-1\cdot} + u_{t2\cdot}, \tag{3.83}$$

so that they are $I(1)$. The implicit assumption in such models is that

$$u_{t\cdot} = (u_{t1\cdot}, u_{t2\cdot}), \quad t \in \mathcal{N}_0 \tag{3.84}$$

is a strictly stationary, and ergodic process. Writing the full model in the more convenient vector form

$$y = (I_q \otimes X)b + u_1, \quad y = \text{vec}(Y), \quad Y = (y_{t\cdot}), \quad X = (x_{t\cdot}), U_1 = (u_{t1\cdot}),$$

$$u_1 = \text{vec}(U_1), \quad b = \text{vec}(B), \tag{3.85}$$

we obtain the least squares estimator as

$$\hat{b} = (I_q \otimes X'X)^{-1}(I_q \otimes X)'y = b + [I_q \otimes (X'X)^{-1}](I_q \otimes X)'u_1. \qquad (3.86)$$

To examine the limiting distribution of the entities in this problem, we first note the novelty it presents. In the models we had examined previously, there was only one basic or fundamental stochastic process, and this process generated **the explanatory variables** as well as the **structural errors**. This was so because the explanatory variables were **only** lagged dependent variables; this implied that the dependent variable was a martingale process and, thus, the structural error was simply the martingale (first) difference. In this problem we have **two** error processes; one, $\{u_{t1.} : t \in \mathcal{N}_0\}$, generates the structural errors, and the other, $\{u_{t2.} : t \in \mathcal{N}_0\}$, generates the explanatory variables (regressors). If we put

$$u_{t.} = (u_{t1.}, \ u_{t2.}), \quad t \in \mathcal{N}_+ \qquad (3.87)$$

and assume the process to be stationary (and ergodic) then all of the previous results will apply, since

$$S_{t.} = S_{t-1.} + u_{t.}, \quad S_{0.} = u_{0.} = 0. \qquad (3.88)$$

In particular, we shall find that by Propositions 7 or 16, etc.(of Chapter 9),

$$\frac{1}{T^2} \sum_{t=1}^{T} S'_{t-1.} S_{t-1.} \xrightarrow{d} P'_0 \left( \int_0^1 B(r)'B(r) \, dr \right) P_0 = \Phi$$

$$\frac{1}{T} \sum_{t=1}^{T} \left( u'_{t.} S_{t-1.} + S'_{t-1.} u_{t.} \right) \xrightarrow{d} P'_0 [B(1)'B(1) - P_0'^{-1} \Sigma_1 P_0^{-1}] P_0 = \Psi^*,$$

and, by the discussion leading to Eq. (3.61), we conclude that *mutatis mutandis*

$$\frac{1}{T} \sum_{t=1}^{T} S'_{t-1.} u_{t.} \xrightarrow{d} \Psi, \quad \Psi = P'_0 \Upsilon P_0 + \Sigma_1^*. \qquad (3.89)$$

The new element in this case is that we wish to treat **separately** certain components of $S_{t.}$ as well as the limiting BM, $B(r)$, and matrices $P'_0 P_0 = \Sigma_0$, and $\Sigma_1^*$. This is so because $u_{t.} = (u_{t1.}, u_{t2.})$ and the first component of the vector pertains to structural errors, while the second component pertains to the behavior of the explanatory variables. We further note that

$$x_{t\cdot} = x_{0\cdot} + \sum_{t=1}^{T} u_{t2\cdot} = x_{0\cdot} + S_{t2\cdot},$$

and moreover

$$\text{vec}(X'U_1) = \sum_{t=1}^{T} x_{t\cdot}' u_{t1\cdot} = x_{0\cdot}' S_{t1\cdot} + \sum_{t=1}^{T} S_{t2\cdot}' u_{t1\cdot}, \quad S_{t1\cdot} = \sum_{j=1}^{t} u_{j1\cdot\cdot}$$

Partitioning $\Phi$ and $\Psi$, conformably with $S_{t\cdot} = (S_{t1\cdot}, S_{t2\cdot})$ and $u_{t\cdot} = (u_{t1\cdot}, u_{t2\cdot})$ we obtain

$$\Phi = \begin{bmatrix} \Phi_{11} & \Phi_{12} \\ \Phi_{21} & \Phi_{22} \end{bmatrix}, \quad \Psi = \begin{bmatrix} \Psi_{11} & \Psi_{12} \\ \Psi_{21} & \Psi_{22} \end{bmatrix}, \tag{3.90}$$

where $\Phi_{11}$, $\Psi_{11}$ are $q \times q$ matrices, and $\Phi_{22}$, $\Psi_{22}$ are $k \times k$ matrices. Noting that

$$T(\hat{b} - b) = \left( I_q \otimes \frac{X'X}{T^2} \right)^{-1} \frac{1}{T} \text{vec}(X'U_1), \quad \text{we find}$$

$$\frac{X'X}{T^2} = \frac{1}{T} x_{0\cdot}' x_{0\cdot} + \frac{x_{0\cdot}'}{\sqrt{T}} \frac{1}{T^{3/2}} \sum_{t=1}^{T} S_{t2\cdot} + \left( \frac{1}{T^{3/2}} \sum_{t=1}^{T} S_{t2\cdot} \right)' \frac{x_{0\cdot}}{\sqrt{T}}$$

$$+ \frac{1}{T^2} \sum_{t=1}^{T} S_{t2\cdot}' S_{t2\cdot}, \quad \text{so that}$$

$$\frac{X'X}{T^2} \xrightarrow{\text{d}} \Phi_{22}; \quad \text{moreover,}$$

$$\frac{1}{T} \text{vec}(X'U_1) = \left( \frac{1}{\sqrt{T}} x_{0\cdot}' \right) \left( \frac{1}{\sqrt{T}} S_{T1\cdot} \right) + \frac{1}{T} \sum_{t=1}^{T} S_{t2\cdot}' u_{t1\cdot} \xrightarrow{\text{d}} \Psi_{21}.$$

Consequently, we obtain

$$T(\hat{b} - b) \xrightarrow{\text{d}} (I_q \otimes \Phi_{22})^{-1} \text{vec}(\Psi_{21}).$$

## 3.7 Instrumental Variables

The discussion of instrumental variable (IV) estimators involves essentially an argument similar to that in the preceding section. Even though such models are conceptually distinct from those we have examined above, from a formal point of view they are identical. The new element introduced by IV is a third stochastic process, so that the fundamental error processes are now $u_{t\cdot} = (u_{t1\cdot}, u_{t2\cdot}, u_{t3\cdot})$ and they pertain, respectively, to the structural errors, explanatory variables and instrumental variables. More precisely the model is

$$y_{t\cdot} = x_{t\cdot}B + u_{t1\cdot}, \quad x_{t\cdot} = x_{t-1\cdot} + u_{t2\cdot}, \quad z_{t\cdot} = z_{t-1\cdot} + u_{t3\cdot},$$

where the variables in $z_{t\cdot}$ represent the instrumental variables.

From the representation above we obtain

$$z_{t\cdot} = z_{0\cdot} + S_{t3\cdot}, \quad x_{t\cdot} = x_{0\cdot} + S_{t2\cdot},$$

$$S_{t\cdot} = (S_{t1\cdot}, S_{t2\cdot}, S_{t3\cdot}), \quad u_{t\cdot} = (u_{t1\cdot}, u_{t2\cdot}, u_{t3\cdot}). \tag{3.91}$$

As in Eq. (3.81) the model is (using the same notation)

$$y = (I_q \otimes X)b + u_1, \quad Z = (z_{t\cdot}), \quad t = 1, 2, \ldots, T.$$

Under any of the conditions given in the previous discussion, we easily establish that

$$\frac{1}{T^2} \sum_{t=1}^{T} S'_{t\cdot} S_{t\cdot} \xrightarrow{d} P'_0 \left( \int_0^1 B(r)' B(r) \, dr \right) P_0. \tag{3.92}$$

This may be justified as follows: Let

$$\frac{1}{T} \sum_{t=1}^{T} E S'_{t\cdot} S_{t\cdot} = \Sigma_T > 0, \quad \Sigma_T = P'_T P_T, \quad P_T \quad \text{nonsingular.} \tag{3.93}$$

This (triangular) decomposition is always possible [see Dhrymes (1984), pp. 68-69]. Moreover, suppose

$$\lim_{T \to \infty} \Sigma_T = \Sigma_0 > 0, \quad \Sigma_0 = P'_0 P_0, \quad P_0 \quad \text{nonsingular.} \tag{3.94}$$

Define

$$X'^T(r) = P'^{-1}_T \frac{S'_{[Tr]}}{\sqrt{T}}, \quad r \in \left[ \frac{t-1}{T}, \frac{t}{T} \right), \quad t = 1, 2, \ldots, T, \tag{3.95}$$

and note that, under a variety of conditions given in Chapter 9,

$$\frac{1}{T^2} \sum_{t=1}^{T} S'_{t\cdot} S_{t\cdot} = P'_T \left( \sum_{t=1}^{T} \int_{\frac{t-1}{T}}^{\frac{t}{T}} X'^T(r) X^T(r) \, dr \right) P_T$$

$$\xrightarrow{d} P'_0 \left( \int_0^1 B(r)' B(r) \, dr \right) P_0 = \Phi > 0. \tag{3.96}$$

Moreover, as in the discussion of the previous section

$$\frac{1}{T}\sum_{t=1}^{T}\left(u_{t\cdot}'S_{t-1\cdot} + S_{t-1\cdot}'u_{t\cdot}\right) = \frac{1}{T}\sum_{t=1}^{T}\left(S_{t\cdot}'S_{t\cdot} - S_{t-1\cdot}'S_{t-1\cdot}\right) - \frac{1}{T}\sum_{t=1}^{T}u_{t\cdot}'u_{t\cdot}$$

$$= \left(\frac{S_{T\cdot}}{\sqrt{T}}\right)'\left(\frac{S_{T\cdot}}{\sqrt{T}}\right) - \frac{1}{T}\sum_{t=1}^{T}u_{t\cdot}'u_{t\cdot};$$

thus we conclude

$$\frac{1}{T}\sum_{t=1}^{T}\left(u_{t\cdot}'S_{t-1\cdot} + S_{t-1\cdot}'u_{t\cdot}\right) \xrightarrow{d} P_0'B(1)'B(1)P_0 - \Sigma_1 = \Psi^*, \tag{3.97}$$

and, by the discussion leading to Eq. (3.85), we obtain

$$\frac{1}{T}S_{t-1\cdot}'u_{t\cdot} \xrightarrow{d} P_0'\left(\int_0^1 B(s)'\,dB(s)\right)P_0 + \Sigma_1^* = \Psi, \tag{3.98}$$

where now we have defined

$$\Sigma_0 = \lim_{T\to\infty}\frac{1}{T}ES_{T\cdot}'S_{T\cdot},$$

$$\Sigma_1 = \lim_{T\to\infty}\frac{1}{T}\sum_{t=1}^{T}Eu_{t\cdot}'u_{t\cdot}, \quad \Sigma_1^* = \frac{1}{T}\sum_{t<s}Eu_{t\cdot}'u_{s\cdot}.$$

Let us now partition the matrices $\Phi$, $\Psi$ conformably with the partition of $S_{t\cdot}$ and $u_{t\cdot}$, thus obtaining

$$\Phi = \begin{bmatrix} \Phi_{11} & \Phi_{12} & \Phi_{13} \\ \Phi_{21} & \Phi_{22} & \Phi_{23} \\ \Phi_{31} & \Phi_{32} & \Phi_{33} \end{bmatrix}, \quad \Psi = \begin{bmatrix} \Psi_{11} & \Psi_{12} & \Psi_{13} \\ \Psi_{21} & \Psi_{22} & \Psi_{23} \\ \Psi_{31} & \Psi_{32} & \Psi_{33} \end{bmatrix}. \tag{3.99}$$

Assuming, for the moment, that $Z$ is of the same dimension as $X$ we may obtain the simple instrumental variable (IV) estimator

$$\hat{b} = [I_q \otimes (Z'X)^{-1}Z']y = b + [I_q \otimes (Z'X)^{-1}](I_q \otimes Z')u_1, \tag{3.100}$$

whose limiting distribution depends only on the limiting distribution of

$$\left(\frac{Z'X}{T^2}\right)^{-1} \quad \text{and} \quad \frac{Z'U_1}{T},$$

since $(I_q \otimes Z')u_1 = \text{vec}(Z'U_1)$. In view of Eqs. (3.92) and (3.94) we conclude that

$$T(\hat{b} - b) \xrightarrow{d} \Phi_{32}^{-1}\text{vec}(\Psi_{31}), \tag{3.101}$$

on the assumption that $\Phi_{32}$ is a.c. **invertible**. While strictly speaking this condition is **not** implied by the fact that $\Phi > 0$, nonetheless this is the standard condition that defines instrumental variables, viz. the condition that they be "correlated" with the variables for which they serve as instruments. This argument is developed as follows. Define the sample version of the coefficient of vector correlation [see Hotelling (1936), or Dhrymes (1970)] by

$$\rho_{z,x} = \frac{|Z'X(X'X)^{-1}X'Z|}{|Z'Z|},$$

and note that, since in all this literature it is assumed that $X'X$ and $Z'Z$ are a.c. invertible, what we must require of instruments is that for $T > q$

$$A_T = \{\omega : |Z(\omega)'X(\omega)| = 0\} \quad \text{obeys} \quad \mathcal{P}(A^*) = 0, \tag{3.102}$$

where $A^* = \overline{\lim}_{T \to \infty} A_T$.

An alternative IV procedure given in Phillips and Hansen (1990) is one that involves two-stage least squares. In the first stage we regress $x_{t.}$ on $z_{t.}$, thereby obtaining the "predictions" $\tilde{x}_{t.}$. In the second stage we simply regress $y_{t.}$ on the "predictions" $\tilde{x}_{t.}$. In such a case the number of instruments, i.e. the **row dimension** of $Z$ must equal or exceed the number of "explanatory" variables, i.e. the **row dimension** of $X$. In this sense it is at least as restrictive a procedure as the simple IV procedure outlined above, and it is akin to the case of **overidentifying restrictions** in a simultaneous equations context. The resulting IV estimator, denoted by $\tilde{b}$ is given by

$$\tilde{b} = [I_q \otimes (\tilde{X}'\tilde{X})^{-1}\tilde{X}']y \tag{3.103}$$

$$= b + [I_q \otimes (X'Z(Z'Z)^{-1}Z'X)^{-1}]\text{vec}[(X'Z(Z'Z)^{-1}Z'U_1], $$

whose limiting distribution is also obtained easily from the results in Eqs. (3.92) and (3.94) above, as

$$T(\tilde{b} - b) \xrightarrow{d} [I_q \otimes (\Phi_{23}\Phi_{33}^{-1}\Phi_{32})^{-1}]\text{vec}[\Phi_{23}\Phi_{33}^{-1}\Psi_{31}]. \tag{3.104}$$

**Remark 7.** This is an opportune juncture to highlight the nature of the so-called spurious regression problem; the latter is a situation in which a set of variables, $y$, is regressed on a set of variables, $x$, producing "significant" results, **even though the two sets of variables are mutually independent**. We can illustrate this fact by obtaining an expression for the matrix $\Phi_{23}$ in Eq. (3.98), which corresponds to the limiting distribution of $Z'X/T^2$. If both processes were stationary and **if** the instruments, $Z$, were uncorrelated with the explanatory variables, $X$,

$$\frac{Z'X}{T} \xrightarrow{\text{d}} 0, \tag{3.105}$$

and, asymptotically, an estimator would not exist in view of the fact that $Z'X/T$ converges to zero in probability and hence in distribution.

Assuming independence among the error processes in Eq. (3.90), the covariance matrix $\Sigma_0$, defined above Eq. (3.98), is of the form

$$\Sigma_0 = \begin{bmatrix} \Sigma_0^{(1)} & 0 & 0 \\ 0 & \Sigma_0^{(2)} & 0 \\ 0 & 0 & \Sigma_0^{(3)} \end{bmatrix}, \quad \Sigma_0^{(i)} = T_i'T_i, \quad i = 1, 2, 3, \tag{3.106}$$

where $T_i$ is the (unique) matrix of the triangular decomposition of $\Sigma_0^{(i)}$ [see Proposition 58 in Dhrymes (1984), pp. 68-69]. Thus,

$$\Phi_{23} = T_2' \left( \int_0^1 B_{(2)}(s)' B_{(3)}(s) \, ds \right) T_3, \tag{3.107}$$

where $B_{(2)}$ is the SMBM corresponding to, or generated by, the explanatory variables ($X$), and $B_{(3)}$ is the SMBM generated by the instrumental variables ($Z$). Since the two sets of variables are mutually independent, evidently $B_{(2)}$ and $B_{(3)}$ are mutually independent $k$-element SMBM. It follows immediately that $\Phi_{23}$ is a.c. **nonsingular**. Thus, when $X$ and $Z$ are generated by $I(1)$ processes the entity analogous to that in the left member of Eq. (3.104) does **not converge to zero** in distribution; in fact, it converges to a **nonsingular matrix**! The reader should note, however, that

$$E\Phi_{23} = 0, \tag{3.108}$$

so that the lack of dependence between explanatory variables and instruments, or actually their mutual independence, is reflected in the expectation above.

This result should also caution the reader that a great deal of the "intuition" that has been acquired in the context of stationary process econometrics does not carry over, without alteration, to the context of nonstationary process econometrics.

# Chapter 4

# Cointegration I

## 4.1 Preliminaries

There are many ways in which one can motivate or explain cointegration; at this juncture it is convenient to present it as an alternative formulation of the problem examined in Chapter 3. As such it is simply a condition in which the regressors, or "explanatory" variables, of a GLM are **integrated processes but the errors are generated by a stationary process**. In the literature of econometrics, however, its origin and motivation are quite different.

In the literature of econometrics, the general notion of (and term) cointegration first occurs in Granger (1981), but was first formally stated in a comprehensive fashion in Engle and Granger (1987) (EG). They presented and motivated cointegration as an indication of a "long-run equilibrium" relationship. The major concept was that even though the elements of a $q$-element vector are each nonstationary processes, there may exist linear combination(s) of them which are stationary, and this was taken to be symptomatic of a long-run equilibrium relation. Upon reflection this is precisely the problem discussed in Chapter 3, where we considered the model $y_t - x_t.\beta = u_t$; if $(y_t, x_t.)$ is taken to be a $q$-element vector of **nonstationary** processes, and $u_t$ is **stationary** we have precisely the cointegration context of EG.

Moreover, in the literature of statistics there have been discussions of properties of parameter estimators in nonstationary autoregressive models since the fifties. The simple unit root problem investigated by Dickey and

Fuller (1979) is an outgrowth of that line of research. The model is

$$y_t = \lambda y_{t-1} + u_t,$$

where the error is an i.i.d. zero mean, finite variance process. Evidently, this represents an instance of cointegration; no matter what $\lambda$ is, a linear combination of the variables of the problem, viz. $y_t - \lambda y_{t-1}$, is stationary. When $\lambda = 1$, the linear combination $a_1 y_t + a_2 y_{t-1}$ is stationary, for $a_1 = 1$, $a_2 = -1$, while each of the two variables is **nonstationary**. Although this is a rather simple model, the same holds true for the general models (discussed in Chapter 3) that involve $I(1)$ "regressors", i.e. models of the form

$$y_{t.} = x_{t.}B + u_{t..}$$

Since the regressors are $I(1)$, the "dependent variables" $y_{t.}$ are **also** $I(1)$, and thus nonstationary. Yet the linear combination

$$z_{t.}B^* = u_{t.}, \quad z_{t.} = (y_{t.}, x_{t.}), \quad B^* = (I, \ -B')',$$

is evidently stationary; hence the variables in $z_{t.}$ are **cointegrated**. This, however, need not indicate any "long-run equilibrium" relation!

Finally, the reader ought to bear in mind the basic research strategy in this branch of the literature, which is to explore two fundamental issues: if we are given a sequence of $I(1)$ vectors, say $\{X_{t.} : t \in \mathcal{N}_+\}$, (1) what does it mean for them to be cointegrated, and (2) if they are cointegrated, what are the implications?

## 4.2    Integrated Processes

In the discussion of the previous section we referred to nonstationary processes in general terms; however, the econometric literature on this topic deals exclusively with a very special type of nonstationary process to be defined below.

**Definition 1.** Let $X = \{X_t : t \in \mathcal{N}_+\}$ be a stochastic sequence defined on the probability space $(\Omega, \mathcal{A}, \mathcal{P})$. The sequence $X$ is said to be **integrated**

**of order** $d$, denoted by $X \sim I(d)$, if and only if

$$Y = \{Y_t : t \in \mathcal{N}, \quad Y_t = (I - L)^d X_t\}, \tag{4.1}$$

is a stationary[1] sequence.

The definition is precisely the same for multivariate processes, i.e. a sequence of random elements (typically $q$-element vectors) $X$, as above, is $I(d)$ if and only if $Y = (I - L)^d X$ is a stationary process.

In this literature distributional assumptions are seldom explicitly stated and it is unclear just what is meant by "stationarity"; we generally mean "strict stationarity", although frequently **covariance stationarity** is an equally acceptable specification. Thus in Definition 1 the phrase "is a stationary sequence" could be replaced by "is a (square integrable) covariance stationary sequence". As a general rule when we use the term "stationary" we mean "strictly stationary", unless otherwise indicated.

**Example 1.** Consider the $MARMA(1, k)$ process $X = \{X_t^{'} : t \in \mathcal{N}\}$, such that

$$X_t^{'} = X_{t-1.}^{'} + \eta_t^{'}, \quad \eta_t^{'} = \sum_{j=0}^{k} A_j \epsilon_{t-j.}^{'}, \quad A_0 = I_q, \tag{4.2}$$

and the $\epsilon$-process is $MWN(\Sigma)$. If we wish to obtain the covariance matrix of the element $X_t.$, we may do so in two steps: First, we obtain an explicit representation of the latter in terms of the entities $\eta_t.$; second, given that representation and the autocovariance kernel of the $\eta$-process, we obtain the desired covariance matrix. Solving Eq. (4.2), with initial conditions $\epsilon_{-j.} = 0$

---

[1] The existence of second moments is not explicitly stated in discussions of integrated processes in the econometrics literature; it is, however, strongly implied by the context. We take this opportunity to note that when we speak of the covariance matrix of an integrated process we always mean, unless otherwise stated to the contrary, that such covariance matrices are **nonsingular**, i.e. we rule out **linear dependencies** among the elements of a sequence.

for $j \geq 0$, we find[2]

$$X_{t\cdot} = \sum_{j=1}^{t} \eta'_{j\cdot}. \tag{4.3}$$

We note that $\eta_{t\cdot} = \sum_{j=0}^{k} A_j \epsilon_{t-j}$. is a multivariate **covariance stationary** process; in fact, it is **strictly stationary as well** because $\epsilon$ is a $MWN(\Sigma)$ process. Since

$$\text{Cov}(X'_{t\cdot}) = \sum_{j=1}^{t} \sum_{s=1}^{t} E\eta'_{j\cdot}\eta_{s\cdot}.$$

the problem is reduced to evaluating the covariance kernel of the right member above. To do so, first note that

$$\text{Cov}(X'_{t\cdot}) = \sum_{j=1}^{t} E\eta'_{j\cdot}\eta_{j\cdot} + \sum_{j\neq j'} E\eta'_{j\cdot}\eta_{j'\cdot}.$$

The evaluation of the first term is straightforward; to evaluate the second, put $j' = j - \tau$, for $1 \leq \tau \leq k$ and note that

$$V(\tau) = E\eta'_{j\cdot}\eta_{j-\tau\cdot} = V(-\tau)' = [E\eta'_{j-\tau\cdot}\eta_{j\cdot}]', \tag{4.4}$$

and moreover that, for $|\tau| \geq k + 1$, $V(\tau) = 0$. When we enforce this condition, we have the representation

$$\text{Cov}(X'_{t\cdot}) = \sum_{j=1}^{t} E\eta'_{j\cdot}\eta_{j\cdot} + \sum_{\tau=1}^{t-1\wedge k} \sum_{j=\tau+1}^{t} \left( E\eta'_{j-\tau\cdot}\eta_{j\cdot} + E\eta'_{j\cdot}\eta_{j-\tau\cdot} \right),$$

$$= C_1(t) + C_2(t), \tag{4.5}$$

where the notation $t - 1 \wedge k$ means the minimum of $t - 1$ and $k$. (Evidently, when $t \geq k + 1$, this notation is unnecessary and we shall dispense with it.) It is interesting to note that while for $t \leq k$ the covariance matrix of $X_t$ changes essentially with $t$, for $t \geq k + 1$ the covariance matrix assumes the form $\Psi_{(0)} + \Psi_{(1)}t$, with **fixed matrices** $\Psi_{(0)}$ and $\Psi_{(1)}$, for all $t \geq k + 1$.[3]

---

[2] In the econometric literature, integrated processes are utilized subject to initial conditions, and it is a common practice to suppose that $X_0 = 0$, a.c., or, what amounts to the same thing, $\epsilon_{-j\cdot} = 0$, a.c., for at least $0 \leq j \leq k$. This is, essentially, the set of initial conditions imposed above. Notice, further, that if such initial conditions **are not imposed** the covariance matrix of the element in question **is not defined!**

[3] To the best of my knowledge this result has not appeared in the econometrics literature; moreover, its derivation is not trivial.

This can be shown by evaluating the expression in Eq. (4.5), which yields

$$\sum_{j=1}^{t} E\eta'_{j\cdot}\eta_{j\cdot} = \sum_{j=1}^{t} E[A_0\epsilon'_{j\cdot} + \ldots + A_k\epsilon'_{j-k\cdot}][A_0\epsilon_{j\cdot} + \ldots + A_k\epsilon'_{j-k\cdot}]'$$

$$= \sum_{r=0}^{k}\sum_{r'=0}^{k} A_r \sum_{j=1}^{t} E[\epsilon'_{j-r\cdot}\epsilon_{j-r'\cdot}]A_{r'}, \quad \text{or}$$

$$C_1(t) = \sum_{r=0}^{k}(t-r)A_r\Sigma A'_r. \tag{4.6}$$

$$\sum_{j=\tau+1}^{t} E\eta'_{j-\tau\cdot}\eta_{j\cdot} = \sum_{j=\tau+1}^{t} E[A_0\epsilon'_{j-\tau\cdot} + \ldots + A_k\epsilon'_{j-\tau-k\cdot}][A_0\epsilon_{j\cdot} + \ldots + A_k\epsilon'_{j-k\cdot}]'$$

$$= \sum_{r=0}^{k}\sum_{r'=0}^{k} A_r[\sum_{j=\tau+1}^{t} E(\epsilon'_{j-\tau-r\cdot}\epsilon_{j-r'\cdot})]A'_{r'}$$

$$= \sum_{r=0}^{k-\tau}(t-\tau-r)A_r\Sigma A'_{r+\tau}, \quad \text{or}$$

$$C_2(t) = \sum_{\tau=1}^{k}\sum_{j=\tau+1}^{t} \left(E\eta'_{j-\tau\cdot}\eta_{j\cdot} + E\eta'_{j\cdot}\eta_{j-\tau\cdot}\right)$$

$$= \sum_{\tau=1}^{k}\sum_{r=0}^{k-\tau}(t-\tau-r)(A_r\Sigma A'_{r+\tau} + A_{r+\tau}\Sigma A'_r)$$

$$= \sum_{s=1}^{k}(t-s)[\sum_{\tau=1}^{s}(A_{s-\tau})\Sigma A'_s + A_s\Sigma(\sum_{\tau=1}^{s} A'_{s-\tau})]. \tag{4.7}$$

The expression in Eq. (4.7) is obtained by putting $s = r + \tau$, $\tau = \tau$ and noting that the range of $s$ is 1 to $t - 1 \wedge k$; moreover, since in this context $\tau = s - r$ **and** $\tau \geq 1$ it follows that the range of $\tau$ is 1 to $s$. The covariance matrix we seek is simply the sum of the rightmost members of Eqs. (4.6) and (4.7). Putting

$$B_0 = A_0\Sigma A'_0 = \Sigma,$$

$$B_1 = A_1\Sigma A'_0 + A_0\Sigma A_1 + A_1\Sigma A'_1, \quad \text{or more generally}$$

$$B_s = A_s\Sigma(\sum_{i=0}^{s} A'_i) + (\sum_{i=0}^{s-1} A_i)\Sigma A'_s, \quad s = 2, 3, \ldots k,$$

we write the covariance matrix in question as

$$\mathrm{Cov}(X'_{t\cdot}) = \sum_{s=0}^{k}(t-s)B_s = (\sum_{s=0}^{k} B_s)t - \sum_{s=1}^{k} s B_s = \Psi_{(1,k)}t + \Psi_{(0,k)}. \qquad (4.8)$$

Can similar results be obtained when $k \to \infty$, i.e. when the forcing function is taken to be $MMA(\infty)$? Proceeding uncritically and letting $k \to \infty$, we obtain

$$\Psi_{(1,\infty)} = \sum_{j=0}^{\infty} B_j, \quad \Psi_{(0,\infty)} = -\sum_{j=1}^{\infty} j B_j, \qquad (4.9)$$

which raises the question of whether or not such matrices are well defined. If they are not, the $I(1)$ process in question **does not possess finite second moments, even if we impose the requisite initial conditions**, which would render this entire topic vacuous. Moreover, certain other results intrinsic to the discussion of cointegration will fail to hold.

It is easy to establish [4]

$$\Psi_{(0,k)} = S_{0,k}\Sigma(\sum_{j=1}^{k} j A_j)' + \sum_{r=1}^{k} S_{r,k}\Sigma S_{0,r-1},$$

$$\Psi_{(1,k)} = S_{0,k}\Sigma S'_{0,k}, \quad S_{j,k} = \sum_{i=j}^{k} A_i. \qquad (4.10)$$

Letting $k \to \infty$ we arrive at Assertion 1.

**Assertion 1.** Consider the $I(1)$ process

$$(I-L)X'_{t\cdot} = \eta'_{t\cdot}, \quad \eta'_{t\cdot} = \sum_{j=0}^{k} A_j \epsilon'_{t-j\cdot},$$

---

[4] To verify the expression above note that

$$\sum_{j=r}^{k} B_j = \sum_{j=r}^{k}\left( A_j\Sigma(\sum_{i=0}^{j} A'_i) + (\sum_{s=0}^{j-1} A_s)\Sigma A'_j \right) = \sum_{j=r}^{k}\sum_{i=0}^{j} A_j\Sigma A'_i + \sum_{j=r}^{k}(\sum_{s=0}^{j-1} A_s)\Sigma A'_j$$

$$= \sum_{i=r}^{k}(\sum_{j=i}^{k} A_j)\Sigma A'_i + \sum_{j=r}^{k}(\sum_{s=0}^{j-1} A_s)\Sigma A'_j + \sum_{i=0}^{r-1}\left(\sum_{j=r}^{k} A_j\right)\Sigma A'_i$$

$$= S_{0,k}\Sigma S'_{r,k} + S_{r,k}\Sigma S_{0,r-1}, \quad \text{where} \quad S_{j,k} = \sum_{i=j}^{k} A_i.$$

of the preceding discussion, subject to the initial conditions $\epsilon_{-j.} = 0$ for $j \geq 0$. A necessary and sufficient condition for it to have well defined second moments, as $k \to \infty$, is that the matrix series

$$\sum_{j=1}^{\infty} j A_j, \quad \text{and thus} \quad \sum_{j=0}^{\infty} A_j,$$

**converge absolutely.**

Proof: We note from the preceding discussion, particularly Eq. (4.10), that

$$\Psi_{(1,k)} = S_{0,k} \Sigma S_{0,k}', \quad \Psi_{(0,k)} = \sum_{r=1}^{k} \sum_{j=r}^{k} B_j = \Psi_{(01,k)} + \Psi_{(02,k)} \quad (4.11)$$

$$\Psi_{(01,k)} = \sum_{r=1}^{k} S_{0,k} \Sigma S_{r,k} = S_{0,k} \Sigma \sum_{r=1}^{k} (r A_r'), \quad \Psi_{(02,k)} = \sum_{r=1}^{k} S_{r,k} \Sigma S_{0,r-1},$$

and moreover that $\Psi_{(02,k)}$ is dominated by $\Psi_{(01,k)}$; thus, it will be sufficient to examine $\Psi_{(1,k)}$ and the first component of $\Psi_{(0,k)}$. Letting $k \to \infty$ we find

$$-\Psi_{(01,\infty)} = S_{0,\infty} \Sigma \left( \sum_{j=1}^{\infty} j A_j \right)', \quad \Psi_{(1,\infty)} = S_{0,\infty} \Sigma S_{0,\infty}',$$

and the result is immediate.

**Corollary 1.** Under the conditions of Assertion 1, if $\sum_{j=1}^{\infty} j A_j$ **converges absolutely**, $\Psi_{(0,\infty)}$ and $\Psi_{(1,\infty)}$ are well defined.

**Remark 1.** Before we take up the discussion of the covariance matrix of $I(2)$ processes, we (1) clarify the role played by initial conditions in the case of an $I(1)$ process whose forcing function is $MMA(\infty)$; (2) establish the nature of the argument in passing to the limit for the case where the forcing function is $MMA(\infty)$ vis-a-vis the case where it is $MMA(k)$ and **show that the previous development is in error**; and finally (3) obtain the proper representation of the covariance matrix for an $I(1)$ process with $MMA(\infty)$ forcing function, and indicate the nature of the problem this creates in the study of cointegration.

To this end, consider the $MARMA(1, \infty)$ process $X = \{X_t : t \in \mathcal{N}_+\}$, such that

$$X_t = X_{t-1} + \sum_{j=0}^{\infty} A_j \epsilon_{t-j}', \quad A_0 = I_q, \quad \sum_{j=0}^{\infty} \| A_j \| < \infty,$$

and the $\epsilon$-process is $MWN(\Sigma)$. No argument advanced earlier is invalidated by this generalization but, as pointed out above, we cannot arrive at the desired result simply by letting $k \to \infty$ in Eq. (4.8). This is so because in the representation of Eq. (4.8) **it is implicitly assumed** that $t \geq k + 1$. Thus, letting $k \to \infty$ necessarily means letting $t \to \infty$ as well.

Returning to the original derivation, consider $X_{t.}' = \sum_{j=1}^{t} \eta_{j.}'$, $\eta_{j.}' = \sum_{s=0}^{\infty} A_s \epsilon_{j-s.}'$. In contrast to the previous context, **no matrix $A_s$ is known a priori to be zero**. Thus,

$$\mathrm{Cov}(X_{t.}') = \sum_{j=1}^{t} E\eta_{j.}'\eta_{j.} + \sum_{j \neq j'} E\eta_{j.}'\eta_{j'.} = C_1(t) + C_2(t),$$

and

$$C_1(t) = \sum_{j=1}^{t}\sum_{s=0}^{\infty}\sum_{s'=0}^{\infty} A_s \left( E\epsilon_{j-s.}'\epsilon_{j-s'.} \right) A_{s'}', \quad \text{nonnull only for } s = s'$$

$$= \sum_{j=1}^{t}\sum_{s=0}^{j-1} A_s \left( E\epsilon_{j-s.}'\epsilon_{j-s.} \right) A_s', \quad \text{changing the order of summation}$$

$$= \sum_{s=0}^{t-1}\sum_{j=s+1}^{t} A_s \left( E\epsilon_{j-s.}'\epsilon_{j-s.} \right) A_s' \;=\; \sum_{s=0}^{t-1}(t - s)A_s \Sigma A_s'.$$

Moreover,

$$C_2(t) = \sum_{j=2}^{t}\sum_{\tau=1}^{j-1} E\left( \eta_{j-\tau.}'\eta_{j.} + \eta_{j.}'\eta_{j-\tau.} \right) = C_{21}(t) + C_{22}(t);$$

Since $C_{22}(t) = C_{21}(t)'$ we need only evaluate $C_{21}(t)$. Thus,

$$C_{21}(t) = \sum_{j=2}^{t}\sum_{\tau=1}^{j-1}\sum_{s=0}^{\infty}\sum_{s'=0}^{\infty} A_s \left( E\epsilon_{j-\tau-s.}'\epsilon_{j-s'.} \right) A_{s'}', \quad \text{nonnull for } s' = s + \tau$$

$$= \sum_{j=2}^{t}\sum_{\tau=1}^{j-1}\sum_{s=0}^{\infty} A_s \left( E\epsilon_{j-\tau-s.}'\epsilon_{j-\tau-s.} \right) A_{s+\tau}', \quad \text{put } \tau + s = i, \; \tau = \tau$$

$$= \sum_{j=2}^{t}\sum_{i=1}^{j-1}\sum_{\tau=1}^{i} A_{i-\tau} \left( E\epsilon_{j-i.}'\epsilon_{j-i.} \right) A_i', \quad \text{change order of summation}$$

$$= \sum_{i=1}^{t-1} S_{0,i-1} \sum_{j=i+1}^{t} \left( E\epsilon_{j-i.}'\epsilon_{j-i.} \right) A_i' \;=\; \sum_{i=1}^{t-1}(t - i)S_{0,i-1}\Sigma A_i'.$$

It follows, therefore, that

$$C(t) = C_1(t) + C_2(t) = \sum_{j=0}^{t-1}(t-j)B_j = \Psi_{(0,t)} + \Psi_{(1,t)}t,$$

where now

$$-\Psi_{(0,t)} = \sum_{j=0}^{t-1} jB_j, \quad \Psi_{(1,t)} = \sum_{j=0}^{t-1} B_j.$$

Comparing this result to Eqs. (4.8) and (4.10), we see that they are basically of the same form, except that the upper limit of the sums is not truncated at $k$. Thus, **we cannot let** $k \to \infty$ as **an independent activity**. This is due to the nature of the initial conditions we have imposed.

Finally, we have the representation

$$\Psi_{(1,t)} = S_{0,t-1} \Sigma S'_{0,t-1}, \quad -\Psi_{(0,t)} = S_{t-1} \Sigma \left(\sum_{r=1}^{t-1} r A_r\right)' + \sum_{r=1}^{t-1} S_{r,t-1} \Sigma S'_{0,r-1}$$

or, alternatively,

$$-\Psi_{(0,t)} = \sum_{r=1}^{t-1} \left( S_{0,t-1} \Sigma S'_{0,t-1} - S_{0,r-1} \Sigma S'_{0,r-1} \right).$$

In the first representation of $\Psi_{0,t}$, we conclude that the two conditions given in Assertion 1 remain valid in this case as well. This so since the mechansim by which the covariance matrix of $X_t$. becomes unbounded is $t \to \infty$, just as in the case where the forcing function is $MMA(k)$. In the second alternative, the covariance matrix becomes

$$\text{Cov}(X'_{t\cdot}) = S_{0,t-1} \Sigma S'_{0,t-1} + \sum_{r=1}^{t-1} S_{0,r-1} \Sigma S'_{0,r-1}.$$

By assumption, the first matrix converges absolutely; thus, we have to rely on the second term to produce the unboundedness, which of course it does. Note that the first term is simply the last term of the partial sum $\sum_{r=0}^{t-1} S_{0,r} \Sigma S'_{0,r}$, whose limit (as $t \to \infty$) is the covariance matrix of $X_t$. as $t \to \infty$.

To avoid such conceptual complexities we generally characterize **integrated processes in terms of forcing functions which are** $MMA(k)$, but we return to these issues later when we examine the cointegration hypothesis.

**Example 2.** Consider the $MARMA(2, k)$ process $X = \{X_t. : t \in \mathcal{N}_+\}$, given by

$$X_t' = 2X_{t-1}' - X_{t-2}' + \sum_{j=0}^{k} A_j \epsilon_{t-j}', \quad A_0 = I_q, \quad \text{Cov}(\epsilon_t') = \Sigma. \tag{4.12}$$

Noting that $X_t. - 2X_{t-1}. + X_{t-2}. = (I - L)^2 X_t.$, we conclude that $X$ is $I(2)$. Moreover, if we assume the initial conditions $X_0. = X_{-1}. = 0$, we may determine the behavior of the covariance matrix of this $I(2)$ sequence by the same method as above, except that here the first step requires a bit more of an explanation.

We begin with the formal expansion [5]

$$(1 - z)^\alpha \sim 1 + (-1)\frac{\alpha}{1!}z + (-1)^2\frac{\alpha(\alpha - 1)}{2!}z^2 + (-1)^3\frac{\alpha(\alpha - 1)(\alpha - 2)}{3!}z^3$$

$$+ \ldots + (-1)^r\frac{\alpha(\alpha - 1)(\alpha - 2)\ldots(\alpha - (r - 1))}{r!}z^r + \ldots.$$

While it is true that this expansion does not converge for $|z| \geq 1$, the coefficients of the various powers of $z$ are valid in the following sense: If we invoke the isomorphism noted in Chapter 1 [see also Dhrymes (1981)], the expansion of the operator $(I - L)^{-2}$ becomes, for $r = t - 1$,

$$(I - L)^{-2} \sim \sum_{s=0}^{t-1}(s + 1)L^s + \ldots = I + 2L + 3L^2 + 4L^3 + \ldots + tL^{t-1} + \ldots,$$

which gives the representation

$$X_t' = \sum_{s=0}^{t-1}(s + 1)L^s\eta_t' = \sum_{s=0}^{t-1}(s + 1)\eta_{t-s}', \quad \eta_t' = \sum_{j=0}^{k} A_j \epsilon_{t-j}'. \tag{4.13}$$

The reader may readily verify (up to $t$ terms) the representation above by solving the difference equation recursively, taking as initial conditions $X_0. = X_{-1}. = 0$. This is the sense in which the partial expansion above is valid.

---

[5] Note that since the argument employed is valid for any $\alpha$, it is a rather trivial matter to deal with fractional differencing, i.e. to examine sequences of the form $(I - L)^\alpha X_t' = \eta_t'$, where $\alpha$ is a fraction, proper or improper.

We can simplify the expression in Eq. (4.13) by putting $j = t - s$ and thus rewriting the representation of the stochastic sequence as

$$X'_{t\cdot} = \sum_{j=1}^{t}(t + 1 - j)\eta_{j\cdot}. \tag{4.14}$$

From Eq. (4.4) $V(\tau) = 0$ for $\tau > k$; thus, we conclude

$$\text{Cov}(X'_{t\cdot}) = C_1(t) + C_2(t),$$

$$C_1(t) = \sum_{j=1}^{t}(t + 1 - j)^2 E\eta'_{j\cdot}\eta_{j\cdot}, \tag{4.15}$$

$$C_2(t) = \sum_{\tau=1}^{k}\sum_{j=\tau+1}^{t}[(t + 1 - j)^2 + \tau(t + 1 - j)]E\left(\eta'_{j-\tau\cdot}\eta_{j\cdot} + \eta'_{j\cdot}\eta_{j-\tau\cdot}\right).$$

Moreover, using the equations immediately preceding and immediately following Eq. (4.6), we obtain the more informative representation

$$C_1(t) = \sum_{r=0}^{k}\sum_{r'=0}^{k} A_r[\sum_{j=1}^{t}(t + 1 - j)^2 E(\epsilon'_{j-r\cdot}\epsilon_{j-r'\cdot})]A_{r'}$$

$$= \sum_{r=0}^{k}\left(\sum_{j=1}^{t-r}j^2\right) A_r\Sigma A'_r, \tag{4.16}$$

$$\phi_{r,\tau} = \sum_{j=r+\tau+1}^{t}[(t + 1 - j)^2 + \tau(t + 1 - j)] = \sum_{j=1}^{t-r-\tau}j^2 + \tau\sum_{j=1}^{t-r-\tau}j,$$

$$C_2(t) = \sum_{\tau=1}^{k}\left[\sum_{r=0}^{k-\tau}\left(\phi_{r,\tau}(A_{r+\tau}\Sigma A'_r + A_r\Sigma A'_{r+\tau})\right)\right], \tag{4.17}$$

which completes the derivation of the covariance matrix of an element of an $I(2)$ process with forcing function $MMA(k)$.

The equation set in Eq. (4.17) and the one preceding it may be easily justified in view of the equation following Eq. (4.6).

While the formal derivation is now complete, the representation above is not particularly informative. To gain some insight into its structure we note that

$$\phi_{r,\tau} = \alpha_{r+\tau} + \tau\beta_{r+\tau}, \tag{4.18}$$

$$\alpha_{r+\tau} = \frac{(t-r-\tau)(t-r-\tau+1)(2t-2r-2\tau+1)}{6},$$

$$\beta_{r+\tau} = \frac{(t-r-\tau)(t-r-\tau+1)}{2},$$

$$\alpha_r = \frac{(t-r)(t-r+1)(2t-2r+1)}{6}, \qquad (4.19)$$

and, consequently, that

$$C_1(t) = \sum_{r=0}^{k} \alpha_r A_r \Sigma A_r',$$

$$C_2(t) = \sum_{\tau=1}^{k}\sum_{r=0}^{k-\tau} \alpha_{r+\tau}(A_{r+\tau}\Sigma A_r' + A_r\Sigma A_{r+\tau}')$$

$$+ \sum_{\tau=1}^{k} \tau \left( \sum_{r=0}^{k-\tau} \beta_{r+\tau}(A_{r+\tau}\Sigma A_r' + A_r\Sigma A_{r+\tau}') \right). \qquad (4.20)$$

Making the change in variable $j = r + \tau$, $\tau = \tau$, we note that the range of the two indices is given by $1 \le j \le k$, $1 \le \tau \le j$, due to the fact that $r = j - \tau$. In this notation we obtain the representation

$$C_1(t) = \sum_{r=0}^{k} \alpha_r A_r \Sigma A_r', \qquad (4.21)$$

$$C_2(t) = \sum_{j=1}^{k} \alpha_j(A_j\Sigma S_{0,j-1}' + S_{0,j-1}\Sigma A_j') + \sum_{j=1}^{k} \beta_j(A_j\Sigma P_{j-1}' + P_{j-1}\Sigma A_j'),$$

$$S_{i,j} = \sum_{s=i}^{j} A_s, \quad P_{j-1} = \sum_{s=0}^{j-1}(j-s)A_s, \quad \text{so that}$$

$$C(t) = \sum_{j=0}^{k} \alpha_j B_j + \sum_{j=1}^{k} \beta_j(A_j\Sigma P_{j-1}' + P_{j-1}\Sigma A_j'). \qquad (4.22)$$

**Example 3.** Here we generalize the discussion above to an arbitrary integrated sequence of order $d$. In the expansion of Example 2, take $\alpha = -d$ and note that the formal expansion therein, terminated at the term for $z^{t-1}$

$$(1-z)^{-d} \sim \sum_{s=0}^{t-1} \binom{d-1+s}{s} z^s, \qquad (4.23)$$

which implies

$$X'_{t\cdot} = \sum_{s=0}^{t-1} \binom{d-1+s}{s} \eta'_{t-s\cdot}; \tag{4.24}$$

the notation $\binom{n}{k}$ represents the binomial coefficient—the number of ways one can choose $k$ out of $n$ distinct objects. Making the change in variable $j = t - s$, we have the canonical representation

$$X'_{t\cdot} = \sum_{j=1}^{t} \binom{d-1+t-j}{t-j} \eta'_{j\cdot}, \tag{4.25}$$

and we can employ the apparatus of the previous examples to obtain a representation of its covariance matrix. By definition

$$\mathrm{Cov}(X'_{t\cdot}) = \sum_{j=1}^{t} \sum_{j'=1}^{t} \binom{d-1+t-j}{t-j}\binom{d-1+t-j'}{t-j} E\eta'_j \eta_{j'}.$$

$$= C_1(t) + C_2(t),$$

$$C_1(t) = \sum_{j=1}^{t} \binom{d-1+t-j}{t-j}^2 E\eta'_j \eta_j, \tag{4.26}$$

$$C_2(t) = \sum_{\tau=1}^{k} \sum_{j=\tau+1}^{t} \gamma_{j,\tau} E(\eta'_{j-\tau\cdot}\eta_{j\cdot} + \eta'_{j\cdot}\eta_{j-\tau\cdot}), \tag{4.27}$$

$$\gamma_{j,\tau} = \binom{d-1+t+\tau-j}{t+\tau-j}\binom{d-1+t-j}{t-j}.$$

Using the results derived immediately above and immediately below Eq. (4.6), we easily establish

$$C_1(t) = \sum_{r=0}^{k} \sum_{r'=0}^{k} A_r \left[ \sum_{j=1}^{t} \binom{d-1+t-j}{t-j}^2 E(\epsilon'_{j-r\cdot}\epsilon_{j-r'\cdot}) \right] A'_{r'} \tag{4.28}$$

$$= \sum_{r=0}^{k} \sum_{s=0}^{t-r-1} \binom{d-1+s}{s} A_r \Sigma A'_r,$$

$$\phi_{r,\tau} = \sum_{j=\tau+r+1}^{t} \binom{d-1+t+\tau-j}{t+\tau-j}\binom{d-1+t-j}{t-j}$$

$$= \sum_{s=0}^{t-\tau-r-1} \binom{d-1+s}{s}\binom{d-1+\tau+s}{\tau+s}, \tag{4.29}$$

$$C_2(t) = \sum_{\tau}^{k} \sum_{r=0}^{k-\tau} \phi_{r,\tau}(A_{r+\tau}\Sigma A_r' + A_r\Sigma A_{r+\tau}'). \tag{4.30}$$

To delineate precisely the nature of the covariance matrix of an $I(d)$, process we need to be able to evaluate sums of powers of the first $n$ integers, as is made very clear from Eq. (4.29). This is done in the appendix to this chapter. We shall simply recall such results as we need them.

We conclude this discussion with Assertion 2.

**Assertion 2.** Let $X = \{X_t : t \in \mathcal{N}_+\}$ be a multivariate $I(d)$ process with suitable initial conditions. If $\mathrm{Cov}(X_{t.}') = \Psi(t)$ then

$$\Psi(t) = \sum_{j=0}^{2d-1} \Psi_i t^i, \tag{4.31}$$

where $\Psi_i$, $0 \le i \le 2d - 1$, are square matrices of order $q$.

Proof: From Eq. (4.29) we have

$$\phi_{r,\tau} = \left(\frac{1}{(d-1)!}\right)^2 \sum_{s=0}^{t-r-\tau-1} \prod_{i=1}^{d-1} [(s+i)(s+\tau+i)], \tag{4.32}$$

which is the sum of a product of $d - 1$ terms, each containing a quadratic term. Hence, it is the sum of $2(d-1)$ powers of integers. By the discussion in the appendix, and particularly Eq. (4.125), it follows that $\phi_{r,\tau}$ contains at least one term which is of degree $2d - 1$.

$$\text{q.e.d.}$$

**Definition 2.** Let $X$ be a multivariate $I(d)$ process as in Definition 1; the components of the random elements $X_{t.}$ are said to be **cointegrated of order** $d, b$, $b \le d$, denoted by $X \sim CI(d,b)$, if and only if there exists a nontrivial vector $\beta$ such that

$$X_{t.}\beta = z_t \sim I(d - b). \tag{4.33}$$

It is said to be **cointegrated of order** $d, b$, and of **cointegrating rank** $r$, $r \le q$, denoted by $CI(d, b, r)$, if and only if there exist exactly $r$ **linearly independent vectors**, say $\beta_{.i}$, such that for $B = (\beta_{.1}, \beta_{.2}, \ldots, \beta_{.r})$

$$X_{t.}B = Z_{t.} \sim I(d-b), \quad Z_{t.} = (z_{ti}), \quad z_{ti} = X_{t.}\beta_{.i}, \quad i = 1, 2, 3, \ldots, r. \tag{4.34}$$

The vector $Z_t$, for want of an appropriate designation in the literature, is henceforth referred to as the **cointegral vector**.

**Remark 2**. In the literature of statistics, as we pointed out in Chapter 3, processes which are individually nonstationary, but a linear combination of which is stationary, are said to be **partially nonstationary** . While there is merit in this terminology, the designation "partially nonstationary" is not sufficiently flexible; for example, a (vector) process which is $CI(d, b, r)$ with $d > b$ is certainly nonstationary but it is **not partially nonstationary**. On the other hand it is certainly cointegrated, in the sense that there exist linear combinations of its elements which are of **a lower order of non-stationarity**. Thus, the cointegration terminology, which originated in the literature of econometrics is to be preferred, and is generally adhered to in our discussion.

In parts of the literature of econometrics, the concept of cointegration is linked to the "long-run equilibrium" relationship among economic (mostly macro) variables. While this could possibly be an explanation as to **why** certain variables may be cointegrated, the concept of cointegration is basically a mathematical and, more precisely, a probabilistic one. In attempting to elucidate the implication of this relationship it is **the mathematical-probabilistic** aspects of the definition that are paramount. Use of only a loosely invoked economic rationalization to determine or judge the existence of cointegration beclouds the issue and could lead to the acceptance of propositions that seem reasonable from an economic theoretic point of view, but cannot be supported by a mathematical argument. In our discussion, we deal solely with the probabilistic aspects of the concept.

To gain some intuition about cointegration, it is useful to compare it with **collinearity**, or **linear dependency**. If we rank the "randomness" of an entity by the magnitude of its variability, say its variance, a zero mean random vector, say $z$, with nonsingular covariance matrix is not "degraded" by a **nontrivial, full rank, linear transformation**. For example, if $A$ is nonsingular, $Az$ still has a nonsingular covariance matrix. If $A_k$ is $k \times q$ of rank $k$, $A_k z$ has a nonsingular covariance matrix as well. On the other hand, **if the covariance matrix is singular** there exists at least one vector, say $\beta$ such that $\mathrm{Var}(\beta'z) = 0$; when this is so, the elements of $z$ are said to be **collinear**, or to exhibit **linear dependencies**, so that the degree of "randomness" of $z$ has been "degraded". The concept of cointegration is similar. For example, if $X$ is $CI(2,1)$, of (cointegrating) rank 1, it means that the components of the vector $X_t$ are each $I(2)$, but there exists a nontrivial vector, say $\beta$, which degrades the randomness of $X_t$. to $I(1)$, since $X_t.\beta \sim I(1)$. Similarly if $X$ is $CI(2,2)$ of (cointegrating) rank 1, there exists a vector, say $\gamma$, such that $X_t.\gamma \sim I(0)$, i.e. the linear combination in question is a square integrable stationary process.

## 4.3   Characterization of Cointegration

In the preceding sections we gave a definition of cointegration; the definition is typically a simple, as well as an easily grasped and extended, statement of the concept we wish to define. In typical mathematical discussions, the definition is quickly followed by an operational criterion which gives succinctly a relatively simple procedure (criterion) by which one may decide whether a given entity conforms to the requirements of the definition. For example, one may define the (column/row) rank of a matrix as the number of **linearly independent** (columns/rows) it contains. This is a simple and easily grasped notion, but does not immediately give rise to a relatively simple procedure for determining the rank of a matrix. In the case of **square matrices** (of order $n$), we have the characterization that such a matrix is of rank $n$ if and only if its determinant is nonnull.

Such a characterization is absent from the econometric literature of cointegration, where much of the published work is devoted to establishing the validity of certain **implications** of cointegration. Our task in this section is to establish a characterization of cointegration. We begin with Proposition 1.

**Proposition 1.** Let $X$ be a (multivariate) $I(d)$ process as in Definition 2;

$X$ is $CI(d, d, r)$ if and only if[6]

$$\mathrm{Cov}(X_{t\cdot}') = \Psi(t) = \Psi_0 + \Psi_1(t), \tag{4.35}$$

such that

i. $\Psi_0$ does not depend on $t$, $\mathrm{rank}(\Psi_0) \geq r$ and the intersection of the null space of $\Psi_0$ and $\Psi_1(t)$ consists only of the zero vector;

ii. $\mathrm{rank}(\Psi_1(t)) = q - r$, i.e. there exists an appropriately dimensioned matrix $B$ of rank, at most, $r$ such that $\Psi_1(t)B = 0$.

Proof: Since for **fixed** $t$, given the initial conditions imposed in the econometric literature, the covariance matrix of $X_{t\cdot}$ is bounded, while it becomes unbounded with $t$, we may without loss of generality write

$$\Psi(t) = \Psi_0 + \Psi_1(t).$$

Necessity: If $X$ is $CI(d, d, r)$, there exists a matrix $B$ of rank $r$ such that for $Z_{t\cdot} = X_{t\cdot}B$, $\mathrm{Cov}(Z_{t\cdot}') = B'\Psi_0 B + B'\Psi_1(t)B = B'\Psi_0 B$.

The expression above defines $Z_{t\cdot}$ as a square integrable stationary process only if $B'\Psi_1(t)B = 0$ and $B'\Psi_0 B > 0$.[7] Thus, $\Psi_1(t)$ is of rank $q - r$ and, since $B$ spans the null space of $\Psi_1(t)$, the intersection of the latter and the null space of $\Psi_0$ consists only of the null vector, which completes the proof of necessity.

Sufficiency: Suppose conditions i and ii are satisfied; then, there exists a suitably dimensioned matrix $B$, of rank at most $r$, such that $Z_{t\cdot} = X_{t\cdot}B$ is a stationary process with fixed covariance matrix $B'\Psi_0 B > 0$, or equivalently $Z_{t\cdot} \sim I(0)$.

q.e.d.

---

[6] Strictly speaking, we should also include a condition on $EX_{t+\tau}'.X_{t\cdot} = \Psi(t + \tau, t) = \Psi_2(\tau) + \Psi_3(t, \tau)$. Frequently, the condition of the proposition having being verified, it is then possible to find an explicit description of the cointegral vector, $Z_{t\cdot}$, and thus establish directly that $EZ_{t+\tau}'.Z_{t\cdot} = C_3(\tau)$. We do so, in Remark 3.

[7] This is another instance in which the cointegration literature is ambiguous. In our characterization we require that $B'\Psi_0 B > 0$; otherwise cointegration becomes **precisely** collinearity. Of course, collinearity is a trivial instance of cointegration.

**Example 4**. Consider the $I(1)$ process $X$ of Example 1, with a forcing function $\eta'_{t.} = \sum_{j=0}^{k} A_j \epsilon'_{t-j}.$. Its covariance matrix is given by

$$\text{Cov}(X'_{t.}) = \Psi_0 + \Psi_1 t, \quad -\Psi_0 = \sum_{s=1}^{k} s B_s, \quad \Psi_1 = \sum_{s=0}^{k} B_s.$$

From the equation just above Eq. (4.8) and using the $S$-notation of Eq. (4.10), we find

$$B_j = A_j \Sigma(\sum_{i=0}^{j} A'_i) + (\sum_{i=0}^{j-1} A_i)\Sigma A'_j = S_{0,j}\Sigma S'_{0,j} - S_{0,j-1}\Sigma S'_{0,j-1}. \qquad (4.36)$$

Since

$$\sum_{j=0}^{k} j B_j = \sum_{r=1}^{k}\sum_{j=r}^{k} B_j, \quad \sum_{j=r}^{k} B_j = S_{0,k}\Sigma S'_{0,k} - S_{0,r-1}\Sigma S'_{0,r-1},$$

we conclude

$$\sum_{j=1}^{k} j B_j = \sum_{r=1}^{k} \left( S_{0,k}\Sigma S'_{0,k} - S_{0,r-1}\Sigma S'_{0,r-1} \right) = k S_{0,k}\Sigma S'_{0,k} - \sum_{r=1}^{k} S_{0,r-1}\Sigma S'_{0,r-1}. \qquad (4.37)$$

Consequently, we have

$$\text{Cov}(X'_{t.}) = (t-k)S_{0,k}\Sigma S'_{0,k} + \sum_{r=1}^{k} S_{0,r-1}\Sigma S'_{0,r-1}. \qquad (4.38)$$

Two important implications arise from Eq. (4.38). First, for the $I(1)$ process above to be $CI(1,1,r)$ there must exist a suitably dimensioned matrix $B$ such that $S'_{0,k}B = 0$ **and** $S'_{0,j}B \neq 0$ for at least one $j = 0, 1, 3, \ldots k-1$.

Second, the process $X$ **cannot possibly be cointegrated** if $k = 0$. This reduces the appeal of referring to cointegration as the property of long-run equilibrium among economic variables since, in principle, there is nothing in economics that precludes equilibrium relations among two series, each of which is a random walk.

**Remark 3**. From the representation of Eq. (4.38) we may infer that the cointegral vector is a **linear transformation** of a process like

$$\xi'_{t.} = \sum_{r=0}^{k-1} S_{0,r}\zeta'_{t-r.}, \quad \text{or} \quad \xi'_{t.} = \sum_{r=1}^{k} S_{r,k}\zeta^{*'}_{t-r.}, \qquad (4.39)$$

where $\zeta$ ( or $\zeta^*$ ) is a $MWN(\Sigma)$; this is so since

$$S_{r,k} = S_{0,k} - S_{0,r-1} \quad \text{and} \quad B' S_{0,k} = 0.$$

We further note that

$$\sum_{r=1}^{k} S_{r,k} = \sum_{j=1}^{k} j A_j. \tag{4.40}$$

For the cointegral vector to be well-defined, as $k \to \infty$, and to have well defined second moments, we require that

$$\lim_{k \to \infty} \sum_{j=1}^{k} j A_j$$

be absolutely summable. A series is **absolutely summable** if and only if for any element $(r, s)$, $\sum_{j=1}^{\infty} |a_{r,s}^{(j)}| < \infty$, which is a condition given in Assertion 1.

As we had pointed out in Remark 1, the operation of letting $k \to \infty$ in the context above, although similar, is not equivalent to assuming **initially** that the forcing function is $MMA(\infty)$. For such a process, it is just not possible to show cointegration since no fixed linear combination of the elements of $X_t$. can possibly have a constant covariance matrix.

Now, with fixed $k$, it is clear that the cointegral vector is **strictly** and, thus, also **covariance stationary** since

$$Z_{t\cdot} = \xi_t.B, \quad \text{Cov}(Z'_{t\cdot}) = B' \left( \sum_{r=0}^{k-1} S_{0,r} \Sigma S'_{0,r} \right) B, \quad \text{Cov}(Z'_{t+\tau\cdot}, Z'_{t\cdot}) = \sum_{r=0}^{k} S_{0,r} \Sigma S'_{0,r-\tau}$$

and $\zeta$ or $\zeta^*$ is $MWN(\Sigma)$.

**Example 5**. We have already established in Example 2 that an $I(2)$ process, has the covariance matrix

$$\text{Cov}(X'_{t\cdot}) = \sum_{j=0}^{k} \alpha_j B_j + \sum_{j=1}^{k} \beta_j (A_j \Sigma P'_{j-1} + P_{j-1} \Sigma A'_j). \tag{4.41}$$

This representation follows from the definitions in Eq. (4.15), adding the first and second equations of Eq. (4.20), and noting that $S_{0,j-1} = 0$ for $j < 1$.

Since from Eq. (4.36)

$$B_j = S_{0,j} \Sigma S_{0,j} - (S_{0,j-1} \Sigma S'_{0,j} - S_{0,j-1} \Sigma A'_j) = S_{0,j} \Sigma S'_{0,j} - S_{0,j-1} \Sigma S'_{0,j-1} \tag{4.42}$$

and

$$\alpha_j - \alpha_{j+1} = (t - j)^2, \quad 0 \le j \le k - 1, \tag{4.43}$$

we can rewrite the covariance matrix of Eq. (4.41) as

$$\mathrm{Cov}(X'_{t.}) = [\alpha_k - (t - k)^2] S_{0,k} \Sigma S'_{0,k} + \sum_{j=0}^{k-1} (t - j)^2 S_{0,j} \Sigma S'_{0,j} \tag{4.44}$$

$$+ \sum_{j=0}^{k} (t - j)^2 \left( S_{0,j} \Sigma S'_{0,j} + \frac{A_j \Sigma P'_{j-1} + P_{j-1} \Sigma A'_j}{2} \right)$$

$$+ \sum_{j=1}^{k} (t - j) \frac{A_j \Sigma P'_{j-1} + P_{j-1} \Sigma A'_j}{2},$$

it being understood that $P_{j-1} = 0$, for $j < 1$.

We now give a representation of the covariance matrix of $I(2)$ processes with forcing function $MMA(k)$, for $k = 0, 1, 2$.

For $k = 0$ we have

$$\mathrm{Cov}(X'_{t.}) = \alpha_0 A_0 \Sigma A_0 = \frac{t(t + 1)(2t + 1)}{6} \Sigma, \tag{4.45}$$

and hence there is **no cointegration except in the form of collinearity**.

For $k = 1$ we find

$$\mathrm{Cov}(X'_{t.}) = \alpha_0 B_0 + \alpha_1 B_1 + \beta_1 (A_1 \Sigma P'_0 + P_0 \Sigma A'_1) \tag{4.46}$$

$$= \frac{t(t + 1)(t - 1)}{3} S_{0,1} \Sigma S'_{0,1} + \frac{t(t + 1)}{2} A_0 \Sigma A'_0 - \frac{t(t - 1)}{2} A_1 \Sigma A'_1,$$

and again there is no cointegration except in the form of collinearity.

For $k = 2$ we obtain from Eq. (4.41)

$$\mathrm{Cov}(X'_{t.}) = \alpha_0 B_0 + \alpha_1 B_1 + \alpha_2 B_2 + \beta_1 (A_1 \Sigma A'_0 + A_0 \Sigma A'_1) \tag{4.47}$$

$$+ \beta_2 (A_1 \Sigma A'_2 + A_2 \Sigma A'_1 + 2 A_0 \Sigma A'_2 + 2 A_2 \Sigma A'_0).$$

To simplify the equation above we note that

$$A_1 \Sigma A'_0 + A_0 \Sigma A'_1 = B_1 - A_1 \Sigma A'_1 \tag{4.48}$$

and moreover that the **coefficient of** $\beta_2$ is given by

$$H = S_{0,2} \Sigma S'_{0,2} - S_{0,1} \Sigma S'_{0,1} + (A_0 + A_1) \Sigma (A_0 + A_1)' - 2 A_2 \Sigma A'_2 - A_0 \Sigma A'_0. \tag{4.49}$$

Noting that $B_2 = S_{0,2}\Sigma S'_{0,2} - S_{0,1}\Sigma S'_{0,1}$, the covariance matrix of Eq. (4.47) may be written as

$$
\begin{aligned}
\operatorname{Cov}(X'_{t.}) &= \alpha_0 B_0 + \alpha_1 B_1 + (\alpha_2 + \beta_2) B_2 + \beta_1 (B_1 - A_1 \Sigma A'_1) \\
&\quad + \beta_2 (A_0 + A_2)\Sigma(A_0 + A_1)' - 2\beta_2 A_2 \Sigma A'_2 - \beta_2 A_0 \Sigma A'_0 \\
&= (\alpha_2 + \beta_2) S_{0,2}\Sigma S'_{0,2} + [(\alpha_1 + \beta_1) - (\alpha_2 + \beta_2)] S_{0,1}\Sigma S'_{0,1} \\
&\quad + \beta_2 (A_0 + A_2)\Sigma(A_0 + A_2)' \qquad\qquad (4.50) \\
&\quad - 2\beta_2 A_2 \Sigma A'_2 - \beta_1 A_1 \Sigma A'_1 + (\alpha_0 - \alpha_1 - \beta_1 - \beta_2) A_0 \Sigma A'_0.
\end{aligned}
$$

Evaluating the coefficients we obtain

$$
\begin{aligned}
\operatorname{Cov}(X'_{t.}) &= \frac{t(t-1)(t-2)}{3} S_{0,2}\Sigma S'_{0,2} + t(t-1) S_{0,1}\Sigma S'_{0,1} \\
&\quad + \frac{(t-1)(t-2)}{2}(A_0 + A_2)\Sigma(A_0 + A_2)' \qquad\qquad (4.51) \\
&\quad - (t-1)(t-2) A_2 \Sigma A'_2 - \frac{t(t-1)}{2} A_1 \Sigma A'_1 + (2t-1) A_0 \Sigma A'_0,
\end{aligned}
$$

which may also be expressed in the more revealing form

$$
\begin{aligned}
\operatorname{Cov}(X'_{t.}) &= \frac{t(t-1)(t-2)}{3} S_{0,2}\Sigma S'_{0,2} + t(t-1)\left( S_{0,1}\Sigma S'_{0,1} - \frac{A_1 \Sigma A'_1}{2} \right) \\
&\quad + (t-1)(t-2)\left( \frac{(A_0 + A_2)\Sigma(A_0 + A_2)'}{2} - A_2 \Sigma A'_2 \right) \\
&\quad + (2t-1) A_0 \Sigma A'_0. \qquad\qquad (4.52)
\end{aligned}
$$

We conclude the example with the following comment. If we write the right member of Eq. (4.52) in canonical form, i.e. by collecting terms corresponding to distinct powers of $t$, the extreme complexity of the conditions under which such a process is $CI(2,1,r)$ becomes quite evident. To see that, let $D_i$, $i = 1, 2, 3, 4$, be the four matrices appearing in the right member of that equation. The canonical representation is

$$
\begin{aligned}
\operatorname{Cov}(X'_{t.}) &= \frac{t^3}{3} D_1 + t^2 (-D_1 + D_2 + D_3) + t \left( \frac{2D_1}{3} - D_2 - 3D_3 + 2D_4 \right) \\
&\quad + (2D_3 - D_4);
\end{aligned}
$$

thus to have cointegration of order one and rank $r$ there must exist a $q \times r$ matrix $B$ such that

$$D_1 B = 0, \ (D_2 + D_3)B = 0, \ \text{and} \ (D_3 - D_4)B \neq 0,$$

and $2D_3 - D_4$ must be positive semidefinite of rank **at least** $r$. For the process to be $CI(2,2,r)$ we must have the conditions

$$D_1 B = 0, \ (D_2 + D_3)B = 0, (D_3 - D_4)B = 0, \ (2D_3 - D_4)B \neq 0,$$

and $BD_3 B' > 0$. This is, however, not possible unless $D_2 = -D_3$, which in general is not the case! Thus the process may be $CI(2,2,r)$ **only through collinearity**. Moreover, for the process to be $CI(2,1,r)$, as noted above, certain very special conditions are required. In particular, the cointegrating matrix, say $B$, must be in the null space of $S_{0,2}$, and contain the common characteristic vectors of the (four) matrices $D_i$! These special conditions reinforce the reservations, earlier expressed, about tying the concept of cointegration too closely to the notion of long-run equilibrium of economic relations.

# 4.4   Implications of Cointegration

## 4.4.1   Cointegration in $MIMA(\infty)$ Processes

Suppose $X$ is a $MIMA(\infty)$, i.e. a multivariate integrated process with a forcing function which is a (multivariate) moving average of infinite order, and we are told that it is $CI(d,b,r)$. What does this imply regarding the parameters of the process and/or its characterization? For convenience, we deal with the case most frequently encountered in applications, viz. $CI(1,1,r)$.

The following result explores the implications of cointegration for $I(1)$ processes, sometimes known as the Granger representation theorem.

**Proposition 2** (Engle and Granger). Let $X = \{X'_t : t \in \mathcal{N}_+\}$ be a stochastic sequence defined on the probability space $(\Omega, \mathcal{A}, \mathcal{P})$, and suppose it is of the form [8]

$$(I - L)X'_{t\cdot} = \sum_{j=0}^{\infty} C_j \epsilon'_{t-j\cdot}, \quad C_0 = I_q, \tag{4.53}$$

such that under the isomorphism $C(L) \leftrightarrow C(z)$, for $z = e^{-i\lambda}$, the series operator $C(L) = \sum_{j=0}^{\infty} C_j L^j$ obeys the conditions: $C(\lambda)$ and $C'(\lambda)$ are well defined and **bounded** at $\lambda = 0$. [9] Moreover, let $X \sim CI(1,1,r)$, $\beta$ be a $q \times r$ matrix containing the $r$ linearly independent cointegrating vectors.

---

[8] The only restriction on the generality of the results here is that the forcing function of the differenced process $(I - L)X_t$ is **a general linear process**, or $MMA(\infty)$. This is merely done out of deference to the original work since, as we have already seen, it is not possible to have cointegration together with an unrestricted forcing function, i.e. one that is $MMA(\infty)$.

[9] The isomorphism between the algebra of lag operators and complex indeterminates is discussed in Section 1.4 of this volume. The requirement that $C(\lambda)$ and $C'(\lambda)$ be **well-defined and bounded at** $\lambda = 0$ is to be understood as follows: $C(0) = \sum_{j=0}^{\infty} C_j^+ - \sum_{j=0}^{\infty} C_j^-$ where, **on an element-by-element basis**, $C_j^+ = \max(0, C_j)$ and $C_j^- = \max(0, -C_j)$. The entity $C(0)$ is not well-defined if and only if $C^+(0) = \sum_{j=0}^{\infty} C_j^+ = \infty$ **and** $C^-(0) = \sum_{j=0}^{\infty} C_j^- = \infty$ in the sense that at least one element of the matrices in question is not finite. Since $C(0) = C^+(0) - C^-(0)$, to be bounded means that **we must have both** $C^+(0) < \infty$ and $C^-(0) < \infty$; alternatively, denoting the $(r,s)$ element of $C_j$ by $c_{r,s}^{(j)}$, $r, s = 1, 2, \ldots q$, we must have $\sum_{j=0}^{\infty} |c_{rs}^{(j)}| < \infty$ so that

$$D = \sum_{j=0}^{\infty} \left( \sum_{r=1}^{q} \sum_{s=1}^{q} |c_{rs}^{(j)}| \right) < \infty.$$

It is evident that

$$S \leq D, \quad \text{where} \quad S = \sum_{j=0}^{\infty} \| C_j \|.$$

If we write $C(z)$, with a **general** complex indeterminate $z$, then what we had termed $C(0)$, for the particular case where $z = e^{i\lambda}$, will be denoted by $C(1)$, i.e. it is $C(z)$ evaluated at $z = 1$. Needless to say, for the case $z = e^{i\lambda}$, $C(\lambda)$ evaluated at $\lambda = 0$ is precisely $C(z)$ evaluated at $z = 1$. In subsequent discussions we generally employ the notation $C(z)$ and $C(1)$, but in this context it is more appealing to use $C(\lambda)$ and $C(0)$.

Similarly, the fact that $C'(\lambda)$ is well defined and bounded at $\lambda = 0$ implies

$$S^* \leq D^* < \infty, \quad \text{where} \quad D^* = \sum_{j=0}^{\infty} j \left( \sum_{r=1}^{q} \sum_{s=1}^{q} |c_{rs}^{(j)}| \right) < \infty, \quad S^* = \sum_{j=0}^{\infty} j \| C_j \|.$$

The condition $S < \infty$ is required so as to ensure the a.c. convergence of $\sum_{j=0}^{\infty} C_j \epsilon'_{t\cdot}$. The condition $S^* < \infty$ is required to ensure that the decomposition in part i is valid, and that the cointegral vector $X_{t\cdot}\beta$ is well-defined. The conditions imposed neither guarantee nor rule out the invertibility of $C(L)$ or $C^*(L)$.

The following statements are true:

i. there exists a representation [10]

$$C(L) = C(1) - (I - L)C^*(L), \; C^*(L) = \sum_{j=0}^{\infty} L^j, \; C_j^* = \sum_{i=1}^{\infty} C_{i+j};$$

ii. $\text{rank}[C(1)] = q - r, \; q > r$;

iii. $X$ has a representation as $MARMA(\infty, \infty)$; specifically, if $d^*(L) = |C(L)|$ and $H^*(L)$ is the adjoint of

$$C(L) = [c_{rs}(L)], \quad c_{rs}(L) = \sum_{j=0}^{\infty} c_{rs}^{(j)} L^j,$$

where $c_{rs}^{(j)}$ is the $r, s$ element of the matrix $C_j$, **then**

$$H(L)X_{t.}' = d(L)I_q \epsilon_{t.}', \quad H(1) \neq 0, \quad d(1) \neq 0, \text{where}$$

$$H^*(L) = (I - L)^{r-1} H(L), \quad d^*(L) = (I - L)^r d(L).$$

iv. the operator $C^*(L) = \sum_{j=0}^{\infty} C_j^* L^j$ does not have a unit root factorization, i.e. $C^*(L) \neq (I - L)C^{**}(L)$;

v. there exist matrices $\beta$, $\Gamma$ of dimension $q \times r$ and rank $r$ such that

$$\text{rank}[H(1)] = r, \quad H(1) = \Gamma\beta', \quad C(1)'\beta = 0, \quad C(1)\Gamma = 0$$

and, moreover, $H(0) = I_q$; [11]

---

The condition $S^* < \infty$, which is somewhat stronger than the one given earlier in Assertion 1 and in the premise of the Proposition 2, was also noted in Stock (1987), and in Phillips and Solo (1992).

[10] The notation $C(1)$ is to be interpreted in the context of the isomorphism $C(L) \leftrightarrow C(z)$ for a (general) complex indeterminate $z$; $C(1)$ means $C(z)$ evaluated at $z = 1$, so that $C(1) = \sum_{j=0}^{\infty} C_j$. Strictly speaking, we should write $C(L) = C(1)I - (I - L)C^*(L)$, but for simplicity of notation the operator $I$ will be omitted.

[11] In the literature of econometrics there is a tendency to interpret $C(0) = C_0$, or $H(0) = H_0$, as we do implicitly here. Strictly speaking, however, this is not correct, since $C(0)$ is the matrix operator $C_0 I$, while $C_0$, or $H_0$ is the matrix (coefficient) which is, by convention, set to $I_q$. We hope that no confusion results from following this common practice.

vi. there exists an "error correction model" (ECM) representation,

$$A(L)(I - L)X'_{t.} = -\Gamma Z'_{t-1.} + d(L)I_q \epsilon'_{t.}, \qquad (4.54)$$

with $A_0 = I_q$;

vii. the cointegral vector $Z_{t.} = X_{t.}\beta$ has the representations

$$Z'_{t.} = -\beta' C^*(L)\epsilon'_{t.}, \quad (I - L)Z'_{t.} = -\beta' \Gamma Z'_{t-1.} + G(L)\epsilon'_{t.},$$

with $|\beta'\Gamma| \neq 0$, $G_0 = \beta' C^*_0$, $G_i = -\beta'(C^*_i + C^*_0 C^*_{i-1})$, for $i = 1, 2, \ldots$.

Proof: By long division, with $C_{(n)}(L) = \sum_{j=0}^{n} C_j L^j$ as the **dividend**, and $L - I$ as the **divisor**, we obtain [12]

$$C_{(n)}(1) = \sum_{j=0}^{n} C_j, \text{ as the } \textbf{remainder},$$

$$C^*_{(n)}(L) = \sum_{j=0}^{n-1} \left( \sum_{s=j+1}^{n} C_s \right) L^j, \text{ as the } \textbf{quotient}, \text{ so that}$$

$$C_{(n)}(L) = C_{(n)}(1) - (I - L)C^*_{(n)}(L).$$

By a limiting process, i.e. letting $n \to \infty$, part i is proved with

$$C(1) = \sum_{j=0}^{\infty} C_j, \quad C^*(L) = \sum_{j=0}^{\infty} C^*_j L^j, \quad C^*_j = \sum_{s=1}^{\infty} C_{j+s}.$$

For the cointegral vector $Z_{t.} = X_{t.}\beta$ to be well defined we must require $C^*(L)\epsilon'_{t.}$ to converge a.c.. This entails the condition $\sum_{j=0}^{\infty} \| C^*_j \| < \infty$, which is **implied** by the condition that $C'(\lambda)$ is well defined and bounded at $\lambda = 0$.

---

[12] In the literature of econometrics this is termed the Beveridge-Nelson decomposition. A variant of it is found in Beveridge and Nelson (1981), but the argument for it, and its interpretation are somewhat misleading. The motivation for this decomposition in the context of Proposition 2 is to determine whether the operator $C(L)$ has a unit root factorization; if it does, and thus the remainder obeys $C(1) = 0$, the process is stationary, possibly with a nonzero mean, provided, minimally, that $\sum_{j=0}^{\infty} \| C^*_j \|^2 < \infty$. If it does not, and thus $C(1) \neq 0$, we have cointegration if $C(1)$ is **singular**; if it is **nonsingular**, we have nonstationarity, i.e. the process is $I(1)$ and **is not cointegrated**.

To prove ii, consider the representation

$$(I - L)X'_{t\cdot} = C(1)\epsilon'_{t\cdot} - (I - L)C^*(L)\epsilon'_{t\cdot}, \qquad (4.55)$$

premultiply by $\beta'$, and let $X_t.\beta = Z_t.$. By the cointegrating assumption, $Z_t.$ is an $r$-element row vector whose elements are jointly **stationary**. Thus, in

$$(I - L)Z'_{t\cdot} = \beta' C(1)\epsilon'_{t\cdot} - (I - L)\beta' C^*(L)\epsilon'_{t\cdot}, \qquad (4.56)$$

we conclude that $\beta' C(1) = 0$, and that $\beta' C^*(L)\epsilon'_{t\cdot}$ represents a **stationary process**; moreover, since $\text{rank}(\beta) = r$, the dimension of the row (and column) null space of $C(1)$ is $r$ and, consequently, $\text{rank}[C(1)] = q - r$, completing the proof of ii. In this connection we also note that this implies that the characteristic equation $|C(z)| = 0$ has $r$ **unit** roots.

To prove iii, we note that $H^*(L)$ has elements which are determinants of $q - 1$-dimensional submatrices of $C(L)$; its typical element, $h^*_{mp}(L) = \sum_{j=0}^{\infty} h^{(j)}_{mp} L^j$, is a sum of products of the form

$$\prod_{i=1}^{q-1} c_{r_i s_i}(L) = \sum_{j_1 \cdots j_{q-1}} \prod_{i=1}^{q-1} c^{j_i}_{r_i s_i} L^{j_i},$$

and thus [13]

$$\sum_{j=0}^{\infty} |h^{(j)}_{mp}| < \left( \sum_{j=0}^{\infty} \| C_j \| \right)^{q-1}.$$

We conclude that the elements of $H^*(L)$ are well defined. Since $d^*(L)$ is **also a determinant**, the same argument yields that $d^*(L)I_q\epsilon'_{t\cdot}$ converges a.c. and is thus well defined. We now show that we have the representation

$$H^*(L) = (I - L)^{r-1}H(L), \quad d^*(L) = (I - L)^r d(L),$$

such that $d(1) \neq 0$ and $H(1) \neq 0$.

---

[13] To illustrate the considerations involved consider a binary product, say $c_{11}(L) \cdot c_{22}(L) = \sum_{j_1, j_2} c^{(j_1)}_{11} c^{(j_2)}_{22} L^{j_1 + j_2}$, which may also be written as $\sum_{k=0}^{\infty} \gamma_k L^k$, with $\gamma_k = \sum_{j_1=0}^{k} c^{(j_1)}_{11} c^{(k-j_1)}_{22}$. Consequently,

$$\sum_{k=0}^{\infty} \sum_{j_1}^{k} |c^{(j_1)}_{11}| \, |c^{(k-j_1)}_{22}| = \left( \sum_{j_1=0}^{\infty} |c^{(j_1)}_{11}| \right) \left( \sum_{k=j_1}^{\infty} |c^{(k-j_1)}_{22}| \right) \leq \left( \sum_{j=0}^{\infty} \| C_j \| \right)^2,$$

where $C_j = [c^{(j)}_{rs}]$, $r, s = 1, 2$.

The second part is evident since, as we have shown in the proof of part ii, $C(L)$ has $r$ unit roots, which implies that $d(1) \neq 0$; as for the first part, we note from part i that $C(L) = C(1) - (I - L)C^*(L)$, and from part ii that $C(1)$ is of rank $q - r$; hence, submatrices of order greater than $q - r$ have zero determinant. By definition

$$h_{ij}^*(L) = (-1)^{i+j}|C_{(j,i)}(1) - (I - L)C^*_{(j,i)}(L)|,$$

so that it is the sum of products of terms one each from each column (and row) of the (sub)matrix in question. The notation $C_{(j,i)}$ denotes the matrix resulting from $C$ after the $j^{th}$ row and the $i^{th}$ column have been eliminated, and it is clear that the sum of terms involving $s$ elements from $C_{(j,i)}(1)$ are null for $s > q - r$, since $C(1)$ is of rank $q - r$. Consequently, the only nonnull terms appearing in $h_{ij}^*(L)$ must involve at least $q - 1 - s$ terms from the component $(I - L)C^*(L)$, where $s \leq q - r$. Since $q - 1 - s \geq r - 1$ we have

$$h_{ij}^*(L) = (I - L)^{r-1}h_{ij}(L), \quad h_{ij}(1) \neq 0, \quad \text{for all } i \text{ and } j,$$

and thus $H^*(L) = (I - L)^{r-1}H(L)$ and $H(1) \neq 0$, as required. Since by definition of an adjoint, $C(L)H^*(L) = H^*(L)C(L) = d^*(L)I_q$, we obtain from Eq. (4.53)

$$H^*(L)(I - L)X'_{t.} = H^*(L)C(L)\epsilon'_{t.} = d^*(L)I_q\epsilon'_{t.}.$$

It follows immediately, by substitution for $H^*(L)$ and $d^*(L)$, that

$$H(L)X'_{t.} = d(L)I_q\epsilon'_{t.}. \tag{4.57}$$

But this is a representation of $X$ as a $MARMA(\infty, \infty)$ process, which concludes the proof of iii.

The proof of iv, follows immediately from the argument above, but we give the details. If $C^*(L)$ has a unit root factorization, we can write $C^*(L) = (I - L)C^{**}(L)$ and, consequently, have the representation $C(L) = C(1) - (I - L)^2 C^{**}(L)$. This means that, if we employ the formulation $(I - L)^2 X'_{t.} = C(L)\epsilon'_{t.}$ we can argue that $X \sim CI(2, 2, r)$, which is a contradiction, thus proving that $C^*(L)$ **does not** have a unit root factorization.

To prove v, we note that by the arguments in the proof of iii and iv, the dimension of the row, as well as column, null space of $C(1)$ is $r$; by the cointegration assumption, $\beta$ spans (is a basis for) the row null space. Moreover, there must exist $r$ linearly independent vectors in the **column null space** which span (form a basis for) that space. Let these vectors be denoted by the $q \times r$ matrix $\Gamma$. To determine the rank of $H(1)$, note that

$$C(L)H^*(L) = (I - L)^r d(L) I_q,$$

$$|C(L)|\,|H^*(L)| = |[(I - L)^r d(L)] I_q| = (I - L)^{rq} d(L)^q.$$

Comparing the determinants of both sides we find $|H(L)| = (I-L)^{q-r} d(L)^{q-1}$, so that $|H(L)| = 0$ has a **unit root of multiplicity** $q - r$. Consequently, $H(1)$ is a **matrix of rank** $r$. Moreover,

$$C(L)H(L) = (I - L)d(L)I_q, \quad \text{and thus} \quad C(1)H(1) = 0,$$

and we see that $H(1)$ **is in the column null space** of $C(1)$. Finally, $C(L)H(L) = H(L)C(L)$ so that $H(1)C(1) = 0$ and thus $H(1)$ is in the **row null space** of $C(1)$ as well. We conclude therefore that $H(1) = \Gamma\beta'$. In view of the standard conventions regarding normalization, $d(0) = 1$ and consequently, from $C(L)H(L) = (I - L)d(L)I_q$, we obtain $C(0)H(0) = d(0)I_q$; since by convention $C(0) = I_q$ we conclude $H(0) = I_q$, completing the proof of v.

To prove vi, we note from Eq. (4.57) that $H(L)X'_{t\cdot} = d(L)I_q\epsilon'_{t\cdot}$. Since $H(1) \neq 0$, we write

$$H(L) = H(1) - (I - L)\bar{H}(L), \quad \bar{H}(L) = \sum_{j=0}^{\infty} \bar{H}_j L^j, \quad \bar{H}_j = \sum_{i=j+1}^{\infty} H_i,$$

and note that $I_q + \bar{H}_0 = H(1)$; thus, Eq. (4.57) may be written as

$$(I - L)X'_{t\cdot} - \sum_{j=1}^{\infty} \bar{H}_j(I - L)X'_{t-j\cdot} = -H(1)X'_{t-1\cdot} + d(L)I_q\epsilon'_{t\cdot} \quad \text{or}$$

$$A(L)(I - L)X'_{t\cdot} = -\Gamma Z'_{t-1\cdot} + d(L)I_q\epsilon'_{t\cdot}, \quad A_0 = I_q, \quad A_j = -\bar{H}_j, \quad j \geq 1.$$

To prove vii, we note that we have the representation

$$(I - L)X'_{t\cdot} = C(1)\epsilon'_{t\cdot} - (I - L)C^*(L)\epsilon'_{t\cdot}.$$

Pre multiplying by $\beta'$, we find $(I - L)Z'_{t.} = -(I - L)\beta' C^*(L)\epsilon'_{t.}$, and canceling the term $(I - L)$ we have the first representation. As for the second, adding and subtracting $\beta' \Gamma Z_{t-1.}$ in the right member of the simplified representation above, and subtracting $Z_{t-1.}$ from both sides, we find

$$(I - L)Z'_{t.} = -\beta' \Gamma Z'_{t-1.} + G(L)\epsilon'_{t.}, \quad G(L) = -\beta'[C^*(L) + C_0^* C^*(L)L].$$

Since $H(1) = \Gamma\beta'$ is of rank $r$, it follows that $\beta'\Gamma$ is also of rank $r$, because the nonzero characteristic roots of the former matrix are exactly the same as the nonzero roots of the latter; since the latter is $r \times r$, it follows that $\beta'\Gamma$ is nonsingular.

q.e.d.

**Corollary 2.** In the context of Proposition 2, $C^*(L)$ need not be an invertible (matrix) operator.

Proof: From the proof of the proposition, we are only able to deduce that $\beta' C^*(1) \neq 0$; this means that the (row) null space of $C^*(1)$ has only the zero vector in common with the null space of $C(1)$, **and in no way implies** that $C^*(1)$ is **nonsingular**, which is required for the invertibility of the operator $C^*(L)$. We complete the proof by giving a counter example.

It will suffice to produce a cointegrated process where the entity above is not an invertible (matrix) operator. To this end consider the example given in Engle and Granger (1987), p. 263, Eqs. (4.7) and (4.8). The (bivariate) model is given by

$$X_{t.} \begin{bmatrix} 1 & 1 \\ \beta & \alpha \end{bmatrix} = u_{t.}, \quad (I - L)u_{t1} = \epsilon_{t1}, \quad u_{t2} = \frac{I}{I - \rho L} \epsilon_{t2}.$$

Manipulating the expression above we find

$$(I - L)X_{t.} \begin{bmatrix} 1 & 1 \\ \beta & \alpha \end{bmatrix} = \begin{pmatrix} \epsilon_{t1} & \frac{I - L}{I - \rho L}\epsilon_{t2} \end{pmatrix}.$$

Further simplification yields

$$(I - L)X'_{t.} = C(L)\epsilon_{t.}, \quad C(L) = \frac{1}{\alpha - \beta} \begin{bmatrix} \alpha I & -\frac{\beta(I - L)}{I - \rho L} \\ -I & \frac{I - L}{I - \rho L} \end{bmatrix},$$

$$C(1) \ = \ \frac{1}{\alpha - \beta} \begin{bmatrix} \alpha & 0 \\ -1 & 0 \end{bmatrix}, \quad C^*(L) \ = \ \frac{1}{\alpha - \beta} \begin{bmatrix} 0 & \frac{\beta I}{I - \rho L} \\ 0 & \frac{I}{I - \rho L} \end{bmatrix},$$

$$C(L) \ = \ C(1) - (I - L)C^*(L).$$

It is clear that $C^*(L)$ is not an invertible operator.

$$\text{q.e.d.}$$

**Remark 4**. The preceding is essentially a restatement of the important result in Engle and Granger (1987) (EG). Our presentation departs from that in EG by providing a number of slightly altered definitions. For example, the definition of integrated processes in EG, p. 252, requires our process $Y$, of Eq. (4.1), to have an ARMA representation. It is not always possible for regular (purely nondeterministic) processes to have a rational representation. The terminology " ..with no deterministic component . ." in that definition is confusing. In fact, the representation in Eq. (3.1) of EG p. 255, has nothing to do with the Wold decomposition; it has to do with Proposition 3 of Chapter 1, which requires the spectral density of $(I - L)x'_{t\cdot}$ to be positive a.e., and to have a unilateral Fourier series representation.

In Proposition 2 we did not include part (6) in EG, since part (6) is nearly vacuous in the $MIMA(\infty)$ context, unless $C(L) = [B(L)^{-1}]D(L)$ where $D(L)$ is a noninvertible finite order (matrix) polynomial. In such a case, evidently,

$$B(L)(I - L)X'_{t\cdot} = D(L)\epsilon'_{t\cdot};$$

if $H_1(L)$ is the adjoint of $D(L)$ and $d_1(L) = |D(L)|$, we obtain

$$A(L)(I - L)X'_{t\cdot} = d_1(L)I_q\epsilon'_{t\cdot},$$

which is of the same form as the representation in part vi of the proposition, except that in this discussion $A(L) = H_1(L)B(L)$ and is thus a **finite order** matrix polynomial. The finite polynomial representation in question is the appropriate outcome in the context of a $MIMA(k)$ process, as we shall see in Proposition 2a below.

## Empirical Implementation

Here we establish certain estimation procedures that are implied by the cointegration assumption. In EG it is argued that the parameters characterizing the cointegrated system of Proposition 2 may be obtained in two steps. In the first step, one obtains the cointegrating matrix $B$; in the second step, one uses the ECM, of part vi Proposition 2, to obtain the remaining parameters of interest. For the first part, EG rely on the characteristic vectors of the empirical covariance matrix of the $X$ process. Denote by [14]

$$X = (X_{t\cdot}), \quad t = 1, 2, 3, \ldots, T \tag{4.58}$$

the data matrix of the problem; EG recommend, *inter alia*, that we consider the characteristic vectors corresponding to the smallest characteristic roots of

$$\hat{M}^{(1)} = \frac{1}{T^2} X' X,$$

on the ground that

$$\frac{1}{T^2} \beta' X' X \beta \xrightarrow{P} 0,$$

"because" $\beta' X' X \beta / T$ is an estimator of the covariance matrix of a well behaved stationary process. Moreover, they argue that this procedure accords "... with the common observation that economic time series data are highly collinear so that moment matrices may be nearly singular even when samples are large" [p. 260 in EG (1987)]. As our discussion in the early part of this chapter makes clear, one needs to distinguish between **cointegration** and **collinearity**, although it is trivially correct to say that collinear $I(1)$ processes are cointegrated. In the face of collinearity, the linear combinations corresponding to the characteristic vectors of zero roots are **constants**, hence trivially $I(0)$. Our earlier discussion also shows that in a well-formulated cointegration context we may conclude that for sufficiently large $t$, $\text{Cov}(\xi'_{t\cdot}) = \Psi_0 + \Psi_1 t$, where $\xi$ is a $CI(1,1,r)$ process. Cointegration then means that there exists a $q \times r$ matrix $B$, of rank $r$, such

---

[14] In this section **only**, the symbol $X$ is used to denote the data matrix defined in Eq. (4.58). The context of the discussion prevents any confusion with the definition of the symbol $X$ hithertofore, which referred to the data generating sequence $X = \{X_{t\cdot} : t \in \mathcal{N}_+\}$.

that

$$z_{t\cdot} = \xi_{t\cdot}B, \quad \text{and} \quad \text{Cov}(z'_{t\cdot}) = B'\Psi_0 B > 0.$$

Otherwise, if we follow the rationale cited above, the cointegral variables (the elements of $z_{t\cdot}$) become, asymptotically, constants. In the EG context, the reasons we may wish to seek out the zero roots is that division by $T^2$ **annihilates** the component $\Psi_0$ in the **expected value** of the matrix $\hat{M}^{(1)}$.

We also observe, in this context, that if the characteristic vectors are normalized by the condition $b'_{\cdot i} b_{\cdot i} = 1$ we have the classic representation of the problem as

$$\min_B F = \text{tr} B' \hat{M}^{(1)} B, \quad \text{subject to} \quad B'B = I_q. \tag{4.59}$$

The minimized value of $F$, subject to the contstraint above, is given by

$$F^* = \text{tr}\Lambda, \quad \Lambda = \text{diag}(\lambda_1, \lambda_2, \cdots, \lambda_q),$$

which yields the conclusion that if the minimizing matrix is to be of rank $r$, **we should choose for its columns** the characteristic vectors of $\hat{M}^{(1)}$ corresponding to the $r$ **smallest roots.**

Placing this development in the context of Proposition 2, we note the following: Matrix $B$ of this discussion corresponds to matrix $\beta$ of Proposition 2; consequently the cointegral vector of this discussion, $z_{t\cdot} = \xi_{t\cdot}B$, corresponds in Proposition 2 to the cointegral vector $Z_{t\cdot} = X_{t\cdot}\beta$ and, moreover, the matrix $\beta$ is defined implicitly by the rank factorization of the $q \times q$ matrix $H(1)$ whose rank is $r$, viz. $H(1) = \Gamma\beta'$. Thus, even though what we seek in the discussion of this section is **uniquely defined** by the normalization **imposed** in Eq. (4.59), what we seek in the larger context of Proposition 2 is **arbitrary to within multiplication on the right by a nonsingular matrix of order** $r$, say $G$. What is meant is that **if** $H(1) = \Gamma B'$ **then we also have** $H(1) = \Gamma^* B^{*'}$, where $\Gamma^* = \Gamma G'^{-1}$, and $B^* = BG$. Thus, to ensure uniqueness it is necessary to impose appropriate restrictions; one convenient set of such restrictions is of the form

$$B = (I_r, \ B'_0)', \quad \text{so that} \quad H(1) = \Gamma(I_r, \ B'_0) \tag{4.60}$$

so that the rank factorization, as well as $B$, are defined uniquely.

Having imposed these restrictions, we write

$$B = \begin{bmatrix} I_r \\ 0 \end{bmatrix} + \begin{bmatrix} 0 \\ I_{q-r} \end{bmatrix} B_0 \;=\; Q_1 + Q_2 B_0, \tag{4.61}$$

and the problem of estimating the (unique) cointegrating matrix, $B$, may be formulated as

$$\min_B F = \mathrm{tr} B' \hat{M}^{(1)} B, \quad \text{subject to} \quad B = Q_1 + Q_2 B_0. \tag{4.62}$$

From Proposition 90 in Dhrymes (1984), pp. 105-106,

$$\mathrm{tr}(A_1' A_2 A_3) = \mathrm{vec}(A_1)'(I \otimes A_2)\mathrm{vec}(A_3),$$

so that the problem of Eq. (4.62) may be reformulated as

$$\min_{b_0} F \;=\; q'(I_r \otimes \hat{M}^{(1)})q, \quad q \;=\; q_1 + (I_r \otimes Q_2)b_0,$$

$$q_1 \;=\; \mathrm{vec}(Q_1), \quad b_0 \;=\; \mathrm{vec}(B_0). \tag{4.63}$$

The first-order conditions imply

$$\hat{b}_0 = -[I_r \otimes (Q_2' \hat{M}^{(1)} Q_2)^{-1} Q_2' \hat{M}^{(1)}]q_1, \tag{4.64}$$

and the estimation problem appears to be resolved.

**Remark 5.** If we rematricize the result in Eq. (4.64) we find

$$\hat{B}_0 \;=\; -(Q_2' \hat{M}^{(1)} Q_2)^{-1} Q_2' \hat{M}^{(1)} Q_1 \;=\; -\hat{M}_{22}^{(1)-1} \hat{M}_{21}^{(1)}, \quad \text{where}$$

$$\hat{M}^{(1)} \;=\; \begin{bmatrix} \hat{M}_{11}^{(1)} & \hat{M}_{12}^{(1)} \\ \hat{M}_{21}^{(1)} & \hat{M}_{22}^{(1)} \end{bmatrix}, \tag{4.65}$$

and $\hat{M}_{11}^{(1)}$ is $r \times r$, $\hat{M}_{22}^{(1)}$ is $q - r \times q - r$ and so on.

An important point regarding this solution to the problem of estimating the cointegrating vectors is that $\hat{B} = (I_r, \; \hat{B}_0')'$ **is not necessarily the matrix whose $r$ columns correspond to the $r$ smallest characteristic roots of $\hat{M}^{(1)}$!** This is so since

$$\hat{M}^{(1)} \hat{B} = \begin{bmatrix} \hat{M}_{11}^{(1)} & \hat{M}_{12}^{(1)} \\ \hat{M}_{21}^{(1)} & \hat{M}_{22}^{(1)} \end{bmatrix} \begin{bmatrix} I_r \\ \hat{B}_0 \end{bmatrix} = \begin{bmatrix} \hat{M}_{11}^{(1)} - \hat{M}_{12}^{(1)} \hat{M}_{22}^{(1)-1} \hat{M}_{21}^{(1)} \\ 0 \end{bmatrix}. \tag{4.66}$$

The conclusion would remain the same even if we replaced $\hat{M}^{(1)}$ by the **limit of its expected value**, say $M$; repeating the operations above we would conclude that the resulting matrix, say $\bar{B}$, obeys

$$M\bar{B} = \begin{bmatrix} M_{11} - M_{12}M_{22}^{-1}M_{21} \\ 0 \end{bmatrix}, \tag{4.67}$$

which is certainly **not** the relation between the matrix $M$ and its characteristic vectors corresponding to the $r$ **zero** roots, **unless** $M_{11} - M_{12}M_{22}^{-1}M_{21} = 0$. From our discussion earlier in this chapter, the matrix $M$ has $r$ zero roots, by the cointegration hypothesis. Without loss of generality suppose the matrix $M_{22} > 0$, i.e. it is nonsingular. We must therefore have

$$|M| = |M_{22}| \, |M_{11} - M_{12}M_{22}^{-1}M_{21}| = 0,$$

so that $M_{11} - M_{12}M_{22}^{-1}M_{21}$ is a **singular matrix**. There is nothing in this context, however, that requires it to be the **zero matrix**. Thus, in moving from the problem of Eq. (4.59) to the probelm of Eq. (4.63) there has been a shift of objectives.

Having obtained an estimator for the matrix $B$ we define the cointegral vector $Z_{t.}$, in the context of Proposition 2 by $X_{t.}\hat{B}$, and, in the context of this discussion, use the ECM representation of part vi of the proposition to obtain

$$(I - L)X'_{t.} = -\Gamma(\hat{B}'X'_{t-1.}) + \sum_{j=1}^{\infty} \bar{H}_j(I - L)X'_{t-j.} + d(L)I_q\epsilon'_{t.}. \tag{4.68}$$

The problem with this representation is that it is **not estimable** owing to the infinite lag structure; thus, as a practical matter, we would have to truncate the lag structure which is equivalent, for all intents and purposes, to having assumed **initially** a $MIMA$ process of finite order.

We close this section by showing that if the forcing function of the process is not of finite order then, strictly speaking, it is not possible to have cointegration as the term is usually defined. This is so since the (second) moments of the cointegral vector depend, instrinsically, on $t$ although they converge to a finite quantity as $t \to \infty$. This dependence is a consequence of **the initial conditions** we are forced to impose in order to obtain finiteness of moments of the underlying $I(1)$ process.

**Remark 6.** In the example given in EG, and utilized in Corollary 2, the cointegral scalar has variance that depends on $t$, and the covariance matrix of the process is of the form $\Psi_t = \Psi_{(0)} + \Psi_{(1)}t$. To see this note that **imposing the initial condition,** $X_0. = 0$, which implies $\epsilon_{-i.} = 0$ for $i \geq 0$, we find

$$X_{t.} = \frac{1}{(\alpha - \beta)}(\epsilon_{t1}^*, \epsilon_{t2}^*)\begin{bmatrix} \alpha & -1 \\ -\beta & 1 \end{bmatrix},$$

where $\epsilon_{t1}^* = \sum_{s=1}^{t} \epsilon_{t1}$ and $\epsilon_{t2}^* = \sum_{j=0}^{t-1} \rho^j \epsilon_{t-j,2}$. It is now easy to establish

$$E\epsilon_{t1}^{*2} = t\phi_{11}, \quad \phi_{ij} = E\epsilon_{ti}\epsilon_{tj}, \quad i, j = 1, 2;$$

$$E\epsilon_{t2}^{*2} = \phi_{22}\rho_2, \quad \rho_2 = \sum_{j=0}^{t-1} \rho^{2j}, \quad \rho_1 = \sum_{j=0}^{t-1} \rho^j;$$

$$E\epsilon_{t1}^*\epsilon_{t2}^* = \phi_{12}\rho_1$$

and that

$$\text{Cov}(X_{t.}') = \frac{1}{(\alpha - \beta)^2}\begin{bmatrix} \alpha & -\beta \\ -1 & 1 \end{bmatrix} \text{Cov}\begin{pmatrix} \epsilon_{t1}^* \\ \epsilon_{t,2}^* \end{pmatrix}\begin{bmatrix} \alpha & -1 \\ -\beta & 1 \end{bmatrix}$$

$$= \Psi_{(0)} + \Psi_{(1)}t,$$

where

$$\Psi_{(0)} = \frac{1}{(\alpha - \beta)^2}\begin{bmatrix} \beta^2\gamma_2 - 2\alpha\beta\gamma_1 & (\alpha + \beta)\gamma_1 - \beta\gamma_2 \\ (\alpha + \beta)\gamma_1 - \beta\gamma_2 & \gamma_2 - 2\gamma_1 \end{bmatrix},$$

$$\gamma_1 = \phi_{12}\rho_1, \quad \gamma_2 = \phi_{22}\rho_2,$$

$$\Psi_{(1)} = \frac{\phi_{11}}{(\alpha - \beta)^2}\begin{bmatrix} \alpha^2 & -\alpha \\ -\alpha & 1 \end{bmatrix}.$$

Note that $\Psi_{(0)}$ depends on $t$ through the entities $\gamma_1$ and $\gamma_2$, while $\Psi_{(1)}$ **does not** depend on $t$. The cointegrating vector is $(1, \alpha)'$, and it is clear that this is a vector in the null space of $\Psi_{(1)}$, **but not** in the null space of $\Psi_{(0)}$.

The cointegral scalar $z_t = X_{t.}(1, \alpha)'$ obeys

$$Ez_t = 0, \quad \text{for all } t, \quad \text{Var}(z_t) = (1, \alpha)\Psi_{(0)}(1, \alpha)' = \gamma_2,$$

so that **it does not have a stationary distribution since its variance
depends on** $t$, through the entity $\rho_2$. However, its variance is essentially
bounded since

$$\lim_{t \to \infty} \rho_2 = \frac{1}{1 - \rho^2}, \quad \lim_{t \to \infty} \gamma_2 = \frac{\phi_{22}}{1 - \rho^2},$$

and thus

$$\lim_{t \to \infty} \text{Var}(z_t) = \frac{\phi_{22}}{1 - \rho^2} < \infty. \tag{4.69}$$

## 4.4.2 Cointegration in $MIMA(k)$ Processes

Here we examine the implications of cointegration in the context of $MIMA(k)$
processes, i.e. processes of the form

$$(I - L)X'_{t \cdot} = \sum_{j=0}^{k} A_j \epsilon'_{t-j \cdot}, \quad A_0 = I_q \tag{4.70}$$

where $\epsilon$ is a $MWN(\Sigma)$, with $\Sigma > 0$. If this process is $CI(1,1,r)$, we may
prove the analog of Proposition 2.

**Proposition 2a.** Let $X = \{X'_{t \cdot} : t \in \mathcal{N}_+\}$ be a stochastic sequence defined
on the probability space $(\Omega, \mathcal{A}, \mathcal{P})$; suppose further that

1. it is of the form

$$(I - L)X'_{t \cdot} = \sum_{j=0}^{k} A_j \epsilon'_{t-j \cdot}, \quad A_0 = I_q; \tag{4.71}$$

2. $X \sim CI(1,1,r)$;

3. $B$ is the $q \times r$ matrix containing the $r$ linearly independent cointegrating vectors.

The following statements are true:

i. there exists a representation

$$A(L) = A(1) - (I - L)A^*(L), \quad A^*(L) = \sum_{j=0}^{k-1} A_j^* L^j, \quad A_j^* = \sum_{i=j+1}^{k} A_i;$$

ii. $\text{rank}[A(1)] = q - r, \ q > r$;

iii. $X$ has a representation as $MARMA(m,n)$; specifically, if $a^*(L) = |A(L)|$ and $H^*(L)$ is the adjoint of

$$A(L) = [a_{rs}(L)], \quad a_{rs}(L) = \sum_{j=0}^{k} a_{rs}^{(j)} L^j,$$

where $a_{rs}^{(j)}$ is the $r,s$ element of the matrix $A_j$ **then**

$$H(L)X'_{t\cdot} = a(L)I_q \epsilon'_{t\cdot}, \quad H(1) \neq 0, \quad a(1) \neq 0, \text{where}$$

$$H^*(L) = (I-L)^{r-1} H(L), \quad a^*(L) = (I-L)^r a(L).$$

iv. the operator $A^*(L) = \sum_{j=0}^{k-1} A_j^* L^j$ does not have a unit root factorization, i.e. $A^*(L) \neq (I-L)A^{**}(L)$;

v. there exist matrices $B$, $\Gamma$ of dimension $q \times r$ and rank $r$ such that

$$\text{rank}[H(1)] = r, \quad H(1) = \Gamma B', \quad A(1)'B = 0, \quad A(1)\Gamma = 0$$

and moreover, $H(0) = I_q$;

vi. there exists an ECM representation,

$$G(L)(I-L)X'_{t\cdot} = -\Gamma Z'_{t-1\cdot} + a(L)I_q \epsilon'_{t\cdot}, \qquad (4.72)$$

with $G_0 = I_q$ and $G_j = -\bar{H}_j$, for $j \geq 1$;

vii. the cointegral vector $Z_{t\cdot} = X_{t\cdot}B$ has the representations

$$Z'_{t\cdot} = -B'A^*(L)\epsilon'_{t\cdot}, \quad (I-L)Z'_{t\cdot} = -B'\Gamma Z'_{t-1\cdot} + B'P(L)\epsilon'_{t\cdot},$$

with $B'\Gamma$ nonsingular, $P(L) = \sum_{j=0}^{k} P_j L^j$, $P_0 = -A_0^*$, and $P_i = -\left[A_i^* + (H(1) - I_r)A_{i-1}^*\right]$.

Proof: To prove i, we divide $A(L)$ by $L-I$, thus obtaining

$$A(L) = A(1) - (I-L)A^*(L), \quad A^*(L) = \sum_{j=0}^{k-1} A_j^* L^j, \quad A_j^* = \sum_{s=j+1}^{k} A_s.$$

To prove ii use result above, and condition 2, i.e. that $X \sim C(1,1,r)$, to obtain

$$(I-L)B'X'_{t\cdot} = B'A(1)\epsilon'_{t\cdot} - (I-L)B'A^*(L)\epsilon_{t\cdot};$$

Since $B$ is the matrix of cointegrating vectors we must have $B'A(1) = 0$, which shows that $A(1)$ is of rank $q - r$; moreover, canceling the term $I - L$ from both sides we obtain

$$B'X'_{t\cdot} = -\sum_{j=0}^{k-1}(B'A_j^*)\epsilon'_{t-j}.$$

which displays the cointegral vector as a stationary process.

To prove iii, let $H^*(L)$ be the adjoint of $A(L)$. Since **every** element of the latter is a polynomial of degree $k$ the elements, $h_{ij}^*(L)$, of the adjoint are polynomials of degree $(q-1)k$; moreover, by the arguments given in the proof of Proposition 2

$$H^*(L) = (I - L)^{r-1}H(L), \quad H(1) \neq 0,$$

and the elements, $h_{ij}(L)$, of $H(L)$ are polynomials of degree $m = qk - k - r + 1$. Similarly, since $a^*(L) = (I - L)^r a(L)$ is a polynomial of degree $qk$ we conclude that $a(L)$ is a polynomial of degree $n = qk - r$. Hence, we obtain the representation $H(L)X'_{t\cdot} = a(L)I_q\epsilon'_{t\cdot}$, which displays $X$ as a $MARMA(m, n)$ process and thus concludes the proof of iii.

To prove iv, we proceed entirely in the manner of the proof of part iv of Proposition 2, so the proof is not repeated.

To prove v, we note that

$$H(L)A(L) = A(L)H(L) = (I - L)I_q a(L), \quad a(0) = 1, \quad a(1) \neq 0.$$

Since

$$|H(L)||A(L)| = (I - L)^q [a(L)]^q \text{ implies } |H(L)| = (I - L)^{q-r}[a(L)]^{q-1},$$

and $a(1) \neq 0$, we conclude that $\text{rank}[H(1)] = r$. Thus, from $H(1)A(1) = A(1)H(1) = 0$ and since $B'A(1) = 0$, with $B$ of full rank, we have that $H(1) = \Gamma B'$; moreover, since $H(0)A(0) = a(0)I_q$ we obtain $H_0 = I_q$, which concludes the proof of v.

To prove vi, we begin with the representation in v and note that since $H(1) \neq 0$ we have

$$H(L) = H(1) - (I - L)\bar{H}(L), \quad \bar{H}(L) = \sum_{j=0}^{m-1}\bar{H}_j L^j, \quad \bar{H}_j = \sum_{s=j+1}^{m-1}H_s,$$

and $H(1) = I_q + \bar{H}_0$. It follows, therefore, that

$$(I_q + \bar{H}_0)X'_{t.} - \bar{H}_0(I - L)X'_{t.} - \sum_{j=1}^{m-1} \bar{H}_j(I - L)X'_{t-j.} = a(L)I_q \epsilon'_{t.}.$$

Adding and subtracting $-LX'_{t.}$ in the left member we find, after some rearrangement,

$$G(L)(I - L)X'_{t.} = -\Gamma Z'_{t-1.} + a(L)I_q \epsilon'_{t.}, \quad G_0 = I_q, \quad G_j = -\bar{H}_j,$$

for $j = 1, 2, \ldots, m - 1$, thus concluding the proof of vi.

To prove vii we note that the first representation was already given in the proof of part ii; for the second representation, add and subtract $B'\Gamma Z'_{t-1.}$ in the right member, and add $-Z'_{t-1.}$ to both sides to obtain

$$(I - L)Z'_{t.} = -B'\Gamma Z'_{t-1.} - B' \left[ A^*(L) + (\Gamma B' - I_r)A^*(L)L \right] \epsilon'_{t.}$$

$$= -B'\Gamma Z'_{t-1.} + B'P(L)\epsilon'_{t.}, \quad P_0 = -A^*_0, \quad P_k = -(\Gamma B' - I_r)A^*_{k-1}$$

$$P_i = -[A^*_i + (\Gamma B' - I_r)A^*_{i-1}], \quad i = 1, 2, \ldots, k - 1.$$

q.e.d.

**Empirical Implementation**

In this section we make use of our discussion in Sections 4.2 and 4.3 to examine systematically the estimation of cointegrating vectors. First, we note that if we are dealing with an $I(1)$ process we are interested in estimating $S_{0,k}\Sigma S'_{0,k}$, and the characteristic vectors corresponding to its zero roots, if any. Evidently, the only information available to us is the data matrix $X$; second, we note that $X'X/T$ does not have a well-defined expectation as $T \to \infty$. It was shown in the previous section that we may define

$$\hat{M}^{(1)} = \frac{1}{T^2}X'X, \quad \text{such that} \quad \lim_{T \to \infty} E\hat{M}^{(1)} \tag{4.73}$$

converges to a well-defined entity and that the latter is finite. In the context of this section, we may be more precise as the following proposition indicates.

**Proposition 3.** Let $\{X_{t.} : t \in \mathcal{N}_+\}$ be a $q$-element vector process defined on the probability space $(\Omega, \mathcal{A}, \mathcal{P})$ and suppose it obeys

$$(I - L)X'_{t.} = \eta'_{t.} = \sum_{j=0}^{\infty} A_j \epsilon'_{t-j.}, \quad \sum_{j=0}^{\infty} \| A_j \| < \infty.$$

Then, for $X = (X_{t.})$, $t = 1, 2, \ldots T$, and

$$\mathbf{M}^{(1)} = \frac{X'X}{T^2}, \quad \lim_{T \to \infty} E\hat{M}^{(1)} = \Psi_\infty$$

where $\Psi_\infty$ is a well defined matrix with finite elements.

Proof: We begin by taking the forcing function to be $MMA(k)$. From Eqs. (4.10) and (4.38) we have, respectively, the representations

$$E\hat{M}^{(1)} = \frac{1}{T^2} \sum_{t=1}^{T} EX'_{t.}X_{t.} = \frac{1}{T^2} \sum_{r=1}^{k} S_{0,r-1} \Sigma S'_{0,r-1}$$

$$+ \frac{T(T - 2k + 1)}{2T^2} S_{0,k} \Sigma S'_{0,k}, \quad S_{ij} = \sum_{s=i}^{j} A_s.$$

It follows immediately that

$$\lim_{T \to \infty} E\hat{M}^{(1)} = \frac{1}{2} S_{0,k} \Sigma S'_{0,k} = \Psi_{(k)}, \qquad (4.74)$$

which is a well-defined matrix; if we let $k \to \infty$ we obtain

$$\lim_{k \to \infty} \lim_{T \to \infty} E\hat{M}^{(1)} = \frac{1}{2} S_{0,\infty} \Sigma S'_{0,\infty} = \Psi_{(\infty)}, \qquad (4.75)$$

which is also a well-defined matrix, given the premise of the proposition. To see whether the limits taken above are permissible in the order stated, consider the forcing function $MMA(\infty)$ and note that, from the discussion following Remark 1,

$$E\hat{M}^{(1)} = \frac{1}{T^2} \sum_{t=1}^{T} C(t) = \frac{1}{T^2} \sum_{t=1}^{T} \sum_{j=0}^{t-1} B_j, \quad B_j = A_j \Sigma S_{0j} + S_{0,j-1} \Sigma A'_j. \quad (4.76)$$

Changing the order of summation we obtain

$$E\hat{M}^{(1)} = \frac{1}{T^2} \sum_{j=0}^{T-1} \sum_{t=j+1}^{T} (t - j) B_j = \frac{1}{T^2} \sum_{j=0}^{T-1} \left( \frac{(T-j)(T-j+1)}{2} \right) B_j$$

$$= \frac{1}{2} \sum_{j=0}^{T-1} B_j + \frac{1}{2T} \sum_{j=0}^{T} B_j - \left( \frac{1}{T} + \frac{1}{2T^2} \right) \sum_{j=0}^{T-1} j B_j + \frac{1}{T^2} \sum_{j=0}^{T-1} j^2 B_j.$$

By the premise of the proposition, the first term of the last member above converges; the second term evidently converges to zero. The third term converges to zero by the requirements of Assertion 1. Since $\sum_{j=0}^{T-1}(j/T)B_j$ converges to zero with $T$ so does the last term (of the last member) above, because $\| (j/T)^2 B_j \| \le \| (j/T)B_j \|$. Thus, in the case where the forcing function is $MMA(\infty)$ as well, we obtain

$$\lim_{T\to\infty} E\hat{M}^{(1)} = \frac{1}{2}S_{0,\infty}\Sigma S_{0,\infty}. \tag{4.77}$$

q.e.d.

**Remark 7.** Note that the limit to which the (expectation of the) estimator of the proposition converges is not the appropriate matrix, but rather a multiple thereof. Precisely, the limit is an **underestimate** of the desired matrix, viz. $S_{0,k}\Sigma S'_{0,k}$ or, in a more general context, of $S_{0,\infty}\Sigma S'_{0,\infty}$.

The preceding discussion suggests, for the $I(1)$ process above, the unbiased estimator

$$\hat{M}^{(2)} = \frac{1}{T}\sum_{t=1}^{T}\left(\frac{X'_{t\cdot}X_{t\cdot}}{t}\right). \tag{4.78}$$

When the forcing function is $MMA(k)$, its expectation is

$$E\hat{M}^{(2)} = \Psi_1 \frac{1}{T}\sum_{t=1}^{T}\left(1 - \frac{k}{t}\right) + \Psi_0\frac{1}{T}\left(\sum_{t=1}^{T}\frac{1}{t}\right), \quad \text{or more appealingly}$$

$$= \Psi_1 + \Psi_0^* \frac{1}{T}\left(\sum_{t=1}^{T}\frac{1}{t}\right), \quad \text{where} \tag{4.79}$$

$$\Psi_1 = S_{0,k}\Sigma S'_{0,k}, \quad \Psi_0 = \sum_{r=1}^{k}S_{0,r-1}\Sigma S'_{0,r-1}, \quad \Psi_0^* = \Psi_0 - k\Psi_1.$$

The estimator above, $\hat{M}^{(2)}$, is a random matrix whose mean is (asymptotically) the matrix (some of) whose characteristic vectors constitute the cointegration matrix, plus a term that goes to zero at a rate between $T^{-1}$ and $T^{-2}$. At a later stage we present a test for cointegration based on the "sample covariance matrix" $\hat{M}^{(2)}$, as defined in Eq. (4.79).

### 4.4.3    Cointegration in $VAR(n)$ Processes

The development of this literature is due mainly to Johansen (J) (1988), (1991), and Ahn and Reinsel (AR) (1990) among others.

In exploring the implications of the cointegration assumption, and thus obtaining a procedure for estimating the cointegrating vector(s), the formulation based on a $VAR(n)$ process is more convenient in estimating the cointegrating vector(s); the reasoning and motivation, however, are essentially those put forth by EG, although the initial formulation is quite different. We begin with the formulation given in J.

The process $X$, as in Definition 1 of this chapter, is $CI(1, 1, r)$ but, in contrast to the EG approach, it is assumed to obey

$$\Pi(L)X'_{t\cdot} = \epsilon'_{t\cdot}, \quad \Pi(L) = \sum_{j=0}^{n} \Pi_j L^j, \quad \Pi_0 = I_q, \tag{4.80}$$

where $\epsilon = \{\epsilon_{t\cdot} : t \in \mathcal{N}_+\}$ is $MWN(\Sigma)$. Evidently, $\Pi(L)$ is **is not** invertible, in the sense that **its characteristic equation** has at least one **unit** root. In fact if it did not, and all its roots were outside the unit circle, we should be able to represent the process as

$$X'_{t\cdot} = [\Pi(L)]^{-1}\epsilon'_{t\cdot},$$

which, under our assumptions, would be a **stationary** process.

**Remark 8.** The two approaches, viz. that of AR and J on one hand and that of EG on the other, have certain similarities and certain differences, which we elucidate before we proceed with our discussion.

In the EG formulation the model is $(I - L)X'_{t\cdot} = C(L)\epsilon'_{t\cdot}$, where $C(L)$ **is not invertible.** Using part iii of Proposition 2, this model has the equivalent representation

$$H(L)X'_{t\cdot} = d(L)I_q\epsilon'_{t\cdot}, \quad d(1) \neq 0. \tag{4.81}$$

Since the scalar lag operator $d(L)$ is invertible, we have the equivalent representation

$$H^*(L)X'_{t\cdot} = \epsilon'_{t\cdot}, \quad H^*(L) = [d(L)]^{-1}H(L), \quad \sum_{i=0}^{\infty} \| H_i^* \| < \infty. \tag{4.82}$$

Thus, the EG formulation is equivalent to a $VAR(\infty)$ model, while the J and AR formulations involve a $VAR(n)$ model. In the EG formulation the cointegrating vectors **reside** in $H(1)$ **which is not directly estimable** [15] while, as we shall see presently, in the J and AR formulations they reside in $\Pi(1)$ **which is directly estimable**. Further, in the EG formulation the cointegration rank $r$ is precisely the **number of unit roots** in $|C(z)| = 0$, where $z$ is a complex indeterminate; as we shall also see presently, in the J and AR formulation the cointegration rank $r$ is given by $r = q - r_0$, where $r_0$ is the number of **unit roots** in $|\Pi(z)| = 0$!

The implications of the cointegration hypothesis in the J and AR context are given in the proposition below.

**Proposition 4.** Let $X = \{X'_{t.} : t \in \mathcal{N}_+\}$ be a stochastic sequence defined on the probability space ($\Omega$, $\mathcal{A}$, $\mathcal{P}$); suppose further that

1. it is of the form

$$\Pi(L)X'_{t.} = \epsilon'_{t.}, \quad \Pi(L) = \sum_{j=0}^{n} \Pi_j L^j, \quad \Pi_0 = I_q;$$

2. $X \sim CI(1,1,r)$;

3. $B$ is the $q \times r$ matrix containing the $r$ linearly independent cointegrating vectors.

The following statements are true:

i. there exists a representation

$$\Pi(L) = \Pi(1) - (I - L)P^*(L), \quad P^*(L) = \sum_{j=0}^{n-1} \Pi_j^* L^j, \quad \Pi_j^* = \sum_{i=j+1}^{n} \Pi_i;$$

ii. the operator $P^*(L) = \sum_{j=0}^{n-1} \Pi_j^* L^j$ does not have a unit root factorization, i.e. $P^*(L) \neq (I - L)P^{**}(L)$;

iii. $\text{rank}[\Pi(1)] = r, \quad q > r$;

---

[15] This would not be the case if we expressed the EG formulation *ab initio* in terms of the representation of Eq. (4.82) in Remark 8.

iv. the characteristic equation of the process, $|\Pi(z)| = 0$, has $r_0 = q - r$ **unit roots**;

v. $X$ has a representation as $MIMA(\infty)$; specifically, if $\pi^*(L) = |\Pi(L)|$, and $H_*(L)$ is the adjoint of

$$\Pi(L) = [\pi_{is}(L)], \quad \pi_{is}(L) = \sum_{j=0}^{n} \pi_{rs}^{(j)} L^j,$$

where $\pi_{is}^{(j)}$ is the $i, s$ element of the matrix $\Pi_j$ **then**

$$(I - L)X'_{t.} = H^*(L)\epsilon'_{t.}, \quad H^*(L) = [\pi(L)]^{-1}H(L), \quad H(1) \neq 0, \quad \text{where}$$

$$H_*(L) = (I - L)^{r_0 - 1}H(L), \quad \pi^*(L) = (I - L)^{r_0}\pi(L), \quad \pi(1) \neq 0.$$

vi. there exist matrices $B$, $\Gamma$ of dimension $q \times r$ and rank $r$ such that

$$\Pi(1) = \Gamma B', \quad B'H(1) = 0, \quad H(1)\Gamma = 0, \quad H(0) = I_q;$$

vii. there exists an ECM representation,

$$(I - L)X'_{t.} = -\Gamma Z'_{t-1.} + \sum_{j=1}^{n-1} \Pi_j^*(I - L)X'_{t-j.} + \epsilon'_{t.},$$

viii. the cointegral vector $Z_{t.} = X_{t.}B$ has the representation

$$Z'_{t.} = G(L)\epsilon'_{t.}, \quad G_0 = B', \quad G_i = (\Gamma'\Gamma)^{-1}\Gamma'\left(\sum_{j=0}^{i} \Pi_j^* H_{i-j}^*\right), \quad i \geq 1.$$

Proof: To prove i, we divide $\Pi(L)$ by $L - I$, thus obtaining

$$\Pi(L) = \Pi(1) - (I - L)P^*(L), \quad P^*(L) = \sum_{j=0}^{n-1} \Pi_j^* L^j, \quad \Pi_j^* = \sum_{s=j+1}^{n} \Pi_s.$$

To prove ii, we note that by the representation above

$$\Pi(1)X'_{t.} = P^*(L)(I - L)X'_{t.} + \epsilon'_{t.},$$

which shows that the left member above must be stationary by condition 2 of the premise; now if $P^*(L) = (I - L)P^{**}(L)$, i.e. if it has a unit root factorization, we should be able to demonstrate that $X$ is cointegrated of order 2. Hence, $P^*(L)$ does not have a unit root factorization and thus $P^*(1) \neq 0$.

To prove iii, we note that by condition 2 of the premise $X \sim C(1,1,r)$, which implies that $\text{rank}[\Pi(1)] = r$. This is so since the representation in ii, exhibits $q$ linear combinations of $X_t$. which are **stationary**. The cointegration hypothesis, on the other hand, implies that only $r$ linear combinations may be linearly independent, which completes the proof of iii.

To prove iv, we note that evaluating $|\Pi(z)| = 0$ at $z = 1$ yields precisely the same result as $|\lambda I_q - \Pi(1)| = 0$ evaluated at $\lambda = 0$. Since $\Pi(1)$ is of rank $r$, the latter has therefore $r_0 = q - r$ **zero roots**, or alternatively, $|\Pi(z)| = 0$ has $r_0$ **unit roots**.

To prove v, note that multiplying on the left, or the right, by the adjoint $H_*(L)$ we obtain

$$H_*(L)\Pi(L) = \Pi(L)H_*(L) = \pi^*(L)I_q = (I - L)^{r_0}\pi(L)I_q, \quad \pi(1) \neq 0.$$

Since the adjoint is the sum of products of terms one each from each row and column, each product contains $s_1$ terms from $\Pi(1)$ and $s_2$ terms from $(I - L)P^*(L)$, such that $s_1 + s_2 = q - 1$. Since $\text{rank}[\Pi(1)] = r$, products that contain $s_1 > r$ terms from $\Pi(1)$ will **vanish** upon summation; hence each term in the adjoint contains $s_2 \geq q - 1 - r = r_0 - 1$ terms from $(I - L)P^*(L)$, which implies that

$$(I - L)^{r_0 - 1}H(L)\Pi(L) = (I - L)^{r_0}\pi(L)I_q.$$

Taking determinants of both sides, we find

$$(I - L)^{(r_0 - 1)q}|H(L)|(I - L)^{r_0}\pi(L) = (I - L)^{r_0 q}[\pi(L)]^q,$$

or $|H(L)| = (I - L)^{q - r_0}[d(L)]^{q - 1}$. This shows that $H(L)$ has $q - r_0 = r$ **unit roots**, and hence it is of rank $q - r = r_0$. It follows that

$$(I - L)X'_{t.} = [\pi(L)]^{-1}H(L)\epsilon'_{t.},$$

which concludes the proof of v.

To prove vi, note that from the argument above

$$H(L)\Pi(L) = \Pi(L)H(L) = (I - L)\pi(L), \quad \pi(1) \neq 0.$$

Consequently, we find

$$H(1)\Pi(1) = 0, \quad \Pi(1)H(1) = 0$$

so that $H(1)$ spans both the row **and** column null space of $\Pi(1)$, and vice versa, i.e. $\Pi(1)$ spans both the row and column null space of $H(1)$ as well. To see this, note that since $\text{rank}[H(1)] = q - r = r_0$, the dimension of its column (as well as row) null space is $r$; we have already shown that $\text{rank}[\Pi(1)] = r$, so that $\Pi(1)$ does, indeed, span the column (and row) null space of $H(1)$. By the rank factorization theorem there exist $q \times r$ matrices $\Gamma$, and $B$ each of rank $r$, such that $\Pi(1) = \Gamma B'$. Since the **rows** of $B'$ are **linearly independent**, $H(1)\Pi(1) = 0$ implies $H(1)\Gamma = 0$. Similarly, $\Pi(1)H(1) = 0$ implies $B'H(1) = 0$, which completes the proof of vi.

To prove vii, we employ the representation of part i, and note that $\Pi_0^* = \Pi(1) - I_q$; adding $-LX_t'$ to both sides we obtain

$$(I - L)X_{t\cdot}' = -\Pi(1)X_{t-1\cdot}' + \sum_{j=1}^{n-1} \Pi_j^*(I - L)X_{t-j\cdot}' + \epsilon_{t\cdot}'$$

which, together with the result in part vi, completes the proof of vii.

To prove viii, we use the representation in parts i and v, so that

$$\Gamma B' X_{t\cdot}' = P^*(L)H^*(L)\epsilon_{t\cdot}' + \epsilon_{t\cdot}'.$$

This exhibits $B$ as the matrix containing the cointegrating vectors. Finally, premultiply by $(\Gamma'\Gamma)^{-1}\Gamma'$ to obtain

$$Z_{t\cdot}' = B'X_{t\cdot}' = G(L)\epsilon_{t\cdot}', \quad G(L) = (\Gamma'\Gamma)^{-1}\Gamma' \left( P^*(L)H^*(L) + I_q \right),$$

where

$$G_0 = (\Gamma'\Gamma)^{-1}\Gamma' \left( I + \Pi_0^* H^*(0) \right) = (\Gamma'\Gamma)^{-1}\Gamma'\Pi(1) = B'.$$

The last equality is valid because $H^*(0) = I_q$ and $\Pi(1) = I_q + \Pi_0^*$.

q.e.d.

## 4.4.4   Empirical Implementation

In the preceding section, we gave a complete characterization of the implications of the cointegration hypothesis in the context of the $VAR(n)$. We now turn to the problem of estimating the matrix of cointegrating vectors, $B$, as well as the other parameters of the model. To give this problem the familiar look of systems of GLM, we depart from the notation in J, by defining

$$y_{t\cdot} = (I - L)X_{t\cdot} + X_{t-1\cdot}\Pi(1)' = \Delta X_{t\cdot} + X_{t-1\cdot}B\Gamma',$$

$$x_{t\cdot} = (\Delta X_{t-1\cdot}, \Delta X_{t-2\cdot}, \Delta X_{t-3\cdot}, \ldots, \Delta X_{t-n+1\cdot}), \quad \epsilon_{t\cdot} = u_{t\cdot},$$

$$\Pi^* = (\Pi_1^*, \Pi_2^*, \ldots, \Pi_{n-1}^*)',$$

$$Y = (y_{t\cdot}), \quad X = (x_{t\cdot}), \quad U = (u_{t\cdot}), \ t = 1, 2, 3, \ldots T,$$

$$y = \text{vec}(Y), \quad \pi^* = \text{vec}(\Pi^*), \quad u = \text{vec}(U), \quad \gamma = \text{vec}(\Gamma'). \quad (4.83)$$

Thus, we may write

$$y_{t\cdot} = x_{t\cdot}\Pi^* + u_{t\cdot}, \quad Y = X\Pi^* + U, \quad \text{or} \ y = (I_q \otimes X)\pi^* + u, \quad (4.84)$$

for a single observation and the entire sample, respectively, and define the estimation problem as follows: Maximize the likelihood function with respect to the unknown parameters **subject to the condition** that $B$ is $q \times r$ of **known rank** $r$. The loglikelihood (LF) of the $T$ observations is given by

$$F_0 = -\frac{qT}{2}\ln(2\pi) - \frac{T}{2}\ln|\Sigma| - \frac{1}{2}\text{tr}(Y - X\Pi^*)\Sigma^{-1}(Y - X\Pi^*)'. \quad (4.85)$$

From Dhrymes (1984), p. 106, we have

$$\text{tr}(Y - X\Pi^*)\Sigma^{-1}(Y - X\Pi^*)' = [y - (I_q \otimes X)\pi^*]'(\Sigma^{-1} \otimes I_T)[y - (I_q \otimes X)\pi^*],$$

and solving the first-order condition $(\partial F_0/\partial \pi^*) = 0$, we find

$$\hat{\pi}^* = [I_q \otimes (X'X)^{-1}X']y. \quad (4.86)$$

Inserting this in Eq. (4.85), we obtain the concentrated LF

$$F_1 = -\frac{qT}{2}\ln(2\pi) - \frac{T}{2}\ln|\Sigma| - \frac{1}{2}y'(\Sigma^{-1}\otimes N)y, \quad N = I_T - X(X'X)^{-1}X'. \quad (4.87)$$

Next, we note that $y = \Delta p + (I_q \otimes P_{-1}B)\gamma$, where

$$P = (X_{t\cdot}), \quad P_{-i} = (X_{t-i\cdot}), \quad t = 1, 2, 3, \ldots, T, \quad i = 1, 2, 3, \ldots n - 1,$$

$$p = \text{vec}(P), \quad p_{-i} = \text{vec}(P_{-i}), \quad \gamma = \text{vec}(\Gamma'), \quad (4.88)$$

and the concentrated LF may be rendered as

$$F_1 = -\frac{qT}{2}\ln(2\pi) - \frac{T}{2}\ln|\Sigma| \tag{4.89}$$

$$-\frac{1}{2}[\Delta p + (I_q \otimes P_{-1}B)\gamma]'(\Sigma^{-1} \otimes N)[\Delta p + (I_q \otimes P_{-1}B)\gamma].$$

Solving the first-order conditions $(\partial F_1/\partial\gamma) = 0$, we find

$$\hat{\gamma} = -[I_q \otimes (B'P'_{-1}NP_{-1}B)^{-1}B'P'_{-1}N]\Delta p. \tag{4.90}$$

Noting that

$$(I_q \otimes N)[\Delta p + (I_q \otimes P_{-1}B)\hat{\gamma}] = (I_q \otimes N^*)(I_q \otimes N)\Delta p, \tag{4.91}$$

$$N^* = I_T - NP_{-1}B(B'P'_{-1}NP_{-1}B)^{-1}B'P'_{-1}N, \tag{4.92}$$

we rewrite the concentrated LF as

$$F_2 = -\frac{qT}{2}\ln(2\pi) - \frac{T}{2}\ln|\Sigma|$$

$$-\frac{1}{2}[(I_q \otimes N^*N)\Delta p]'(\Sigma^{-1} \otimes I_T)[(I_q \otimes N^*N)\Delta p]. \tag{4.93}$$

Using the results in Dhrymes (1984), p. 106, and "rematricizing" the last expression in the LF, we rewrite it as

$$F_2 = -\frac{qT}{2}\ln(2\pi) - \frac{T}{2}\ln|\Sigma| - \frac{T}{2}\text{tr}\Sigma^{-1}S, \quad S = \frac{1}{T}(N\Delta P)'N^*(N\Delta P). \tag{4.94}$$

Maximizing $F_2$ with respect to the elements of $\Sigma^{-1}$, we obtain

$$\frac{\partial F_2}{\partial \text{vec}(\Sigma^{-1})} = \frac{T}{2}\text{vec}(\Sigma)' - \frac{T}{2}\text{vec}(S) = 0, \quad \text{or} \quad \hat{\Sigma} = S. \tag{4.95}$$

Inserting this in Eq. (4.95), the (ultimately) concentrated LF becomes

$$F_3 = -\frac{qT}{2}[\ln(2\pi) + 1] - \frac{T}{2}\ln|S|, \tag{4.96}$$

which is now to be **maximized with respect to** $B$.

To continue, it is convenient to define

$$W = N\Delta P = N(P - P_{-1}), \quad V = NP_{-1}, \tag{4.97}$$

and to omit the term $(1/T)$ from the definition of $S$; in this notation, the maximization of $F_3$ in Eq. (4.96) is **equivalent to the minimization of**

$$D(B) = |S| = |W'W - W'VB(B'V'VB)^{-1}B'V'W| \qquad (4.98)$$

$$= |W'W||I_q - J|, \quad J = W'VB(B'V'VB)^{-1}B'V'W(W'W)^{-1}.$$

From Dhrymes (1984) p. 39, we have that

$$|I_q - J| = |B'V'VB|^{-1}|B'[V'V - V'W(W'W)^{-1}W'V]B|. \qquad (4.99)$$

Since the function $D(B)$ is **homogeneous of degree zero in** $B$, we need to impose a normalization, otherwise the problem is not well-defined, in other words there is an infinite number of solutions. A convenient normalization, under the circumstances, is $B'V'VB = I_r$, so that the quantity to be minimized, with respect to $B$, is

$$D(B) = |W'W||B'[V'V - V'W(W'W)^{-1}W'V]B|. \qquad (4.100)$$

To do so we proceed in a roundabout fashion. Consider the characteristic roots of $V'W(W'W)^{-1}W'V$ in the metric of $V'V$:

$$|\lambda V'V - V'W(W'W)^{-1}W'V| = 0. \qquad (4.101)$$

Since the matrix $V'V - V'W(W'W)^{-1}W'V$ is **positive definite,** all such characteristic roots are (positive and) less than unity [see Dhrymes (1984), p. 75. Let the roots $\{\lambda_j : j = 1, 2, 3, \ldots, q\}$ be arranged in **decreasing magnitude,** let the corresponding characteristic vectors be $\{g_{\cdot j} : j = 1, 2, 3, \ldots, q\}$, define

$$\Lambda = \text{diag}(\lambda_1, \lambda_2, \lambda_3, \ldots, \lambda_q), \quad G = (g_{\cdot 1}, g_{\cdot 2}, g_{\cdot 3}, \ldots, g_{\cdot q}), \qquad (4.102)$$

and impose the normalization $G'V'VG = I_q$. Since

$$V'W(W'W)^{-1}W'VG = V'VG\Lambda, \qquad (4.103)$$

we see that

$$D(G) = |W'W||G'[V'V - V'W(W'W)^{-1}W'V]G| = |W'W||I_q - \Lambda|. \quad (4.104)$$

Thus, the estimator we seek, $\hat{B}$, must be a subset of the columns of the matrix $G$, it must be of rank $r$, and must minimize $D(G)$. The solution is then obvious: **We must choose $\hat{B}$ to be the submatrix $G_{(r)}$ containing the $r$ characteristic vectors corresponding to the $r$ largest roots.** Thus, the maximized value of the LF is given by

$$F_4(\hat{B}) = \frac{qT}{2}[\ln(2\pi) + 1] - \frac{T}{2}\left|\frac{W'W}{T}\right| - \frac{T}{2}\sum_{j=1}^{r}\ln(1 - \lambda_j). \qquad (4.105)$$

We have therefore proved Proposition 5.

**Proposition 5.** Consider the $VAR(n)$ model of Proposition 4,

$$\Pi(L)X'_{t.} = \epsilon'_{t.}, \quad \text{or} \quad \sum_{j=0}^{n}\Pi_j X'_{t-j.} = \epsilon'_{t.}, \quad \Pi_0 = I_q,$$

and suppose $\{\epsilon'_{t.} : t \in \mathcal{N}\}$ is a sequence of i.i.d. $N(0, \Sigma)$ vectors, with $\Sigma > 0$. **If the sequence $\{X_t : t \in \mathcal{N}_+\}$ is cointegrated of known rank** $r$, the following statements are true:

i. the (log) likelihood function of the problem is given, in the notation of the equation set Eq. (4.84), by

$$F_0 = -\frac{qT}{2}\ln(2\pi) - \frac{T}{2}\ln|\Sigma| - \frac{1}{2}\text{tr}(Y - X\Pi^*)\Sigma^{-1}(Y - X\Pi^*)';$$

ii. the ML estimators of the unknown parameters in $\Pi^*$ and $\Gamma'$ are given by

$$\hat{\pi^*} = [I_q \otimes (X'X)^{-1}X']y, \quad \hat{\gamma} = [I_q \otimes (B'P'_{-1}NP_{-1}B)^{-1}B'P'_{-1}N]\Delta p;$$

iii. defining $N$ as in Eq. (4.87), $N^*$ as in Eq. (4.92), $V$ and $W$ as in Eq. (4.97), the ML estimator of $B$ is given by the characteristic vectors corresponding to the $r$ **largest** roots of

$$|\lambda V'V - V'W(W'W)^{-1}W'V| = 0,$$

and the ML estimator of $\Sigma$ is given by

$$\hat{\Sigma} = \frac{1}{T}W'N^*W.$$

**Remark 9.** Even though the preceding discussion has accomplished the objective of obtaining an estimator for the cointegrating vectors, we should note two aspects. First, in expressing $\Pi(1) = \Gamma B'$, there is an ambiguity, since the choice $\Gamma^* = \Gamma Q'^{-1}$, $B^* = BQ$, **for an arbitrary nonsingular matrix** $Q$, will do just as well. Thus, what we have obtained is an estimator for **a basis of the space spanned by the "true" cointegrating vectors**.

Second, there is no intuition as to the meaning to this solution, nor is it clear how this solution is connected with the notion of "long-run equilibrium".

In the popular interpretation of this particular procedure for obtaining cointegrating vectors, much is made of its relation to the process of defining **canonical variates**. We give a brief discussion of this topic below; for a more extended discussion the reader may wish to consult Dhrymes (1970), p. 42. Thus, suppose

$$x \sim N(0, \Sigma), \quad x = (x'_{(1)}, x'_{(2)})', \quad \Sigma = \begin{bmatrix} \Sigma_{11} & \Sigma_{12} \\ \Sigma_{21} & \Sigma_{22} \end{bmatrix} \tag{4.106}$$

such that $x$ is an $n$-element vector, $x_{(1)}$ contains the first $k$ elements, and $\Sigma$ has been partitioned **conformably**. Suppose, further, that we wish to form linear combinations $x'_{(1)}g$, and $x'_{(2)}h$, **such that the correlation between these two linear combinations is maximized**, given that their **variance is normalized to unity**. Thus, we deal with the Lagrangian

$$F = g'\Sigma_{12}h + \frac{1}{2}\lambda_1(1 - g'\Sigma_{11}g) + \frac{1}{2}\lambda_2(1 - h'\Sigma_{22}h) \tag{4.107}$$

and the first-order conditions imply

$$-\lambda_1\Sigma_{11}g + \Sigma_{12}h = 0, \quad \Sigma_{21}g - \lambda_2\Sigma_{22}h = 0. \tag{4.108}$$

Pre multiplying the two equations above by $g'$ and $h'$, respectively, we find

$$g'\Sigma_{12}h = \lambda_1, \quad h'\Sigma_{21}g = \lambda_2,$$

which shows that $\lambda_1 = \lambda_2 = \lambda$. Inserting this in Eq. (4.108), and solving (the second equation) for $h$, we find $h = (1/\lambda)\Sigma_{22}^{-1}\Sigma_{21}g$. Inserting this in the first equation of Eq. (4.108), and noting that $\lambda = 0$ is not relevant, we conclude that the first equation of Eq. (4.109) is equivalent to

$$[\lambda^2\Sigma_{11} - \Sigma_{12}\Sigma_{22}^{-1}\Sigma_{21}]g = 0. \tag{4.109}$$

The solution we are seeking is simply the **characteristic vector corresponding to the largest root**, say $\lambda_1^2$, of $\Sigma_{12}\Sigma_{22}^{-1}\Sigma_{21}$ **in the metric of** $\Sigma_{11}$. Moreover, the vector in question, say $g._1$, obeys

$$\Sigma_{12}\Sigma_{22}^{-1}\Sigma_{21}g._1 = \lambda_1^2\Sigma_{11}g._1.$$

We also obtain

$$h._1 = \frac{1}{\lambda_1}\Sigma_{22}^{-1}\Sigma_{21}g._1,$$

and verify that, given the normalizations above,

$$\mathrm{Corr}(g._1'x_{(1)}, h._1'x_{(2)}) = \frac{1}{\lambda_1}g._1'\Sigma_{12}\Sigma_{22}^{-1}\Sigma_{21}g._1 = \lambda_1. \qquad (4.110)$$

In fact, it may be shown that if $G_{(r)}$ contains the $r$ characteristic vectors corresponding to the $r$ largest roots in Eq. (4.109), the two sets of linear combinations, $G_{(r)}'x_{(1)}$ and $H_{(r)}'x_{(2)}$, **are pairwise uncorrelated and obey**

$$E[G_{(r)}'x_{(1)}x_{(2)}'H_{(r)}] = \Lambda_{(r)}, \quad r \leq k. \qquad (4.111)$$

This means that corresponding elements of the two linear combinations have unit variance and exhibit **maximal correlation, given that they are uncorrelated with other pairs.**

The connection to the estimation problem in Proposition 5 is made as follows: If we associate

$$x_{(1)}' \sim V, \quad x_{(2)}' \sim W, \quad \Sigma_{11} \sim V'V, \quad \Sigma_{12} = V'W, \quad \Sigma_{22} \sim W'W,$$

the last phase of the estimating procedure therein represents a rough analog to the procedure for obtaining canonical variates. However, in the context of the $VAR(n)$ model dealt with in Proposition 5, there is **no intention, or requirement**, that we obtain canonical variates; in fact, the analog of $H_{(r)}$ **never arises therein.** It is just that the estimation procedure in the context of that model **resembles, in part, the procedure for obtaining canonical variates**, but there is no **instrinsic connection between them.**

**An Alternative Estimator**

The estimation procedure examined in this section was suggested by Ahn and Reinsel (1990) and employs the same approach employed in an earlier section, particularly in connection with Eq. (4.60). Since the matrix of cointegrating vectors is defined implicitly by the rank factorization of $\Pi(1)$, **it is arbitrary to within left multiplication by an arbitrary nonsingular matrix** $Q$, i.e.

$$\Pi(1) = \Gamma B' \quad \text{and} \quad \Pi(1) = \Gamma^* B^{*'}, \quad B^* = QB, \quad \Gamma^* = \Gamma Q'^{-1}.$$

Thus, we have an "identification" problem, which may be "resolved" by the conditions imposed in Eq. (4.60), viz.

$$B = (I_r, B_0')' = Q_1 + Q_2 B_0, \quad Q_1 = \begin{bmatrix} I_r \\ 0 \end{bmatrix}, \quad Q_2 = \begin{bmatrix} 0 \\ I_{q-r} \end{bmatrix}. \qquad (4.112)$$

In the discussion of the Johansen formulation, the estimator of $B$ contains the characteristic vectors corresponding to the $r$ **largest roots** of

$$|\lambda V'V - V'W(W'W)^{-1}W'V| = 0. \qquad (4.113)$$

Now, characteristic vectors are arbitrary to within normalization so that in order to gain uniqueness we impose on the estimator of $B$ $r$ restrictions, **without loss of generality**. Our objective, however, is not to **estimate uniquely** certain characteristic vectors. It is, rather, to estimate **uniquely** the rank factorization component $B$. This requires, for uniqueness, $r^2$ restrictions! The two approaches may not be compatible, in the sense that the estimator of $B$, as restricted by Eq. (4.112), is a simple transformation of the estimator of $B$, obtained as the charateristic vectors corresponding to the $r$ largest roots of Eq. (4.113), and normalized by the condition $B'V'VB = I_r$.

To estimate parameters in this formulation, we proceed as before through Eq. (4.96), but **we do not seek to obtain an estimator for** $B$, **in closed form.** Instead, we seek to estimate $B$ by iterative methods, i.e. we return to the form of the LF as in Eq. (4.89) and write it as

$$F_1^* = -\frac{qT}{2}\ln(2\pi) - \frac{T}{2}\ln|\hat{\Sigma}| \qquad (4.114)$$

$$-\frac{1}{2}[\Delta p - (I_q \otimes P_{-1}B)\hat{\gamma}]'(\hat{\Sigma}^{-1} \otimes N)[\Delta p - (I_q \otimes P_{-1}B)\hat{\gamma}].$$

However, taking into account Eq. (4.112) we obtain

$$(I_q \otimes P_{-1}B)\hat{\gamma} = (\hat{\Gamma} \otimes P_{-1})\text{vec}(B) = (\hat{\Gamma} \otimes P_{-1})q_1 + (\hat{\Gamma} \otimes P_{-1})(I_r \otimes Q_2)b,$$

where $q_1 = \text{vec}(Q_1)$ and $b = \text{vec}(B_0)$. It follows that the LF of Eq. (4.114) may be written more conveniently as

$$F_1^* = -\frac{qT}{2}\ln(2\pi) - \frac{T}{2}\ln|\hat{\Sigma}| \tag{4.115}$$

$$-\frac{1}{2}[\Delta p^* - (\hat{\Gamma} \otimes P_{-1}Q_2)b]'(\hat{\Sigma}^{-1} \otimes N)[\Delta p^* - (\hat{\Gamma} \otimes P_{-1}Q_2)b],$$

where $\Delta p^* = \Delta p - (\hat{\Gamma} \otimes P_{-1})q_1$. If we treat $\hat{\Sigma}$, $\hat{\Gamma}$, and $\Delta p^*$ (as well as $\hat{\Pi}^*$) as having been determined by a **prior** iteration, we may obtain an estimator for $b$ by maximizing $F_1^*$; the first-order conditions yield, at the $s^{th}$ iteration,

$$\hat{b}_{(s)} = H^{-1}(\hat{b}_{(s-1)})g(\hat{b}_{(s-1)}), \quad \text{where} \tag{4.116}$$

$$H(\hat{b}_{(s-1)}) = (\hat{\Gamma}'\hat{\Sigma}^{-1}\hat{\Gamma} \otimes Q_2'V'VQ_2), \quad g(\hat{b}_{(s-1)}) = [\hat{\Gamma}'\hat{\Sigma}^{-1} \otimes Q_2'V']\Delta p^*.$$

Since we had already determined

$$\hat{\Gamma}'_{(s)} = (\hat{B}'_{(s-1)}V'V\hat{B}_{(s-1)})^{-1}\hat{B}'_{(s-1)}V'W, \quad \Delta p^*_{(s)} = \Delta p - (\hat{\Gamma}_{(s-1)} \otimes P_{-1})q_1,$$

$$\hat{\Sigma}_{(s)} = \frac{1}{T}W'[I_T - V\hat{B}_{(s-1)}(\hat{B}'_{(s-1)}V'V\hat{B}_{(s-1)})^{-1}\hat{B}'_{(s-1)}V']W, \tag{4.117}$$

the estimation problem is completely resolved upon convergence.

It is worth pointing out that all estimators, with the exception of $\hat{b}$, are **exclusively** functions of **stationary** entities.

## 4.5 $VAR(n)$ with a Constant Term

In Chapter 3 we briefly examined the model

$$y_t = \lambda y_{t-1} + \beta + u_t,$$

and we noted that this representation, with initial conditions $y_t = 0$ for $t \leq 0$ implies, in the case of a unit root ($\lambda = 1$), that the process $y_t$ obeys

$$y_t = \beta t + \eta_t, \quad \eta_t = \sum_{s=1}^{t} u_s.$$

More generally, if we take the position that this process came into existence only at "time" $t = 0$ the equation describing the process yields $\Delta y_t = \beta + u_t$, which is consistent with the view that its "solution" or "integral" is given by

$$y_t = \alpha + \beta t + \eta_t, \quad \eta_t = \sum_{s=1}^{t} u_s,$$

where $\alpha$ is a constant akin to the constant of integration. Thus, even if $u$ is a $WN(\sigma^2)$ process, what we have in $y$ is a process whose mean is a **time trend** and its random component is $I(1)$. If $\beta = 0$, $y$ is generally an $I(1)$ with a **nonzero constant** mean. This is, perhaps, a more appropriate view in econometric applications since it is rare that economic time series have mean zero; moreover, in this context, we ought to deal with entities of the form $y_t - \bar{y}$ rather than $y_t$.

In this section we investigate the situation where a cointegrated $VAR(n)$ model has a constant term, i.e. we investigate the model

$$\Pi(L)X'_{t\cdot} = \mu + \epsilon'_{t\cdot}, \quad t = 1, 2, 3, \ldots. \tag{4.118}$$

The questions we wish to answer are as follows:

i. Is it always the case that the representation in Eq. (4.118) implies that the $X$-process is one with (nonrandom) time trends?

ii. Whether it is or not, how does the presence of the constant term affect the estimation of parameters and the limiting distribution of the resulting estimators? More properly the answer to this last question will be dealt with in the next chapter when limiting distribution issues are discussed.

Putting the model in its ECM form yields

$$(I - L)X'_{t\cdot} = -\Pi(1)X'_{t-1\cdot} + \sum_{i=1}^{n-1} \Pi_i^*(I - L)X'_{t-i\cdot} + \mu + \epsilon'_{t\cdot}$$

As it stands, the ECM representation does not suggest how the cointegration hypothesis may be used to shed light on the issue of whether a **cointegrated** $VAR(n)$, with a constant term, implies that the underlying $X$-process invariably contains nonrandom time trends. In what follows we show that this need not be the case, if certain conditions hold.

To bring the cointegration assumption to bear on this issue we utilize part v of Proposition 4 of this chapter. Thus, in Eq. (4.118), if we premultiply by the adjoint of $\Pi(L)$, and cancel the term $(I - L)^{r_0 - 1}$ from both sides of the equation, we find

$$\pi(L)(I - L)X'_{t.} = H(1)\mu + H(L)\epsilon'_{t.}. \qquad (4.119)$$

Since by part vi of Proposition 4, the row (and column) null space of $H(1)$ is of rank [16] $q - r$, by the rank factorization theorem there exist $q \times q - r$ matrices $C_1$, $C_2$ such that $H(1) = C_1 C'_2$. Consequently, from

$$0 = \Pi(1)H(1) = \Gamma B' C_1 C'_2, \quad 0 = H(1)\Pi(1) = C_1 C'_2 \Gamma B', \qquad (4.120)$$

and the fact that the columns of $C_1$ are linearly independent, we conclude that $C'_2 \Gamma B' = 0$; since the rows of $B'$ are linearly independent, we also conclude that $C'_2 \Gamma = 0$. A similar argument will show that $B' C_1 = 0$, so that we may write

$$\Pi(1) = \Gamma B', \quad H(1) = C_1 C'_2, \quad B' C_1 = 0 \quad C'_2 \Gamma = 0.$$

Thus, if $\mu$ lies in the space spanned by $\Gamma$, and $C'_2$ lies in the orthogonal complement of that space it follows that $H(1)\mu = 0$. For such a case, solving Eq. (4.119), we conclude that the underlying $X$-process is an $I(1)$ process **without nonrandom time trends**, but possibly with a constant term, which would disappear upon differencing. If this condition does not hold, the underlying $X$-process will be the sum of an $I(1)$ stochastic process, a nonrandom time trend, and possibly a constant term.

---

[16] It might appear from part vi that we are only guaranteed that the rank of $H(1)$ is **at most** $q - r$, rather than precisely $q - r$. We now show that the latter is the true situation. To this end suppose that the rank of $H(1)$ is less than $q - r$, say $q - r - 1$. Then there exists a vector $b^*$, linearly independent of the columns of $B$, such that $B^* = (B, b^*)$ [which is $q \times (r + 1)$ matrix of rank $r + 1$] is a **basis** of the row null space of $H(1)$. In Eq. (4.119) suppose $\mu = 0$, and "divide" $H(L)$ by $L - I$ to obtain $H(L) = H(1) - (I - L)H^*(L)$, where $H^*$ is a matrix polynomial lag operator of order one less than $H(L)$. Premultiplying Eq. (4.119) by $B^{*'}$ we obtain

$$\pi(L)(I - L)B^{*'}X'_{t.} = B^{*'}H(1)\epsilon'_{t.} - (I - L)B^{*'}H^*(L)\epsilon'_{t.} = -(I - L)B^{*'}H^*(L)\epsilon'_{t.}.$$

Canceling the term $(I - L)$ from both sides we conclude that $X_{t.}$ is **cointegrated of rank** $r + 1$, which is a contradiction.

The preceding discussion may be summarized in Proposition 6.

**Proposition 6.** Consider the $VAR(n)$ model of Proposition 4, and suppose we modify it by adding the constant term $\mu$ as in Eq. (4.118). The following statements are true, on the assumption that the process came into existence at "time" zero [17] and is cointegrated:

i. If $\mu$ does not lie in the space spanned by $\Gamma$, the underlying $X$-process consists of an $I(1)$ stochastic process and a linear time trend of the form $\mu_0 + \mu t$, i.e.

$$X'_{t.} = \mu_0 + \mu t + \sum_{s=1}^{t} u'_{s.}.$$

subject to the usual assumptions on the $u$-process.

ii. If $\mu$ **does** lie in the space spanned by $\Gamma$ then the underlying $X$-process obeys

$$X'_{t.} = \mu_0 + \sum_{s=1}^{t} u'_{s.},$$

where possibly $\mu_0 = 0$.

---

[17] The same result would hold if we asserted that the initial conditions are $X_r. = 0$, for $r < 0$, and $X_t. = \mu_0$, for $t = 0$.

# Appendix to Chapter 4

In this appendix we derive a general expression for the sum of the $s^{th}$ powers of the first $n$ integers, or more precisely for

$$\sum_{j=1}^{n} j^s, \quad s = 0, 1, 2, \ldots.$$

We begin by stating the well-known results

$$\sum_{j=1}^{n} j^0 = n, \quad \sum_{j=1}^{n} j = \frac{n(n+1)}{2}. \qquad (4.121)$$

It is interesting that given the results above we can obtain **recursively** the sums $\sum_{j=1}^{n} j^s$, **for any** $s$. To do so we note that

$$2\sum_{i=1}^{j} i = j^2 + j, \quad \text{or} \quad j^2 = 2\sum_{i=1}^{j} i - j,$$

and summing over $j$ we find

$$\sum_{j=1}^{n} j^2 = 2\sum_{j=1}^{n}\sum_{i=1}^{j} i - \frac{n(n+1)}{2}.$$

Changing the order of summation,

$$\sum_{j=1}^{n} j^2 = 2\sum_{i=1}^{n}\sum_{j=i}^{n} i - \frac{n(n+1)}{2} = 2\sum_{i=1}^{n}(n+1-i)i - \frac{n(n+1)}{2},$$

whence we obtain

$$3\sum_{j=1}^{n} j^2 = n(n+1)^2 - \frac{n(n+1)}{2} = \frac{n(n+1)(2n+1)}{2}.$$

Thus,

$$\sum_{j=1}^{n} j^2 = \frac{n(n+1)(2n+1)}{6}. \qquad (4.122)$$

For $s = 3$, we note that

$$6 \sum_{i=1}^{j} i^2 = 2j^3 + 3j^2 + j, \quad \text{or} \quad 2j^3 = 6 \sum_{i=1}^{j} i^2 - 2j^2 - j.$$

Summing over $j$ and changing the order of summation as above, we find

$$2 \sum_{j=1}^{n} j^3 = 6(n+1) \sum_{i=1}^{n} i^2 - 6 \sum_{i=1}^{n} i^3 - \frac{n(n+1)(2n+1)}{2} - \frac{n(n+1)}{2}.$$

Finally, simplifying and rearranging terms we obtain

$$\sum_{j=1}^{n} j^3 = \frac{n^2(n+1)^2}{4}. \tag{4.123}$$

For $s = 4$, we find

$$4 \sum_{i=1}^{j} i^3 = j^4 + 2j^3 + j^2, \quad \text{or} \quad j^4 = 4 \sum_{i=1}^{j} i^3 - 2j^3 - j^2.$$

Summing over $j$ and changing the order of summation as above, we find

$$\sum_{j=1}^{n} j^4 = n^2(n+1)^3 - 4 \sum_{i=1}^{n} i^4 - \frac{n^2(n+1)^2}{2} - \frac{n(n+1)(2n+1)}{6}.$$

Finally, simplifying and rearranging terms we obtain

$$\sum_{j=1}^{n} j^4 = \frac{n(n+1)(2n+1)}{30} [3n(n+1) - 1]. \tag{4.124}$$

In this fashion one may compute as many sums as one wishes, by the routine procedure we have described above. The significance of the developments in this appendix becomes evident when we discuss the structure of the covariance matrix of integrated processes. Finally, we note the obvious fact that the sum of the $s^{th}$ power of the first $t$ integers is a polynomial of degree $s + 1$ in $t$, i.e.

$$\sum_{j=1}^{t} j^s = a_0 t^{s+1} + a_1 t^s + a_2 t^{s-1} + \ldots + a_{s+1}. \tag{4.125}$$

# Chapter 5

# Cointegration II

## 5.1  Introduction

In the preceding chapter we obtained, *inter alia*, the implications of cointegration in $MIMA(\infty)$, $MIMA(k)$ and $VAR(n)$ processes, and derived the specifications from which the underlying parameters may be estimated.

The estimating formulations were given, respectively, in Eqs. (4.73), (4.77) and (4.88), which are reproduced below for convienience.

$$\Delta X'_{t\cdot} = -\Gamma B' X'_{t-1\cdot} + \sum_{j=1}^{\infty} \bar{H}_j \Delta X'_{t-j\cdot} + d(L) I_q \epsilon'_{t\cdot}, \quad d(L) = \sum_{j=0}^{\infty} d_j L^j;$$

$$\Delta X'_{t\cdot} = -\Gamma B' X'_{t-1\cdot} + \sum_{j=1}^{m-1} \bar{H}_j \Delta X'_{t-j\cdot} + a(L) I_q \epsilon'_{t\cdot}, \quad a(L) = \sum_{j=0}^{n} a_j L^j;$$

$$\Delta X'_{t\cdot} = -\Gamma B' X'_{t-1\cdot} + \sum_{j=1}^{n-1} \Pi_j^* \Delta X'_{t-j\cdot} + \epsilon'_{t\cdot}. \tag{5.1}$$

From a pragmatic point of view there is no material difference between the first two specifications, which result from the assertion that the underlying process is $MIMA(\infty)$ and $MIMA(k)$, respectively. Thus, we shall make no distinction between them in examining the limiting distribution of their parameter estimators; rather, we deal exclusively with the last two cases and the models that generate them. These models are, specifically,

$$(I - L)X'_{t\cdot} = A(L)\epsilon_{t\cdot}, \quad \Pi(L)X'_{t\cdot} = \epsilon'_{t\cdot}, \quad A(L) = \sum_{j=0}^{k} A_j L^j, \quad \Pi(L) = \sum_{j=0}^{n} \Pi_j L^j,$$

232

where $A_0 = \Pi_0 = I_q$ .[1]

**Remark 1**. The normality assumption in Johansen (1988), (1991), (J) and Ahn and Reinsel (1990) (AR) does not play any essential role beyond motivating the form of the function we need to maximize in order to obtain estimators for the unknown parameters. We can certainly use the same procedure to obtain estimators **even though the underlying distribution is not normal**. In the simultaneous equations literature this is referred to as a "quasi"- or "pseudo-maximum likelihood" procedure.

**Remark 2**. The crucial difference in the $MIMA(k)$ and $VAR(n)$ formulations, as they are exhibited in the last two equations of Eq. (5.1), lies in the specification of the error process. In the $MIMA(k)$ case both the error process and the "systematic" part of the relation **are functions of the same parametric set**, viz. the parameters in the matrix operator $A(L)$. The error process contains such parameters through the determinant of $A(L)$, while the "systematic" part depends on these parameters through the simplified adjoint $H(L)$; the latter is a matrix polynomial operator of degree $m_1$, where $m_1 = (q-1)k - (r-1)$, as determined in the previous chapter. [2] Thus, a "fully efficient" procedure that takes these relations into account is too complex to be seriously considered for implementation. Even more important, the matrices $\bar{H}_j$ **cannot be estimated consistently** by least squares, while the entity $\Gamma B'$ may be. On the other hand, in the initial stages of this literature the parameters in the matrix polynomial $A(L)$ were of no particular interest; interest was focused **solely** on the cointegrating vectors, which were thought to constitute the "long-run equilibrium" relation

---

[1] The reader might wonder why we write the stochastic sequences as **rows** rather than **columns**. As we pointed out earlier, in the absence of this convention data matrices would become the transpose of what they usually are in the literature of the general linear model and related discussions. Thus, for example, the data matrix $X$ in the context of Eq. (4.89) of Chapter 4, would lead to the respresentation of the estimator of $\Pi^*$ in Eq. (4.90) as $\hat{\Pi}^* = Y'X(X'X)^{-1}$ , which is rather awkward and inconvenient. The convention we follow consistently throughout this volume preserves the customary representation for estimators as we find them in other aspects of econometrics such as the general linear model, and the simultaneous equations literature. The "price" we pay is that we often have to use the transposed form of the stochastic sequence in question.

[2] Incidentally, note that $m < n$ .

among the elements of the vector $X_{t.}$. Thus, if we adhere to this point of
view we may simply ignore the terms containing the $\bar{H}_j$, and obtain by least
squares a consistent estimator of $\Gamma B'$.

If we refer to the discussion in Remark 8 of Chapter 4, a certain similarity
between the two formulations becomes apparent. Specifically, from part iii
of Proposition 2a (in Chapter 4) we have the representation

$$H(L)X'_{t.} = a(L)I_q \epsilon'_{t.},$$

where $H(L)$ is a matrix polynomial operator of degree $m_1 = (q-1)k - r + 1$,
$r$ is the cointegration rank, and $a(L)$ is an **invertible** (scalar) polynomial
operator of degree $m_2 = qk - r$. Hence, we may write the $MIMA(k)$ model
as

$$H^*(L)X'_{t.} = \epsilon'_{t.}.$$

Since $H^*(L) = [a(L)]^{-1}H(L)$, it follows that in the expansion

$$H^*(L) = \sum_{j=0}^{\infty} H^*_j L^j, \quad \sum_{j=0}^{\infty} j \parallel H^*_j \parallel < \infty,$$

since

$$H^*_k = \sum_{i=0}^{k} b_{k-i} H_i, \ k \in \mathcal{N}_0, \quad [a(L)]^{-1} = \sum_{j=0}^{\infty} b_j L^j, \quad |b_j| < j|\lambda_0| \qquad (5.2)$$

where $\lambda_0$ is the largest (in modulus) inverse root of $a(z) = 0$, and obeys
$|\lambda_0| < 1$. Thus, the matrices $H^*_j$ converge to zero exponentially. Noting,
further, that

$$H^*(1) = \frac{H(1)}{a(1)} \neq 0, \quad H^*(L) = H^*(1) - (I - L)\bar{H}^*(L),$$

we may write the second equation of Eq. (5.1) as

$$\Delta X'_{t.} = -\Gamma B' X'_{t-1.} + \sum_{j=1}^{\infty} H^*_j \Delta X'_{t-j.} + \epsilon'_{t.}. \qquad (5.3)$$

In Eq. (5.3) $\Gamma B' = H^*(1)$, and $\Gamma$ is unspecified beyond its dimension
and rank; thus, apart from the properties of the error process, the notation
therein is not in substantial conflict with the representation in the second
equation of Eq. (5.1), where we had also written $\Gamma B' = H(1)$, since $H(1)$

differs from $H^*(1)$ by a nonzero scalar only! Moreover, it is evident from the representation of the $MIMA(k)$ model in Eq. (5.3) that its differences relative to the $VAR(n)$ formulation are miniscule. Indeed, we may think of the $VAR(n)$ formulation of Eq. (5.1) as beginning initially with the $MIMA(k)$ formulation, and then **arbitrarily truncating the operator** $H^*(L)$ **at the term** $H_n^* L^n$.

For parameter estimation we have, in both cases, the option of obtaining (1) **unrestricted** estimators of the underlying parameters, $\bar{H}(1)$, $\bar{H}_j$, in the $MIMA(k)$ case, and $\Pi(1), \Pi_j^*$ in the $VAR(n)$ case, or (2) **restricted** estimators, restricted by the cointegration hypothesis that imposes the **rank factorization**

$$H(1) = \Gamma B', \quad \text{or} \quad \Pi(1) = \Gamma B'.$$

In the $MIMA(k)$ case there is also the option, suggested by Engle and Granger (1987) (EG), of obtaining parameter estimators **asymmetrically**. This involves estimating the parameters in $\Gamma$ and $\bar{H}_j$, $j = 1, 2, \ldots, m-1$ by least squares methods, with $\hat{B}$ given by a prior consistent estimator.

In nearly all of the cases above we also have the option of obtaining $B$ uniquely, by imposition of ( $r^2$ ) "just identifying" conditions, say,

$$B = Q_1 + Q_2 B_0, \quad Q_1 = \begin{bmatrix} I_r \\ 0 \end{bmatrix}, \quad Q_2 = \begin{bmatrix} 0 \\ I_{q-r} \end{bmatrix}, \tag{5.4}$$

where $B_0$ is a $q-r \times r$ matrix of **free parameters**; or, alternatively, by estimating $B$ as a matrix of characteristic vectors subject only to normalization, which imposes $r$ restrictions.

**Remark 3**. From an inference point of view, the preceding discussion raises the issue of whether the cointegrating vectors are indeterminate. In the literature, this issue is avoided by the observation that the (charateristic) vectors in question **span the space of cointegrating vectors**. If our objective, however, is to establish and **measure** a "long-run" equilibrium relationship among economic variables, this intepretation of the estimated parameters is rather unsatisfactory. It is akin to the statement that if we are interested in estimating the $m$-element vector $b^* = (b', 0)$, where $b$ is an $r$-element column vector, $r \leq m$, we should be content with knowing only the vectors $e_{.i}$, $i = 1, 2, \ldots, r$ where $e_{.i}$ is an $m$-element **column** vector, all of whose elements are zero, save the $i^{th}$ which is unity, since the latter span the appropriate subspace of $R^m$ !

Ultimately, the issue is this: What is the controlling inference from the cointegration hypothesis? Is it that the cointegral vector is stationary? Or is it that $\Pi(1)$ or $H(1)$ is of **reduced** rank? If we take the former point of view, and operate as we did in the previous chapter, i.e. we obtain the characteristic vectors corresponding to the $r$ smallest roots of $\hat{M}^{(2)}$ then, on the assumption that the cointegration hypothesis is correct, we would have estimated, subject only to trivial normalization, the cointegrating vectors. If we do not take this position, and operate only through the rank factorization of $\Pi(1)$, (or $H(1)$), it is not entirely clear that what we have obtained **are** the cointegrating vectors.

This putative resolution of the problem, however, is more apparent than real. While it is true that the characteristic vectors corresponding to the smallest roots of $\hat{M}^{(2)}$ are quite well determined in finite samples, they suffer from the same ambiguities in the limit. This is so since, in the limit, the matrix $B$ of such charateristic vectors obeys $\bar{M}B = 0$, where $\bar{M} = \lim_{T \to \infty} E\hat{M}_2$; but **characteristic vectors corresponding to zero roots** are **arbitrary** to within multiplication by a nonsingular matrix! Thus, uniqueness and lack of ambiguity in determining cointegrating vectors may be obtained **only** by imposing "prior" restrictions. If such restrictions produce only "just identification", the economic implications of the empirical findings that result from some **particular set** of prior restrictions cannot be said to dominate other empirical findings that result from a **different set** of "just identifying" prior restrictions. Only the imposition of "overidentifying" prior restrictions on the matrix $B$ may be subjected to test!

## 5.2   Limiting Distributions

### 5.2.1   Notation and Preliminaries

In this section we retain the extensive notational scheme we developed in the previous chapter, and expand it as necessary. Moreover, since the major difference between that formulation and the $MIMA(k)$ case, as exhibited in Eq. (5.1), lies in the specification of the error process we continue to operate in the notational context of the previous chapter; we make the differences in the implications of the two formulations clear only as such differences actually arise.

Retaining the notation in Eq. (4.89) define

$$x_{t\cdot}^* = (X_{t-1\cdot}, x_{t\cdot}), \quad J = (-\Pi(1), \ \Pi^*)', \tag{5.5}$$

where $x_{t\cdot}$, $\Pi^*$ are as also defined in Eq. (4.89). We may, thus, write

$$\Delta X_{t\cdot} = x_{t\cdot}^* J + u_{t\cdot}, \quad t = 1, 2, \dots, T, \quad \Delta P = X^* J + U, \tag{5.6}$$

for the $t^{th}$ observation on the model and for the entire sample, respectively.

The least squares estimator of the (unrestricted) parameters is given by

$$\hat{J} = (X^{*'} X^*)^{-1} X^{*'} \Delta P = J + (X^{*'} X^*)^{-1} X^{*'} U. \tag{5.7}$$

Using the partitioned inverse form [Proposition 31 in Dhrymes (1984), p. 37], we find

$$-[\hat{\Pi}(1) - \Pi(1)]' = (V'V)^{-1} V' U, \quad V = N P_{-1}, \quad N = I_T - X(X'X)^{-1} X',$$

$$(\hat{\Pi}^* - \Pi^*)' = F_{22} X' N_* U, \quad N_* = I_T - P_1(P_{-1}' P_{-1})^{-1} P_{-1}',$$

$$F_{22}^{-1} = X'X - X' P_{-1}(P_{-1}' P_{-1})^{-1} P_{-1}' X. \tag{5.8}$$

Since $X_{t-1\cdot}$ is $I(1)$, while $x_{t\cdot} = (\Delta X_{t-1\cdot}, \cdots, \Delta X_{t-n+1\cdot})$ is $I(0)$, the discussion in Chapter 3 implies

$$\frac{P_{-1}' P_{-1}}{T^2} \xrightarrow{d} S_0, \quad \frac{X' P_{-1}}{T} \xrightarrow{d} S_1.$$

Here, $S_0$ is an a.c. finite random matrix consisting of (Lebesgue) integrals of Brownian motion (BM), while $S_1$ is an a.c. finite random matrix consisting of stochastic integrals involving BM. It follows, therefore, that

$$\frac{1}{T} F_{22}^{-1} \xrightarrow{P} M_{xx}, \quad M_{xx} = \underset{T \to \infty}{\text{plim}} \frac{1}{T} X'X. \tag{5.9}$$

Later we verify that the probability limits of Eq. (5.9) exist, and we produce an expression for them, but for the moment let us take their existence as given. Moreover, since

$$K_{T1} = \frac{1}{T} \left[ X' P_{-1} (P'_{-1} P_{-1})^{-1} P'_{-1} X \right]$$

$$= \frac{1}{T} \left[ \left( \frac{X' P_{-1}}{T} \right) \left( \frac{P'_{-1} P_{-1}}{T^2} \right)^{-1} \left( \frac{P'_{-1} X}{T} \right) \right] \xrightarrow{P} 0, \quad \text{and}$$

$$K_{T2} = \frac{1}{T} \left[ X' P_{-1} (P'_{-1} P_{-1})^{-1} P'_{-1} U \right]$$

$$= \frac{1}{T} \left[ \left( \frac{X' P_{-1}}{T} \right) \left( \frac{P'_{-1} P_{-1}}{T^2} \right)^{-1} \left( \frac{P'_{-1} U}{T} \right) \right] \xrightarrow{P} 0,$$

it is evident that $T^\alpha K_{Ti} \xrightarrow{P} 0$, for $\alpha \in [0,1)$, $i = 1, 2$. We conclude, therefore, that

$$\sqrt{T}(\hat{\Pi}^* - \Pi^*)' \sim \left( \frac{X'X}{T} \right)^{-1} \frac{1}{\sqrt{T}} X' U. \tag{5.10}$$

If $E X' U = 0$, its limiting distribution may be obtained through a standard CLT, which we do in the next section. If not, however, the limiting distribution would not be well defined or, more appropriately, it would have to be centered not about $\Pi^*$, but about another entity reflecting the apparent inconsistency of the estimator as exhibited above.

To examine the limiting distribution of $[\hat{\Pi}(1) - \Pi(1)]'$, we need to examine the behavior of

$$V'V = P'_{-1} P_{-1} - P'_{-1} X (X'X)^{-1} X' P_{-1}, \quad \text{and}$$

$$V'U = P'_{-1} U - P'_{-1} X (X'X)^{-1} X' U.$$

For $V'V$ to be well behaved, it must be normalized by $T^{-2}$, thus obtaining

$$\frac{V'V}{T^2} = \frac{P'_{-1} P_{-1}}{T^2} - \frac{1}{T} \left[ \left( \frac{P'_{-1} X}{T} \right) \left( \frac{X'X}{T} \right)^{-1} \left( \frac{X'U}{T} \right) \right];$$

in addition,

$$\frac{V'U}{T} = \frac{P'_{-1} U}{T} - \left[ \left( \frac{P'_{-1} X}{T} \right) \left( \frac{X'X}{T} \right)^{-1} \left( \frac{X'U}{T} \right) \right].$$

**Provided** $(X'U/T) \overset{\text{P}}{\to} 0$, we may conclude that

$$- T[\hat{\Pi}(1) - \Pi(1)]' \sim \left( \frac{P'_{-1}P_{-1}}{T^2} \right)^{-1} \frac{P'_{-1}U}{T}, \qquad (5.11)$$

whose limiting distribution is nonstandard, and may be determined by the methods developed in Chapter 3.

If $(X'U/T)$ **does not converge to zero**, but instead converges to a well-defined, finite, constant matrix, say $M_{xu}$, we shall again conclude that

$$- T[\hat{\Pi}(1) - \Pi(1)]' \sim \left( \frac{P'_{-1}P_{-1}}{T^2} \right)^{-1} \left( \frac{P'_{-1}U}{T} - \frac{P'_{-1}X}{T} M_{xx}^{-1} M_{xu} \right), \qquad (5.12)$$

whose limiting distribution is nonstandard but may also be determined by the methods developed in Chapter 3.

We now take up the convergence issue left pending.

**Assertion 1.** In the context of the discussion above the following statements are true:

i. for the $MIMA(k)$ model (of Proposition 2a in Chapter 4)

$$\frac{X'U}{T} \overset{\text{a.c.}}{\to} \Sigma_{xu}, \qquad \Sigma_{xu} = D^*\Sigma,$$

$$D^* = [\, D'_1, \quad D'_2, \quad \cdots, \quad D'_{m-1} \,]', \qquad D_i = \sum_{j=i}^{k+i} a_j A_{j-i},$$

where $U = (u_{t\cdot})$, $u'_{t\cdot} = a(L)I_q \epsilon'_{t\cdot}$;

ii. for the $VAR(n)$ model

$$\frac{X'U}{T} \overset{\text{a.c.}}{\to} 0, \qquad U = (\epsilon_{t\cdot}).$$

iii. for both models

$$\frac{X'X}{T} \overset{\text{a.c.}}{\to} M_{xx} > 0;$$

iv. for the $MIMA(k)$ model

$$\frac{H(1)P'_{-1}U}{T} \overset{\text{a.c.}}{\to} G^*\Sigma, \qquad G^* = -H(1) \sum_{j=0}^{k-1} a_{j+1} A_j^*;$$

v. for the $VAR(n)$ model

$$\frac{\Pi(1)P'_{-1}U}{T} \overset{\text{a.c.}}{\to} 0.$$

Proof: To prove the assertion, we need certain representations which are used repeatedly below. To this end, note that from the formulation of the $MIMA(k)$ model, and Eq. (4.76) of Chapter 4, we obtain

$$x_{t\cdot} = \epsilon_{t\cdot}[\,LA(L)',\quad L^2A(L)',\quad \cdots,\quad L^{m-1}A(L)'\,] \tag{5.13}$$

$$u'_{t\cdot} = a(L)I_q\epsilon'_{t\cdot},\quad a(L) = \sum_{s=0}^{n} a_s L^s,\quad n = qk - r.$$

For the $VAR(n)$ formulation, we recall from Proposition 4 in Chapter 4

$$\pi^*(L) = |\Pi(L)| = (I - L)^{r_0}\pi(L),\quad \pi(1) \neq 0, \tag{5.14}$$

so that $\pi(L)$ is **invertible**. Specifically, if $z_i$ are the roots $\pi(z) = 0$, where $z$ is a complex indeterminate, they all lie outside the unit circle; consequently, we have the representation

$$\frac{I}{\pi(L)} = \prod_{i=1}^{k}\sum_{j=0}^{\infty} \nu_i^j L^j,\quad \nu_i = \frac{1}{z_i},\quad |\nu_i| < 1.$$

Thus, for the $VAR(n)$ formulation, we obtain the relations

$$(I - L)X'_{t\cdot} = H^*(L)\epsilon'_{t\cdot},\; H^*(L) = [\pi(L)]^{-1}H(L) = \sum_{i=0}^{\infty} H_i^* L^i,$$

$$x'_{t\cdot} = [\,LH^*(L)',\quad L^2H^*(L)',\quad \cdots,\quad L^{n-1}H^*(L)'\,]'\,\epsilon'_{t\cdot},$$

$$\Pi(1)X'_{t\cdot} = \sum_{j=0}^{n-1}\Pi_j^*(I - L)X'_{t-j\cdot} + \epsilon'_{t\cdot} = \sum_{k=0}^{\infty} G_s\epsilon'_{t-s\cdot},\; \sum_{j=0}^{\infty}\|\,H_i^*\,\| < \infty$$

$$G_s = \sum_{j=0}^{(n-1)\wedge s}\Pi_j^* H_{s-j}^*,\; s \geq 1,\; G_0 = \Pi(1). \tag{5.15}$$

Finally, in Example 1 of Chapter 9 we show, in effect, that the entities of interest in this discussion, viz., $\Delta X_{t-i\cdot}$, $H(1)X'_{t\cdot}$, $\Pi(1)X'_{t\cdot}$ and $u_{t\cdot}$ are all **stationary and ergodic**, have bounded second moments, and each is representable as a square integrable GLP (general linear process).

To prove i we note, from Eq. (5.13), that $X'U/T$ is a **block** matrix, the $i^{th}$ block of which obeys

$$\frac{1}{T}\sum_{t=1}^{T} E\Delta X'_{t-i\cdot}u_{t\cdot} = \frac{1}{T}\sum_{s=0}^{n}\sum_{j=i}^{k+i} A_{j-i}a_j \sum_{t=1}^{T} E\epsilon'_{t-j\cdot}\epsilon_{t-s\cdot}. \tag{5.16}$$

$$= D_i\Sigma, \quad D_i = \sum_{j=i}^{k+i} a_j A_{j-i}, \quad i = 1, 2, \cdots, m-1.$$

Since the sequences

$$\{(x_{t\cdot}, u_{t\cdot}) : t \in \mathcal{N}_+\}, \quad \{(x_{t\cdot}, u_{t\cdot})'(x_{t\cdot}, u_{t\cdot}) : t \in \mathcal{N}_+\}$$

are stationary and ergodic,[3] we conclude by Theorem 2 in Hannan (1970) p. 203, or Proposition 31 (Mean Ergodic Theorem) in Dhrymes (1989), p. 357,

$$\frac{1}{T}\sum_{t=1}^{T}\begin{bmatrix} x'_{t\cdot}.x_{t\cdot} & x'_{t\cdot}.u_{t\cdot} \\ u'_{t\cdot}.x_{t\cdot} & u'_{t\cdot}.u_{t\cdot} \end{bmatrix} \overset{\text{a.c.}}{\to} \begin{bmatrix} \Sigma_{xx} & \Sigma_{xu} \\ \Sigma_{ux} & \Sigma_{uu} \end{bmatrix}.$$

From the definition of $x_{t\cdot}$ and $u_{t\cdot}$ we further establish that

$$\frac{X'U}{T} \overset{\text{a.c.}}{\to} \Sigma_{xu} = D^*\Sigma,$$

which concludes the proof of part i.

To prove part ii, we note that here $u_{t\cdot} = \epsilon_{t\cdot}$, and from the first two equations in Eq. (5.15) it is clear that the $i^{th}$ block of $X'U/T$ obeys

$$E\left(\frac{\Delta X'_{t-i\cdot}\epsilon_{t\cdot}}{T}\right) = \frac{1}{T}\sum_{j=0}^{\infty}\sum_{t=1}^{T} H^*_j E\epsilon'_{t-i-j\cdot}\epsilon_{t\cdot} = 0,$$

which, by the discussion above, shows that in the $VAR(n)$ model

$$\frac{X'U}{T} \overset{\text{a.c.}}{\to} 0.$$

To prove part iii for the $MIMA(k)$ case, we note from the first equation in Eq. (5.13) that the $(i, j)$ block of $x'_{t\cdot}.x_{t\cdot}$ obeys

$$E(\Delta X'_{t-i\cdot}.\Delta X_{t-j\cdot}) = \sum_{r=0}^{k}\sum_{s=0}^{k} A_s E(\epsilon'_{t-i-s\cdot}.\epsilon_{t-j-r\cdot})A'_r$$

---

[3] This is, again, slightly inaccurate since we assume certain initial conditions; however, for $t$ sufficiently large, their deviation from these properties may be made arbitrarily small.

$$= \sum_{r=0}^{k-(j-i)} A_{j-i+r} \Sigma A_r' \quad \text{for } j \geq i,$$

$$= \sum_{s=0}^{k-(i-j)} A_s \Sigma A_{i-j+s}' \quad \text{for } j < i$$

$$= 0, \quad \text{for } |i - j| > k. \tag{5.17}$$

For the $VAR(n)$ model, using the first equation of Eq. (5.15) we have that

$$E(\Delta X_{t-i.}' \Delta X_{t-j.}) = \sum_{r=0}^{\infty} \sum_{s=0}^{\infty} H_s^* E(\epsilon_{t-i-s.}' \epsilon_{t-j-r.}) H_r^{*'}$$

$$= \sum_{r=0}^{\infty} H_{j-i+r}^* \Sigma H_r^{*'} \quad \text{for } j \geq i,$$

$$= \sum_{s=0}^{\infty} H_s^* \Sigma H_{i-j+s}^{*'}, \quad \text{for } j < i. \tag{5.18}$$

The discussion above has established the following:

i. For the $MIMA(k)$ model, the matrix $M_{xx}$ is positive definite, non-singular, and (block) symmetric; its diagonal blocks are **all of the form** $\sum_{r=0}^{k} A_r \Sigma A_r'$ and its $s^{th}$ **super-diagonal** block is of the form $\sum_{r=0}^{k-s} A_{s+r} \Sigma A_r'$, for $s \leq k$, and **zero** for $s > k$.

ii. For the $VAR(n)$ model, the matrix $M_{xx}$ is positive definite, non-singular, and (block) symmetric; its diagonal blocks are **all of the form** $\sum_{r=0}^{\infty} H_r^* \Sigma H_r^{*'}$, and its $s^{th}$ **super diagonal block** is of the form $\sum_{r=0}^{\infty} H_{s+r}^* \Sigma H_r^{*'}$.

This completes the proof of iii.

To prove part iv we begin with the representation (part vii of Proposition 2a of Chapter 4)

$$B'X_{t.}' = -B'A^*(L)\epsilon_{t.}', \quad \text{so that} \quad H(1)X_{t.}' = -H(1)A^*(L)\epsilon_{t.}', \quad H(1) = \Gamma B'.$$

Since $\Gamma Z_{t-1.}' = H(1)X_{t-1.}'$,

$$E\left(\frac{H(1)P_{-1}'U}{T}\right) = -H(1)\frac{1}{T}\sum_{j=0}^{k-1}\sum_{s=0}^{n}\sum_{t=1}^{T} a_s A_j^* E\epsilon_{t-1-j.}' \epsilon_{t-s.}.$$

$$= G^*\Sigma, \quad G^* = -H(1)\sum_{j=0}^{k-1} a_{j+1} A_j^*. \tag{5.19}$$

By the same arguments given earlier we have, in the $MIMA(k)$ model,

$$\frac{H(1)P'_{-1}U}{T} \overset{\text{a.c.}}{\to} \Gamma\Sigma_{zu} = G^*\Sigma,$$

thus proving iv.

For part v, using the third equation in the equation set Eq. (5.15) we find

$$E\left(\frac{\Pi(1)P'_{-1}U}{T}\right) = \frac{1}{T}\sum_{s=0}^{\infty} G_s\left(\sum_{t=1}^{T} E\epsilon'_{t-1-s}.\epsilon_{t\cdot}\right) = 0.$$

By the same arguments given earlier, we conclude that in the $VAR(n)$ model

$$\frac{\Pi(1)P'_{-1}U}{T} \overset{\text{a.c.}}{\to} \Gamma\Sigma_{zu} = 0,$$

which concludes the proof of v.

q.e.d.

## 5.2.2   Unrestricted Estimators

We now examine the problems encountered in obtaining the limiting distribution of the unrestricted estimators obtained above. We do so first in the case of the $VAR(n)$ model. Since

$$\sqrt{T}(\hat{\Pi}^* - \Pi^*)' \sim \left(\frac{X'X}{T}\right)^{-1}\xi_T, \quad \xi_T = \frac{1}{\sqrt{T}}\sum_{t=1}^{T} x'_{t\cdot}u_{t\cdot}, \tag{5.20}$$

it is evident that the limiting distribution in question depends on the behavior of the random matrix, $\xi_T$; vectorizing it, we find

$$\text{vec}(\xi_T) = \frac{1}{\sqrt{T}}\sum_{t=1}^{T}(I_q \otimes x'_{t\cdot})u'_{t\cdot}. \tag{5.21}$$

In the $VAR(n)$ context, $u_{t\cdot} = \epsilon_{t\cdot}$ and it is clear that $\text{vec}(\xi_T)$ is a **martingale difference** process which obeys a **Lindeberg** condition. Hence, by Proposition 21 in Dhrymes (1989), p. 337, we conclude that

$$\sqrt{T}\text{vec}[(\hat{\Pi}^* - \Pi^*)'] \overset{\text{d}}{\to} N(0, \Sigma \otimes M_{xx}^{-1}). \tag{5.22}$$

To obtain the limiting distribution of $T[\hat{\Pi}(1) - \Pi(1)]$ we note that

$$\frac{P'_{-1}P_{-1}}{T^2} = \frac{1}{T^2}\sum_{t=1}^{T} X'_{t-1\cdot}X_{t-1\cdot}, \quad \frac{1}{T}P'_{-1}U = \frac{1}{T}\sum_{t=1}^{T} X'_{t-1\cdot}\epsilon_{t\cdot}. \qquad (5.23)$$

To evaluate the limiting behavior of these entities, we first obtain, from the equation set Eq. (5.15), the representations

$$X'_{t\cdot} = \sum_{s=1}^{t} \eta^{*'}_{t\cdot}, \quad X_{0\cdot} = 0, \quad \eta^{*'}_{t\cdot} = H^*(L)\epsilon'_{t\cdot}. \qquad (5.24)$$

Next, define the **fictitious process**, $\nu$, by $(I - L)\nu'_{t\cdot} = \epsilon'_{t\cdot}$ so that

$$\nu'_{t\cdot} = \sum_{s=1}^{t} \epsilon'_{s\cdot}, \qquad (5.25)$$

and consider the $2q$-element sequence

$$(I - L)\zeta_{t\cdot} = \phi_{t\cdot}, \quad \phi_{t\cdot} = (\eta^*_{t\cdot}, \epsilon_{t\cdot}). \qquad (5.26)$$

When solved, Eq. (5.26) yields

$$\zeta'_{t\cdot} = \sum_{s=1}^{t} \phi'_{s\cdot} = S'_{t\cdot}, \quad S_{t\cdot} = (S_{t1\cdot}, \ S_{t2\cdot}). \qquad (5.27)$$

The sequence $\phi_{t\cdot}$ satisfies the conditions of Proposition 16 of Chapter 9 with $p = q = 2$, and thus obeys a FCLT. We, therefore, conclude

$$\frac{1}{T^2}\sum_{t=1}^{T} S'_{t\cdot}S_{t\cdot} \overset{d}{\to} P'_0 \left(\int_0^1 B'(s)B(s)\,ds\right) P_0 = \Phi,$$

$$\frac{1}{T}\sum_{t=1}^{T} S'_{t-1\cdot}\phi_{t\cdot} \overset{d}{\to} P'_0 \left(\int_0^1 B'(s)\,dB(s)\right) P_0 + \Sigma^*_1 = \Psi + \Sigma^*_1, \qquad (5.28)$$

$$P'_0 P_0 = \Sigma_0 = \lim_{T\to\infty} \frac{1}{T}\sum_{t=1}^{T} ES'_{t\cdot}S_{t\cdot}, \quad \Sigma^*_1 = \lim_{T\to\infty} \frac{1}{T}\sum_{t<s} E\phi'_{t\cdot}\phi_{s\cdot}.$$

The discussion surrounding Eqs. (3.57) through (3.61) in Chapter 3 ensures the validity of the second equation set in Eq. (5.28). For the case under consideration, we note that $\phi$ is a **strictly stationary** process; thus, by the discussion surrounding Eq. (3.46) (in Chapter 3) the matrix $\Sigma_0$ has the representation [4]

$$\Sigma_0 = E\phi'_{1.}\phi_{1.} + \sum_{\tau=1}^{\infty}\left(E\phi'_{1.}\phi_{1+\tau.} + E\phi'_{1+\tau.}\phi_{1.}\right) = \Sigma_1 + \Sigma_1^* + \Sigma_1^{*'}, \qquad (5.29)$$

where

$$\Sigma_1 = \begin{bmatrix} \sum_{i=0}^{\infty} H_i^*\Sigma H_i^{*'} & \Sigma \\ \Sigma & \Sigma \end{bmatrix}, \quad \Sigma = \mathrm{Cov}(\epsilon'_{1.})$$

$$\Sigma_1^* = \begin{bmatrix} \sum_{\tau=1}^{\infty}\sum_{i=0}^{\infty} H_i^*\Sigma H_{i+\tau}^{*'} & 0 \\ \Sigma H^{*'} & 0 \end{bmatrix}, \quad H^* = \sum_{j=1}^{\infty} H_j^*. \qquad (5.30)$$

In the preceding discussion what **is of particular interest to us** is the (1,1) block in the (naturally) partitioned matrix $\Phi$, and the (1,2) block in the (naturally) partitioned matrix $\Psi$, both as defined in Eq. (5.28) above.

We now turn to the limiting distribution of the unrestricted parameters in the context of the $MIMA(k)$ model. Assertion 1 shows that the least squares estimator of the coefficient matrices of the lagged differences is **inconsistent**; thus, strictly speaking, the limiting distribution of the analog of the entity in Eq. (5.22) does not exist. What we may obtain is the limiting distribution of

$$\sqrt{T}(\hat{\Pi}^* - \Pi^* - \Sigma_{xu}) = \left(\frac{X'X}{T}\right)^{-1} \frac{1}{\sqrt{T}} \sum_{t=1}^{T}\left(x'_{t.}u_{t.} - \Sigma_{xu}\right),$$

which is of little, if any, interest and will thus be omitted.

To obtain the limiting distribution of $T[\hat{H}(1) - H(1)]$, in the context of the $MIMA(k)$ model, we note that

$$\frac{P'_{-1}P_{-1}}{T^2} = \frac{1}{T^2}\sum_{t=1}^{T} X'_{t-1.}X_{t-1.}, \quad \frac{1}{T}P'_{-1}U = \frac{1}{T}\sum_{t=1}^{T} X'_{t-1.}u_{t.}. \qquad (5.31)$$

Again making use of the equation set Eq. (5.15), we find from the definition of the process,

$$X'_{t.} = \sum_{s=1}^{t} \eta_{s.} \quad X_{0.} = 0, \quad \eta_{t.} = A(L)\epsilon'_{t.}. \qquad (5.32)$$

---

[4] This representation is slightly inaccurate since the initial conditions set $X_{-i.} = 0$ for $i \leq 0$.

Define the **fictitious process**, $\nu$, by [5] $(I - L)\nu'_{t\cdot} = u'_{t\cdot}$ so that

$$\nu'_{t\cdot} = \sum_{s=1}^{t} u'_{s\cdot}, \quad u'_{t\cdot} = \sum_{s=0}^{n} a_s \epsilon'_{t-s\cdot}, \quad n = qk - r. \tag{5.33}$$

Next, consider the $2q$-element sequence

$$(I - L)\zeta_{t\cdot} = \phi_{t\cdot}, \quad \phi_{t\cdot} = (\eta_{t\cdot}, u_{t\cdot}), \tag{5.34}$$

which, when solved, yields

$$\zeta'_{t\cdot} = \sum_{s=1}^{t} \phi'_{s\cdot} = S'_{t\cdot}, \quad S_{t\cdot} = (S_{t1\cdot}, S_{t2\cdot}). \tag{5.35}$$

As before, what is of interest to us is the $(1,1)$ block in the (naturally) partitioned matrix $S'_{t\cdot}S_{t\cdot}$, and the $(1,2)$ block in the (naturally) partitioned matrix $S'_{t\cdot}\phi_{t\cdot}$. Since the sequence $\phi_{t\cdot}$ satisfies the conditions of Proposition 16 of Chapter 9 with $p = q = 2$, it obeys a FCLT. Consequently, we have that

$$\frac{1}{T^2} \sum_{t=1}^{T} S'_{t\cdot}S_{t\cdot} \overset{\mathrm{d}}{\to} P'_0 \left( \int_0^1 B'(s)B(s)\, ds \right) P_0,$$

$$\frac{1}{T} \sum_{t=1}^{T} S'_{t-1\cdot}\phi_{t\cdot} \overset{\mathrm{d}}{\to} P'_0 \left( \int_0^1 B'(s)\, dB(s) \right) P_0 + \Sigma_1^*,$$

$$P'_0 P_0 = \Sigma_0 = \lim_{T\to\infty} \frac{1}{T} \sum_{t=1}^{T} E S'_{t\cdot}S_{t\cdot}. \quad \Sigma_1^* = \lim_{T\to\infty} \frac{1}{T} \sum_{t<s} E \phi'_{t\cdot}\phi_{s\cdot}.$$

To particularize these results to the $MIMA(k)$ case, we note that $\phi$ is a **strictly stationary** process; thus, by the discussion surrounding Eq. (3.46) (in Chapter 3) the matrix $\Sigma_0$ has the representation

$$\Sigma_0 = E\phi'_{1\cdot}\phi_{1\cdot} + \sum_{\tau=1}^{\infty} \left( E\phi'_{1\cdot}\phi_{1+\tau\cdot} + E\phi'_{1+\tau\cdot}\phi_{1\cdot} \right) = \Sigma_1 + \Sigma_1^* + \Sigma_1^{*'}, \tag{5.36}$$

where

$$\Sigma_1 = \begin{bmatrix} \sum_{i=0}^{k} A_i \Sigma A'_i & F\Sigma \\ \Sigma F' & \alpha\Sigma \end{bmatrix}, \quad \Sigma = \mathrm{Cov}(\epsilon_{1\cdot}), \quad F = \sum_{j=0}^{k} a_j A_j,$$

---

[5] Note that this process is **not the same** as the one defined in Eq. (5.25).

$$\Sigma_1^* = \begin{bmatrix} \sum_{\tau=1}^{k} \sum_{i=0}^{k-\tau} A_i \Sigma A_{i+\tau}' & \left(\sum_{j=0}^{k} \beta_j A_j\right) \Sigma \\ \Sigma \left(\sum_{j=0}^{k} a_j A_j^*\right)' & \alpha^* \Sigma \end{bmatrix}, \tag{5.37}$$

$$\beta_j = \sum_{i=j+1}^{n} a_i, \quad \alpha = \sum_{s=0}^{n} a_s^2, \quad \alpha^* = \sum_{\tau=1}^{n} \sum_{j=0}^{n-\tau} a_j a_{\tau+j}, \quad A_j^* = \sum_{i=j+1}^{k} A_i.$$

We have therefore proved Proposition 1.

**Proposition 1.** In the context of the $CI(1,1,r)$ model examined in the previous chapter consider the (last two) estimating equations in Eq. (5.1), corresponding to the $MIMA(k)$ and $VAR(n)$ specifications. The following statements are true:

i. the unrestricted least squares parameter estimators are given, generically, by

$$-[\hat{\Pi}(1) - \Pi(1)]' = (V'V)^{-1}V'U, \quad V = NP_{-1}, \quad W = N\Delta P$$

$$(\hat{\Pi}^* - \Pi^*)' = F_{22}X'N_*U, \quad N = I_T - X(X'X)^{-1}X'$$

$$N_* = I_T - P_1(P_{-1}'P_{-1})^{-1}P_{-1}'$$

$$F_{22}^{-1} = X'X - X'P_{-1}(P_{-1}'P_{-1})^{-1}P_{-1}'X.$$

ii. their limiting behavior is determined, respectively, by

$$\left(\frac{P_{-1}'P_{-1}}{T^2}\right)^{-1} \left(\frac{P_{-1}'U}{T} - \frac{P_{-1}'X}{T}M_{xx}^{-1}M_{xu}\right), \quad \left(\frac{X'X}{T}\right)^{-1}\frac{X'U}{\sqrt{T}};$$

iii. in both specifications $M_{xx} > 0$; in the $VAR(n)$ specification $M_{xu} = 0$; in the $MIMA(k)$ specification $M_{xu} \neq 0$;

iv. in the $MIMA(k)$ specification the least squares estimator of $\Pi^*$ is **inconsistent**, and to obtain its limiting distribution we need to center the estimator **not about** $\Pi^*$ **but about** $\Pi^* + M_{xu}$. The least squares estimator of $H(1)$, however, is consistent of order $T^\alpha$, $\alpha \in [0,1)$;

v. in the $VAR(n)$ specification the limiting distribution of $\hat{\Pi}(1)$ is determined by the (1,1) block and (1,2) block, respectively, in

$$P_0' \left(\int_0^1 B'(s)B(s)\,ds\right) P_0, \quad P_0' \left(\int_0^1 B'(s)\,dB(s)\right) P_0 + \Sigma_1^*,$$

where

$$P_0'P_0 \;=\; \Sigma_0 \;=\; \Sigma_1 + \Sigma_1^* + \Sigma_1^{*\prime}, \quad \Sigma_1 = \lim_{T\to\infty} \frac{1}{T} \sum_{t=1}^{T} \phi_{t\cdot}' \phi_{t\cdot},$$

$$\Sigma_1^* \;=\; \lim_{T\to\infty} \frac{1}{T} \sum_{t<s} E \phi_{t\cdot}' \phi_{s\cdot},$$

the matrices having been defined in Eq. (5.30); the limiting distribution of $\hat{\Pi}^*$ is given by

$$\sqrt{T}\,\mathrm{vec}[(\hat{\Pi}^* - \Pi^*)'] \xrightarrow{\mathrm{d}} N(0, \Sigma \otimes M_{xx}^{-1}).$$

The estimator of $\Pi(1)$ is consistent of order $T^\alpha$, for $\alpha \in [0,1)$; the estimator of $\Pi^*$ is consistent of order $T^\alpha$, for $\alpha \in [0, 1/2)$.

## 5.2.3   Unrestricted Estimators in $VAR(n)$

As the preceding discussion makes quite clear, the $MIMA(\infty)$ [or, for that matter the $MIMA(k)$] formulation of the cointegration problem is not very felicitous since it leads to great complications, and **inconsistent estimators**. Thus, in the sequel we concentrate solely on the $VAR(n)$ formulation which, as noted particularly in Remark 9 of Chapter 4, may be thought of as an approximation to the $MIMA(k)$ formulation. The discussion in the preceding two sections still remains quite useful, however, even though it does not go sufficiently far in exploiting the implications of the cointegration hypothesis.

Return to Eq. (5.15), and note that the first equation therein is derived from

$$(I - L)\pi(L)I_q X_{t\cdot}' = H(L)\epsilon_{t\cdot}' = H(1)\epsilon_{t\cdot}' - (I - L)H^{**}(L)\epsilon_{t\cdot}', \qquad (5.38)$$

where $H^{**}(L)$, $\pi(L)$ are, respectively, a matrix polynomial of degree $m - (n-1)$, and a scalar polynomial of degree $m = nq - r_0$, in the lag operator $L$. Hence, **if** there exists a matrix $A$ such that $A'H(1) = 0$ we conclude that $A'X_{t\cdot}'$ is a **stationary, ergodic** process. From Eq. (5.14) and the definition of adjoints and determinants, we have

$$\Pi(L)H(L) \;=\; H(L)\Pi(L) = (I - L)\pi(L)I_q, \quad \text{which implies}$$

$$\Pi(1)H(1) \;=\; 0, \quad H(1)\Pi(1) = 0. \qquad (5.39)$$

To exploit these results, we recall the discussion of Chapter 3, and note that $\Pi(1) = I_q + \Pi_0^*$. Consequently, $-\Pi_0^*$ has $r_0$ **unit roots**, because $|\lambda I_q + \Pi_0^*| = 0$ **evaluated at** $\lambda = 1$ yields $|\Pi(1)| = 0$, and $|\Pi(z)| = 0$ is known to have $r_0$ **unit roots**. For simplicity of exposition, assume that $\Pi_0^*$ is **diagonalizable**.[6] and let $D$ be the (nonsingular) matrix of its characteristic vectors so that

$$D^{-1}\Pi_0^* D = - \begin{bmatrix} I_{r_0} & 0 \\ 0 & \Lambda_1 \end{bmatrix},$$

where $\Lambda_1$ is a **diagonal matrix** of order $r = q - r_0$ containing the roots which are **less than one** in absolute value. It follows, therefore, that

$$\Pi(1) = D \begin{bmatrix} 0 & 0 \\ 0 & I_r - \Lambda_1 \end{bmatrix} D^{-1} = D_2(I_r - \Lambda_1)D_2^{*'},$$

$$D = (D_1, D_2), \quad D^{-1} = (D_1^*, D_2^*)', \tag{5.40}$$

where $D_1$ is $q \times r_0$ and contains the characteristic vectors corresponding to the **unit** roots. Similarly, $D_2^{*'}$ is $(q - r_0) \times q$ and so on. Using the representation in Eq. (5.40), and the condition $\Pi(1)H(1) = 0$, we obtain

$$D_2^{*'} H(1) = 0, \quad \text{and thus that} \quad D_2^{*'} X_t' \text{ is stationary and ergodic.}$$

---

[6] In any event, the case where the matrix is not necessarily diagonalizable has been discussed adequately in Chapter 3, in the section on cointegration. It would disrupt the flow of our discussion to reintroduce this complication now. Moreover, this assertion entails little loss of generality, or relevance. Note that, in this context, we assume that $\Pi(1)$ has $r_0$ zero roots, and hence it is of rank $r = q - r_0$. Thus, the dimension of its (column) null space is $r_0$, which means that there exists a nonsingular matrix, say $A_{r_0}$, such that $\Pi(1)A = 0$. In turn, this implies that $-\Pi_0^* A_{r_0} = A_{r_0}$, so that $A_{r_0}$ contains the $r_0$ **linearly independent** characteristic vectors corresponding to the **unit roots** of $-\Pi_0^*$. To assert that $-\Pi_0^*$ is diagonalizable, in this context, it is sufficient to assert that its remaining $r$ roots are distinct. This is a rather insignificant restriction on the range of applicability of such results. What we gain is the property that $\Pi(1)$ is diagonalizable, and has the representation $A^{-1}\Pi(1)A = \text{diag}(\Lambda_r, 0_{r_0})$, where $\Lambda_r$ is a diagonal matrix of dimension $r$ containing the stationary roots, and $A = (A_*, A_{r_0})$, $A_*$ being the $q \times r$ matrix corresponding to the other (distinct) roots of $-\Pi_0^*$. Since we now have

$$-A^{-1}\Pi_0^* A = \begin{bmatrix} \Lambda_r^* & 0 \\ 0 & I_{r_0} \end{bmatrix}, \quad \text{it follows that } A^{-1}\Pi(1)A = \begin{bmatrix} \Lambda_r & 0 \\ 0 & 0_{r_0} \end{bmatrix}, \quad \Lambda_r = I_r - \Lambda_r^*.$$

Alternatively, if we asserted that $\Pi(1)$ is diagonalizable, this will be sufficient to render $-\Pi_0^*$ **diagonalizable**.

In fact, from the representation in Eq. (5.38) we have

$$D_2^{*'} X_{t.}' = [\pi(L)]^{-1} H^{**}(L)\epsilon_{t.}', \tag{5.41}$$

which demonstrates the stationarity and ergodicity of the left member above. This is so since the roots of $\pi(z) = 0$ are outside the unit circle, and thus the expansion of the inverse converges exponentially, while $H^{**}(L)$ is a matrix polynomial of finite degree $m - (n - 1) = (q - 1)n - r_0 + 1$ .

Return now to Eq. (5.6), and define the $nq \times nq$ nonsingular matrices

$$D_*^{-1} = \begin{bmatrix} T^{-1} D_1^{*'} & 0 \\ T^{-(1/2)} D_2^{*'} & 0 \\ 0 & T^{-(1/2)} I_{(n-1)q} \end{bmatrix}, \quad R = D_*^{-1} X^{*'} X^* D_*^{'-1}, \tag{5.42}$$

$$R = \begin{bmatrix} T^{-2} D_1^{*'} P_{-1}' P_{-1} D_1^* & T^{-(3/2)} D_1^{*'} P_{-1}' P_{-1} D_2^* & T^{-(3/2)} D_1^{*'} P_{-1}' X \\ T^{-(3/2)} D_2^{*'} P_{-1}' P_{-1} D_1^* & T^{-1} D_2^{*'} P_{-1}' P_{-1} D_2^* & T^{-1} D_2^{*'} P_{-1}' X \\ T^{-(3/2)} X' P_{-1} D_1^* & T^{-1} X' P_{-1} D_2^* & T^{-1} X' X \end{bmatrix}.$$

From the results obtained in Chapter 3, we conclude

$$D_*^{-1} X^{*'} X^* D_*^{'-1} \to \begin{bmatrix} \Phi_{11}^* & 0 & 0 \\ 0 & M_{zz} & M_{zx} \\ 0 & M_{xz} & M_{xx} \end{bmatrix}, \tag{5.43}$$

$$\Phi_{11}^* = D_1^{*'} \Phi_{11} D_1^*,$$

where $\Phi_{11}$ is the (1,1) block of the matrix $\Phi$ defined in Eq. (5.28). The convergence to $\Phi_{11}^*$ is convergence in distribution, while the convergence to the other matrices of the right member is convergence a.c. by virtue of the mean ergodic theorem. Utilizing the notation of Eq. (5.42), we have

$$D_*'(\hat{J} - J) = \left( D_*^{-1} X^{*'} X^* D_*^{'-1} \right)^{-1} D_*^{-1} X^{*'} U, \quad U = (\epsilon_{t.}), \quad t = 1, 2, \dots, T.$$

Therefore

$$D_*^{-1} X^{*'} U = \begin{bmatrix} T^{-1} D_1^{*'} P_{-1}' U \\ W_T U \end{bmatrix}, \quad W_T U = \frac{1}{\sqrt{T}} \sum_{t=1}^{T} \psi_{t.}' \epsilon_{t.}, \quad \psi_{t.}' = \begin{pmatrix} z_{t-1.}' \\ x_{t.}' \end{pmatrix},$$

where $z_{t.} = X_{t.} D_2^*$. The limiting distribution of the entity $\text{vec}(W_T U)$ is easily obtained as follows. Write

$$\text{vec}(W_T U) = \frac{1}{\sqrt{T}}(I_q \otimes \psi_{t.}')\epsilon_{t.}', \tag{5.44}$$

and note that the right member above is a **martingale difference** sequence with respect to the stochastic basis $\{\mathcal{A}_t : \mathcal{A}_t = \sigma(\epsilon_{s\cdot}, \; 1 \le s \le t)\}$. In addition, it satisfies a **Lindeberg condition**. Hence by Proposition 21 in Dhrymes (1989), p. 337, we conclude

$$\text{vec}(W_T U) \overset{\text{d}}{\to} N(0, \Sigma \otimes M^*), \quad M^* = \begin{bmatrix} M_{zz} & M_{zx} \\ M_{xz} & M_{xx} \end{bmatrix}. \tag{5.45}$$

We must now find the limiting distribution of $D_1^{*\prime} P_{-1}^\prime U / T$; in effect, this distribution was already determined as the (1,2) block of the matrix $\Psi + \Sigma_1^*$ in Eq. (5.28). Noting, from Eq. (5.30), that the (1,2) block of $\Sigma_1^*$ is **null**, we recall from Eqs. (5.28) and (5.29) that

$$P_0^\prime P_0 = \Sigma_0 = \begin{bmatrix} \Omega_* & H^*(1)\Sigma \\ \Sigma H^{*\prime}(1) & \Sigma \end{bmatrix}, \quad \Omega_* = \Omega_1 + \Omega_1^* + \Omega_1^{*\prime}, \tag{5.46}$$

$$\Omega_1 = \sum_{i=0}^{\infty} H_i^* \Sigma H_i^{*\prime}, \quad \Omega_1^* = \sum_{\tau=1}^{\infty} \sum_{i=0}^{\infty} H_i^* \Sigma H_{i+\tau}^{*\prime}, \quad \Omega_* = T_1^\prime T_1.$$

Next, consider the (unique) triangular decomposition [see Proposition 58 in Dhrymes (1984), p. 68]:

$$\Sigma_0 = \begin{bmatrix} T_1^\prime & 0 \\ T_2^\prime & T_3^\prime \end{bmatrix} \begin{bmatrix} T_1 & T_2 \\ 0 & T_3 \end{bmatrix} = P_0^\prime P_0, \tag{5.47}$$

such that

$$T_2^\prime = \Sigma H(1)^\prime T_1^{-1}, \quad T_3^\prime T_3 = \Sigma[\Sigma^{-1} - H(1)^\prime \Omega_*^{-1} H(1)]\Sigma. \tag{5.48}$$

In the first convergence representation of Eq. (5.28), the standard multivariate Brownian motion (SMBM) appearing therein is of dimension $2q$, so that we may write

$$B(s) = (B_{(1)}(s), B_{(2)}(s)), \tag{5.49}$$

and **each of the constituent elements of the SMBM**, $B_{(i)}(s)$, is of dimension $q$. Hence,

$$\begin{aligned} \Phi_{11} &= (I_q, 0) P_0 \left( \int_0^1 \begin{bmatrix} B_{(1)}(s)^\prime B_{(1)}(s)\,ds & B_{(1)}(s)^\prime B_{(2)}(s)\,ds \\ B_{(2)}(s)^\prime B_{(1)}(s)\,ds & B_{(2)}(s)^\prime B_{(2)}(s)\,ds \end{bmatrix} \right) P_0(I_q, 0)^\prime \\ &= T_1^\prime \left( \int_0^1 B_{(1)}(s)^\prime B_{(1)}(s)\,ds \right) T_1, \end{aligned}$$

$$\Psi_{12} = (I_q, 0) P_0' \left( \int_0^1 \begin{bmatrix} B_{(1)}(s)' \, dB_{(1)}(s) & B_{(1)}(s)' \, dB_{(2)}(s) \\ B_{(2)}(s)' \, dB_{(1)}(s) & B_{(2)}(s)' \, dB_{(2)}(s) \end{bmatrix} \right) P_0(0, I_q)'$$

$$= T_1' \left( \int_0^1 B_{(1)}(s)' \, dB(s) \right) \begin{bmatrix} T_2 \\ T_3 \end{bmatrix}. \tag{5.50}$$

In obtaining the limiting distribution of $P_{-1}' U / T$, and in deriving the full implications of the cointegration hypothesis, we cannot operate with the representation of $\Psi_{12}$ as in Eq. (5.50). We need to simplify it. It is shown in Remark 2 of Chapter 7 that a general MBM with covariance matrix $\Upsilon$ may always be written as $BP$, where $B$ is a SMBM and $P'P = \Upsilon$ is the (unique) triangular decomposition of the covariance matrix. Viewed in this light, what we have in $\Psi_{12}$ is the integration of the general MBM $B_{(1)}(s)T_1$, with respect to the general MBM $B_{(1)}(s)T_2 + B_{(2)}T_3$. The covariance matrix of the latter is thus

$$s(T_2' T_2 + T_3' T_3) = s\Sigma, \quad \text{in view of Eq. (5.48)}.$$

Hence, again following the practice of Remark 2 in Chapter 6, we may write this general MBM in the equivalent representation

$$B_{(1)}(s)T_2 + B_{(2)}(s)T_3 = B_{(4)}(s)T_4, \quad T_4' T_4 = \Sigma, \tag{5.51}$$

where $B_{(4)}(s)$ is a ent SMBM and $T_4$ is the triangular matrix of the decomposition of $\Sigma$. Hence, $\Psi_{12}$ may be written in the canonical form

$$\Psi_{12} = T_1' \left( \int_0^1 B_{(1)}'(s) \, dB_{(4)}(s) \right) T_4. \tag{5.52}$$

The MBM involved in the integral above, however, are not independent, since their covariance matrix is, by Eq. (5.48),

$$\text{Cov}(T_1' B_{(1)}'(s), \ B_{(1)}(s)T_2 + B_{(2)}(s)T_3) = s \, T_1' T_2 = s \, H^*(1)\Sigma \neq 0.$$

Combining the results of Eqs. (5.46), (5.47), and (5.52), we obtain

$$\frac{1}{T} D_1^{*'} P_{-1}' U \xrightarrow{d} D_1^{*'} T_1' \left( \int_0^1 B_{(1)}(s)' \, dB_{(4)}(s) \right) T_4.$$

Employing the same argument, we note that $D_1^{*'} \Psi_{12}$ yields the result

$$\frac{1}{T} D_1^{*'} P_{-1}' U \xrightarrow{d} T^{*'} \left( \int_0^1 B_*(s)' \, dB_{(4)}(s) \right) T_4, \quad B_*(s)T^* = B_{(1)}(s)T_1 D_1^*, \tag{5.53}$$

where $T^{*'} T^* = D_1^{*'} \Omega_* D_1^*$ is the triangular decomposition of the covariance matrix of the MBM $B_{(1)}(s) T_1 D_1^*$. Note that $T^*$ is $r_0 \times r_0$, and $B_*(s)$ is an $r_0$-element (row) SMBM. It is important to note that in the representation above, the covariance between the two MBM, $B_*(s) T^*$ and $B_{(4)}(s) T_4$, is given by

$$\mathrm{Cov}(T^{*'} B_*(s)', \ B_{(4)}(s) T_4) = \mathrm{Cov}(D_1^{*'} T_1' B_{(1)}(s)', \ B_{(1)}(s) T_2 + B_{(2)}(s) T_3)$$

$$= s\, D_1^{*'} T_1' T_2 = D_1^{*'} H^*(1) \Sigma. \qquad (5.54)$$

This result will play an important role at a later stage.

By a similar argument on $\Phi_{11}^*$ of Eq. (5.43), we find

$$\frac{1}{T^2} D_1^{*'} P_{-1}' P_{-1} D_1^* \xrightarrow{\mathrm{d}} T^{*'} \left( \int_0^1 B_*(s)' B_*(s)\, ds \right) T^*. \qquad (5.55)$$

This completes the derivation of the (canonical form of the) limiting distribution of the estimators of parameters in the $VAR(n)$ specification. We have therefore proved Proposition 2.

**Proposition 2.** In the context of Proposition 1, consider the estimation of parameters in the $VAR(n)$ model, $\Delta P = X^* J + U$. The following statements are true:

i. the least squares estimator obeys

$$\hat{J} - J = (X^{*'} X^*)^{-1} X^{*'} U;$$

ii. let $D$ be the matrix of characterstic vectors of $-\Pi_0^*$, partition $D = (D_1, \ D_2)$ such that $D_1$ is $q \times r_0$ and corresponds to its $(-\Pi_0^*) \ r_0$ **unit roots**, and define

$$D_* = \begin{bmatrix} T D_1 & T^{(1/2)} D_2 & 0 \\ 0 & 0 & T^{(1/2)} I_{(n-1)q} \end{bmatrix};$$

then

$$D_*'(\hat{J} - J) \xrightarrow{\mathrm{d}} \begin{bmatrix} \zeta_{(0)} \\ \xi_{(1)} \\ \xi_{(2)} \end{bmatrix},$$

$$\zeta_0 = \left[ T^{*-1} \left( \int_0^1 B_*(s)' B_*(s)\, ds \right)^{-1} T^{*'-1} \right] L_*,$$

$$L_* = T^{*'} \left( \int_0^1 B_*(s)'\, dB_{(4)}(s) \right) T_4,$$

$$\text{vec}(\xi) \sim N(0, \Sigma \otimes M^{*-1}), \quad \xi = \begin{bmatrix} \xi_{(1)} \\ \xi_{(2)} \end{bmatrix}. \tag{5.56}$$

**Remark 4**. Notice that in the canonical representation of Eq. (5.56) the only integral involving stochastic measure is $\int_0^1 B_*(s)'\, dB_{(4)}(s)$; since $B_*(s)$ is an $r_0$-element SMBM, and $B_{(4)}(s)$ is a ent SMBM, the matrix in question is $r_0 \times q$. In the preceding section, where the implications of cointegration were not fully exploited, the corresponding integral with stochastic measure as given in part v of Proposition 1 entailed a $q \times q$ matrix involving stochastic integration.

Notice, further, that in Eq. (5.56) we **do not** exhibit the limiting distribution of the estimator of $J$, but rather a certain transformation of it, which is given by

$$T D_1^{*'}[\hat{\Pi}(1) - \Pi(1)]', \quad T^{(1/2)} D_2^{*'}[\hat{\Pi}(1) - \Pi(1)]', \quad T^{(1/2)}(\hat{\Pi}^* - \Pi^*)'.$$

The limiting distributions of these entities, in the notation of Eq. (5.56), correspond to those of $\zeta_{(0)}$, $\xi_{(1)}$ and $\xi_{(2)}$, respectively. In addition, note that this means that a certain transform of the estimator of $\Pi(1)$ has "components" that converge to $\Pi(1)$ at different rates; thus, for the transformation $D^{-1}[\hat{\Pi}(1) - \Pi(1)]$, its first component, $D_1^{*'}[\hat{\Pi}(1) - \Pi(1)]$, is consistent of order $T^\alpha$, for any $\alpha \in [0,1)$, while the second component, $D_2^{*'}[\hat{\Pi}(1) - \Pi(1)]$, is consistent of order $T^\alpha$, for any $\alpha \in [0, 1/2)$. Hence, the asymptotic behavior ascribed to the estimator above in part v of Proposition 1 of the preceding section did not fully exploit the cointegration properties of the model.

Finally, note that the matrix $B$ in the rank factorization $\Pi(1) = \Gamma B'$ is of dimension $q - r$ **with** $r = q - r_0$. Thus, the matrix $B_0$ in Eq. (5.4), containing the "free parameters", is of dimension $r_0 \times r$, where $r_0$ is the number of **unit roots** of $|\Pi(z)| = 0$, and $r = q - r_0$ is the **cointegration rank**.

## 5.2.4   Restricted Estimators

The bulk of the work needed to establish the limiting distribution of restricted estimators was done in the previous section, where we derived the canonical representation of the limiting distribution of functions of the unrestricted estimators. Moreover, we also obtained the ML estimator of parameters obeying the restriction of Eq. (5.4), using a fixed point algorithm. Here we discuss the Newton-Raphson (NR) procedure, by which Ahn and Reinsel (1990) obtain ML estimators obeying such restrictions. For this topic we need to revise slightly the notation employed in Eq. (5.6). To this end, rewrite the model as

$$\Delta P = -P_{-1}\Pi(1)' + X\Pi^{*'} + U = -P_{-1}B\Gamma' + X\Pi^* + U,$$

and define

$$X^* = (P_{-1}B, \ X), \quad J = (-\Gamma, \ \Pi^*)', \quad \delta = (b_0', \ j^{*'})',$$

$$b_0 = \text{vec}(B_0), \quad j^* = \text{vec}(J). \tag{5.57}$$

The matrices $X^*$ and $J$ differ from the corresponding entities in the previous section **only in their first component**; in particular, $X^*$ **now depends on the unknown parameter** $B$.

The LF of the observations may be written, with $c_* = -\frac{qT}{2}\ln(2\pi)$, as

$$F_0 = c_* - \frac{T}{2}\ln|\Sigma| - \frac{1}{2}\text{tr}(\Delta P - X^*J)\Sigma^{-1}(\Delta P - X^*J)' \tag{5.58}$$

$$= c_* - \frac{T}{2}\ln|\Sigma| - \frac{1}{2}[\Delta p - \text{vec}(X^*J)]'(\Sigma^{-1} \otimes I_T)(\Delta p - \text{vec}(X^*J)).$$

The NR procedure is based on the mean value theorem expansion of the gradient of the LF and thus involves **both the gradient and the Hessian of the LF**; by contrast, the fixed point procedure given in the previous section involves only the gradient. The mean value theorem expansion, for any two points $\hat{\delta}$ and $\tilde{\delta}$, is

$$\left(\frac{\partial F_0}{\partial \delta}(\hat{\delta})\right)' = \left(\frac{\partial F_0}{\partial \delta}(\tilde{\delta})\right)' + \frac{\partial^2 F_0}{\partial \delta \partial \delta}(\bar{\delta})(\hat{\delta} - \tilde{\delta}), \quad \text{or}$$

$$(\hat{\delta} - \tilde{\delta}) = -\left(\frac{\partial^2 F_0}{\partial \delta \partial \delta}(\bar{\delta})\right)^{-1}\left(\frac{\partial F_0}{\partial \delta}(\tilde{\delta})\right)', \quad |\bar{\delta} - \tilde{\delta}| \leq |\hat{\delta} - \tilde{\delta}|. \tag{5.59}$$

In the NR procedure we evaluate the Hessian at $\tilde{\delta}$, interpret the latter as the estimator at the $s-1$ iteration, $\delta_{(s-1)}$, and $\hat{\delta}$ as the $s^{th}$ iterate, $\delta_{(s)}$, thus obtaining

$$\delta_{(s)} = \delta_{(s-1)} - \left(\frac{\partial^2 F_0}{\partial\delta\partial\delta}(\delta_{(s-1)})\right)^{-1}\left(\frac{\partial F_0}{\partial\delta}(\delta_{(s-1)})\right)'. \qquad (5.60)$$

To apply this procedure in our context we need to obtain expressions for the gradient and the Hessian of the LF in Eq. (5.59). Recall that

$$B = Q_1 + Q_2 B_0, \quad Q_1 = \begin{bmatrix} I_r \\ 0 \end{bmatrix}, \quad Q_2 = \begin{bmatrix} 0 \\ I_{r_0} \end{bmatrix} \qquad (5.61)$$

and note that, taking into account Eq. (5.58), we have the two (equivalent) representations

$$\mathrm{vec}(X^*J) = (I_{(n-1)q+r} \otimes X^*)j^*, \quad X^{**} = (P_{-1}Q_1, \; X),$$

$$\mathrm{vec}(X^*J) = (I_{(n-1)q+r} \otimes X^{**})j^* + (I_{r_0} \otimes P_{-1}Q_2)b_0. \qquad (5.62)$$

The first representation is useful when differentiating with respect to $j^*$, the second when differentiating with respect to $b_0$. Finally, note that $X^{**}$ **does not depend on any unknown parameters**.

The gradient and Hessian of the LF are easily found to be

$$\left(\frac{\partial F_0}{\partial b_0}\right)' = C_1'(\Sigma^{-1} \otimes I_T)[\Delta p - \mathrm{vec}(X^*J)], \quad C_1 = -\Gamma \otimes P_{-1}Q_2,$$

$$\left(\frac{\partial F_0}{\partial j^*}\right)' = C_2'(\Sigma^{-1} \otimes I_T)[\Delta p - \mathrm{vec}(X^*J)], \quad C_2 = I_q \otimes X^*,$$

$$\frac{\partial^2 F_0}{\partial b_0 \partial b_0} = -C_1'(\Sigma^{-1} \otimes I_T)C_1, \quad \frac{\partial^2 F_0}{\partial b_0 \partial j^*} = -C_1'(\Sigma^{-1} \otimes I_T)C_2,$$

$$\frac{\partial^2 F_0}{\partial j^* \partial j^*} = -C_2'(\Sigma^{-1} \otimes I_T)C_2, \quad C = (C_1, \; C_2). \qquad (5.63)$$

Thus, the NR procedure entails the iteration

$$\delta_{(s)} = \delta_{(s-1)} + \left[C_{(s-1)}'(\Sigma^{-1} \otimes I_T)C_{(s-1)}\right]^{-1} C_{(s-1)}'(\Sigma^{-1} \otimes I_T)c_{(s-1)},$$

$$c_{(s-1)} = \Delta p - \mathrm{vec}(X^*J)_{s-1}. \qquad (5.64)$$

Since the ML estimator of $\Sigma$ is given by

$$\hat{\Sigma} = \frac{1}{T}[\Delta P - \hat{X}^* \hat{J}]'[\Delta P - \hat{X}^* \hat{J}], \tag{5.65}$$

we may include the covariance matrix $\Sigma$ in the iteration process; to do so, denote its estimator after the $s-1$ iteration by $\Sigma_{s-1}$, and note that it may be obtained by evaluating $X^* J$ at $\delta_{s-1}$. We can write the NR procedure of Eq. (5.64) as

$$\delta_s = \delta_{s-1} + \left[ C'_{(s-1)} (\Sigma_{s-1}^{-1} \otimes I_T) C_{(s-1)} \right]^{-1} C'_{(s-1)} (\Sigma_{s-1}^{-1} \otimes I_T) c_{(s-1)},$$

$$c_{(s-1)} = \Delta p - \text{vec}(X^* J)_{(s-1)}. \tag{5.66}$$

**Remark 5. For the converging iterate to be a ML estimator**, the initial estimator of $\delta$, the zeroth iterate, must be consistent of order at least $T^\alpha$ for $\alpha \in [0, 1/2)$. Otherwise, we cannot be assured of having found, in the converging iterate, a consistent root of the ML equations and, thus, we cannot be assured of having found the ML estimator. Hence, to complete the NR procedure we need to supply an initial consistent estimator of $\delta$.

## An Initial Consistent Estimator

In part v of Proposition 1, we have shown that the least squares estimator of $\Pi(1)$ is consistent of order $T^\alpha$, $\alpha \in [0, 1)$, while the estimator of $\Pi^*$ is consistent of order $T^\alpha$, $\alpha \in [0, 1/2)$. Thus, we need only obtain consistent estimators for $\Gamma$ and $B$ as restricted by Eq. (5.4). Since in the restricted context in which we operate $\Pi(1) = (\Gamma, \Gamma B'_0)$, an initial consistent estimator for $\Gamma$ is easily, and obviously, available as $\tilde{\Gamma} = G_1$, where we have partitioned the least squares estimator $\tilde{\Pi}(1) = (G_1, G_2)$. The initial consistent estimator for $B_0$ is also easily obtained as $\tilde{B}'_0 = (G'_1 G_1)^{-1} G'_1 G_2$, since $\tilde{\Pi}(1)$ is consistent and, thus,

$$\text{plim}_{T \to \infty} \tilde{B}'_0 = (\Gamma' \Gamma)^{-1} \Gamma' \Gamma B'_0 = B'_0.$$

We have therefore proved Proposition 3.

**Proposition 3.** Consider the model in Eq. (5.6), obeying the restrictions in Eq. (5.4). In the context of the notation in Eq. (5.57), let $\tilde{\Pi}(1) = (G_1, G_2)$,

$\tilde{\Pi}^*$ be the (unretricted) least squares estimators of the corresponding parameters. The following statements are true:

i. the estimators

$$\tilde{\Gamma} = G_1, \quad \tilde{B}_0 = (G_1'G_1)^{-1}G_1'G_2,$$

are consistent of order $T^\alpha$, $\alpha \in [0,1)$, for the corresponding parameters $\Gamma$, $B_0$;

ii. the estimator $\tilde{\Pi}^*$ is consistent of order $T^\alpha$, $\alpha \in [0,1/2)$, for the
parameter $\Pi^*$;

iii. the estimator

$$\tilde{\delta} = (\tilde{b}_0', \tilde{j}^{*'})',$$

is a consistent estimator for $\delta$, **and is defined as the initial consistent estimator of that parameter**, where

$$\tilde{b}_0 = \text{vec}(\tilde{B}_0), \quad \tilde{j}^* = \text{vec}(\tilde{J}), \quad \tilde{J} = (-\tilde{\Gamma}, \tilde{\Pi}^*)'.$$

**Limiting Distribution of the Restricted Estimators**

Examining the expression for the Hessian of the LF the following become
apparent, relative to the matrix $C'(\hat{\Sigma}^{-1} \otimes I_T)C$ in the converging iterate of
the NR procedure:

i. the (1,1) block of that matrix converges in distribution **only** if normalized by $T^{-2}$; in fact, it converges to the same entity as the (1,1) block
of the matrix $R$ of Eq. (5.41), with $D_1^*$ therein **replaced** by $Q_2$;

ii. the (1,2) and (2,1) blocks of that matrix, when normalized by $T^{-(3/2)}$,
converge to zero in probability;

iii. when normalized by $T^{-1}$, the (2,2) block of that matrix converges, at
least in probability, to $M^*$ as defined in Eq. (5.45); in the present case
$z_{t-1.} = X_{t-1.}B$, while in the definition of Eq. (5.45) it was defined as
$z_{t-1.} = X_{t-1.}D_2^*$. In either case, the entity in question is **stationary
and ergodic** and represents the cointegral vector.

Since the parameter matrices $\Sigma$, $J$, and $B_0$ are all consistently estimated, we may employ again the mean value theorem expansion of the LF relative to $\hat{\delta}$, the converging iterate of the NR procedure, and $\delta^\circ$, the true parameter vector (restricted to $J$ and $B_0$) to obtain

$$0 = \left(\frac{\partial F_0}{\partial \delta}(\hat{\delta})\right)' = \left(\frac{\partial F_0}{\partial \delta}(\delta^\circ)\right)' + \left(\frac{\partial^2 F_0}{\partial \delta \partial \delta}(\bar{\delta})\right)(\hat{\delta} - \delta^\circ), \quad |\bar{\delta} - \delta^\circ| \leq |\hat{\delta} - \delta^\circ|.$$

Since $\hat{\delta}$ converges to $\delta^\circ$ at least in probability, so does $\bar{\delta}$. Defining the normalizing matrix

$$D_T = \begin{bmatrix} I_r \otimes T^{-1}I_{r_0} & 0 \\ 0 & I_q \otimes T^{-(1/2)}I_{q(n-1)+r} \end{bmatrix}, \tag{5.67}$$

we may write

$$D_T(\hat{\delta} - \delta^\circ) \sim \left(D_T C_\circ'(\Sigma_\circ^{-1} \otimes I_T)C_\circ D_T\right)^{-1} D_T C'(\Sigma_\circ^{-1} \otimes I_T)u, \quad u = \text{vec}(U), \tag{5.68}$$

and note that

$$D_T C_\circ'(\Sigma_\circ^{-1} \otimes I_T)C_\circ D_T \xrightarrow{\text{d}} \begin{bmatrix} \Gamma_\circ' \Sigma_\circ^{-1} \Gamma_\circ \otimes K & 0 \\ 0 & \Sigma_\circ^{-1} \otimes M^* \end{bmatrix}, \tag{5.69}$$

where the notation $C_\circ$, for example, indicates that the entity in question is evaluated at the **true parameter point**. Hence, the limiting distribution of the ML estimator of $\delta$ is essentially determined by

$$D_T C_\circ'(\Sigma_\circ^{-1} \otimes I_T)u = \begin{bmatrix} \text{vec}(A_1) \\ \text{vec}(A_2) \end{bmatrix}, \quad A_1 = Q_2'\left(\frac{P_{-1}'U}{T}\right)\Sigma_\circ^{-1}\Gamma_\circ, \quad A_2 = \frac{X^{*\prime}U}{\sqrt{T}}. \tag{5.70}$$

The entities in Eq. (5.70) involve sequences whose limiting distribution was established earlier. In fact, from Eqs. (5.50) and (5.52), we easily establish that

$$T(\hat{b}_0 - b_0) \xrightarrow{\text{d}} \text{vec}\left[K^{-1}H\Sigma_\circ^{-1}\Gamma_\circ(\Gamma_\circ'\Sigma_\circ^{-1}\Gamma_\circ)^{-1}\right],$$

$$K = Q_2'T_1'\left[\int_1^1 B_{(1)}(s)' B_{(1)}(s)\,ds\right]T_1 Q_2, \tag{5.71}$$

$$\sqrt{T}(\hat{j}^* - j^*) \xrightarrow{\text{d}} N(0, \Sigma \otimes M^{*-1}), \quad H = Q_2'T_1'\left[\int_0^1 B_{(1)}(s)' dB_{(4)}(s)\right]T_4.$$

The limiting distribution of the ML estimator of the covariance matrix is easily obtained from its definition,

$$\hat{\Sigma} = \frac{1}{T}\hat{U}'\hat{U}, \quad \hat{U} = \Delta P - \hat{X}^*\hat{J}. \tag{5.72}$$

Adding and subtracting $\hat{X}^*J$ in the right member below, we find

$$\hat{U} = U - \hat{X}^*(\hat{J} - J) - (\hat{X}^* - X^*)J, \quad \hat{X}^* = (P_{-1}\hat{B}, \ X),$$

and thus

$$\frac{\hat{U}'\hat{U}}{\sqrt{T}} = \frac{U'U}{\sqrt{T}} + \frac{1}{T^{3/2}}\sqrt{T}(\hat{J} - J)'[\hat{X}^*\hat{X}^*]\sqrt{T}(\hat{J} - J)$$

$$+ \frac{1}{T^{5/2}}[T\Gamma(\hat{B}_0 - B_0)']P'_{-1}P_{-1}[T(\hat{B}_0 - B_0)\Gamma'] + \text{cross products}.$$

Since all terms, other than the first, converge to zero at least in probability, we obtain

$$\frac{\hat{U}'\hat{U}}{\sqrt{T}} \sim \frac{U'U}{\sqrt{T}}. \tag{5.73}$$

The limiting distribution of the latter, however, is found by standard techniques. Thus, set

$$\xi_{(T)} = \frac{1}{\sqrt{T}}\text{vec}(U'U) = \frac{1}{\sqrt{T}}\sum_{t=1}^{T}\xi'_{t\cdot}, \quad \xi_{t\cdot} = (\epsilon_{t\cdot} \otimes \epsilon_{t\cdot}). \tag{5.74}$$

The sequence $\{\xi_{t\cdot} : t \geq 1\}$ is one of i.i.d. random vectors with

$$E\xi_{t\cdot} = E\xi_{1\cdot} = \text{vec}(\Sigma) = \sigma, \quad \text{Cov}(\xi'_{1\cdot}) = E(\epsilon'_{1\cdot}\epsilon_{1\cdot} \otimes \epsilon'_{1\cdot}\epsilon_{1\cdot}) = \Sigma^*. \tag{5.75}$$

Since it evidently obeys a Lindeberg condition we conclude, from Proposition 45 and Remark 25 in Dhrymes (1989), pp. 271-275,

$$\sqrt{T}\text{vec}(\hat{\Sigma} - \Sigma) \xrightarrow{d} N[0, \Sigma^*]. \tag{5.76}$$

To evaluate $\Sigma^*$, we first note that the vector $\xi_{1\cdot}$ contains redundancies, and hence that $\Sigma^*$ is **singular**. In particular, the $i, j$ element of the $r, s$ block of $E(\epsilon'_{1\cdot}\epsilon_{1\cdot} \otimes \epsilon'_{1\cdot}\epsilon_{1\cdot})$ is given by

$$E(\epsilon_{1r}\epsilon_{1r}\epsilon_{1i}\epsilon_{1j}) = \sigma_{rs}\sigma_{ij} + \sigma_{ri}\sigma_{sj} + \sigma_{rj}\sigma_{si}, \quad \text{so that} \quad \Sigma^* = \Sigma \otimes \Sigma + 2\sigma\sigma',$$

which is evidently a **singular** matrix.

The preceding discussion has proved Proposition 4.

**Proposition 4.** In the context of Proposition 3, let $\hat{b}_0$, $\hat{j}^*$, $\hat{\sigma}$ be the ML estimators of the parameters in the corresponding matrices, $B_0$, $J$ and $\Sigma$, obtained either by the fixed point or the NR algorithm. The following statements are true:

i. $T(\hat{b}_0 - b_0) \overset{\text{d}}{\to} \zeta$, where

$$\zeta = \text{vec} \left[ K^{-1} H \Sigma_\circ^{-1} \Gamma_\circ (\Gamma_\circ' \Sigma_\circ^{-1} \Gamma_\circ)^{-1} \right],$$

$$K = Q_2' T_1' \left[ \int_1^1 B_{(1)}(s)' B_{(1)}(s) \, ds \right] T_1 Q_2,$$

$$H = Q_2' T_1' \left[ \int_0^1 B_{(1)}(s)' \, dB_{(4)}(s) \right] T_4;$$

ii. the limiting distribution of the ML estimator of the unknown elements in $J$ is given by

$$\sqrt{T}(\hat{j}^* - j^*) \overset{\text{d}}{\to} N(0, \, \Sigma \otimes M^{*-1});$$

iii. the limiting distribution of the ML estimator of the unknown parameters in $\Sigma$ is given by

$$\sqrt{T} \text{vec}(\hat{\Sigma} - \Sigma) \overset{\text{d}}{\to} N[0, \, \Sigma^*].$$

Although, in principle, the preceding competely resolves the inference problem for cointegrated $VAR(n)$ models, we have not exhausted the simplification potential of the cointegration hypothesis. The basis for the discussion below is Example 5 in Chapter 8, where it is shown that, under certain circumstances, stochastic integrals of the form $\int_0^t B(s)' \, dV(s)$, properly normalized, have a standard normal distribution, where $B$ and $V$ are MBM. The condition under which this holds is that $B$ and $V$ are mutually independent. Since BM are Gaussian systems, lack of correlation is a sufficient basis for this result. To that end, consider again the matrix

$$H_{(*)} = Q_2' T_1' \left[ \int_0^1 B_{(1)}(s)' \, dB_{(4)}(s) \right] T_4 \Sigma_\circ^{-1} \Gamma_\circ (\Gamma_\circ' \Sigma_\circ^{-1} \Gamma_\circ)^{-1},$$

which multiplies $K^{-1}$ (part i of Proposition 4) in the representation of the limiting distribution of $T(\hat{b}_0 - b_0)$. We note that it involves the stochastic integration of the MBM $V_{(1)} = B_{(1)}(s)T_1 Q_2$ relative to the MBM

$$V_{(2)} = B_{(4)}(s)T_4 \Sigma_o^{-1} \Gamma_o (\Gamma_o' \Sigma_o^{-1} \Gamma_o)^{-1} = [B_{(1)}T_2 + B_{(2)}T_3] \Sigma_o^{-1} \Gamma_o (\Gamma_o' \Sigma_o^{-1} \Gamma_o)^{-1},$$

where the terms are as defined in Eqs. (5.47), (5.48), and (5.49). The covariance of these two MBM was in part obtained in Eq. (5.54); repeating the derivation here, we find the covariance matrix to be

$$\mathrm{Cov}[V_{(1)}(s),\ V_{(2)}(s)] = s\, Q_2' H^*(1) \Sigma_o \Sigma_o^{-1} \Gamma_o (\Gamma_o' \Sigma_o^{-1} \Gamma_o)^{-1}$$

$$= s Q_2' H^*(1) \Gamma_o (\Gamma_o' \Sigma_o^{-1} \Gamma_o)^{-1} = 0.$$

The last result is valid because of Eq. (5.39), the fact that $H^*(1) = \alpha H(1)$, with $\alpha = [1/\pi(1)]$, and the restrictions of Eq. (5.4) which imply that $\Pi(1) = (\Gamma_o,\ \Gamma_o B_o')$. Consequently, the two MBM $V_{(1)}$ and $V_{(2)}$ are **mutually independent**. We note, further, that the entities in question are, respecively, $r_0$-element and $r$-element MBM with covariance matrices

$$\mathrm{Cov}[V_{(1)}(s)] = s\, Q_2' T_1' T_1 Q_2 = s\, \Omega_{*22}, \quad \mathrm{Cov}(V_{(2)}) = s\, (\Gamma_o' \Sigma_o^{-1} \Gamma_o)^{-1}, \quad (5.77)$$

where $\Omega_{*22}$ is the principal submatrix of $\Omega_*$, as defined in Eq. (5.46), consisting of the last $r$ rows and columns. Putting

$$B_{(5)}(s)T_5 = B_{(1)}(s)T_1 Q_2, \quad T_5' T_5 = \Omega_{*22}, \quad T_6' T_6 = (\Gamma_o' \Sigma_o^{-1} \Gamma_o)^{-1},$$

$$B_{(6)} T_6 = [B_{(1)}(s)T_2 + B_{(2)}T_3] \Sigma_o^{-1} \Gamma_o (\Gamma_o' \Sigma_o^{-1} \Gamma_o)^{-1}, \quad (5.78)$$

we may express the limiting distribution of $T(\hat{b}_0 - b_0)$ in the more parsimonious notation

$$T(\hat{b}_0 - b_0) \overset{d}{\to} \mathrm{vec}\left\{ \left[ T_7' T_7 \right]^{-1} \left[ T_5' \left( \int_0^1 B_{(5)}(s)' dB_{(6)}(s) \right) T_6 \right] \right\}$$

$$T_7' T_7 = \left[ T_5' \left( \int_0^1 B_{(5)}(s)' B_{(5)}(s)\, ds \right) T_5 \right] \quad (5.79)$$

$$(I_r \otimes T_7)[T(\hat{b}_0 - b_0)] = (I_r \otimes T_7'^{-1}) \mathrm{vec}\left\{ \left[ T_5' \left( \int_0^1 B_{(5)}(s)' dB_{(6)}(s) \right) T_6 \right] \right\}.$$

Now, consider the $i^{th}$ subvector (column) of the right member of the last equation above; it is given by

$$\phi_{\cdot i} = [T_7'^{-1}]T_5 \int_0^1 B_{(5)}(s)\, dB_{(6)} t_{\cdot i}^{(6)},$$

where $t_{\cdot i}^{(6)}$ is the $i^{th}$ column of $T_6$. Since $B_{(5)}T_5$ is **independent** of $B_{(6)}T_6$, and thus $B_{(5)}$ is **independent** of $B_{(6)}$, the arguments given in Example 5 of Chapter 8 imply that the conditional distribution of $\phi_{\cdot i}|B_{(5)}$ is normal with mean zero and covariance matrix

$$
\begin{aligned}
C_{\phi_i} &= T_7'^{-1} T_5' \left( \int_0^1 \int_0^1 B_{(5)}(s_1)' B_{(5)}(s_2) t_{\cdot i}^{(6)'} E[dB_{(6)}(s_1)' dB_{(6)}(s_2)] t_{\cdot i}^{(6)} \right) T_5 T_7^{-1} \\
&= t_{\cdot i}^{(6)'} t_{\cdot i}^{(6)} T_7'^{-1} \left[ T_5 \left( \int_0^1 B_{(5)}(s)' B_{(5)}(s)\, ds \right) T_5 \right] T_7^{-1} \\
&= t_{\cdot i}^{(6)'} t_{\cdot i}^{(6)} I_r = \text{Cov}(\phi_{\cdot i}|B_{(5)}).
\end{aligned}
\tag{5.80}
$$

Consequently, the conditional distribution of $\phi_{\cdot i}|B_{(5)}$ is $N(0, \psi_{ii} I_{r_0})$; since it **does not depend on** $B_{(5)}$, the result above holds **unconditionally** as well, i.e.

$$\phi_{\cdot i} \sim N(0,\ \psi_{ii} I_{r_0}), \quad (\psi_{ij}) = \Psi = (\Gamma_o' \Sigma_o^{-1} \Gamma_o)^{-1}. \tag{5.81}$$

We may similarly establish that

$$\text{Cov}(\phi_{\cdot i},\ \phi_{\cdot j}) = \psi_{ij} I_{r_0}, \tag{5.82}$$

and consequently that

$$(I_r \otimes T_7')[T(\hat{b}_0 - b_0)] \sim N(0,\ \Psi \otimes I_{r_0}), \quad \Psi = (\Gamma_o' \Sigma_o^{-1} \Gamma_o)^{-1}. \tag{5.83}$$

We have therefore proved Proposition 5.

**Proposition 5.** In the context of Proposition 4, there exists a transformation of $T(\hat{b}_0 - b_0)$ such that the resulting entity has a standard distribution. More precisely

$$(I_r \otimes T_7')[T(\hat{b}_0 - b_0)] \sim N(0,\ \Psi \otimes I_{r_0}), \quad \Psi = (\Gamma_o' \Sigma_o^{-1} \Gamma_o)^{-1}.$$

where the entities involved have been defined in Eqs. (5.78) and (5.79).

**Remark 6.** Proposition 5 is not merely of theoretical interest, but has quite significant implications for empirical work. This is so since the matrix $K$, which is decomposed by the triangular matrix $T_7$ of the proposition is the limit of

$$\frac{1}{T^2}Q_2'P_{-1}'P_{-1}Q_2 = \frac{1}{T^2}\sum_{t=1}^{T}X_{t-1.}^{(2)'}X_{t-1.}^{(2)},$$

where $X_{t-1.}^{(2)}$ contains the last $r_0$ elements of the vector $X_{t-1.}$. If we obtain the triangular decomposition of this matrix, say,

$$\hat{T}_7'\hat{T}_7 = \frac{1}{T^2}\sum_{t=1}^{T}X_{t-1.}^{(2)'}X_{t-1.}^{(2)} = \hat{S}_{22},$$

it follows that

$$(I_r \otimes \hat{T}_7)[T(\hat{b}_0 - b_0)] \xrightarrow{d} N(0, \ \Psi \otimes I_{r_0}).$$

Using the results in Eq. (5.78) we may also obtain the consistent estimator of $\hat{T}_6$ given by

$$\hat{T}_6'\hat{T}_6 = (\hat{\Gamma}'\hat{\Sigma}^{-1}\hat{\Gamma})^{-1};$$

consequently,

$$(\hat{T}_6'^{-1} \otimes \hat{T}_7[T(\hat{b}_0 - b_0)]' \xrightarrow{d} N(0, \ I_r \otimes I_{r_0}),$$

and thus

$$[T(\hat{b}_0 - b_0)]'(\hat{\Psi}^{-1} \otimes \hat{S}_{22}[T(\hat{b}_0 - b_0)] \xrightarrow{d} \chi_{rr_0}.$$

From these results we may obtain tests of significance on individual parameters, or groups of parameters in $B_0$; consequently, Proposition 5 affords us very useful inferential tools in empirical applications.

## 5.3   Tests for Cointegration Rank

### 5.3.1   Model-free Tests Based on the Covariance Matrix

Since in applications we do not generally know the cointegration rank, it is important to determine procedures by which hypotheses on the rank of cointegration may be tested. Before we proceed, however, let us note that if we are going to test whether the elements of a vector are **cointegrated**, it is only appropriate that we should rule out the possibility that the variables in question are simply **collinear**. [7] We analyze this issue first in the context of the $MIMA(k)$ model. Thus, let $X$ be the stationary sequence

$$X_{t\cdot} = \sum_{j=0}^{k} u_{t-j\cdot} A_j', \quad A_0 = I_q, \tag{5.84}$$

and consider the sample covariance matrix

$$M_T^{(0)} = \frac{1}{T} \sum_{t=1}^{T} X_{t\cdot}' X_{t\cdot}, \tag{5.85}$$

whose expectation is given by

$$EM_T^{(0)} = \sum_{j=0}^{k} A_j \Sigma A_j' = \Omega.$$

Testing for collinearity is equivalent to a test of the hypothesis that the **smallest** characteristic root of $\Omega$ is zero. [8] If we **accept** the hypothesis that the smallest root is zero, evidently there is no point in proceeding with a cointegration test since the relevant estimators in that context will produce roots which are uniformly **smaller than those** of $M_T^{(0)}$. Even though this does not, *ipso facto*, guarantee that we shall find cointegration if we have **already found** collinearity (because of differing distributions under the null),

---

[7] In this connection, it should be noted that Richard Stone writing in 1947, observed that in dealing with 17 US macro variables over the period 1922-1938, **three principal components** accounted for over 97% of their variation. Without going into the merits of Stone's research, his finding would imply that there are 14 linear combinations of these 17 variables whose variation over the sample period is neglible. Perhaps on the basis of his data we might even test, and accept, the hypothesis that the 14 smallest roots of the covariance matrix of these 17 variables are zero. In a contemporary context, a similar procedure applied somewhat differently may well have led to the conclusion that these 17 variables are cointegrated. While from a formal point point of view collinearity **is** cointegration, in the sense that the linear combination in question is constant, and thus stationary, the substantive economic and econometric implications are rather strikingly different.

[8] In the context of the model as we have specified it in Chapter 4, collinearity implies that there exists at least one vector, say $\beta$, which is in the null space of all the matrices $A_j \Sigma A_j'$; a little reflection, however, will show that under the **normalization** we have imposed in Eq. (5.84) this is possible only if the matrix $\Sigma$ itself is **singular**.

it is wise to proceed to a cointegration test **only if collinearity has been rejected**; otherwise, the meaning and interpretation of the cointegration test results are problematic. Thus, having satisfied ourselves that we are **not** dealing with "collinear" $I(0)$ processes, we may proceed with cointegration tests employing the "sample covariance" matrix estimators given in Eqs. (4.78) and (4.83) of Chapter 4, which are reproduced here for convenience:

$$\hat{M}^{(1)} \;=\; \frac{1}{T^2} \sum_{t=1}^{T} X_t' . X_t. \quad \text{and} \tag{5.86}$$

$$\hat{M}^{(2)} \;=\; \frac{1}{T} \sum_{t=1}^{T} \frac{X_t' . X_t.}{t}. \tag{5.87}$$

Of course dealing with estimators of the component of the covariance matrix that contains the information relevant for cointegration issues has certain practical problems. First, for the covariance matrix to "recover" from the imposition of the usual initial conditions at least $k$ periods must elapse; this is so since before this "time" the relevant component of the covariance matrix **changes with** "time".[9] To see just why such tests based on the estimator in Eq. (5.87) are appropriate, and why they yield information on the desired entities, consider the sequence of Eq. (5.84) and note that for $t$ sufficiently large we can argue, heuristically,

$$X_t^{*'} \;=\; \frac{1}{\sqrt{t}} \sum_{j=1}^{t} \eta_j' . \;=\; \sum_{r=0}^{k} A_r \zeta_r,$$

$$\zeta_r \;=\; \frac{1}{\sqrt{t}} \sum_{j=1}^{t} \epsilon_{j-r.}', \quad \zeta_r \xrightarrow{\text{d}} \zeta \sim N(0, \Sigma), \tag{5.88}$$

and consequently that, asymptotically,

$$X_t^{*'} \xrightarrow{\text{d}} \left( \sum_{r=0}^{k} A_r \right) \zeta \sim N(0, \Phi), \quad \Phi = S_{0,k} \Sigma S_{0,k}. \tag{5.89}$$

Thus, the matrix whose singularity **defines the property of cointegration** may be thought of as the covariance matrix of the limiting distribution of

---

[9] Incidentally, this is one of the reasons why the general $MIMA(\infty)$ formulation in EG has been replaced by the $MIMA(k)$ formulation.

a suitably normalized $X$-sequence. [10] Thus, whether we have defined our estimator to be the "sample covariance" matrix of Eq. (5.86) or Eq. (5.87), we can be assured that as the sample increases more and more information accrues. The key aspect to be understood, is that these entities have a well-defined limiting distribution, and a number of cointegration related tests **are simply tests regarding certain aspects of their mean!**

To further motivate this approach, we may put forth the following argument: if we are convinced that the sequences we are dealing with **are** $I(1)$, our problem is to determine whether there exist linear combinations $X_{t.}\beta$ that, in some sense, have "small variance" and, if so, how many such **linearly independent** combinations there are. Noting that

$$\text{Var}\left(\frac{X_{t.}\beta}{\sqrt{t}}\right) = \beta'\Phi\beta + \frac{1}{t}\beta'D\beta$$

we conclude that a reasonable formulation of the problem is the following: Find a linear combination such that

$$\beta'\Phi\beta \quad \text{is minimized subject to} \quad \beta'\beta = 1.$$

Setting up the Lagrangian

$$L_1 = \beta'\Phi\beta + \lambda(1 - \beta'\beta),$$

the first-order conditions lead to

$$\Phi\beta = \lambda\beta, \quad \beta'\beta = 1,$$

which, therefore, define the solution, say $\beta_{.1}$, as the characteristic vector corresponding to the **smallest** root of $\Phi$. If more than one such linear combination is desired, we may formulate the problem as follows: Minimize

$$\text{tr}B_r'\Phi B_r, \quad \text{subject to} \quad B_r'B_r = I_r.$$

---

[10] Notice that a similar argument cannot be made, at least not as easily, in the $MIMA(\infty)$ case. For example, in Eq. (5.88), the $MIMA(\infty)$ context would require that the argument be applied to an infinite sequence of $\zeta_r$; moreover, the limiting distribution of $X_{t.}^{*'}$, if one could establish such result, would have covariance matrix given by

$$\lim_{t\to\infty} \frac{1}{t}\sum_{r=0}^{t-1} S_{0,r}\Sigma S_{0,r}',$$

whose existence would have to be separately established.

Or, if the number of zero roots **is not known** *a priori*, we may set up the problem as follows: Find a matrix of **maximal rank** such that

$$\operatorname{tr} B' \Phi B \text{ is minimized subject to } B' B = I.$$

The nature of the problem indicates that the maximal rank possible is $q$, since the (column) dimension of $X_{t\cdot}$ is $q$. Let the columns of $B$ be denoted by $\beta_{\cdot i}$, $i = 1, 2, 3, \ldots, q$, at most. The Lagrangian of this problem is

$$L_2 = \operatorname{tr} B' \Phi B + \sum_{i=1}^{q} \lambda_i (1 - \beta'_{\cdot i} \beta_{\cdot i}),$$

with first-order conditions,

$$\Phi \beta_{\cdot i} = \lambda_i \beta_{\cdot i}, \quad \beta'_{\cdot i} \beta_{\cdot i} = 1, \quad i = 1, 2, 3, \ldots, q. \tag{5.90}$$

If $b_{\cdot i_0}$ is a solution, then this is the characteristic vector **corresponding to the** $i_0^{th}$ root; moreover since the trace for a positive semidefinite matrix is a **nonnegative function**, the minimand is bounded below by zero. Hence, the solution must be: Choose $B$ to contain all characteristic vectors **corresponding to zero roots**. If the number of such roots is unknown *a priori*, having arranged the roots in **increasing order**, and having inserted the corresponding characteristic vectors in the minimand, we find that the concentrated form of the latter becomes

$$\operatorname{tr} B' \Phi B = \sum_{i=1}^{q} \lambda_{(i)}, \tag{5.91}$$

where $\lambda_{(i)}$ are the **ordered** roots of $\Phi$, arranged in **increasing order**. From the representation above it is clear that if the **first nonnull ordered root** is the $(r+1)^{st}$ root **then** $\lambda_{(r)} = 0$, the minimand above attains the value zero, and the solution to the problem is given by

$$\min_{B \text{ of rank } r} \operatorname{tr} B'_r \Phi B_r = \sum_{i=1}^{r} \lambda_{(i)} = 0,$$

where $B_r = (\beta'_{\cdot 1}, \beta'_{\cdot 2}, \ldots, \beta'_{\cdot r})'$ and $\beta_{\cdot i}$ is the characteristic vector corresponding to $\lambda_{(i)}$. On the other hand, if we accept the hypothesis that $\lambda_{(r+1)} = 0$, we may take $B$ to be of rank $r + 1$, i.e. add the characteristic vector corresponding to the $(r+1)^{st}$ root as the last column of $B$ and so on.

In an inference context, i.e. operating with an estimator of $\Phi$, the question of whether or not we have cointegration may be formulated as:

$H_0 : \lambda_{(1)} = 0$,

as against the alternative

$H_1 : \lambda_{(1)} > 0$.

This test is equivalent to

$H_0 : \text{rank}(B) = 1$,

as against the alternative

$H_1 : \text{rank}(B) = 0$,

where $\lambda_{(1)}$ is the **smallest root**.

We now turn to the general issue of the rank of cointegration. Here the question concerns the rank of $B$, and more particularly the determination of the **order** of the **first nonnull** root. This is easily determined by extracting all roots of the estimator of $\Phi$ and finding the **largest** $r$ for which the hypothesis $H_0 : \lambda_{(r)} = 0$ is accepted; this is equivalent to the formal test

$H_0 : \text{rank}(B) = r$,

as against the alternative

$H_1 : \text{rank}(B) < r$.

Such tests may be based on the limiting distribution of the roots. Even though the distribution of these roots is not standard, the basic idea of such tests is quite analogous to what the reader might consider the standard $F$- or $\chi^2$-tests in ordinary regression. For example, if we wish to test for the presence of a group of variables in a regression context, we obtain the sum of squared residuals in the regression **with and without** the group of variables in question. The **difference** in the two sums of squared residuals (properly normalized) is the basis for testing whether the coefficients of such variables are or are not significantly different from zero. The (first) basic test given above is the test for cointegration, and it involves the resolution of the simple question of whether the smallest root of $\Phi$ is zero. That this is, indeed, **the** test for cointegration is evident since **a necessary and sufficient condition** for cointegration is that the matrix $\Phi$ be **singular**. Let us now expand the scope of our discussion by determining the rank of cointegration and thus present the analog of the regression issue alluded to above. Suppose we pose the problem

$H_0$ : we have cointegration of rank $r + s$

as against the alternative

$H_1$ : we have cointegration of rank $r$ .

In this problem the minimand, under the null, assumes the value

$$L_{H_0} = \operatorname{tr} B' \Phi B = \sum_{i=1}^{r+s} \lambda_{(i)},$$

while under the alternative it assumes the value [11]

$$L_{H_1} = \operatorname{tr} B' \Phi B = \sum_{i=1}^{r} \lambda_{(i)}.$$

If the null is true, $L_{H_0} - L_{H_1} \approx 0$, a situation quite analogous to what we have in a regression context.

For the tests discussed above to become operational, we require an **estimator** for $\Phi$, and (at least) its limiting distribution. We begin with the estimator in Eq. (5.86), even though this is a **biased** estimator of $\Phi$; thus, consider

$$M_T^{(1)} = \frac{1}{T^2} \sum_{t=1}^{T} X'_{t\cdot} X_{t\cdot} = \frac{1}{T} \sum_{t=1}^{T} \left( \frac{X'_{t\cdot}}{\sqrt{T}} \right) \left( \frac{X'_{t\cdot}}{\sqrt{T}} \right)',$$

and define the stochastic process

$$H_T(\tau, \omega) = \frac{1}{\sqrt{T}} \Lambda^{-(1/2)} Q' X'_{t-1}. \tag{5.92}$$

$$= \left( \frac{[\tau T]}{T} \right)^{1/2} \left( \frac{1}{[\tau T]} \right)^{1/2} \Lambda^{-(1/2)} Q' X'_{[\tau T]}, \quad \tau \in \left[ \frac{t-1}{T}, \frac{t}{T} \right),$$

for $t = 1, 2, 3, \ldots, T$, where $\Phi = Q \Lambda Q'$ and $\Lambda$, $Q$ are, respectively, the matrices of its characteristic roots, and corresponding characteristic vectors. [12] Note that, for $t$, $T \to \infty$ with $\tau$ **fixed**, we may conclude that

$$H_T(\tau, \omega) \overset{\mathrm{d}}{\to} \sqrt{\tau} \zeta, \quad \zeta \sim N(0, I),$$

---

[11] Notice that since we are dealing with a minimization problem the signs appear to be the opposite of what one might expect in a ML context.

[12] If we were to entertain, as a **maintained hypothesis**, that $\Phi$ were singular of rank $q - r$, the matrix $\Lambda^{-(1/2)}$ would be replaced by $\Lambda_g^{-(1/2)} = \operatorname{diag}(\Lambda_{q-r}^{-(1/2)}, 0)$.

provided certain conditions hold, which ware explored extensively in Chapters 3 and 9. One such condition may be

$$\sup_t E|\epsilon_t.|^\beta < \infty, \quad \text{for some } \beta \in (2, \infty]. \tag{5.93}$$

This property may be easily asserted for the $MWN(\Sigma)$ process of this problem with hardly any loss of relevance. Moreover, by Propositions 18, or 22, 23, 24, 25, 27, 28 or 29 of Chapter 9, and Corollary 5 in Dhrymes (1989), p. 243, we may further conclude that

$$\Lambda^{-(1/2)} Q' M_T^{(2)} Q \Lambda^{-(1/2)} = \sum_{t=1}^{T} \int_{\frac{t-1}{T}}^{\frac{t}{T}} H_T(\tau, \omega) H_T(\tau, \omega)' d\tau$$

$$\xrightarrow{d} \int_0^1 B(\tau)' B(\tau) d\tau, \quad \text{or} \tag{5.94}$$

$$M_T^{(1)} \xrightarrow{d} Q \Lambda^{(1/2)} \left( \int_0^1 B(\tau)' B(\tau) \, d\tau \right) \Lambda^{(1/2)} Q',$$

where $B(\tau)$ is a SMBM process, evaluated at $\tau$.

The distribution of the unbiased estimator $M_T^{(2)}$ may be obtained by noting that

$$\Lambda^{-(1/2)} Q' M_T^{(2)} Q \Lambda^{-(1/2)} = \sum_{t=1}^{T} \int_{\frac{t-1}{T}}^{\frac{t}{T}} \left( \frac{1}{\tau} H_T(\tau, \omega) H_T(\tau, \omega)' \right) d\tau,$$

$$\xrightarrow{d} \int_0^1 \left( \frac{1}{\tau} B(\tau)' B(\tau) \right) d\tau, \quad \text{or} \tag{5.95}$$

$$M_T^{(2)} \xrightarrow{d} Q \Lambda^{(1/2)} \left[ \int_0^1 \left( \frac{1}{\tau} B(\tau)' B(\tau) \right) \, d\tau \right] \Lambda^{(1/2)} Q',$$

it being understood that we define $(B(0)/0) = 0$. Given the fact that the characteristic roots and vectors of a matrix are **continuous functions** of its elements, it follows again from Corollary 5 in Dhrymes (1989), p. 243, that their respective distributions are easily obtainable from the distributions in Eqs. (5.94) and (5.95), **although not in closed form**. Incidentally, notice from the last expressions in the two equation sets above, that the expected value of the limiting entity to which $M^{(1)}$ and $M^{(2)}$ is given by $(1/2)\Phi$ and $\Phi$, respectively, agreeing with the earlier observation that $M^{(1)}$ is **biased** estimator of $\Phi$, where the latter is as defined in Eq. (5.89). This conclusion is due to the fact that

$$E \int_0^1 B(\tau)' B(\tau) \, d\tau \;=\; I_q \int_0^1 \tau \, d\tau \;=\; \frac{1}{2} I_q, \quad \text{while}$$

$$E \int_0^1 \frac{1}{\tau} B(\tau)' B(\tau) \, d\tau \;=\; I_q \int_0^1 d\tau \;=\; I_q.$$

Finally, we note that a similar approach to the problem of testing for the presence of cointegration is given in Phillips and Ouliaris (1990).

## 5.3.2   Rank Tests Based on the $VAR(n)$ Formulation

Tests of the presence and/or rank of cointegration based on characteristic roots were introduced into the literature by Johansen (1988), (1991), (J) and were alluded to at an earlier section. J's test for the presence of cointegration of rank $r$, operates in the context of the characteristic equation in iv of Proposition 5 of Chapter 4, and involves a test of the null that there are **only r nonnull roots** as against the alternative that **there are q nonnull roots**. The precise test is based on the likelihood ratio (LR) and is thus a LRT, even though the LF exists **only because** of the special initial conditions. When the cointegration rank is **unknown** *a priori*, one may set up the hypothesis

$H_0 : \text{rank}(B) = r$,

as against the alternative

$H_1 : \text{rank}(B) = q$.

To explore the rationale and the properties of such tests, return to Eq. (4.104) of Chapter 4 and note that what is required is to minimize the determinant

$$D(B) \;=\; |S| \;=\; |W'W - W'VB(B'V'VB)^{-1}B'V'W|$$

$$=\; |W'W| \, |I_q - W'V(B'V'VB)^{-1}B'V'W(W'W)^{-1}|$$

$$=\; |W'W| \, |B'V'VB|^{-1} \, |B'[V'V - V'W(W'W)^{-1}W'V]B| \quad \text{where}$$

$$W \;=\; N\Delta P, \;\; V = NP_{-1}, \;\; N = I_T - X(X'X)^{-1}X'. \tag{5.96}$$

Under the null hypothesis $B$ is $q \times r$ of rank $r$. In this context we **impose the normalization** that its estimator, say $\hat{B}$, obeys $B'V'VB = I_r$; thus, in order to **minimize** $D(B)$ we need only operate on

$$D^*(B) = |B'[V'V - V'W(W'W)^{-1}W'V]B|. \qquad (5.97)$$

As we noted in Chapter 4, the entity above is minimized by choosing the columns of $B$ to be the characteristic vectors corresponding to the largest characteristic roots of

$$|\lambda V'V - V'W(W'W)^{-1}W'V| = 0. \qquad (5.98)$$

If it is known *a priori* that the rank of cointegration is $r$, we evidently choose the $r$ **largest** characteristic roots, and their corresponding characteristic vectors constitute the columns of the estimated matrix, say $\hat{B}$. If $r$ is not known *a priori* the J procedure consists of taking as the columns of this estimator the characteristic vectors corresponding to the roots which are not judged to be **insignificantly different from zero**.

We also note, parenthetically, that if some roots are zero, the characteristic vectors corresponding to them are **arbitrary** subject only to the condition that they are linearly independent of the characteristic vectors corresponding to positive roots.

If we require that the rank of cointegration be $r$ then, by the rank factorization theorem [Dhrymes (1984), p. 23], the representation $\Pi(1) = \Gamma B'$ **implies** that $B$ is of rank $r$. The question that needs to be addressed, however, is this: If, under the null, we have cointegration of rank $r$, does it follow that the characteristic roots of Eq. (5.98), in the limit, consist of $r$ positive roots and $q - r$ zero roots? If the answer to this question is yes, the test procedure implied in the discussion above is fully justified. If the answer is no, it is not entirely clear just what the testing procedure amounts to. The problem, as in the classic identification tests of simultaneous equations theory, is that we do not approach the problem directly, but rather indirectly. For example, we may estimate $\Pi(1)$ without restrictions and test the hypothesis that its rank is $r$, as against some suitable alternative. If the alternative is that there is no cointegration, i.e. the $VAR(n)$ has **only** stationary roots (and thus $X$ is a GLP), we must have that $\Pi(1)$ is of rank $q$ and we return to the **unrestricted** estimator examined in a previous discussion. Instead of doing this, however, we examine the characteristic roots of some other matrix and we make a judgment on the rank of cointegration depending on how many roots may be judged to be nonzero in this other context.

If we operate under the null hypothesis that the cointegration rank is $r$, the maximum of the LF under the null is

$$F_4(\hat{B}) = \frac{qT}{2}[\ln(2\pi) + 1] - \frac{T}{2}\left|\frac{W'W}{T}\right| - \frac{T}{2}\sum_{j=1}^{r}\ln(1 - \hat{\lambda}_j). \qquad (5.99)$$

while under the alternative it is

$$F_4(\hat{B}) = \frac{qT}{2}[\ln(2\pi) + 1] - \frac{T}{2}\left|\frac{W'W}{T}\right| - \frac{T}{2}\sum_{j=1}^{q}\ln(1 - \hat{\lambda}_j). \qquad (5.100)$$

Hence, the log of the LRT statistic is given by

$$\sup_{H_0} F_4(B) - \sup_{H_1} F_4(B) = \frac{T}{2}\sum_{j=r+1}^{q}\ln(1 - \hat{\lambda}_j), \qquad (5.101)$$

whose limiting distribution is examined below.

**Remark 7**. Note that, in this context, the cointegration rank $r$ is linked, in the limit, to the number of nonnull roots in Eq. (5.98), which is given by $r = q - r_0$, $r_0$ being the number of **unit roots of** $|\Pi(z)| = 0$, as made clear in Remark 9 of Chapter 4.

For a test on the "existence" of cointegration take $r = q - 1$ so that the LRT statistic is given by

$$\sup_{H_0} F_4(\hat{B}) - \sup_{H_1} F_4(\hat{B}) = \frac{T}{2}\ln(1 - \hat{\lambda}_q).$$

If this test rejects the null, we conclude that there is no cointegration, since the result would imply that the smallest characteristic root is significantly different from zero and thus that $B$ and $\Pi(1)$ are of rank $q$. If the presence of cointegration is not in doubt, and the only testable hypothesis pertains to the **rank of cointegration**, the test is of the form

$H_0 : \text{rank}(B) = r$,

as against the alternative

$H_1 : \text{rank}(B) = r + s, \quad 1 \leq s < q - r$.

An entirely similar argument yields the conclusion that the logarithm of the LRT statistic is given by

$$\sup_{H_0} F_4(\hat{B}) - \sup_{H_1} F_4(\hat{B}) = \frac{T}{2} \sum_{j=r+1}^{s} \ln(1 - \hat{\lambda}_j). \tag{5.102}$$

In either case, the test is based on the supposition that the rank of cointegration **is equal to the number of nonnull roots**. Thus, in the first case we will accept $H_0$ if the LRT statistic leads to the conclusion that the entity

$$\sum_{j=r+1}^{q} \ln(1 - \lambda_j)$$

is **insignificantly different from zero**. In the second case we will accept the null for the **smallest** $r$ such that

$$\sum_{j=r+1}^{r+s} \ln(1 - \lambda_j)$$

is **insignificantly different from zero for any** $s \geq 1$. Note that since the roots are non-negative and are arranged in decreasing order, if the null is accepted for $s = 1$ then it will be accepted for $1 < s \leq q - r$. It bears repeating that the alternative, that the sequence in question is $I(1)$ **but not cointegrated**, is not an admisible one. Rejecting cointegration in this context means that all roots are not null and thus the process in question is **stationary**.

At least two issues need to be discussed in connection with the preceding. The first issue has to do with the manipulation of determinants in Eq. (5.96). Note that in the initial representation $D(B) = |W'W - W'VB(B'V'VB)^{-1}B'V'W|$, so that if we divide each matrix by $T$ we deal, under the cointegration assumption, with entities which all converge to well-defined constant matrices, provided $B$ represents the "true" matrix of cointegrating vectors, or a suitable consistent estimator of it. However, in the final representation of $D^*(B)$, Eq. (5.97), we are led to consider the characteristic roots in Eq. (5.98), i.e.

$$|\lambda V'V - V'W(W'W)^{-1}W'V| = 0.$$

Even if we divide both matrices above by $T$, we seem to have a serious problem in that the first matrix, $V'V/T$, **will diverge** while the second matrix will converge in distribution. The question is whether this is a substantive problem or merely one of appearances.

The second issue is whether we can connect directly the question of cointegration rank, or even the existence of cointegration, to the nature of the roots in Eq. (5.98), or its asymptotic equivalent; certainly, the implication of the procedure is that the number of nonnull roots is connected to the number of unit roots of $|\Pi(z)| = 0$, as indicated in Remark 7 above.

In the absence of such a connection, if we do not assert that we know that cointegration is present, and that its rank is precisely $r$, **we do not have a rationale for choosing the number of characteristic vectors to be retained**, based on the development so far. For example if, as in the standard maximum likelihood problem, we choose estimates that **maximize the LF**, it is clear that we should choose **all** characteristic vectors, provided the corresponding roots obey $0 \leq \hat{\lambda}_{(i)} < 1$; even if several roots are "nearly" zero **there is no basis in the estimation procedure for declining the zero** roots; all there is, is relative flatness of the LF, a situation commonly encountered in many nonlinear problems in which the choice of widely varying values for the underlying parameters **yields substantially the same value for the LF**. For example, if the last $q - r$ roots are "nearly" null then

$$F_4(\hat{B}; q) = \frac{qT}{2}[\ln(2\pi) + 1] - \frac{T}{2}\left| \frac{W'W}{T} \right| - \frac{T}{2}\sum_{j=1}^{q} \ln(1 - \hat{\lambda}_j) \qquad (5.103)$$

$$\approx \frac{qT}{2}[\ln(2\pi) + 1] - \frac{T}{2}\left| \frac{W'W}{T} \right| - \frac{T}{2}\sum_{j=1}^{r} \ln(1 - \hat{\lambda}_j) = F_4(\hat{B}; r),$$

and the investigator is not able to determine whether $B$ should be chosen to be of rank $r$ or of rank $q$. Indeed, in the canonical variates literature it is precisely the specification of **how many** canonical variate sets we wish to obtain that allows us to choose a number smaller than the maximal possible. The same is true of principal components. In the test we discussed in the previous section it is the desire to test the hypothesis that $\Phi$ is singular, and thus that the $X$-sequence is both $I(1)$ and cointegrated, that leads us to test the hypothesis that the smallest (ordered) root is null. **There is no similar conceptual basis in the J estimation context**, unless additional argument is produced to justify this practice. Thus, it is important that the two issues raised above be fully examined.

The first issue is relatively easily disposed of. Thus, observe that

$$D^{**}(B) = \left| \lambda \frac{V'V}{T} - \frac{V'W}{T} \left( \frac{W'W}{T} \right)^{-1} \frac{W'V}{T} \right|$$

$$= \left| \rho \frac{V'V}{T^2} - \frac{V'W}{T} \left( \frac{W'W}{T} \right)^{-1} \frac{W'V}{T} \right|, \quad \rho = T\lambda. \quad (5.104)$$

The result above implies that **all** characteristic roots of Eq. (5.98), when normalized by $T$, have **a well defined limiting distribution** in view of the earlier discussion in this chapter. This disposes of the first issue.

The resolution of the second issue is somewhat more complex, in that we must utilize in our argument the fact that the $I(1)$ stochastic process $\{X_t : t \in \mathcal{N}_+\}$ is cointegrated of rank $r$. To this end, suppose we have cointegration of rank $r$. Let $B$ be the $q \times r$ matrix of the true cointegrating vectors,[13] and consider the matrix

$$S(\lambda) = \lambda \frac{V'V}{T} - \frac{V'W}{T} \left( \frac{W'W}{T} \right)^{-1} \frac{W'V}{T}.$$

Let $G_2$ be an **arbitrary** $q \times q - r$ matrix whose columns are linearly independent, such that $(B, G_2)$ is a nonsingular matrix, and define

$$C_T = \left( B, \frac{1}{\sqrt{T}} G_2 \right), \quad S^*(\lambda) = C_T' S(\lambda) C_T. \quad (5.105)$$

It is easily verified that the characteristic roots of interest in our discussion are the roots of $|S^*(\lambda)| = 0$ and, moreover,

$$S^*(\lambda) = \begin{bmatrix} S_{11}^*(\lambda) & S_{12}^*(\lambda) \\ S_{21}^*(\lambda) & S_{22}^*(\lambda) \end{bmatrix}, \quad S_{21}^*(\lambda) = S_{12}^{*\prime}(\lambda) \quad \text{where} \quad (5.106)$$

$$S_{11}^*(\lambda) = \lambda \frac{B' P_{-1}' N P_{-1} B}{T} - \frac{B'V'W}{T} \left( \frac{W'W}{T} \right)^{-1} \frac{W'VB}{T},$$

$$S_{12}^*(\lambda) = \lambda \frac{B'V'VG_2}{T^{3/2}} - \frac{B'V'W}{T} \left( \frac{W'W}{T} \right)^{-1} \frac{W'VG_2}{T^{3/2}},$$

---

[13] Actually any element of the space spanned by the true cointegrating vectors will do, for example $BH$, where $H$ is an arbitrary nonsingular matrix of order $r$.

$$S_{22}^*(\lambda) = \lambda G_2' \frac{V'V}{T^2} G_2 - \frac{G_2'V'W}{T^{3/2}} \left(\frac{W'W}{T}\right)^{-1} \frac{W'VG_2}{T^{3/2}}.$$

In view of the discussion in earlier sections, $S_{12}^*(\lambda) \xrightarrow{d} 0$ and, asymptotically,

$$\frac{G_2'V'VG_2}{T^2} \sim \frac{G_2'P_{-1}'P_{-1}G_2}{T^2} \xrightarrow{d} G_2'P_0' \left(\int_0^1 B(s)'B(s)\,ds\right) P_0 G_2, \qquad (5.107)$$

where $P_0$ is a triangular decomposition of the matrix $\Sigma_0$, i.e.

$$P_0'P_0 = \Sigma_0 = \Sigma_1 + \Sigma_1^* + \Sigma_1^{*\prime} = E\eta_{1\cdot}'\eta_{1\cdot} + \sum_{\tau=1}^{\infty} E\eta_{1\cdot}'\eta_{1+\tau\cdot} + \sum_{\tau=1}^{\infty} E\eta_{1+\tau\cdot}'\eta_{1\cdot}, \quad (5.108)$$

$B(s)$ being a SMBM of dimension $q$. We also note

$$S_{11}^*(\lambda) = \lambda \frac{Z_{-1}'NZ_{-1}}{T} - \frac{Z_{-1}'N\Delta P}{T} \left(\frac{(\Delta P)'N\Delta P}{T}\right)^{-1} \frac{(\Delta P)'NZ_{-1}}{T}.$$

It follows, therefore, that

$$\frac{Z_{-1}'NZ_{-1}}{T} = \frac{Z_{-1}'Z_{-1}}{T} - \frac{Z_{-1}'X}{T} \left(\frac{X'X}{T}\right)^{-1} \frac{X'Z_{-1}}{T}$$

$$\xrightarrow{\text{a.c.}} M_{zz} - M_{zx}M_{xx}^{-1}M_{xz} = M_{zz}^*, \qquad (5.109)$$

where $M_{zz}^*$ denotes the (conditional) covariance matrix of the cointegral vector $z_{t-1}$. **given** $x_{t\cdot}$. In a similar fashion we may show that

$$\frac{Z_{-1}'N\Delta P}{T} \sim -\left[\frac{Z_{-1}'Z_{-1}}{T} - \frac{Z_{-1}'X}{T}\left(\frac{X'X}{T}\right)^{-1}\frac{X'Z_{-1}}{T}\right]\Gamma' \xrightarrow{\text{a.c.}} -M_{zz}^*\Gamma'$$

$$\frac{(\Delta P)'N\Delta P}{T} \xrightarrow{\text{a.c.}} \Sigma + \Gamma M_{zz}^*\Gamma' = M_{ww}, \quad \text{so that} \qquad (5.110)$$

$$S_{11}^*(\lambda) \xrightarrow{\text{a.c.}} \lambda M_{zz}^* - M_{zz}^*\Gamma' M_{ww}^{-1}\Gamma M_{zz}^*,$$

$$S_{22}^*(\lambda) \xrightarrow{d} \lambda G_2'P_0'K P_0 G_2 = \lambda \bar{S}_{22}, \quad K = \int_0^1 B(s)'B(s)\,ds.$$

We therefore conclude that the characteristic roots of $|S(\lambda)| = 0$ converge, in distribution or a.c., to the characteristic roots of $|\bar{S}^*(\lambda)| = 0$, where

$$\bar{S}^*(\lambda) = \begin{bmatrix} \lambda M_{zz}^* - M_{zz}^*\Gamma' M_{ww}^{-1}\Gamma M_{zz}^* & 0 \\ 0 & \lambda \bar{S}_{22} \end{bmatrix}, \qquad (5.111)$$

and it is evident that **there are at least** $q - r$ **zero roots**; to complete the argument we must show that the roots of

$$|\lambda M_{zz}^* - M_{zz}^* \Gamma' M_{ww}^{-1} \Gamma M_{zz}^*| = 0$$

are **nonnull**, so that the number of nonzero roots corresponds to the number of cointegrating vectors. Since the cointegral vector $z_{t-1}$ and $x_{t\cdot}$ are evidently **not linearly dependent**, $M_{zz}^* > 0$; moreover, since equally evidently $M_{zz}^* \Gamma' M_{ww}^{-1} \Gamma M_{zz}^* > 0$, we conclude from Proposition 62 in Dhrymes (1984) p. 74 that the roots in question obey $0 < \lambda_j < 1$, for $j = 1, 2, \ldots, r$ and $\lambda_j = 0$ for $j = r + 1, r + 2, \ldots, q$.

In view of the fact that the characteristic roots of $S(\lambda)$ are continuous functions of its elements, its ordered characteristic roots obey, asymptotically, the relations implied by Eq. (5.111), i.e. the largest $r$ roots have positive a.c. limits, while the smallest $q - r$ roots have zero limits (in distribution).

**Remark 8.** Since we have employed the normalization $\hat{B}'V'V\hat{B}/T = I_r$, we are led to the conclusion that $M_{zz}^* = I_r$. Thus, the limiting form of the characteristic equation yielding the characteristic roots is given by

$$0 = |\lambda I_r - \Gamma'(\Sigma + \Gamma\Gamma')^{-1}\Gamma| \, |\lambda G_2' P_0' K P_0 G_2|, \tag{5.112}$$

which clearly shows that the $r$ positive roots come from the first factor, while the $q - r$ zero roots come from the second factor. Finally, using the results in Dhrymes (1984) p. 39, we conclude that the $r$ largest characteristic roots of relevance to our discussion, $\hat{\lambda}_j$, $j = 1, 2, \ldots, r$ converge to entities, $\lambda_j$, which are related to the roots of

$$0 = |(1 - \lambda)I_r - (I_r + \Gamma'\Sigma^{-1}\Gamma)^{-1}| = |\mu I_r - \Gamma'\Sigma^{-1}\Gamma|, \tag{5.113}$$

by $\mu_j = \lambda_j/(1 - \lambda_j)$. In view of the fact that $\Gamma'\Sigma^{-1}\Gamma > 0$, $\mu_j > 0$, for all $j$, and thus $0 < \lambda_j < 1$. Since $\Gamma'\Sigma^{-1}\Gamma$ is a **nonsingular matrix** of dimension equal to the **cointegration rank**, we see that the number of the characteristic roots of $S(\lambda)$ which, in the limit, are positive is precisely equal to the rank of cointegration. We have therefore proved Proposition 6.

**Proposition 6.** Consider the model in Eq. (5.6); suppose we employ the normalization $\hat{B}'(V'V/T)\hat{B} = I_r$ and the stochastic sequence $\{X_{t\cdot} : t \in \mathcal{N}_+\}$ is cointegrated of rank $r$. Then the following statements are true:

i. the characteristic roots of

$$0 = |S(\lambda)| = \left| \lambda \frac{V'V}{T} - \frac{V'W}{T} \left( \frac{W'W}{T} \right)^{-1} \frac{W'V}{T} \right|$$

are precisely those of

$$|S^*(\lambda)| = 0, \quad S^*(\lambda) = C_T' S(\lambda) C_T, \quad C_T = \left( B, \frac{1}{\sqrt{T}} G_2 \right),$$

where $B$ is the matrix containing the $r$ "true" cointegrating vectors and $G_2$ is a $q \times q - r$ matrix such that $(B, G_2)$ is nonsingular;

ii. the matrix $S^*(\lambda)$ obeys

$$S^*(\lambda) \to \begin{bmatrix} \lambda M_{zz}^* - M_{zz}^* \Gamma' M_{ww}^{-1} \Gamma M_{zz}^* & 0 \\ 0 & \lambda G_2' P_0' K P_0 G_2 \end{bmatrix} = \bar{S}^*(\lambda),$$

where convergence of the $(1,1)$ block is convergence a.c., and convergence of the other blocks is convergence in distribution,

$$K = \int_0^1 B(s)' B(s) \, ds, \quad P_0' P_0 = \Sigma_0 = \Sigma_1 + \Sigma_1^* + \Sigma_1^{*'},$$

as these entities were defined in Eq. (5.108);

iii. the matrix $\bar{S}^*(\lambda)$ has a number of **positive roots equal to the cointegration rank**, $r$, and $q - r$ zero roots, i.e. it has as many **zero** roots as $|\Pi(z)| = 0$ has **unit** roots;

iv. if the cointegration rank is $r$, the $r$ **largest** characteristic roots of $S(\lambda)$ converge a.c. to **positive** limits, and the $q - r$ **smallest** roots converge in distribution to null limits.

**Remark 9.** Under the null that cointegration is of rank $r$, Proposition 6 answers completely the issues raised in the discussion immediately preceding, and fully justifies the procedure of choosing for the columns of $\hat{B}$ **only those characteristic vectors** corresponding to the largest $r$ roots of $|S(\lambda)| = 0$.

## 5.3.3 Limiting Distribution of Characteristic Roots

We now turn to the question of the limiting distribution of the smallest $q - r$ characteristic roots; as we have observed in the discussion of the previous section, the various tests for the existence and/or rank of cointegration depend precisely on this distribution. As we have also noted earlier, $T\hat{\lambda}_j = \hat{\rho}_j$ has a well-defined limiting distribution, if it corresponds to one of the roots that converge to zero. To this end, note first that

$$S_{22}^*(T^{-1}\rho) - S_{21}^*(T^{-1}\rho)[S_{11}^*(T^{-1}\rho)]^{-1}S_{12}^*(T^{-1}\rho) \overset{d}{\to} 0, \quad (5.114)$$

$$S^{**}(\rho) = TS_{22}^*(T^{-1}\rho) - TS_{21}^*(T^{-1}\rho)[S_{11}^*(T^{-1}\rho)]^{-1}S_{12}^*(T^{-1}\rho) \overset{d}{\to} \Upsilon(\rho),$$

where $\Upsilon(\rho)$ is a well-defined matrix to be determined below, and moreover

$$S_{11}^*(T^{-1}\rho) = \frac{\rho B'V'VB}{T^2} - \frac{B'V'W}{T}\left(\frac{W'W}{T}\right)^{-1}\frac{W'VB}{T}$$

$$\overset{a.c.}{\to} -M_{zz}^*\Gamma'M_{ww}^{-1}\Gamma M_{zz}^*,$$

$$TS_{22}^*(\rho) = \rho\frac{G_2'V'VG_2}{T^2} - \frac{G_2'V'W}{T}\left(\frac{W'W}{T}\right)^{-1}\frac{W'VG_2}{T},$$

$$\sim \rho G_2'P_0'KP_0G_2 - \frac{G_2'V'W}{T}M_{ww}^{-1}\frac{W'VG_2}{T}$$

$$\sqrt{T}S_{12}^*(T^{-1}\rho) \sim -M_{zz}^*\Gamma'M_{ww}^{-1}\frac{W'VG_2}{T}. \quad (5.115)$$

Consequently,

$$S^{**}(\rho) = \rho G_2'P_0'KP_0G_2 - \left(\frac{G_2'V'W}{T}\right)F\left(\frac{W'VG_2}{T}\right),$$

$$F = M_{ww}^{-1} - M_{ww}^{-1}\Gamma(\Gamma'M_{ww}^{-1}\Gamma)^{-1}\Gamma'M_{ww}^{-1}. \quad (5.116)$$

Since $V'W = P_{-1}'[NU - NP_{-1}B\Gamma']$ and $\Gamma'F = 0$ , we need only be concerned about the limiting distribution of $P_{-1}'NU/T$ , which obeys

$$\frac{P_{-1}'NU}{T} \overset{d}{\to} P_0'K_1T_3, \quad K_1 = \int_0^1 B(s)'\,dB_2(s), \quad T_3'T_3 = \Sigma,$$

where $B(s)$ and $B_{(1)}(s)$ are each SMBM of dimension $q$. This derivation was extensively discussed earlier in connection with Eqs. (5.46) through (5.52). Finally, we obtain

$$S^{**}(\rho) \xrightarrow{d} \Upsilon(\rho) = \rho G_2' P_0' K P_0 G_2 - G_2' P_0' K_1 T_1 F T_1' K_1' P_0 G_2, \qquad (5.117)$$

and thus the limiting distribution of the ordered roots $T\hat{\lambda}_j$, $j = r+1, r+2, \ldots, q$ of $|S(\lambda)| = 0$ is given by the distribution of the ordered roots $\rho_j$, $j = 1, 2, \ldots, q - r$, of

$$0 = \left| \rho G_2' P_0' \left( \int_0^1 B(s)' B(s)\, ds \right) P_0 G_2 - G_2' P_0' \left( \int_0^1 B(s)'\, dB_{(2)}(s) \right) T_3 F T_3' \right.$$

$$\left. \left( \int_0^1 dB_{(2)}(s)' B(s) \right) P_0 G_2 \right|. \qquad (5.118)$$

It may further be shown that, with some rearrangement, the characteristic equation yielding the roots on which we may base tests for the existence, or rank, of cointegration may be represented as

$$0 = \left| \rho I_{r_0} - \left( \int_0^1 B_4(s)'\, dB_3(s) \right)' \left( \int_0^1 B_4(s)' B_4(s)\, ds \right)^{-1} \left( \int_0^1 B_4(s)'\, dB_3(s) \right) \right|,$$

$$\qquad (5.119)$$

where $B_4$ and $B_3$ are $r_0$-element SMBM, which are not independent of each other.

Details of the preceding derivation are given in the appendix to this chapter. We have therefore proved Proposition 7.

**Proposition 7.** In the context of Proposition 6, let $\rho = T\lambda$ and consider the $r_0 = q - r$ smallest roots of $|S(\lambda)| = 0$. The limiting distribution of such roots is the distribution of the roots of

$$|\Upsilon(\rho)| = |\rho G_2' P_0' K P_0 G_2 - G_2 P_0' K_1 T_1 F T_1' K_1' P_0 G_2| = 0, \qquad (5.120)$$

which is equivalent to the distribution of the roots of

$$\left| \rho I_{r_0} - \left( \int_0^1 B_4(s)'\, dB_3(s) \right)' \left( \int_0^1 B_4(s)' B_4(s)\, ds \right)^{-1} \left( \int_0^1 B_4(s)'\, dB_3(s) \right)' \right| = 0,$$

$$\qquad (5.121)$$

where $B_4$ and $B_3$ are $r_0$-element SMBM; thus, the limiting distribution of such roots is **free of nuisance parameters**. A test of the hypothesis that the rank of cointegration is $r$ may thus be formulated as

$H_0 : \operatorname{tr} \Upsilon(\rho) = 0$

as against the alternative

$H_1 : \operatorname{tr} \Upsilon(\rho) > 0$ .

This is equivalent to obtaining the $r_0 = q - r$ **smallest** characteristic roots of

$$\left| \lambda V'V - V'W(W'W)^{-1}W'V \right| = 0, \tag{5.122}$$

say $\hat{\lambda}_{r+i}$, $i = 1, 2, \ldots, r_0$ and comparing the test statistic

$$JCT(r_0) = T \sum_{i=1}^{r_0} \hat{\lambda}_{r+i} \tag{5.123}$$

with the critical value, as specfied by the size of the test.

Similarly, a test of the hypothesis that the rank of cointegration is $r$ as against the alternative that it is $r + 1$ , is equivalent to the hypothesis that $\Upsilon(\rho)$ is of dimension $r_0 - 1$ and may be tested by comparing the test statistic

$$JCT(r_0 - 1) = T \hat{\lambda}_{r+1} \tag{5.124}$$

with the critical value, as specified by the size of the test.[14]

---

[14] Tabulations of the distribution of the roots of Eq. (5.119) are given in Johansen (1988) as well as Osterwald-Lenum (1992).

# Appendix to Chapter 5

## Simpler Representation of $P'_{-1}NP_{-1}$ and $P'_{-1}NU$

In this appendix we provide the detailed argument in the transition from Eq. (5.115) to Eq. (5.118).

We begin by noting that the $VAR(n)$ model of Eq. (5.1) may also be represented (in row form) as

$$X_{t.} - X_{t-1.} = -X_{t-1.}.\Pi(1)' + x_{t.}.\Pi^{*'} + \epsilon_{t.},$$

thus obtaining

$$X_{t.} = X_{t-1.}.\Pi'_* + x_{t.}.\Pi^{*'} + \epsilon_{t.}, \quad \Pi_* = I_q - \Pi(1). \tag{5.125}$$

A formal solution is given by

$$X_{t.} = \sum_{j=0}^{\infty} x_{t-j.}.\Pi^{*'}(\Pi'_*)^j + \sum_{j=0}^{\infty} \epsilon_{t-j.}.(\Pi'_*)^j. \tag{5.126}$$

To examine the nature of the representation in (5.180), consider the roots of $|I_q + \Pi_1 z + \Pi_2 z^2 + \dots, +\Pi_n z^n| = 0$. In the context of our discussion, this equation has the roots $z_i = 1$, for $i = 1, 2, \dots r_0$, and $|z_i| > 1$, for $i = r_0 + 1, r_0 + 2, \dots, qn$. It follows therefore, on the assumption that $\Pi(1)$ is diagonalizable, that $|\lambda I_q - \Pi(1)| = 0$ has $r_0$ **zero** roots, and $r = q - r_0$ roots obeying $\lambda_i \in (0,1)$, $i = r_0 + 1, r_0 + 2, \dots, q$. Hence there exists a nonsingular matrix $A$ such that

$$\Pi(1) = A \begin{bmatrix} 0 & 0 \\ 0 & \Lambda_1 \end{bmatrix} A^{-1}. \tag{5.127}$$

Moreover, since $\Pi_* = I_q - \Pi(1)$, we have

$$0 = |\mu I_q - \Pi_*| = |(1 - \mu)I_q - \Pi(1)|, \tag{5.128}$$

which implies that the roots of $\Pi_*$, say $\mu_i$, are related to the roots of $\Pi(1)$, say $\lambda_i$, as $\mu_i = 1 - \lambda_i$, $i = 1, 2, \ldots, q$. Consequently, the roots of $\Pi_*$ obey: $\mu_i = 1$, for $i = 1, 2, \ldots, r_0$, $\mu_i \in (0, 1)$, for $i = r_0 + 1, r_0 + 2, \ldots, q$. Thus, we have the representation

$$\Pi_* = A \begin{bmatrix} I_{r_0} & 0 \\ 0 & I - \Lambda_1 \end{bmatrix} A^{-1}, \quad \text{so that for } \textbf{any } i \geq 1,$$

$$\Pi(1)\Pi_*^i = A \begin{bmatrix} 0 & 0 \\ 0 & \Lambda_1(I_r - \Lambda_1)^i \end{bmatrix} A^{-1}. \tag{5.129}$$

The sample representation is formally given by

$$P_{-1} = \sum_{j=0}^{\infty} X_{-1-j} \Pi^{*'} (\Pi_*')^j + \sum_{j=0}^{\infty} U_{-1-j} (\Pi_*')^j,$$

where $X_{-s} = (x_{t-s.})$, $U_{-s} = (\epsilon_{t-s.})$. Thus

$$\frac{P_{-1}' N P_{-1}}{T^2} = P_1^* + P_2^*,$$

$$P_1^* = \frac{1}{T^2} \sum_{j=0}^{\infty} \sum_{j'=0}^{\infty} \Pi_*^j U_{-1-j}' N U_{-1-j'} (\Pi_*')^{j'},$$

$$P_2^* = P_{21}^* + P_{21}^{*'} + P_{22}^*,$$

$$P_{22}^* = \frac{1}{T^2} \sum_{j=0}^{\infty} \sum_{j'=0}^{\infty} \Pi_*^j \Pi^* X_{-1-j}' N X_{-1-j'} \Pi^{*'} (\Pi_*')^{j'},$$

$$P_{21}^* = \frac{1}{T^2} \sum_{j=0}^{\infty} \sum_{j'=0}^{\infty} \Pi_*^j \Pi^* X_{-1-j}' N U_{-1-j'} (\Pi_*')^{j'}.$$

Since for every $j, j'$,

$$\frac{1}{T^2} (\Pi_*')^j \Pi^* X_{-1-j}' N X_{-1-j'} \Pi^{*'} (\Pi_*')^{j'} \xrightarrow{\text{P}} 0,$$

$$\frac{1}{T^2} \Pi_*^j \Pi^* X_{-1-j}' N U_{-1-j'} (\Pi_*')^{j'} \xrightarrow{\text{P}} 0, \tag{5A.6}$$

we conclude that $P_2^* \xrightarrow{\text{P}} 0$. The justification for this claim is as follows. The matrices $X_{-1-j}' N X_{-1-j'}$ are of the form

$$X_{-1-j}' X_{-1-j'} - X_{-1-j}' X (X'X)^{-1} X' X_{-1-j'},$$

and $X'_{-1-j}X_{-1-j'}$ consists of blocks of the form $\Delta X'_{t-2-j-s.}\Delta X_{t-2-j'-s'..}$
Similarly, $X'_{-1-j}X$ consists of blocks of the form $X'_{t-2-j-s.}\Delta X_{t-1-r..}$ The
expected value of such blocks was determined in Eq. (5.18) to be of the
generic form $\sum_{s=0}^{\infty} H_s^* \Sigma H_{\tau+s}^{*'}$, where $\tau$ is the difference in the index of the
left factor minus the index of the right factor. The elements of such blocks
are **stationary**, and the matrices $H_j^*$ converge to zero exponentially, a fact
we shall demonstrate below. Thus, it may shown that the blocks converge
a.c. to their expected value. The summation

$$\sum_{\tau=1}^{\infty}\sum_{s=0}^{\infty} H_s^* \Sigma H_{\tau+s}^{*'} \quad \text{converges absolutely;}$$

since we divide by $T^2$, the matrices of Eq. (5A.6) converge to zero, at least
in probability.

We now demonstrate that the matrices $H_k^*$ converge to zero exponentially. Recall that

$$H^*(L) = \frac{H(L)}{\pi(L)},$$

where $H(L)$ is the adjoint of $\Pi(L)$ and, thus, a matrix polynomial operator
of degreee $m_1 = n(q-1) - r_0 + 1$, and $\pi(L)$ is a **scalar, stationary**,
polynomial operator of degree $m = nq - r_0$. The inverse of $\pi(L)$ is given
by

$$\prod_{i=1}^{m}(I - \lambda_i L)^{-1}, \quad |\lambda_i| < 1.$$

Expanding, we obtain

$$\prod_{i=1}^{m}(I - \lambda_i L)^{-1} = \sum_{i=1}^{\infty} \gamma_k L^k, \quad \gamma_k = \sum_{s_1=0}^{k} \cdots \sum_{s_{m-1}=0}^{s_{m-2}} \lambda_1^{k-s_1} \lambda_2^{s_1-s_2} \lambda_3^{s_2-s_3} \cdots \lambda_m^{s_{m-1}}.$$

Carrying out this summation, we have

$$\gamma_k = \sum_{i=1}^{m} c_{m,i} \lambda_i^k,$$

where the constants $c_{m,i}$ are functions of the roots, and the degree of the
polynomial alone, and the roots are arranged in **decreasing** order of absolute
value. It is evident from the preceding that

$$|\gamma_{k+1}| \leq |\gamma_k|.$$

If we write $H^*(L) = \sum_{r=0}^{\infty} H_r^* L^r$, the matrix $H_k^*$ is given by

$$H_k^* = \sum_{i=0}^{m_1} \gamma_{k-i} H_i, \quad \text{for} \quad k \geq m_1.$$

Consequently,

$$\| H_k^* \| \leq K^* |\gamma_{k-i}| \leq K |\lambda_1|^k,$$

for finite constant $K$, and the matrices $H_k^*$ converge to zero (in norm) at least at the rate at which powers of the largest root converge to zero. We conclude, therefore,

$$\frac{P_{-1}' N P_{-1}}{T^2} \sim \frac{U_{-1}^{*\prime} U_{-1}^*}{T^2}, \quad \text{where} \quad U_{-1}^* = \sum_{j=0}^{\infty} U_{-1-j} (\Pi_*')^j.$$

A similar argument will show that

$$\frac{P_{-1}' N U}{T^2} \sim \frac{U_{-1}^{*\prime} U}{T^2}, \quad U = (\epsilon_{t\cdot}).$$

## Limiting Behavior

To find the limiting behavior of $P_{-1}' N P_{-1}/T^2$ and $P_{-1}' N U/T$, consider

$$\zeta_{t\cdot} = u_{t\cdot}^* - u_{t-1\cdot}^* = \epsilon_{t\cdot} - \sum_{j=0}^{\infty} \epsilon_{t-1-j\cdot} (\Pi_*')^j \Pi(1)', \quad \phi_{t\cdot} = (\zeta_{t-1\cdot}, \epsilon_{t\cdot}),$$

define

$$S_{t1\cdot} = \sum_{s=1}^{t} \zeta_{t-1\cdot}, \quad S_{t2\cdot} = \sum_{s=1}^{t} \epsilon_{s\cdot}, \quad S_{t\cdot} = (S_{t1\cdot}, S_{t2\cdot}),$$

and note that

$$\frac{1}{T^2} P_{-1}' N P_{-1} \text{ behaves like the (1,1) block of } \frac{1}{T^2} \sum_{t=1}^{T} S_{t\cdot}' S_{t\cdot}, \tag{5A.7}$$

and

$$\frac{1}{T} P_{-1}' N U \text{ behaves like the (1,2) block of } \frac{1}{T} \sum_{t=1}^{T} S_{t\cdot}' \phi_{t\cdot}. \tag{5A.8}$$

Following the procedures employed in Chapter 3, we easily establish that

$$\frac{1}{T^2} \sum_{t=1}^{T} S'_{t\cdot} S_{t\cdot} \xrightarrow{d} \Phi^* = S'_0 \left( \int_0^1 B(s)' B(s) \, ds \right) S_0, \quad \text{and}$$

$$\frac{1}{T} \sum_{t=1}^{T} S'_{t\cdot} \phi_{t\cdot} \xrightarrow{d} \Psi^* = S'_0 \left( \int_0^1 B(s)' \, dB(s) \right) S_0 + S_1^*,$$

where $B = (B_1, B_2)$, each $B_i$, $i = 1, 2$, is $q$-element SMBM, and $S_0$ is the (lower) triangular decomposition of the "long run" covariance matrix

$$S'_0 S_0 = S_1 + S_1^* + S_1^{*'}, \tag{5A.9}$$

$$S_1 = E\phi'_{t\cdot} \phi_{t\cdot}, \quad \text{and} \quad S_1^* = \sum_{\tau=1}^{\infty} E\phi'_{t\cdot} \phi_{t+\tau\cdot}.$$

We now derive the precise form of the covariance matrices above. We note that, in view of the stationarity [15] of the process $\{\phi_{t\cdot} : t \in \mathcal{N}\}$ we obtain

$$E\phi'_{t\cdot} \phi_{t\cdot} = E \begin{bmatrix} \zeta'_{t-1\cdot} \zeta_{t-1\cdot} & \zeta'_{t-1\cdot} \epsilon_{t\cdot} \\ \epsilon'_{t\cdot} \zeta_{t-1\cdot} & \epsilon'_{t\cdot} \epsilon_{t\cdot} \end{bmatrix} = \begin{bmatrix} \Sigma_{11} & 0 \\ 0 & \Sigma \end{bmatrix},$$

where $\Sigma = E\epsilon'_{t\cdot} \epsilon_{t\cdot}$, and

$$\Sigma_{11} = \Sigma + \Pi(1)\Phi_{00}\Pi(1)', \quad \Phi_{00} = \sum_{i=0}^{\infty} \Pi_*^i \Sigma (\Pi'_*)^i.$$

Moreover,

$$S_1^* = \begin{bmatrix} S_{11}^* & 0 \\ S_{21}^* & 0 \end{bmatrix}, \quad S_{21}^* = \Sigma - \Sigma D' \Pi(1)', \quad D = \sum_{i=0}^{\infty} \Pi_*^i, \tag{5A.10}$$

$$S_{11}^* = -\Sigma D' \Pi(1)' + \Pi(1) \left( \sum_{\tau=1}^{\infty} \Phi_{0\tau} \right) \Pi(1)', \quad \Phi_{0\tau} = \sum_{i=0}^{\infty} \Pi_*^i \Sigma (\Pi'_*)^{i+\tau}.$$

This is so since

$$E\zeta'_{t-1\cdot} \epsilon_{t+\tau\cdot} = 0, \quad E\epsilon'_{t\cdot} \epsilon_{t+\tau\cdot} = 0, \quad \text{for } \tau \geq 1,$$

---

[15] There is a slight problem here in that by the initial conditions we have assumed the process is not stationary in the strict sense of the term. On the other hand, for $t$ sufficiently large, the manner in which it differs from a statinary sequence may be made arbitrarily small.

$$E\epsilon'_{t.}\zeta_{t+\tau-1.} = \Sigma, \text{ for } \tau = 1,$$

$$= -\Sigma(\Pi'_*)^{\tau-2}, \text{ for } \tau \geq 2,$$

$$E\zeta'_{t-1.}\zeta_{t+\tau-1.} = -\Sigma(\Pi'_*)^{\tau-1}\Pi(1)' + \Pi(1)\Phi_{0\tau}\Pi(1)'.$$

**Remark 1A.** The notation $\Phi_{00}$ and $\Phi_{0\tau}$ is somewhat infelicitous since these matrices represent **divergent** series. None the less it is quite useful in giving succinct representation to $\Sigma_{11}$, $S^*_{11}$, and other similar entities. Needless to say the latter are well behaved, and represent **convergent series**. We ilustrate this with $\Pi(1)\Phi_{00}\Pi(1)'$. To this end, refer to Eq. (5.183), partition conformably

$$A = \begin{bmatrix} A_{11} & A_{12} \\ A_{21} & A_{22} \end{bmatrix}, \quad A^{-1} = \begin{bmatrix} A^{11} & A^{12} \\ A^{21} & A^{22} \end{bmatrix},$$

and note that

$$\Pi(1)\Pi^i_* = A_*\Lambda_1(I_r - \Lambda_1)^i A^*, \quad A_* = \begin{pmatrix} A_{12} \\ A_{22} \end{pmatrix}, \quad A^* = (A^{21}, \ A^{22}).$$

In this notation

$$\Sigma_{11} = \Sigma + A_*\Lambda_1 \left[ \sum_{i=0}^{\infty} (I_r - \Lambda_1)^i A^*\Sigma A^{*'}(I_r - \Lambda_1)^i \right] \Lambda_1 A'_*,$$

which is evidently a **convergent series**, since the (diagonal) elements of $I_r - \Lambda_1$ are less than one in absolute value.

To determine the limiting distribution of $P'_{-1}NP_{-1}/T$ and $P'_{-1}NU/T$ we must obtain the (1,1) block of $\Phi^*$ and the (1,2) block of $\Psi^*$. To this end, write the (lower) triangular matrix $S_0$ as

$$S_0 = \begin{bmatrix} T_1 & 0 \\ T_2 & T_3 \end{bmatrix}$$

where $T_1$ and $T_3$ are (lower) triangular matrices; in addition, note that from Eqs. (5A.9) and (5A.10) we obtain

$$T'_3 T_3 = \Sigma$$

$$T'_2 T_3 = \Sigma - \Pi(1)D\Sigma, \text{ so that}$$

$$T_2 = T_3 - T_3 D'\Pi(1)', \text{ and}$$

$$T'_1 T_1 = \Pi(1)[\Phi_{00} + \sum_{\tau=1}^{\infty}(\Phi_{0\tau} + \Phi_{\tau 0}) - D\Sigma D']\Pi(1)'.$$

Consequently,

$$\frac{P'_{-1}NP_{-1}}{T^2} \xrightarrow{\mathrm{d}} \int_0^1 [T'_1 B_1(s)' + T'_2 B_2(s)'][B_1(s)T_1 + B_2(s)T_2]\, ds$$

and

$$\frac{P'_{-1}NU}{T} \xrightarrow{\mathrm{d}} \int_0^1 [T'_1 B_1(s)' + T'_2 B_2(s)']\, dB_2(s)T_3.$$

To simplify the representation above, let $T'_1 T_1 + T'_2 T_2 = P'_0 P_0$ where $P_0$ is a lower triangular matrix, and define

$$BP_0 = B_1 T_1 + B_2 T_2,$$

so that $B$ is a $q$-element SMBM and $BP_0$ is a MBM with covariance matrix $P'_0 P_0$. In this notation

$$\frac{P'_{-1}NP_{-1}}{T^2} \xrightarrow{\mathrm{d}} P'_0 \left( \int_0^1 B(s)' B(s)\, ds \right) P_0$$

$$\frac{P'_{-1}NU}{T} \xrightarrow{\mathrm{d}} P'_0 \left( \int_0^1 B(s)'\, dB_2(s) \right) T_3;$$

thus, Eq. (5.117) may be rendered as

$$\Upsilon(\rho) = G'_2 P'_0 \left( \rho \int_0^1 B(s)' B(s)\, ds \right) P_0 G_2 - G'_2 P'_0 \left( \int_0^1 B(s)'\, dB_2(s) \right)$$

$$\times T_3 F T'_3 \times \left( \int_0^1 B(s)'\, dB_2(s) \right)' P_0 G_2.$$

Define $B_4 T_4 = BP_0 G_2$, with $T'_4 T_4 = G'_2 P'_0 P_0 G_2$, and rewrite $\Upsilon(\rho)$ as

$$\Upsilon(\rho) = T'_4 \left[ \rho K - K_1 T_3 F T'_3 K'_1 \right] T_4,$$

$$K = \int_0^\infty B_4(s)' B_4(s)\, ds, \quad K_1 = \int_0^1 B_4(s)'\, dB_2(s).$$

We recall from Eq. (5.116) that

$$F = M_{ww}^{-1} - M_{ww}^{-1} \Gamma (\Gamma' M_{ww}^{-1} \Gamma)^{-1} \Gamma' M_{ww}^{-1}.$$

From Corollary 5 in Dhrymes (1984) p. 39 we find

$$M_{ww}^{-1} = \Sigma^{-1} - \Sigma^{-1} \Gamma (I_q + \Gamma' \Sigma^{-1} \Gamma)^{-1} \Gamma' \Sigma^{-1}, \quad M_{ww}^{-1}\Gamma = \Sigma^{-1}\Gamma (I_q + \Gamma' \Sigma^{-1}\Gamma)^{-1},$$

so that

$$F = \Sigma^{-1} - \Sigma^{-1}\Gamma(\Gamma'\Sigma^{-1}\Gamma)^{-1}\Gamma'\Sigma^{-1}, \quad T_3 F T_3' = I_q - T_3'^{-1}\Gamma(\Gamma'\Sigma^{-1}\Gamma)^{-1}\Gamma'T_3^{-1}.$$

The matrix $F^* = T_3 F T_3'$ is symmetric idempotent of rank $r_0$ and, consequently, it has the representation $F^* = Q_1 Q_1'$, where $Q_1$ is $q \times r_0$, with **orthonormal columns**. Thus, $B_2 Q_1 = B_3$ is MBM with covariance matrix $Q_1'Q_1 = I_{r_0}$, so that $B_3$ is a SMBM; it is, however, **not distinct from** $B_4 T_4 = (B_1 T_1 + B_2 T_2) G_2$. It follows therefore that in Eq. (5.118), $|\Upsilon(\rho)| = 0$ may be written as

$$0 = \left| \rho I_{r_0} - \left( \int_0^1 B_4(s)' \, dB_3 \right)' \left( \int_0^1 B_4(s)' B_4(s) \, ds \right)^{-1} \left( \int_0^1 B_4(s)' \, dB_3(s) \right) \right|.$$

Thus, the distribution of the roots involved in tests for the rank of cointegration **contains no nuisance parameters**.

# Chapter 6

# Cointegration III

## 6.1  Introduction

In the previous chapter we examined at great length cointegration issues chiefly in the context of the standard $VAR(n)$ specification, by fully exploiting the implication of the cointegration hypothesis. In particular, the LR procedure for testing for the presence and rank of cointegration relies on estimators obtained under the null of cointegration, and is thus fully reflective of the relations the latter imposes.

In this chapter we examine a number of additional topics bearing on the general issue of cointegration, such as tests for the presence and rank of cointegration in $VAR(n)$ or $MIMA(k)$ system in a context in which the cointegration hypothesis has not been imposed at the estimation phase. In the literature this is apt to be termed a "Wald test", although the propriety of this term is ambiguous. In our discussion this will be termed a Conformity Test, (CT) in that we are asking whether certain entities estimated without imposing the restrictions implied by the cointegration hypothesis, nonetheless conform to these restrictions. In addition, we extend the methods discussed extensively in the previous chapter vis-a-vis $VAR(n)$ models to $ARMA(n, m)$ models.

## 6.2 CT for Cointegration in $VAR(n)$ Models

The formal tests for the existence and rank of cointegration generally prevailing in the applied literature assume a $VAR(n)$ and are essentially likelihood ratio (LR) procedures.

Relying on the basic definition of cointegration, we can devise a relatively simple, direct, and straight forward test (conformity test) for the presence and rank of cointegration that relies on the **unrestricted** estimator of $\Pi(1)$ examined in the previous chapter. The procedure entails estimation of $\Pi(1)$ **without restrictions**, the formation of a certain second moment matrix, which is a transformation of the "sample covariance matrix" of the usual cointegral vector, and a **test** for its rank. This is precisely the essence of the cointegration assertion, viz. that there are certain linearly independent linear combinations, but less than the maximal possible number, that exhibit stationarity, even though the underlying vector is one of $I(1)$ elements.

To implement this procedure return to Proposition 1 of Chapter 5, and note from part i that the **unrestricted** estimator for $\Pi(1)$ is given by

$$\hat{\Pi}(1)' = \Pi(1)' - (V'V)^{-1}V'U. \tag{6.1}$$

In keeping with the framework of this analysis we redefine the cointegral vector in this section by $z_{t\cdot} = X_{t\cdot}\Pi(1)'$, and set

$$\hat{M} = \hat{\Pi}(1)\left(\frac{P'_{-1}P_{-1}}{T}\right)\hat{\Pi}(1)' = \left(\frac{\Pi(1)P'_{-1}P_{-1}\Pi(1)'}{T}\right) + A_{12} + A'_{12} + A_{22}$$

$$A_{12} = -\frac{1}{T}\left(\frac{U'NP_{-1}}{T}\right)\left(\frac{P'_{-1}NP_{-1}}{T^2}\right)^{-1}\left(\frac{P'_{-1}P_{-1}\Pi(1)'}{T}\right), \tag{6.2}$$

$$A_{22} = \frac{1}{T}\left(\frac{U'NP_{-1}}{T}\right)\left(\frac{P'_{-1}NP_{-1}}{T^2}\right)^{-1}\left(\frac{P'_{-1}P_{-1}}{T^2}\right)\left(\frac{P'_{-1}NP_{-1}}{T^2}\right)^{-1}\left(\frac{P'_{-1}NU}{T}\right).$$

Under the null of cointegration, $\Pi(1) = \Gamma B'$, and we see that $\hat{M}$ is, indeed, a transformation of the "second moment matrix" of the cointegral vector, viz. $T\hat{M} = \Gamma(\hat{B}'P_{-1}'P_{-1}\hat{B})\Gamma'$ .[1]  In addition,

$$A_{11} = \left(\frac{\Pi(1)P_{-1}'P_{-1}\Pi(1)'}{T}\right) \overset{a.c.}{\to} M_{zz} = \Gamma M_{zz}^{*}\Gamma', \quad A_{12} \overset{d}{\to} 0, \quad A_{22} \overset{d}{\to} 0,$$

$$\lambda I_q - \hat{M} \overset{d}{\to} \lambda I_q - M_{zz}, \quad \frac{B'P_{-1}'P_{-1}B}{T} \overset{a.c.}{\to} M_{zz}^{*}. \tag{6.3}$$

Hence, the ordered characteristic roots of $\hat{M}$ converge to the (ordered) characteristic roots of $M = M_{zz}$, and it is evident that, as expected, the number of zero roots of $\Pi(1)$, and thus of $M$, corresponds to the number of unit roots of $|\Pi(z)| = 0$, and its rank corresponds to the cointegration rank.

To examine the limiting distribution of such roots, we first note that $A_{12}$, $A_{22}$ both converge to zero at the rate of $T^{\alpha}$, for $\alpha \in [0, 1)$. Thus, the limiting distribution of the roots depends entirely on the first term, and we obtain

$$\sqrt{T}\left(\lambda I_q - \hat{M}\right) \sim \sqrt{T}\left[\lambda I_q - \left(\frac{1}{T}\Pi(1)'P_{-1}'P_{-1}\Pi'\right)\right], \tag{6.4}$$

owing to the fact that

$$\sqrt{T}A_{12} \overset{d}{\to} 0, \quad \sqrt{T}A_{22} \overset{d}{\to} 0. \tag{6.5}$$

Since under the null of cointegration $\Pi(1) = \Gamma B'$, we have

$$\frac{1}{\sqrt{T}}\Pi(1)P_{-1}'P_{-1}\Pi' - \sqrt{T}M_{zz} = \frac{1}{\sqrt{T}}\sum_{t=1}^{T}\left(z_{t-1.}'z_{t-1.} - M_{zz}\right), \tag{6.6}$$

and

$$\left\{\zeta_t = \mathrm{vec}(z_{t.}'z_{t.} - M_{zz}) : z_{t.} = X_{t.}\Pi(1)', \quad t \in \mathcal{N}_0\right\}$$

is a zero mean strictly stationary ergodic process. Since the $MWN(\Sigma)$ that defines the $VAR(n)$ model is assumed in this context to be normal, the entities of Eq. (6.6) obey

$$E \parallel \zeta_t \parallel^{2+\alpha} < \infty, \quad \alpha > 0, \tag{6.7}$$

---

[1] In general, the cointegral vector is defined by $z_{t.} = X_{t.}B$, and therefore its sample covariance matrix is given by $B'Z'ZB/T$, where $Z = (z_{t.})$. However, since in this discussion we do not impose the hypothesis of cointegration it is much more convenient to define it by $Z_{t.} = X_{t.}\Pi(1)'$, and its sample covariance matrix by $\Pi(1)Z'Z\Pi(1)'/T$. Thus, in this discussion, the term $M_{zz}$ will denote the limit of $\Pi(1)Z'Z\Pi(1)'/T$.

and it follows, from Corollary 1a in Chapter 9, that

$$\psi_T^* = \text{vec}\left(\frac{1}{\sqrt{T}}\Pi P_{-1}' P_{-1}\Pi' - \sqrt{T}M_{zz}\right) = \frac{1}{\sqrt{T}}\sum_{t=1}^{T}\zeta_t, \tag{6.8}$$

obeys a CLT. Consequently, its limiting distribution is given by

$$\psi_T^* \xrightarrow{d} \psi^* \sim N(0, \Psi^*), \quad \text{where}$$

$$\Psi^* = \lim_{T\to\infty} E\psi_T^*\psi_T^{*'}.$$

Evidently, the matrix $\Psi^*$ is **singular** since the vector $\zeta_t.$ contains redundancies such as $z_{t-1,i}z_{t'-1,j} - m_{ij}(0)$ and $z_{t-1,j}z_{t'-1,i} - m_{ji}(0)$. Still, it is preferable to proceed in this formal fashion rather than consider only the distinct elements of $\psi_T^*$. The precise form of $\Psi^*$ may be obtained as follows. By definition

$$\Psi_T^* = E\psi_T^*\psi_T^{*'} = \sum_{t=1}^{T}\sum_{t'=1}^{T} E\zeta_t'.\zeta_{t'}.,$$

$$\zeta_t'. = z_{t-1}'. \otimes z_{t-1}'. - \text{vec}(M_{zz}). \tag{6.9}$$

To proceed, we need to modify our notation so that it becomes more flexible. To this end, let

$$M(\tau) = Ez_{t+\tau}'.z_t., \quad \text{so that} \quad M(-\tau) = M(\tau)'. \tag{6.10}$$

In this notation, what we had called $M_{zz}$ above becomes $M(0)$ and we further define

$$\text{vec}[M(\tau)] = m(\tau), \quad \tau = 0, \pm 1, \pm 2, \ldots, \tag{6.11}$$

so that

$$E\zeta_t'.\zeta_{t'}. = E\left(z_{t-1}'.z_{t'-1}. \otimes z_{t-1}'.z_{t'-1}\right) - m(0)m(0)'. \tag{6.12}$$

To evaluate the first expression in the right member above, we note that we are dealing with a block matrix whose $(i,j)$ block is $z_{t-1,i}z_{t'-1,j}z_{t-1}'.z_{t'-1}$; hence the $(r,s)$ element in the $(i,j)$ block is $z_{t-1,i}z_{t'-1,j}z_{t-1,r}z_{t'-1,s}$. If the underlying $WN(\Sigma)$ is normal, or if the fourth-order cumulants of this distribution are appropriately behaved [see e.g. Hannan (1970), pp. 208-229 or Anderson (1971), pp. 250-271], we find that the $(r, s)$ element of the $(i, j)$ block of $\Psi_T^*$ is given by

$$\Psi_{T,(i,j),(r,s)}^* = \frac{1}{T} \sum_{t=1}^{T} \sum_{t'=1}^{T} \left[ m_{ij}(t - t')m_{rs}(t - t') + m_{is}(t - t')m_{jr}(t' - t) \right].$$

(6.13)

Letting $t - t' = \tau$, we note that, for fixed $\tau = \tau_0 \geq 0$, the term in square brackets in Eq. (6.13) is repeated $T - \tau_0$ times for $t' = 1, 2, \dots, T - \tau_0$; similarly for $\tau_0 < 0$, the term in square brackets is repeated $T + \tau_0$ times for $t' = -\tau_0 + 1, -\tau_0 + 2, \dots, T$. Moreover, since $m_{jr}(-\tau) = m_{rj}(\tau)$, we may write

$$\Psi_{T,(i,j),(r,s)}^* = \sum_{\tau=-T+1}^{T-1} \left( 1 - \frac{|\tau|}{T} \right) [m_{ij}(\tau)m_{rs}(\tau) + m_{is}(\tau)m_{rj}(\tau)].$$

(6.14)

If the series above converges,[2] as it will under the standard assumptions of this literature, we conclude

$$\Psi_T^* = \sum_{\tau=-T+1}^{T-1} \left( 1 - \frac{|\tau|}{T} \right) [M(\tau) \otimes M(\tau) + m_{\cdot j}(\tau)m_{\cdot i}(\tau)']$$

$$\to \Psi^* = \sum_{\tau=-\infty}^{\infty} [M(\tau) \otimes M(\tau) + m_{\cdot j}(\tau)m_{\cdot i}(\tau)'].$$

(6.15)

Noting that

$$(m_{\cdot j}(\tau)m_{\cdot i}(\tau)') = \begin{bmatrix} M(\tau) \otimes m_{\cdot 1}(\tau)' \\ M(\tau) \otimes m_{\cdot 2}(\tau)' \\ \vdots \\ M(\tau) \otimes m_{\cdot q}(\tau)' \end{bmatrix},$$

(6.16)

---

[2] The convergence of the series above, i.e. the fact that

$$\left\| \lim_{T \to \infty} \sum_{\tau=-T+1}^{T-1} [m_{ij}(\tau)m_{rs}(\tau) + m_{is}(\tau)m_{rj}(\tau)] \right\| < \infty,$$

may be justified either in terms of Proposition 23 of Chapter 9, in view of the requirement in Eq. (6.7), or in terms of Proposition 4 of Chapter 1, and the requirement that the spectral matrix of the distinct elements of $\zeta_t$ is positive definite a.e. This enables us to write it as a GLP that converges exponentially.

we may verify that each term

$$M(\tau) \otimes M(\tau) + \begin{bmatrix} M(\tau) \otimes m_{.1}(\tau)' \\ M(\tau) \otimes m_{.2}(\tau)' \\ \vdots \\ M(\tau) \otimes m_{.q}(\tau)' \end{bmatrix}$$

contains certain row redundancies; for example, row $j$ is the same as row $qj$ since the first corresponds to the covariance of the element $z_{t-1,1}z_{t'-1,j} - m_{1j}(0)$, with respect to all other elements, while the second corresponds to the covariance of the element $z_{t-1,j}z_{t'-1,1} - m_{j1}(0)$, with respect to all other elements, and the two are identical! In addition, since we are dealing with real processes, $M(\tau) = M(-\tau)$. Consequently, we find

$$\Psi^* = M(0) \otimes M(0) + [m_{.j}(0)m_{.i}(0)' + m_{.i}(0)m_{.j}(0)'] \qquad (6.17)$$

$$+2\sum_{\tau=1}^{\infty}[M(\tau) \otimes M(\tau)] + \sum_{\tau=1}^{\infty}\{[m_{.j}(\tau)m_{.i}(\tau)'] + [m_{.i}(\tau)m_{.j}(\tau)']\}.$$

A simpler representation of the covariance matrix is obtained by noting that since the $\zeta$-sequence is strictly stationary and square integrable

$$\Psi^* = 2\pi f(0), \qquad (6.18)$$

where, provided it is continuous at zero, $f(\lambda)$ is the spectral matrix of the process[3] $\{\zeta_t : t \geq 1\}$, as the latter is defined in Eq. (6.9).

We have thus proved Proposition 1.

**Proposition 1.** In the context of the discussion above, consider the **unrestricted** estimator $\hat{\Pi}(1)$ and the matrix

$$\hat{M} = \frac{1}{T}\hat{\Pi}(1)P'_{-1}P_{-1}\hat{\Pi}(1)'.$$

The following statements are true:

i. $\hat{M} \xrightarrow{d} M_{zz} = M(0)$,[4] and thus in probability as well, by Proposition 40, p. 263, in Dhrymes (1989);

---

[3] Strictly speaking this is required of the spectral matrix of the $\zeta$-process, after redundancies have been eliminated.

[4] For simplicity of notation, in the remainder of the chapter, we use $M(0)$ to denote $M_{zz}$.

ii. $\mathrm{vec}\,\{\sqrt{T}[\hat{M}-M(0)]\}\xrightarrow{\mathrm{d}} N(0,\Psi^*)$, where $\Psi^*$ is as defined in Eq. (6.15), or Eq. (6.18), $f$ being the spectral matrix of the sequence $\{\zeta_t. : t \geq 1\}$, as the latter is defined in Eq. (6.9).

**Remark 1.** If the $X$-sequence is **stationary** and all roots of $M$ are positive (and assumed distinct), the covariance matrix $\Psi^*$ of the limiting distribution above, after removal of redundancies, is positive definite. If, on the other hand, some roots of $M(0)$ are zero, as would be the case under cointegration, we note a certain complication. Let $\Lambda$ and $Q$ be, respectively, the matrices of characteristic roots and characteristic vectors of $M(0)$. If $r_0$ of the characteristic roots are null, we have the following representation

$$Q'M(0)Q = \begin{bmatrix} \Lambda_1 & 0 \\ 0 & \Lambda_2 \end{bmatrix}, \tag{6.19}$$

where $\Lambda_2 = 0$, i.e. it contains the $r_0$ zero roots, and $\Lambda_1 > 0$. Partitioning conformably, $Q = (Q_1, Q_2)$, we have that

$$Q_2'M(0)Q_2 = 0; \quad \text{moreover, since} \quad \|\,M(\tau)\,\|\leq\|\,M(0)\,\|, \quad Q_2'M(\tau)Q_2 = 0, \tag{6.20}$$

for all $\tau$. Thus, the suitably modified matrix $\Psi^*$ is **singular**.

We have thus proved Corollary 1.

**Corollary 1.** Under the null of cointegration, the limiting distribution of $\sqrt{T}\mathrm{vec}[\hat{M} - M(0)]$ has a **singular covariance matrix**.

We are now in a position to formulate a (conformity) test for the presence of cointegration based on the result above which imply the validity of Proposition 2.

**Proposition 2.** Under the conditions of Proposition 1, let $\gamma$ be a small number, say $\gamma = .001$, then the conformity test for cointegration (CCT) may be formulated as

$H_0:\ \mathrm{tr}\,M(0) = \gamma$,

as against the alternative,

$H_1:\ \mathrm{tr}M(0) > \gamma$.

Since

$$\sqrt{T}[\operatorname{tr}\hat{M} - \operatorname{tr}M(0)] \overset{\mathrm{d}}{\to} N(0, \phi^2),\tag{6.21}$$

the conformity test statistic may be formulated as

$$CCTS: \frac{\sqrt{T}(\operatorname{tr}\hat{M} - \gamma)}{\phi} \overset{\mathrm{d}}{\to} N(0,1).\tag{6.22}$$

If the hypothesis $H_0$ above is accepted, we should conclude that cointegration does not exist, and the $X$-process is $I(1)$ and **not cointegrated**. If it is rejected, we conclude that **it is cointegrated**.

**Remark 2.** One might ask why we inserted the parameter $\gamma$ in this discussion. From a practical point of view, it makes no particular difference since in the presence of cointegration $\operatorname{tr}\hat{M}$ overwhelms $\gamma$. The reason for its insertion, however, is simply to uphold the applicability of the limiting distribution to the null. Notice that if we state the null as $\operatorname{tr}M(0) = 0$, it would mean that this **is not an admissible hypothesis**, given the conditions employed in deriving the limiting distribution. On the other hand $\operatorname{tr}M(0) = \gamma$ **is, strictly speaking, admissible**. Consequently, this device preserves the formal niceties and at the same time gives us a particularly simple procedure in testing for the presence of cointegration.

We next turn to the question of the rank of cointegration. This necessitates the extraction of the characteristic roots of $\hat{M}$ and, through them, tests of certain hypotheses.

Thus, we shall be dealing with tests for cointegration and/or stationarity. To rule out stationarity we test the null

$$H_0: \lambda_{\min} = 0$$

as against the alternative

$$H_1: \lambda_{\min} > 0.$$

Acceptance indicates that the sequence is $I(1)$, and cointegrated of rank at most $q-1$. More generally, if the cointegration rank is $r$, the $q-r$ smallest characterstic roots of $M(0)$ must be zero. Therefore, to devise such a test we need to obtain the limiting distribution of the characteristic roots of $\hat{M}$.

Such a result is not available in the literature. We provide it in the form of Proposition 3.

**Proposition 3.** In the context of Proposition 1, the (ordered) characteristic roots of $\hat{M}$, contained in the diagonal matrix $\tilde{\Lambda}$ obey

$$\sqrt{T}d^*[(\tilde{\Lambda} - \Lambda)] \sim d^*[\sqrt{T}Q'[\hat{M} - M(0)]Q],$$

where the notation $d^*[A]$ indicates the **diagonal elements of the matrix** $A$, and $Q$ is the orthogonal matrix of the decomposition $M(0) = Q\Lambda Q'$, as given in Eq. (6.19).

Proof: Since $\hat{M}$ and $M(0)$ are at least positive semidefinite, by Proposition 52, in Dhrymes (1984) pp. 61-62, they have the (orthogonal) decomposition

$$\hat{M} = \tilde{Q}\tilde{\Lambda}\tilde{Q}', \quad M(0) = Q\Lambda Q'. \tag{6.23}$$

Moreover, by the results of Theorem 1 of Chapter 5 and Proposition 28, Corollary 5 in Dhrymes (1989), pp. 242-244, $\tilde{Q}, and \tilde{\Lambda}$ converge, respectively, to $Q$ and $\Lambda$; in addition, $\sqrt{T}(\tilde{Q} - Q)$, and $\sqrt{T}(\tilde{\Lambda} - \Lambda)$ have well-defined limiting distributions. Next, consider

$$\sqrt{T}[Q'(\hat{M} - M(0))Q] = \sqrt{T}(\tilde{M}^* - \Lambda), \quad \tilde{M}^* = Q'\tilde{Q}\tilde{\Lambda}\tilde{Q}'Q, \tag{6.24}$$

and put

$$\sqrt{T}(\tilde{Q}'Q - I_q) = C, \quad \sqrt{T}(\tilde{\Lambda} - \Lambda) = D, \quad \sqrt{T}(\tilde{M}^* - \Lambda) = G. \tag{6.25}$$

Note that by Proposition 1, $C$, $D$, $G$ are all a.c. finite random elements, in the sense that they have well-defined limiting distributions, and thus may assume the values $\pm\infty$ only on a set of measure zero. It follows immediately that

$$\sqrt{T}(\tilde{M}^* - \Lambda) = G \sim D + \Lambda C - C\Lambda. \tag{6.26}$$

Since

$$g_{ii} = d_{ii}, \quad g_{ij} = (\lambda_i - \lambda_j)c_{ij}, \quad \text{or} \quad c_{ij} = g_{ij}/(\lambda_i - \lambda_j), \quad i \neq j, \tag{6.27}$$

this concludes the proof of the proposition, for the case where all characteristic roots are different so that the elements $c_{ij}$ are well-defined. We now examine the case where

$$\Lambda = \begin{bmatrix} \Lambda_1 & 0 \\ 0 & \Lambda_2 \end{bmatrix}, \quad \Lambda_2 = 0,$$

and $\Lambda_1$ is a diagonal matrix containing the $(r = q - r_0)$ positive roots under the null of cointegration; evidently, $\Lambda_2$ contains the $r_0$ zero roots. Partitioning the other matrices conformably we determine

$$\sqrt{T}(\tilde{M}^* - \Lambda) = G \sim \begin{bmatrix} D_{11} & 0 \\ 0 & D_{22} \end{bmatrix} + \begin{bmatrix} \Lambda_1 C_{11} - C_{11}\Lambda_1 & \Lambda_1 C_{12} \\ -C_{21}\Lambda_1 & 0 \end{bmatrix}. \quad (6.28)$$

Consequently, we have again

$$g_{ii} = d_{ii}, \quad i = 1, 2, \ldots, q, \quad g_{ij} = (\lambda_i - \lambda_j)c_{ij}, \quad i, j = 1, 2, \ldots, r, \quad i \neq j;$$

$$g_{ij} = \lambda_i c_{ij}, \quad i = 1, 2, \ldots, r, \quad j = r+1, r+2, \ldots, q;$$

$$= -\lambda_j c_{ij}, \quad i = r+1, r+2, \ldots, q, \quad i = 1, 2, \ldots, r;$$

$$= 0, \quad i, j = r+1, r+2, \ldots, q, \quad i \neq j. \quad (6.29)$$

From the construction above it follows that if we partition $G$ conformably

$$G = \begin{bmatrix} G_{11} & G_{12} \\ G_{21} & G_{22} \end{bmatrix},$$

we must conclude that $G_{22} = D_{22}$ and, moreover, that $D_{22} \xrightarrow{d} 0$. To verify this claim we note that, using the construction of Eq. (6.19), the covariance matrix of the limiting distribution of $G_{22}$ may be obtained from the result

$$\sqrt{T}(Q' \otimes Q')\mathrm{vec}[\hat{M} - M(0)] \xrightarrow{d} N(0, \Psi), \quad \Psi = (Q' \otimes Q')\Psi^*(Q \otimes Q),$$

where

$$\Psi = \begin{bmatrix} \Psi_{11} & \Psi_{12} \\ \Psi_{21} & \Psi_{22} \end{bmatrix} = \begin{bmatrix} [Q_1' \otimes Q']\Psi^*[Q \otimes Q_1] & [Q_2' \otimes Q']\Psi^*[Q \otimes Q_2] \\ [Q_1' \otimes Q']\Psi^*[Q \otimes Q_2] & [Q_2' \otimes Q']\Psi^*[Q \otimes Q_2] \end{bmatrix}. \quad (6.30)$$

Thus, the **marginal** limiting distribution of $G_{22}$ obeys

$$\mathrm{vec}(G_{22}) \xrightarrow{d} N(0, \Psi_{22}),$$

or what is equivalent,

$$\sqrt{T}\mathrm{vec}[(\hat{\Lambda}_2 - \Lambda_2)] \xrightarrow{\mathrm{d}} N(0, \Psi_{22}), \quad \Psi_{22} = [Q_2' \otimes Q']\Psi^*[Q \otimes Q_2]. \qquad (6.31)$$

Since $\Psi_{22} = 0$, we have that

$$\sqrt{T}(\tilde{\Lambda}_2 - \Lambda_2) \xrightarrow{\mathrm{d}} 0, \quad \text{and hence that} \quad \sqrt{T}(\tilde{\Lambda}_2 - \Lambda_2) \xrightarrow{\mathrm{P}} 0.$$

q.e.d.

**Remark 3.** Since

$$\sqrt{T}\mathrm{tr}\tilde{\Lambda}_2 \quad \text{and} \quad \hat{\Psi}_{22} = [\hat{Q}_2' \otimes \hat{Q}']\hat{\Psi}^*[\hat{Q} \otimes \hat{Q}_2]$$

**both** converge to zero, the usual $t$-ratio applied to estimators of **zero roots alone** involves at least an indeterminate form. Consequently, we cannot base any formal tests on such constructs. To deal with this problem, we shall presently derive the limiting distribution of zero root estimators, on which tests for the rank of cointegration may be based. Note, however, that in the framework of the conformity test for cointegration the presence of zero roots is almost self evident by visual inspection. The separation of roots is striking, especially when compared to the separation of characteristic roots in the context of the likelihood ratio test framework given in Johansen (1988), (1991). This is illustrated in Table 1, where PC1* denotes a $VAR(3)$ with one unit root and moderate stationary roots and PC2* denotes a $VAR(3)$ with one unit root and large stationary roots of the characteristic equation $|\Pi(z)| = 0$.

Table 1

Root Separation, Conformity and Johansen (LR) Tests
Trivariate $VAR(3)$, 3,000 Replications

| | Test Type | | | |
|---|---|---|---|---|
| **Experiment** | **Conformity** | | | **Johansen (LR)** |
| | Mean Root | True Root | Bias | Mean Root |
| **PC1** *, T = 100 | | | | |
| $\lambda_1$ | 6.609 | 5.739 | 0.910 | .458 |
| $\lambda_2$ | 2.456 | 2.406 | 0.050 | .258 |
| $\lambda_3$ | 0.010 | 0.000 | 0.010 | .012 |
| | | | | |
| **PC1** *, T = 300 | | | | |
| $\lambda_1$ | 6.039 | 5.739 | 0.300 | .443 |
| $\lambda_2$ | 2.436 | 2.406 | 0.030 | .247 |
| $\lambda_3$ | 0.003 | 0.000 | 0.003 | .004 |
| | | | | |
| **PC2** *, T = 100 | | | | |
| $\lambda_1$ | 681.711 | 709.528 | -27.817 | 0.852 |
| $\lambda_2$ | 10.899 | 8.989 | 1.910 | 0.283 |
| $\lambda_3$ | 0.060 | 0.000 | 0.060 | 0.015 |
| **PC2** *, T = 300 | | | | |
| | | | | |
| $\lambda_1$ | 698.108 | 709.528 | -11.420 | 0.849 |
| $\lambda_2$ | 9.481 | 8.989 | 0.492 | 0.258 |
| $\lambda_3$ | 0.008 | 0.000 | 0.008 | 0.005 |

An immediate consequence of the proposition and the preceding discussion are the following:

**Corollary 2.** The characteristic roots of $\hat{M}$, and hence their limiting distributions, are exactly those of $\tilde{M}^*$.

Proof: Obvious since $Q$ is a fixed orthogonal matrix.

**Corollary 3.** The distribution of the (associated) characteristic vectors is given by the distribution of $QC'$.

**Corollary 4.** A test on the rank of cointegration cannot be carried out in this context since the usual "t-ratio" converges to an indeterminate form.

The remaining problem is to determine the matrix $\Psi_q$ corresponding to the covariance matrix of the limiting distribution of the roots. Denoting the elements of the $q^2 \times q^2$ matrix $\Psi$ by $(\psi_{ij})$, $i,j = 1,2,\ldots,q^2$, we give below the elements in the upper triangular positions of the matrix $\Psi_q$.

$$\Psi_q = \begin{bmatrix} \psi_{11} & \psi_{1,q+2} & \psi_{1,2q+3} & \psi_{1,3q+4} & \cdots & \psi_{1,q^2} \\ * & \psi_{q+2,q+2} & \psi_{q+2,2q+3} & \psi_{q+2,3q+4} & \cdots & \psi_{q+2,q^2} \\ * & * & \psi_{2q+3,2q+3} & \psi_{2q+3,3q+4} & \cdots & \psi_{2q+3,q^2} \\ \vdots & \vdots & \vdots & \vdots & \vdots & \vdots \\ * & * & * & * & * & \psi_{q^2,q^2} \end{bmatrix} \qquad (6.32)$$

**Remark 4.** The major computational burden entailed by this procedure is the estimation of the spectral matrix of the vector process $\hat{z}'_{t-1.} \otimes \hat{z}'_{t-1.}$, where $\hat{z}_{t.} = X_{t-1.}\hat{\Pi}(1)$ is the estimated cointegral vector; or, equivalently, the estimation of the covariance matrix, $\Psi$, of the limiting distribution above.

In connection with the matrix in Eq. (6.32), and more generally in connection with the limiting distribution of sample "covariance matrices", note that the limiting distribution of the **distinct** elements of the latter may be obtained routinely through the relation

$$\sqrt{T}d[\text{vec}(\hat{M}-M(0))] = \sqrt{T}P\text{vec}(\hat{M}-M(0)), \quad P = \begin{bmatrix} I_q & 0 & 0 & \cdots & 0 \\ 0 & I_{q-1} & 0 & \cdot & 0 \\ \vdots & \vdots & \vdots & \vdots & \vdots \\ 0 & 0 & \cdots & I_2 & 0 \\ 0 & 0 & 0 & \cdots & I_1 \end{bmatrix},$$

where the notation $d[\cdot]$ denotes the distinct elements of the (symmetric) matrix appearing (in vectorized form) within the brackets, $P$ is of dimension $q(q+1)/2 \times q^2$ and, evidently, $I_1$ is the identity matrix of order **one**, or otherwise the **scalar** unity! Thus, the covariance matrix of the limiting distribution of the **distinct elements** is given by $P\Psi^*P'$, where $P$ is as defined in Eq. (5.141).

## 6.2.1   Conformity Cointegration Tests
##         for $MIMA(k)$ Processes

In this section we show that the results above are almost identically applicable, even if we assume that the $X$-process is $MIMA(k)$, $k < \infty$, i.e.

$$(I - L)X'_{t.} = A(L)\epsilon'_{t.}, \quad A(L) = \sum_{j=0}^{k} A_j L^j, \quad A_0 = I_q, \tag{6.33}$$

and $|A(z)| = 0$ has $r_0$ **unit roots**. We begin by noting that in Proposition 2a of Chapter 4 it is shown that the $X$-process has the representation

$$H(L)X'_{t.} = a(L)I_q\epsilon'_{t.}, \tag{6.34}$$

where $H(L)$ is a matrix polynomial lag operator, and $a(L)$ is an invertible scalar polynomial lag operator, respectively, of degrees

$$m_1 = (q - 1)(k - 1) + r_0, \quad m_2 = q(k - 1) + r_0. \tag{6.35}$$

There are two differences between this formulation and the $VAR(n)$ formulation. First, $MIMA(k)$ is **equivalent to an** $ARMA(m_1, m_2)$ process. Thus, the properties of the error term in Eq. (6.34) are considerably more complex than in the $VAR(n)$ formulation. Second, in the $MIMA(k)$ formulation the parameter $k$ need not be specified explicitly; but even for moderate values like $k = 10$ and $q = 6$, the autoregressive component of the resulting ARMA process is of order $(q - 1)(k - 1) + r_0 = 54 + r_0$! As we show below, however, an accurate specification of the lags is not essential for establishing the limiting distribution of the conformity cointegration test statistics. Thus, Propositions 1, 2, 3 and Corollaries 1, through 4, remain valid if

$$(I - L)X'_{t.} = \sum_{j=0}^{k} A_j\epsilon'_{t-j.}, \tag{6.36}$$

the $\epsilon$-process is $MWN(\Sigma)$, and is required to obey a condition analogous to that in Eq. (5.131). We formalize this in Proposition 4.

**Proposition 4.** Consider the $MIMA(k)$ model of Eq. (6.33) and suppose the moment condition $E \parallel \epsilon_{t.} \parallel^{2+\alpha} < \infty$ holds, for $\alpha > 0$, where the cointegral vector is defined by $z_{t.} = X_{t.}H(1)'$, the entity $H(1)$ being as implicitly defined in Eq. (6.34). Then the conclusions of Propositions 1, 2, 3 and Corollaries 1 through 4 (of this chapter) remain valid, whether or not the ARMA representation of the $X$-process in Eq. (6.34) is correctly specified.

Moreover, if the matrix of characteristic roots of $M$ is given by

$$\Lambda = \begin{bmatrix} \Lambda_1 & 0 \\ 0 & \Lambda_2 \end{bmatrix}, \quad \Lambda_2 = 0,$$

then

$$\sqrt{T}(\hat{\Lambda}_2 - \Lambda_2) \xrightarrow{d} 0, \quad \text{and hence} \quad \sqrt{T}(\hat{\Lambda}_2 - \Lambda_2) \xrightarrow{P} 0.$$

Proof: From Eq. (6.34) we may obtain the equivalent representation

$$(I - L)X_{t\cdot} = -X_{t-1\cdot}H(1)' + \sum_{j=1}^{m_2-1} (I - L)X_{t-j\cdot}H_j^{*'} + u_{t\cdot}, \qquad (6.37)$$

$$H(L) = H(1) - (I - L)H^*(L), \qquad H_j^* = \sum_{i=j+1}^{m_2} H_i.$$

To maintain maximal correspondence with the preceding discussion, let

$$X_1 = (\Delta X_{t-1\cdot}, \Delta X_{t-2\cdot}, \ldots, \Delta X_{t-n+1\cdot}), \qquad (6.38)$$

$$X_2 = (\Delta X_{t-n\cdot}, \Delta X_{t-n-1\cdot}, \ldots, \Delta X_{t-m_2+1\cdot}),$$

and, in the obvious notation, write the observations on this model as

$$\Delta P = -P_{-1}H(1)' + X_1 H_{(1)}^{*'} + X_2 H_{(2)}^{*'} + U, \quad U = (u_{t\cdot}), \quad u_{t\cdot} = a(L)\epsilon_{t\cdot}. \quad (6.39)$$

Note that since $a(L)$ is invertible, the $u$-process is strictly stationary. Suppose in the estimation of $H(1)$ **we neglect** the variables in $X_2$, thus running a regression of $\Delta P$ on $(P_{-1}, X_1)$. The least squares estimator of $H(1)$ is given by

$$\hat{H}(1)' = H(1)' - (V'V)^{-1}V'U + (V'V)^{-1}V'X_2 H_{(2)}^{*'}. \qquad (6.40)$$

Moreover, we note that for any $\alpha \in [0, 1)$

$$T^\alpha[\hat{H}(1) - H(1)] \xrightarrow{P} 0. \qquad (6.41)$$

If we now define the matrix

$$\hat{M} = \frac{1}{T}\hat{H}(1)P'_{-1}P_{-1}\hat{H}(1)',$$ (6.42)

we have, by precisely the same arguments employed earlier, that

$$\sqrt{T}(\hat{M} - M_{zz}) \sim \frac{1}{\sqrt{T}}\sum_{t=1}^{T}(z'_{t-1.}z_{t-1.} - M_{zz}),$$ (6.43)

where now

$$z_{t.} = X_{t.}H(1)', \quad \frac{H(1)P'_{-1}P_{-1}H(1)'}{T} \overset{\text{a.c.}}{\to} M_{zz} \geq 0.$$ (6.44)

In addition, note that in deriving the limiting distribution of the entity in Eq. (6.8) **we did not make use of the fact that** the $\epsilon$-process was one of i.i.d. random variables. We only made use of the fact that it was a **strictly stationary process**. Since the same condition holds in this case as well, the conclusions of Propositions 1, 2, 3 and Corollaries 1 through 4 continue to be valid.

q.e.d.

Before we leave this subject, it is useful to explore two aspects: First, what is the intuitive content of the derivation of the limiting distribution of the matrix $\hat{M}$ and, second, what is the connection between the roots of $\hat{M}$ and the roots that appear in the LR procedure (Johansen). We develop these issues in the two remarks below.

**Remark 5.** Since by definition

$$\hat{\Pi}(1)' = -(V'V)^{-1}V'W, \quad W = N\Delta P,$$

we might as well have written

$$\hat{M} = \frac{1}{T}W'V(V'V)^{-1}P'_{-1}P_{-1}(V'V)^{-1}V'W.$$

Doing so, leads us to observe that since

$$(V'V)^{-1}P'_{-1}P_{-1} \overset{\text{d}}{\to} I_q, \quad \text{and thus} \quad (V'V)^{-1}P'_{-1}P_{-1} \overset{\text{P}}{\to} I_q,$$

we have

$$\hat{M} \sim \frac{1}{T} W'V(V'V)^{-1}V'W. \qquad (6.45)$$

If the representation above were appropriate, it would imply that the theory developed in this section can equally well be developed on the basis of the entity

$$\hat{M}^{(1)} = \frac{1}{T}\hat{\Pi}(1)V'V\hat{\Pi}(1)' = \frac{1}{T}W'V(V'V)^{-1}V'W. \qquad (6.46)$$

This is, however, incorrect since $\hat{M}^{(1)}$ need not have the same probability limit as $\hat{M}$, and need not have the same limiting distribution. To see this, expand its representation to obtain

$$\hat{M}^{(1)} = \frac{1}{T}(B_{11} + B_{12} + B'_{12} + B_{22}),$$

$$B_{11} = \Pi(1)V'V\Pi(1)', \quad B_{12} = \Pi(1)P'_{-1}V'U,$$

$$B_{22} = U'V(V'V)^{-1}V'U.$$

It is clear that

$$\frac{1}{\sqrt{T}}B_{22} \xrightarrow{\text{P}} 0, \quad \frac{1}{\sqrt{T}}B_{12} \xrightarrow{\text{P}} 0,$$

and moreover

$$\frac{1}{T}B_{11} = \frac{1}{T}\Pi(1)P'_{-1}P_{-1}\Pi(1)' - \frac{\Pi(1)P'_{-1}X}{T}\left(\frac{X'X}{T}\right)^{-1}\frac{X'P_{-1}\Pi(1)'}{T}.$$

Evidently,

$$\frac{1}{T}B_{11} \xrightarrow{\text{ac.}} M^*_{zz} = M_{zz} - M_{zx}M_{xx}^{-1}M_{xz}. \qquad (6.47)$$

**In this context**, $M^*_{zz}$ is the **conditional covariance matrix** of the cointegral vector $z_t = X_{t-1}.\Pi(1)'$, conditioned on the $\sigma$-field induced by the elements of the matrix $X$, i.e. the differences $\Delta X_{t-i\cdot}$, $i = 1, 2, \ldots, n-1$. Moreover, the limiting distribution is obtained from

$$\sqrt{T}(\hat{M}^{(1)} - M^*_{zz}) \sim \frac{1}{\sqrt{T}}\sum_{t=1}^{T}(z'_{t-1\cdot}z_{t-1\cdot} - M_{zz}) \qquad (6.48)$$

$$-M_{zx}M_{xx}^{-1}\left(\frac{1}{\sqrt{T}}\sum_{t=1}^{T}[x'_{t\cdot}z_{t-1\cdot} - M_{xz}]\right).$$

Evidently, the behavior of $\hat{M}$ and $\hat{M}^{(1)}$ will be **identical if** $M_{zx} = 0$!

**Remark 6.** We now turn to the connection between the (characteristic) roots encountered in the conformity test, and those encountered in the LR test. If we concentrate the likelihood function [see Eq. (4.98) in Chapter 4], we ultimately find that we need to **minimize** with respect to $B$, which is a $q \times r$ matrix of rank $r$, the determinant

$$D(B) = |W'W - W'VB(B'V'VB)^{-1}B'V'W|. \qquad (6.49)$$

After considerable manipulation, we determine that this requires us to obtain the (characteristic) roots and vectors of

$$|\lambda V'V - V'W(W'W)^{-1}W'V| = 0, \qquad (6.50)$$

The solution to the problem is to select the $r$ largest roots, and their associated characteristic vectors. The latter serve as the estimator of the matrix $B$ in the rank factorization $\Pi(1) = \Gamma B'$. Note further that, under the null of cointegration, the remaining roots are zero. Hence, in the LR procedure the rank of cointegration is simply the number of positive roots in the limit version of Eq. (6.50). In the conformity test context, the rank of cointegration is determined by the positive roots of the limit of $\hat{M}^{\cdot}$. An important question then is this: How are the roots of Eq. (6.50) related to the roots of $\hat{M}$? Using a number of properties of determinants, particularly Proposition 43 in Dhrymes (1984), p. 51, we have that the characteristic roots of Eq. (6.50) are precisely those of

$$|\lambda W'W - W'V(V'V)^{-1}V'W| = 0. \qquad (6.51)$$

Thus, basically, the LR procedure determines the rank of cointegration by the number of positive roots of the limit of $\hat{M}^{(1)}$, **in the metric of** $M^*_{x_0 x_0}$. The latter is simply the conditional covariance matrix of $\Delta X_{t\cdot}$, conditioned on the same $\sigma$-field as above. It is this feature that renders such roots **less than unity**, and thus impedes the effective separation of roots in empirical applications. By contrast, the conformity approach determines the rank of cointegration by the positive characteristic roots of the limit of $\hat{M}$, which is the **unconditional covariance** matrix of the cointegral vector, **in the metric of the identity matrix**. Thus, the roots are not compressed by measuring them in terms of "units" of a possibly large covariance matrix, and this contributes to a very effective separation of the zero roots in an empirical context.

## 6.2.2  Limiting Distribution of Zero Root Estimators, Case i: Standard VAR, no Constant Term

The preceding discussion was designed to obtain a test for the presence of cointegration whose limiting distribution was standard. The consequence was that while that objective was achieved, we had no reliable test for the **rank** of cointegration, due to the fact that the "t-ratio" of test statistics **involving estimators of zero roots alone** converges to an indeterminate form. In this section we remedy this condition by obtaining such a result; the resulting limiting distributions, however, are nonstandard.

We begin by considering the problem of obtaining the characteristic roots of $\hat{M}$ **in the metric of** $\Sigma$, which is the covariance matrix of the "structural error" of the VAR. The reason for proceeding in this fashion is to produce a limiting distribution for the zero root estimators that is free of nuisance parameters. Thus, we consider

$$|\lambda\Sigma - \hat{M}| = |\Sigma||\lambda I_q - T_0^{'-1}\hat{M}T_0^{-1}| = 0, \quad \Sigma = T_0^{'}T_0, \qquad (6.52)$$

where $T_0$ is the triangular matrix of the decomposition of $\Sigma$. Evidently, the rank of $\hat{M}$ is not changed by this operation, and all previous results regarding the **nonzero roots** hold, *mutatis mutandis*. Precisely, if we redefine $\hat{M}$ to be $\hat{T}_0^{'-1}\hat{M}\hat{T}_0^{-1}$ all previous results will continue to hold as stated, where $\hat{T}_0$ is the triangular matrix of the decomposition of

$$\hat{\Sigma} = \frac{1}{T}\hat{U}'\hat{U}, \quad \hat{U} = \Delta P + P_{-1}\hat{\Pi}(1)^{'} - X\hat{\Pi}^*, \qquad (6.53)$$

$\hat{\Pi}(1)$ and $\hat{\Pi}^*$ being the least squares estimators of the corresponding parameter matrices. This is so since $\hat{\Sigma} \overset{\text{P}}{\to} \Sigma$, as well as a.c.

In the first representation of Eq. (6.52), pre- and post-multiply by $D'$ and $D$, respectively, where

$$\Pi(1)^{'} = D\Lambda D^{-1}, \ \Lambda = \begin{bmatrix} \Lambda_1 & 0 \\ 0 & 0 \end{bmatrix}, \ D^{-1} = \begin{bmatrix} D_1^* \\ D_2^* \end{bmatrix}, \ D = (D_1, D_2), \qquad (6.54)$$

to obtain

$$|\lambda D'\Sigma D - D'\hat{M}D| = 0. \tag{6.55}$$

We note that since the rank of $\hat{M}$ (and its limit) is not affected by this operation, the number of nonzero and zero roots is preserved.

By the cointegration hypothesis, $\Lambda_1$ in Eq. (6.52) contains the $r$ nonzero roots along its diagonal; $D$ is the (nonsingular) matrix of characteristic vectors, $D_1$ is $q \times r$, $D_1^*$ is $r \times q$ and, moreover,

$$D_1^*D_1 = I_r, \quad D_2^*D_2 = I_{r_0}, \quad D_1^*D_2 = 0, \quad D_2^*D_1 = 0. \tag{6.56}$$

Next consider the matrix

$$\lambda I_*D'\Sigma DI_* - I_*D'\hat{M}DI_*, \quad I_* = \begin{bmatrix} I_r & 0 \\ 0 & \sqrt{T}I_{r_0} \end{bmatrix}, \tag{6.57}$$

and note that setting the determinant of the first matrix in Eq. (6.57) to zero yields precisely the same roots as Eq. (6.55). Moreover,

$$\lambda I_*D'\Sigma DI_* - I_*D'\hat{M}DI_* \tag{6.58}$$

$$\sim \begin{bmatrix} \lambda D_1'\Sigma D_1 - \tilde{M}_{11}^* & \sqrt{T}\lambda D_1'\Sigma D_2 \\ \\ \sqrt{T}\lambda D_2'\Sigma D_1 & T\lambda D_2'\Sigma D_2 - \tilde{M}_{22}^* \end{bmatrix}$$

$$\tilde{M}_{11}^* = \Lambda_1\left(\frac{1}{T}D_1'P_{-1}'P_{-1}D_1\right)\Lambda_1, \quad \tilde{M}_{22}^* = \left(\frac{D_2'U'V}{T}\right)\left(\frac{V'V}{T^2}\right)^{-1}\left(\frac{V'UD_2}{T}\right).$$

The operator $\sim$ in Eq. (6.58) signifies that the difference between the two sides of the equation converges to zero in probability.[5] The determinant of the right member of that relation, set to zero, yields the characteristic roots of interest, from the point of view of their limiting behavior. We have

$$0 = |\lambda D_1'\Sigma D_1 - \tilde{M}_{11}^*| \tag{6.59}$$

$$\times \ |T\lambda D_2'\Sigma D_2 - \tilde{M}_{22}^* - \sqrt{T}\lambda D_2'\Sigma D_1(\lambda D_1'\Sigma D_1 - \tilde{M}_{11}^*)^{-1}\sqrt{T}\lambda D_1'\Sigma D_2|.$$

---

[5] More generally we shall use this operator a bit more loosely to mean that what is on the left behaves essentially like what is on the right in matters that are relevant for the problem at hand.

The estimators of the nonzero roots are obtained from

$$|\lambda D_1' \Sigma D_1 - \tilde{M}_{11}^*| = 0, \tag{6.60}$$

and evidently converge to constants with probability one, by previous discussion. In particular, note that the matrix whose roots we take converges to a constant matrix with probability one. The estimators of the zero roots are obtained from the second determinant above, i.e. from the equation

$$|T\lambda D_2' \Sigma D_2 - \tilde{M}_{22}^* - T\lambda^2 D_2' \Sigma D_1 (\lambda D_1' \Sigma D_1 - \tilde{M}_{11}^*)^{-1} D_1' \Sigma D_2| = 0. \tag{6.61}$$

Earlier, in Proposition 3, we established that estimators of zero roots converge to zero (in probability), so that the inverse matrix in Eq. (6.59) remains well behaved as $T \to \infty$; moreover, it was also established that for an estimator of a zero root $\sqrt{T}\lambda \xrightarrow{\text{P}} 0$, which implies that in Eq. (6.61) $T\lambda^2 \xrightarrow{\text{P}} 0$. Putting $\rho = T\lambda$, we conclude that the limiting distribution of the zero roots of $\hat{M}$ **in the metric of** $\Sigma$ is given by the distribution of the roots of

$$0 = |\rho D_2' \Sigma D_2 - \bar{M}_{22}^*| = |\rho I_{r_0} - J' \bar{M}_{22}^* J|, \quad J'^{-1} J^{-1} = D_2' \Sigma D_2, \tag{6.62}$$

where $\bar{M}_{22}^*$ is the limit (in distribution) of $\tilde{M}_{22}^*$, i.e.

$$\tilde{M}_{22}^* \xrightarrow{\text{d}} \bar{M}_{22}^*. \tag{6.63}$$

**Remark 7.** Note that while the behavior of estimators of nonnull roots depends essentially on the limiting behavior of $\sqrt{T}A_{11}$, centered appropriately, **the (limiting) distribution of zero root estimators depends exclusively on the limiting behavior of** $TA_{22}$.

To determine the form of the limit in Eq. (6.63), we observe that since the error process is one of i.i.d. vectors [6]

$$\frac{1}{T}V'UD_2 \xrightarrow{\text{d}} P_0'\Upsilon T_0 D_2, \quad \Upsilon = \int_0^1 B(r)' d\,B_2(r), \tag{6.64}$$

---

[6] The results in Eqs. (6.64) and (6.65) are obtained as follows: Define $\phi_{t\cdot} = (\eta_{t\cdot}, \epsilon_{t\cdot})$, $S_{t\cdot} = \sum_{s=1}^t \phi_{s\cdot}$, where $\eta_{t\cdot}$ is as defined in Eq. (6.66). From Chapter 3,

$$\frac{1}{T^2}\sum_{t=1}^T S_{t\cdot}' S_{t\cdot} \xrightarrow{\text{d}} P_0^{*'}\left(\int_0^1 B(s)' B(s)ds\right) P_0^*, \quad \frac{1}{T}S_{t-1\cdot}' \phi_{t\cdot} \xrightarrow{\text{d}} P_0^{*'}\left(\int_0^1 B(s)' dB(s)\right) P_0^*,$$

where $P_0^*$ is the triangular decomposition of the "long-run" covariance matrix of the $\phi$-process, and $B$ is a $2q$-element SMBM, say $B = (B_1, B_2)$. What is of relevance to this

and moreover

$$\frac{1}{T^2}V'V \xrightarrow{d} P_0'KP_0, \quad K = \int_0^1 B'(r)B(r)\,dr, \tag{6.65}$$

where $P_0$ is the matrix of the triangular decomposition of what is termed in the literature the long-run covariance matrix of $(I-L)X_{t\cdot}$, and is more precisely the sum of the autocovariance matrices of the process as follows: Write the $VAR(n)$ as

$$(I-L)X_{t\cdot} = \epsilon_{t\cdot}\frac{H(L)'}{\pi(L)} = \eta_{t\cdot}, \tag{6.66}$$

where $\eta_{t\cdot}$ is strictly stationary, and define the autocovariance matrix sequence as

$$C(\tau) = E\eta_{t-\tau}'\eta_{t\cdot}, \quad \tau \in \mathcal{N}, \quad C(-\tau) = C(\tau)'; \tag{6.67}$$

the "long run covariance matrix" is obtained as

$$C = C(0) + \sum_{\tau=1}^{\infty}[C(\tau) + C(\tau)'], \quad \text{and} \quad C = P_0'P_0. \tag{6.68}$$

Consequently, the entity in Eq. (6.63) is given by

$$\tilde{M}_{22}^* \xrightarrow{d} D_2'T_0'\left(\int_0^1 B(r)'dB_2(r)\right)' P_0(P_0'KP_0)^{-1}P_0'\left(\int_0^1 B(r)'d\,B_2(r)\right)T_0D_2$$

$$= D_2'T_0'\Upsilon'K^{-1}\Upsilon T_0D_2. \tag{6.69}$$

---

discussion is the (1,1) block of the first limit, and the (1,2) block element of the second limit above. If we write

$$P_0^* = \begin{bmatrix} T_1 & 0 \\ T_2 & T_3 \end{bmatrix},$$

the (1,1) block is given by

$$\int_0^1 B_3^*(s)'B_3^*ds = P_0'\left(\int_0^1 B_3(s)'B_3(s)ds\right)P_0,$$

where $B_3^* = B_1T_1 + B_2T_2$, i.e. it is a MBM with covariance matrix $T_1'T_1 + T_2'T_2 = P_0'P_0$. Thus, we may write $B_3^* = B_3P_0$, where $B_3$ is a SMBM, i.e. a MBM whose covariance matrix is $I_q$. The (1,2) block of the second limit is given by

$$\int_0^1 B_3^{*'}(s)dB_2(s)T_3 = P_0'\left(\int_0^1 B_3(s)'dB_2(s)\right)T_3, \quad T_3'T_3 = \Sigma.$$

In the discussion of the main text $P_0$ has the same meaning as in this footnote, $B$ corresponds to the SMBM $B_3$ of this footnote, $B_2$ has the same meaning as in this footnote, and $T_0$ corresponds to $T_3$ in this footnote.

From the second representation in Eq. (6.69), we further conclude that the distribution of the estimators of the zero roots of $\hat{M}$ in the metric of $\Sigma$ converges to the distribution of the roots of $J'\bar{M}^*_{22}J$. Examining the stochastic integral involved in the limit, we obtain

$$J'\bar{M}^*_{22}J = J'D'_2T'_0\left(\int_0^1 dB_2(r)'B(r)\right)K^{-1}\left(\int_0^1 B'(r)d\,B_2(r)\right)T_0D_2J, \quad (6.70)$$

so that $B_2T_0D_2J$ is a $MBM$ with covariance matrix

$$J'D'_2T'_0T_0D_2J = J'D'_2\Sigma D_2J = I_{r_0}. \qquad (6.71)$$

Thus, we may define the $r_0$-element SMBM

$$B_1(r) = B_2(r)P_0D_2J \qquad (6.72)$$

and finally conclude that the limiting distribution of the estimators of the zero characteristic roots of (the limit of) $\hat{M}$ in the metric of $\Sigma$ is the distribution of the characteristic roots of

$$0 = \left|\rho I_{r_0} - \left(\int_0^1 dB_1(r)'B(r)\right)\left(\int_0^1 B(r)'B(r)\,dr\right)^{-1}\left(\int_0^1 B(r)'\,dB_1(r)\right)\right|.$$
$$(6.73)$$

We have therefore proved Proposition 5.

**Proposition 5**. Consider the $VAR(n)$ model under the conditions of Proposition 1, and the characteristic roots of

$$|\lambda\Sigma - \hat{M}| = 0.$$

The following statements are true:

i. If in Proposition 3 we redefine $\hat{M}$ to be $\hat{T}'^{-1}_0\hat{M}\hat{T}^{-1}_0$, the limiting distribution of the **nonzero roots** remains as given therein;

ii. a test for the rank of cointegration, $r$, i.e. a test of the null hypothesis that there are $r_0 = q - r$ **zero roots**, may be based on

$$T\sum_{i=r+1}^q \hat{\lambda}_i \xrightarrow{d} \mathrm{tr}\,\Upsilon'K^{-1}\Upsilon, \quad K = \int_0^1 B(r)'B(r)dr, \quad \Upsilon = \int_0^1 B(r)'dB_1(r),$$

where $B, B_1$ are SMBM, represented as row vectors with $q$ and $r_0$ elements, respectively.

iii. a test that the rank of cointegration is $r$ as against the alternative that it is $r+1$, i.e. a test that there are $r_0$ zero roots as against the alternative that there are $r_0 - 1$ zero roots, may be carried out through the statistic $T\hat{\lambda}_r$.

**Remark 8.** Note that this distribution is **not the same** as that given in Johansen and Juselius (1990), in Osterwald-Lenum (1992), or in Saikkonen (1992), although it is certainly of the same genre. In Johansen, the equivalent of $B$ and $B_1$ **are of the same dimension**, while in the conformity context the first is of dimension $q$ and the second of dimension $r_0$. Thus, in the Johansen context the test statistic for a unit root in a two equation system has precisely the same distribution as a test statistic for a unit root in a $q$-equation system. Or, put in slightly different terms, a test statistic for **three unit** roots in a four-equation system has the same distribution as the test statistic for three unit roots in a ten-equation system. Thus, in empirical applications it is more likely that in larger systems we would find a higher incidence of cointegration than is perhaps inherent in the data. The distribution in Proposition 5, however, has two parameters, the dimension of the system, $q$, and the number of unit roots, $r_0$.

# 6.3 $VAR(n)$ with a Constant Term

## 6.3.1 Preliminaries and Implications

When the (cointegrated) $VAR(n)$ model is stated as

$$\Pi(L)X'_{t.} = \mu' + \epsilon'_{t.}, \tag{6.74}$$

there are certain implications for the behavior of the $X$-sequence as well as for the estimation of parameters. In the context of Eq. (6.74) we distinguish three cases:

i. $\mu = 0$, a case we had considered above;

ii. $\mu \neq 0$ but it lies in the row space, $\mathcal{R}(\Gamma')$, of $\Gamma'$, i.e. there exists a nonnull (row) vector $c$ such that $\mu = c\Gamma'$;

iii. $\mu \neq 0$ **and** $\mu \notin \mathcal{R}(\Gamma')$.

When we examined case i, we implicitly assumed that the underlying process $X$ had **mean zero**. Although this is not necessarily the case, it accords with the general tenor of the VAR without a constant term. To draw more fully the implications here we note that if we pre- and post-multiply the $VAR(n)$ by the adjoint of $\Pi(L)$, say $H_*(L)$, we find, respectively,

$$H_*(L)\Pi(L) = \Pi(L)H_*(L) = \pi^*(L)I_q. \tag{6.75}$$

By assumption, however, the system has $r_0$ **unit** roots. Hence we have

$$H_*(L)\Pi(L) = (I - L)^{r_0}\pi(L)I_q, \quad \pi(1) \neq 0, \tag{6.76}$$

where $\pi(L)$ is an invertible scalar polynomial operator. It may also be shown that

$$H_*(L) = (I - L)^{r_0-1}H(L), \quad \text{such that} \quad \text{rank}[H(1)] = q - r = r_0, \tag{6.77}$$

where $H(L)$ is a matrix polynomial lag operator of degree at most $(q - 1)n - r_0$. It follows that

$$H(L)\Pi(L) = (I - L)\pi(L), \quad \pi(1) \neq 0, \tag{6.78}$$

and we see that

$$\Pi(1)H(1) = H(1)\Pi(1) = 0, \tag{6.79}$$

which, by the cointegration hypothesis, implies that $H(1)$ is of rank $q - r = r_0$. By the rank factorization theorem, Proposition 15 in Dhrymes (1984), p. 23, there exist $q \times r_0$ matrices, $C_1, C_2$, each of rank $r_0$, such that $H(1) = C_2 C_1'$. Consequently, we find

$$0 = H(1)\Pi(1) = C_2 C_1' \Gamma B'. \tag{6.80}$$

Since the **rows** of $B'$ are linearly independent, Eq. (6.80) implies that

$$C_2 C_1' \Gamma = 0; \tag{6.81}$$

moreover, since the **the columns** of $C_2$ are **linearly independent** we conclude that Eq. (6.81) implies

$$C_1' \Gamma = 0, \quad \text{or} \quad \Gamma' C_1 = 0. \tag{6.82}$$

Finally, we note that premultiplying Eq. (6.74) by $H(L)$ yields

$$(I - L)X'_{t\cdot} = \left(\frac{H(1)}{\pi(1)}\right)\mu' + \eta'_{t\cdot}, \quad \eta_{t\cdot} = \epsilon_{t\cdot}\frac{H(L)'}{\pi(L)}, \tag{6.83}$$

where $\eta$ is a **strictly stationary** process.

**If** $\mu \in \mathcal{R}(\Gamma')$, $\mu H(1)' = 0$ and the equation above becomes $(I - L)X_{t\cdot} = \eta_{t\cdot}$. We may solve this to obtain

$$X_{t\cdot} = \mu_0 + S_{t\cdot}, \quad S_{t\cdot} = \sum_{s=1}^{t} \eta_{s\cdot}, \tag{6.84}$$

where $\mu_0$ is "the constant of integration".

For case iii, we conclude that [7]

$$X_{t\cdot} = \mu_0 + \mu_1 t + S_{t\cdot}, \quad \text{for} \quad \mu \notin \mathcal{R}(\Gamma'), \quad \mu_1 = \mu\frac{H(1)'}{\pi(1)}. \tag{6.85}$$

## 6.3.2 Case ii: Constant in VAR; No Trends in Levels

In this section we examine the behavior of the conformity test in the last two cases, and determine whether and how the limiting distribution in cases ii

---

[7] We do not consider linear time trends in the specification of the $VAR(n)$, since this would imply that the underlying $X$ process has **quadratic** time trends. In the author's view this does not correspond to the behavior of economic time series, i.e. it is unlikely that economic time series involve unit roots **and** quadratic time trends. On the other hand, introducing linear time trends in the specification of the VAR is perfectly routine in this context, and the results we obtain below will remain valid, *mutatis mutandis*.

The model examined here allows for unit roots and **linear** time trends in the levels. It is easy to verify from Eq. (6.84) that the cointegral vector $X_{t\cdot}\Pi(1)'$ **does not have** linear time trends, although it may have a nonzero mean. Moreover, reconstituting the original $VAR(n)$ model we obtain

$$\Pi(L)X'_{t\cdot} = \Pi(1)\mu'_0 - \sum_{i=1}^{n} i\Pi_i\mu'_1 + \Pi(1)\mu'_1 t + \Pi(L)S'_{t\cdot}.$$

Since $\mu_1 = \mu H(1)'/\pi(1)$, and $\mu_1 = c\Gamma'$ it follows that $\mu_1\Pi(1)' = c\Gamma'C_1C'_2 = 0$, by Eq. (6.82). Thus, **no trends are present** in the "levels". Moreover,

$$\Pi(L)S'_{t\cdot} = \sum_{s=1}^{t}\left(\frac{H(L)\Pi(L)}{\pi(L)}\right)\epsilon'_{s\cdot} = (I - L)\sum_{s=1}^{t}\epsilon'_{s\cdot} = \epsilon'_{t\cdot},$$

so that we get back the original $VAR(n)$ model with a constant term.

and iii differs from that in i, as determined in Proposition 5. This is the case where the $VAR(n)$ is stated with a constant term and the underlying $X$-process is asserted to have a (constant) nonzero mean and **no time trends**. From the error correction model (ECM) representation we have

$$\Delta P = -P_{-1}\Pi(1)^{'} + X\Pi^* + e\mu + \epsilon_{t\cdot}. \tag{6.86}$$

The least squares estimator of $\Pi(1)$ may be obtained from

$$(\Delta P)^* = -P^*_{-1}\Pi(1) + X^*\Pi^* + U^*, \quad X^* = (I - \frac{ee^{'}}{T})X, \quad P^*_{-1} = (I - \frac{ee^{'}}{T})P_{-1}, \tag{6.87}$$

and similarly for all other starred quantities. Specifically, the estimator is given by

$$\hat{\Pi}(1)^{'} = \Pi(1)^{'} - (P^{*'}_{-1}N_x P^*_{-1})^{-1}P^{*'}_{-1}N_x U^*, \tag{6.88}$$

which is, *mutatis mutandis*, the same representation of the unrestricted estimator obtained in Eq. (6.1), except that now

$$N = N_x = I_T - X^*(X^{*'}X^*)^{-1}X^{*'}. \tag{6.89}$$

Since the intuition of the conformity test is that we obtain the (estimator of the) covariance matrix of the cointegral vector and test for its rank, we are led to define the matrix $\hat{M}$ in this case by

$$\hat{M} = \frac{1}{T}\hat{\Pi}(1)P^{*'}_{-1}P^*_{-1}\hat{\Pi}(1)^{'}. \tag{6.90}$$

It is shown in Appendix I of this chapter that the discussion surrounding Eqs. (6.2) and (6.3) remains valid. Thus, we may now examine the limiting distribution of the relevant (zero) roots in the case $\mu \in \mathcal{R}(\Gamma^{'})$. From the discussion of Chapter 3,

$$\frac{1}{T^{3/2}}e^{'}P_{-1} = \frac{1}{T^{3/2}}\sum_{t=1}^{T}S_{t-1\cdot} \xrightarrow{d} P^{'}_0\int_0^1 B(r)^{'}dr, \quad \frac{e^{'}U}{T} \xrightarrow{d} 0,$$

where $e$ is a $T$-element vector of unities and, moreover,

$$\frac{1}{T}e^{'}U \xrightarrow{a.c} 0. \tag{6.91}$$

It follows, therefore that

$$\sqrt{T}A_{12}, \quad \sqrt{T}A_{22} \xrightarrow{P} 0.$$

Thus, the conclusions of Eqs. (6.4) and (6.5) remain valid. Moreover, the argument in Eqs. (6.54) through (6.63) also remains valid. We need only examine the limiting distributions in Eqs. (6.64) and (6.65). We note that

$$\frac{P_{-1}^{*'} N_x P_{-1}^{*}}{T^2} = \frac{P_{-1}^{*'} P_{-1}^{*}}{T^2} - \left(\frac{P_{-1}^{*'} X^*}{T}\right) \left(\frac{X^{*'} X^*}{T}\right)^{-1} \left(\frac{X^{*'} P_{-1}}{T}\right) \sim \frac{P_{-1}^{*'} P_{-1}^{*}}{T^2} \tag{6.92}$$

Similarly,

$$\frac{P_{-1}^{*'} N_x U}{T} \sim \frac{P_{-1}^{*'} U}{T}, \tag{6.93}$$

whose limits are obtained in Appendix I of this chapter.

We have therefore proved Proposition 6.

**Proposition 6**. Under the conditions of Proposition 1, write the $VAR(n)$ model as

$$\Pi(L) X_{t.}^{'} = \mu^{'} + \epsilon_{t.}^{'}, \quad \mu \in \mathcal{R}(\Gamma^{'}).$$

Let $\hat{\lambda}_i$, $i = r + 1, r + 2, \ldots, r + r_0$ be the ($r_0$) smallest roots of $\hat{M}$ in the metric of $\hat{\Sigma}$. Their limiting distribution is the distribution of the roots of

$$|\rho I_{r_0} - \Upsilon^{'} K^{-1} \Upsilon| = 0, \quad \Upsilon = \Upsilon_0 - \Upsilon_1, \quad K = K_0 - K_1,$$

where $\Upsilon_1 = A_1^{'} A_4$, $K_1 = A_1^{'} A_1$, and

$$K_0 = \int_0^1 B(r)^{'} B(r) \, dr, \quad \Upsilon_0 = \int_0^1 B(r)^{'} dB_1(r), \quad A_1 = \int_0^1 B(r) dr,$$

$$A_2 = \int_0^1 r B(r) dr, \quad A_3 = \int_0^1 B_1(r) dr, \quad A_4 = B_1(1). \tag{6.94}$$

Moreover, the following statements are true.

   i. a test for the rank of cointegration, $r$, i.e. a test of the null that there are $r_0 = q - r$ **zero roots,** may be based on

$$T \sum_{i=r+1}^{q} \hat{\lambda}_i \xrightarrow{d} \mathrm{tr} \Upsilon^{'} K^{-1} \Upsilon,$$

   where $B, B_1$ are SMBM, represented as row vectors with $q$ and $r_0$ elements, respectively, and $K$, $\Upsilon$ are as defined in the statement of the proposition.

ii. a test that the rank of cointegration is $r$ as against the alternative that it is $r+1$, i.e. a test that there are $r_0$ zero roots as against the alternative that there are $r_0 - 1$ zero roots, may be carried out through the statistic $T\hat{\lambda}_r$.

### 6.3.3    Case iii: Constant in VAR, Trends in Levels

This is the case where the $VAR(n)$ is stated with a constant term and the underlying $X$-process is asserted to have a nonzero mean and **linear time trends**. From the ECM representation of the model we have

$$\Delta P = -P_{-1}\Pi(1)' + X\Pi^* + e\mu + U, \quad U = (\epsilon_{t.}). \tag{6.95}$$

It is convenient to write the equation from which we shall obtain the least squares estimator of $\Pi(1)'$, as

$$\Delta P = -P_{-1}^*\Pi(1)' + X\Pi^* + U^*, \quad U^* = U - G(G'G)^{-1}G'P_{-1}\Pi(1)',$$

$$P_{-1}^* = N_g P_{-1}, \tag{6.96}$$

$$G = (e, t^*), \quad t^* = (0, 1, 2, \ldots, T-1)' \quad N_g = I_T - G(G'G)^{-1}G'.$$

We note that $N_g e = 0$, so that $N_g(I_T - ee'/T) = N_g$. The least squares estimator of $\Pi(1)'$ is given by

$$\hat{\Pi}(1)' = \Pi(1)' - (P_{-1}^{*'} N_x P_{-1}^*)^{-1} P_{-1}^{*'} N_x U^*, \tag{6.97}$$

which is, *mutatis mutandis*, the same representation of the unrestricted estimator obtained in Eqs. (6.1) and (6.88), except that now

$$N = N_x = I_T - X^*(X^{*'}X^*)^{-1}X^{*'}, \text{ and } P_{-1}^* = N_g P_{-1}. \tag{6.98}$$

Since the intuition of the conformity test is that we obtain the (estimator of the) covariance matrix of the cointegral vector and test for its rank, we are led to define the matrix $\hat{M}$ in this case by

$$\hat{M} = \frac{1}{T}\hat{\Pi}(1)' P_{-1}' N_g P_{-1}\hat{\Pi}(1). \tag{6.99}$$

It is shown in Appendix I of this chapter that the discussion surrounding Eqs. (6.2) and (6.3) remains valid, so that the limiting distribution of the zero root estimators depends only on

$$A_{22} = \frac{1}{T} \left( \frac{U^{*\prime} N_x P_{-1}^*}{T} \right) \left( \frac{P_{-1}^{*\prime} N_x P_{-1}^*}{T^2} \right)^{-1} \left( \frac{P_{-1}^{*\prime} P_{-1}^*}{T^2} \right)$$

$$\times \left( \frac{P_{-1}^{*\prime} N_x P_{-1}^*}{T^2} \right)^{-1} \left( \frac{P_{-1}^{*\prime} N_x U^*}{T} \right). \tag{6.100}$$

It is also shown therein that

$$\frac{P_{-1}^{*\prime} N_x U^*}{T} \sim \frac{P_{-1}^{\prime} N_g U}{T}, \quad \text{and} \quad \frac{P_{-1}^{*\prime} N_x P_{-1}^*}{T^2} \sim \frac{P_{-1}^{\prime} N_g P_{-1}}{T^2},$$

so that

$$T A_{22} \sim \left( \frac{P_{-1}^{\prime} N_g U}{T} \right)^{\prime} \left( \frac{P_{-1}^{\prime} N_g P_{-1}}{T^2} \right)^{-1} \frac{P_{-1}^{\prime} N_g U}{T}. \tag{6.101}$$

By the same argument as in case ii, we may thus establish that the estimators of zero roots converge in distribution to the roots of

$$|\rho I_{r_0} - \Upsilon^{\prime} K^{-1} \Upsilon| = 0, \quad K = K_0 - K_1, \quad \Upsilon = \Upsilon_0 - \Upsilon_1. \tag{6.102}$$

where now

$$K_0 = \int_0^1 B(r)^{\prime} B(r) \, dr, \quad \Upsilon_0 = \int_0^1 B(r)^{\prime} dB_1(r),$$

$$\Upsilon_1 = 4 A_1^{\prime} A_4 - 6(A_1^{\prime} A_4 - A_1^{\prime} A_3 + A_2^{\prime} A_4) + 12(A_2^{\prime} A_4 - A_2^{\prime} A_3),$$

$$K_1 = 4 A_1^{\prime} A_1 - 6(A_1^{\prime} A_2 + A_2^{\prime} A_1) + 12 A_2^{\prime} A_2. \tag{6.103}$$

We have, therefore, proved Proposition 7.

**Proposition 7.** Under the conditions of Proposition 1, write the $VAR(n)$ model as

$$\Pi(L)X_{t.}^{\prime} = \mu^{\prime} + \epsilon_{t.}^{\prime}, \quad \mu \notin \mathcal{R}(\Gamma^{\prime}).$$

Let $\hat{\lambda}_i$, $i = r + 1, r + 2, \ldots, r + r_0$ be the ($r_0$) smallest roots of $\hat{M}$ in the metric of $\hat{\Sigma}$. Their limiting distribution is the distribution of the roots of

$$|\rho I_{r_0} - \Upsilon^{\prime} K^{-1} \Upsilon| = 0, \quad \Upsilon = \Upsilon_0 - \Upsilon_1, \quad K = K_0 - K_1,$$

where $\Upsilon, \Upsilon_1$ and $K, K_1$ are as defined in Eqs. (6.103) and (6.104). Moreover, the following statements are true:

i. A test for the rank of cointegration, $r$, i.e. the null that there are $r_0 = q - r$ **zero roots**, may be based on

$$T \sum_{i=r+1}^{q} \hat{\lambda}_i \xrightarrow{\mathrm{d}} \mathrm{tr}\Upsilon' K^{-1}\Upsilon,$$

where $B, B_1$ are SMBM, represented as row vectors with $q$ and $r_0$ elements, respectively, and $K$, $\Upsilon$ are as defined in Eqs. (6.106) and (6.107).

ii. a test that the rank of cointegration is $r$ as against the alternative that it is $r + 1$, i.e. a test that there are $r_0$ zero roots as against the alternative that there are $r_0 - 1$ zero roots, may be carried out through the statistic $T\hat{\lambda}_r$.

## 6.4   Extension to MARMA Processes

### 6.4.1   Unrestricted Estimators

The estimation of **stationary** MARMA processes was extensively discussed in Chapter 2 (Section 2.5.2). In this section we examine the case of **cointe-grated** $MARMA(n, k)$. In so doing we utilize the results of Section 2.5.2. We begin with the model

$$\Pi(L)X'_{t.} = A(L)\epsilon'_{t.},  \quad A(L) = \sum_{j=1}^{k} A_j L^j, \quad t = 1, 2, \ldots, T. \tag{6.104}$$

We assume that the characteristic equation $|A(z)| = 0$ has all roots outside the unit circle, that $|\Pi(z)| = 0$ has $r_0$ **unit roots**, and the remaining $(nq - r_0)$ roots lie outside the unit circle. Moreover, the error process is assumed to be a Gaussian $MWN(\Sigma)$, where the term here means i.i.d. Writing the model above in ECM form, we have

$$\Delta X'_{t.} = -\Pi(1)X'_{t-1.} + \sum_{i=1}^{n-1} \Pi_i^* \Delta X'_{t-i.} + \sum_{j=1}^{k} A_j \epsilon'_{t-j.} + \epsilon'_{t.},$$

or, in conformity with the notation in previous discussions,

$$y'_{t.} = -\Pi(1)p'_{t-1.} + \sum_{i=1}^{n-1} \Pi_i^* y'_{t-i.} + \sum_{j=1}^{k} A_j \epsilon'_{t-j.} + \epsilon'_{t.} = JV'_{t.} + \epsilon'_{t.}.$$

$$J = (-\Pi(1), \Pi_1^*, \Pi_2^*, \ldots, \Pi_{n-1}^*, A_1, A_2, \ldots, A_k), \quad p_{t-1\cdot} = X_{t-1\cdot}.$$

$$V_{t\cdot} = (p_{t-1\cdot}, y_{t-1\cdot}, y_{t-2\cdot}, \ldots, y_{t-n+1\cdot}, \epsilon_{t-1\cdot}, \epsilon_{t-2\cdot}, \ldots, \epsilon_{t-k\cdot}), \quad \text{so that}$$

$$\epsilon_{t\cdot}' = y_{t\cdot}' - (V_{t\cdot} \otimes I_q)\gamma, \quad \gamma = \text{vec}(J). \tag{6.105}$$

The log-likelihood function is given by [8]

$$L(\Pi, A, \Sigma) = -\frac{Tq}{2}\ln(2\pi) - \frac{T}{2}\ln|\Sigma| - \frac{1}{2}\sum_{t=1}^{T} \epsilon_{t\cdot} \Sigma^{-1} \epsilon_{t\cdot}'. \tag{6.106}$$

The first-order conditions are

$$\left(\frac{\partial L}{\partial \gamma}\right)' = -\sum_{t=1}^{T} \left(\frac{\partial \epsilon_{t\cdot}'}{\partial \gamma}\right)' \Sigma^{-1} \epsilon_{t\cdot}' = 0. \tag{6.107}$$

As noted in Section 2.5.2 the LF and its derivative are nonlinear in the parameters of the problem, and may be written entirely in terms of observables. We not repeat these arguments here, but instead operate with Eqs. (6.106) and (6.107). To this end, we note that

$$\frac{\partial \epsilon_{t\cdot}'}{\partial \gamma} = -(V_{t\cdot} \otimes I_q) - \sum_{j=1}^{k} A_j \frac{\partial \epsilon_{t-j\cdot}'}{\partial \gamma}, \quad \text{or} \quad A(L)\frac{\partial \epsilon_{t\cdot}'}{\partial \gamma} = -(V_{t\cdot} \otimes I_q). \tag{6.108}$$

The right member of the last equation above contains the **nonstationary** component $p_{t-1\cdot}$, but all other components are **stationary**. Since $A(L)$ is a **stationary operator**, it is convenient to separate the components of the derivative. Noting that $\gamma = (\gamma_{\cdot 1}', \gamma_{\cdot 2}')'$, and $\gamma_{\cdot 1} = \pi(1) = \text{vec}(\Pi(1))$, the derivative of Eq. (6.108) may be written as

$$A(L)\left(\frac{\partial \epsilon_{t\cdot}'}{\partial \gamma}\right) = -\left[(A(L))\left(\frac{\partial \epsilon_{t\cdot}'}{\partial \pi(1)}\right), \ A(L)\left(\frac{\partial \epsilon_{t\cdot}'}{\partial \gamma_2}\right)\right]. \tag{6.109}$$

Employing the usual decomposition of polynomial lag operators we may write

$$A(L) = A(1) - (I - L)\sum_{j=0}^{k-1} A_j^* L^j, \tag{6.110}$$

$$A(L)\left(\frac{\partial \epsilon_{t\cdot}'}{\partial \pi(1)}\right) = -(p_{t-1\cdot} \otimes I_q), \quad \text{or} \quad \frac{\partial \epsilon_{t\cdot}'}{\partial \pi(1)} = -[A(L)]^{-1}(p_{t-1\cdot} \otimes I_q),$$

---

[8] We note that in Eq. (6.105), the vector $V_{t\cdot}$ is $1 \times (n+k)q$ and $\gamma$ is $(n+k)q^2 \times 1$.

$$(I - L)\left(\frac{\partial \epsilon'_{t\cdot}}{\partial \pi(1)}\right) = -[A(L)]^{-1}(\Delta p_{t-1\cdot} \otimes I_q), \quad \text{so that}$$

$$\frac{\partial \epsilon'_{t\cdot}}{\partial \pi(1)} = -[A(1)]^{-1}(p_{t-1\cdot} \otimes I_q)$$

$$+[A(1)]^{-1} \sum_{j=0}^{k-1} A_j^* \left(\frac{\partial \epsilon'_{t-j\cdot}}{\partial \pi(1)} - \frac{\partial \epsilon'_{t-1-j\cdot}}{\partial \pi(1)}\right).$$

The first component of the (right member of the) last equation above is **non-stationary**, while the second is **stationary**; thus the stationary component may be neglected in limiting distribution arguments, in which case we would operate only with [9]

$$\frac{\partial \epsilon'_{t\cdot}}{\partial \pi(1)} = -[A(1)]^{-1}(p_{t-1\cdot} \otimes I_q). \tag{6.111}$$

Since the iterative procedure is based on the expansion

$$\left(\frac{\partial L}{\partial \gamma}(\hat{\gamma})\right)' = \left(\frac{\partial L}{\partial \gamma}(\tilde{\gamma})\right)' + \frac{\partial^2 L}{\partial \gamma \partial \gamma}(\gamma^*)(\hat{\gamma} - \tilde{\gamma}),$$

where $\tilde{\gamma}$ is an initial consistent estimator (ICE), we require an expression for the Hessian of the LF. To this end note that

$$\frac{\partial^2 L}{\partial \gamma \partial \gamma} = -\sum_{t=1}^{T} \left\{ \left(\frac{\partial \epsilon'_{t\cdot}}{\partial \gamma}\right)' \Sigma^{-1} \left(\frac{\partial \epsilon'_{t\cdot}}{\partial \gamma}\right) + (\epsilon_{t\cdot}\Sigma^{-1} \otimes I_{(n+k)q^2}) \frac{\partial}{\partial \gamma} \text{vec}\left[\left(\frac{\partial \epsilon'_{t\cdot}}{\partial \gamma}\right)'\right] \right\}.$$

As in Chapter 2, the last term in the right member above converges in probability to zero and may thus be neglected. Consequently, we shall operate with the expression

$$\frac{\partial^2 L}{\partial \gamma \partial \gamma} = -\sum_{t=1}^{T} \left(\frac{\partial \epsilon'_{t\cdot}}{\partial \gamma}\right)' \Sigma^{-1} \left(\frac{\partial \epsilon'_{t\cdot}}{\partial \gamma}\right) = -\sum_{t=1}^{T} \begin{bmatrix} \Phi_{11t} & \Phi_{12t} \\ \Phi_{21t} & \Phi_{22t} \end{bmatrix},$$

$$\Phi_{11t} = p'_{t-1\cdot}p_{t-1\cdot} \otimes (A(1)\Sigma A(1)')^{-1},$$

---

[9] A more extensive discussion of the negligibility issue is given in Yap and Reinsel (1995), which forms the basis of this discussion. Note, however, that in the computations for obtaining estimators for $\gamma$ we **do not** utilize this simplification. Doing so will force us to use two different procedures in computing the required derivatives, and this entails additional complications in carrying out the iterative estimation process.

$$\Phi_{12t} = (p'_{t-1\cdot} \otimes A(1)'^{-1}\Sigma^{-1})\frac{\partial \epsilon'_{t\cdot}}{\partial \gamma_{\cdot 2}}, \quad \Phi_{21t} = \Phi'_{12t},$$

$$\Phi_{22t} = \left(\frac{\partial \epsilon'_{t\cdot}}{\partial \gamma_{\cdot 2}}\right)' \Sigma^{-1} \left(\frac{\partial \epsilon'_{t\cdot}}{\partial \gamma_{\cdot 2}}\right), \quad \text{in view of the fact that}$$

$$\frac{\partial \epsilon'_{t\cdot}}{\partial \gamma} = \left(\frac{\partial \epsilon'_{t\cdot}}{\partial \pi(1)}, \frac{\partial \epsilon'_{t\cdot}}{\partial \gamma_{\cdot 2}}\right) = \left(-A(1)^{-1}(p_{t-1\cdot} \otimes I_q), \frac{\partial \epsilon'_{t\cdot}}{\partial \gamma_{\cdot 2}}\right). \quad (6.112)$$

The equations defining the iterative precedure are

$$\tilde{\gamma}_{(s)} = \tilde{\gamma}_{(s-1)} - \left[\sum_{t=1}^{T} \left(\frac{\partial \epsilon'_{t\cdot}}{\partial \gamma}\right)' \Sigma^{-1} \left(\frac{\partial \epsilon'_{t\cdot}}{\partial \gamma}\right)\right]^{-1}_{(s-1)} \left[\sum_{t=1}^{T} \left(\frac{\partial \epsilon'_{t\cdot}}{\partial \gamma}\right)' \Sigma^{-1} \epsilon'_{t\cdot}\right]_{(s-1)},$$

$$\tilde{\Sigma}_{(s)} = \frac{1}{T}\tilde{U}'_{(s)}\tilde{U}_{(s)}, \quad \tilde{U}_{(s)} = (\tilde{\epsilon}_{t\cdot,(s)}), \quad \text{and}$$

$$\tilde{\epsilon}'_{t\cdot,(s)} = y'_{t\cdot} + \tilde{\Pi}(1)_{(s-1)}p'_{t-1\cdot} - \sum_{i=1}^{n-1} \tilde{\Pi}^*_{i,(s-1)}y'_{t-i\cdot} - \sum_{j=1}^{k} \tilde{A}_{j,(s-1)}\tilde{\epsilon}'_{t-j\cdot,(s)}, \quad (6.113)$$

where the notation $[\cdot]_{(s)}$ means that the parameters appearing in the square bracket are evaluated at the $s^{th}$ iterate, and the last equation is computed recursively given the initial conditions $\epsilon_{-t\cdot} = 0$, for $t \geq 0$. The order of iteration is as follows: Given an ICE, compute the $\epsilon$'s of the last equation of Eq. (6.113); then $\Sigma$ from the next to the last equation; then compute the derivative, recursively using the same initial conditions as above, from Eq. (6.108). This enables the computation of the first iterate from the first equation in the equation set Eq. (6.113) and so on. The converging iterate, say $\hat{\gamma}$, is the (unrestricted) ML estimator of $\gamma$, and obeys $(\partial L/\partial \gamma)(\hat{\gamma}) = 0$.

**Initial Consistent Estimator**

Obtaining an ICE for this model is more difficult than for the VAR model. However, the following is a feasible procedure: Choose $(n-1)$ instrumental variables (IV) from among the set $\{y_{t-j\cdot} : j > k\}$ and one instrumental variable from the set $\{X_{t-i\cdot} : i \geq 1\}$. Using the specific instrumental set, say $\{X_{t-1\cdot}, y_{t-k-1\cdot}, \ldots, y_{t-k-n+1\cdot}\}$, the resulting IV estimators are consistent for the VAR parameters, $\Pi(1), \Pi_1^*, \ldots, \Pi_{n-1}^*$. Thereafter, we compute

$$\tilde{U} = Y - P_{-1}\tilde{\Pi}(1)' - \sum_{i=1}^{n-1} Y_{-i}\tilde{\Pi}_i^{*'}, \qquad (6.114)$$

which is, evidently, a matrix containing $T$ "observations" on the process $\sum_{j=0}^{k} \epsilon_{t-j}.A_j'$. Consistent estimators for the MA parameters, $A_j$, may be found by the factorization methods given in Tunnicliffe-Wilson (1972), or by regressing $\tilde{U}$ on $\tilde{U}_{-j}$, $j = 1, 2, \ldots, k$, **without a constant term**.

**Limiting Distribution**

We now consider the limiting distribution of the ML estimator. By the mean value theorem,

$$0 = \left(\frac{\partial L}{\partial \gamma}(\hat{\gamma})\right)' = \left(\frac{\partial L}{\partial \gamma}(\gamma^\circ)\right)' + \left(\frac{\partial^2 L}{\partial \gamma \partial \gamma}(\gamma^*)\right)(\hat{\gamma} - \gamma^\circ), \quad |\gamma^* - \gamma^\circ| \leq |\hat{\gamma} - \gamma^\circ|,$$

so that we have the representation

$$\hat{\gamma} - \gamma^\circ = -\left(\frac{\partial^2 L}{\partial \gamma \partial \gamma}(\gamma^*)\right)^{-1}\left(\frac{\partial L}{\partial \gamma}(\gamma^\circ)\right)'.$$

To establish the limiting distribution of the estimator $\hat{\gamma}$, we need to show that the matrix in the right member of the equation above, properly normalized, converges to a limit at least in probability, and that the vector therein, properly normalized, obeys a CLT or a FCLT. The major blocks of the matrix in question are partially displayed in Eq. (6.112), but we need a more explicit expression for the derivative with respect to $\gamma_{.2}$. Define

$$V_{t2.} = (y_{t-1.}, y_{t-2.}, \ldots, y_{t-n+1.}, \epsilon_{t-1.}, \epsilon_{t-2.}, \ldots, \epsilon_{t-k.})$$

and note, from Eq. (6.108) that [10]

$$\frac{\partial \epsilon_{t.}'}{\partial \gamma_{.2}} = -[A(L)]^{-1}(V_{t2.} \otimes I_q) = -V_{t2.}^* \otimes I_q. \qquad (6.115)$$

---

[10] Even though the representation below is useful for theoretical purposes, **it is not** for computational purposes; in fact computing the derivative in question by recursion in Eq. (6.113) will give us precisely what is required.

Completing the representation of the entities in Eq. (6.112), we have

$$\Phi_{12t} = (p'_{t-1.}V^*_{t2.} \otimes A(1)'^{-1}\Sigma^{-1}), \quad \Phi_{21t} = \Phi'_{21t}, \quad \Phi_{22t} = V^{*'}_{t2.}V^*_{t2.} \otimes \Sigma^{-1},$$

$$\tilde{\Phi}_{11} = \sum_{t=1}^{T} \Phi_{11t}, \quad \tilde{\Phi}_{12} = \sum_{t=1}^{T} \Phi_{12t}, \quad \tilde{\Phi}_{22} = \sum_{t=1}^{T} \Phi_{22t}, \quad \text{and}$$

$$\tilde{\Phi} = \begin{bmatrix} \tilde{\Phi}_{11} & \tilde{\Phi}_{12} \\ \tilde{\Phi}_{21} & \tilde{\Phi}_{22} \end{bmatrix} = -\frac{\partial^2 L}{\partial\gamma\partial\gamma}, \tag{6.116}$$

which gives an explicit representation of the Hessian matrix in Eq. (6.112) exclusive of terms which, when properly normalized, converge in probability to zero. Since

$$\left(\frac{\partial L}{\partial\gamma}(\gamma^\circ)\right)' = -\sum_{t=1}^{T} \begin{bmatrix} p'_{t-1.} \otimes A(1)'^{-1} \\ V^{*'}_{t2.} \otimes I_q \end{bmatrix} \Sigma^{-1}\epsilon'_{t.},$$

we may rewrite the mean value expansion above as

$$\begin{bmatrix} \tilde{\Phi}_{11} & \tilde{\Phi}_{12} \\ \tilde{\Phi}_{21} & \tilde{\Phi}_{22} \end{bmatrix} \begin{pmatrix} -(\hat{\pi}(1) - \pi(1)) \\ \hat{\gamma}_{(2)} - \gamma^\circ_{(2)} \end{pmatrix} = -\sum_{t=1}^{T} \begin{bmatrix} p'_{t-1.} \otimes A(1)'^{-1} \\ V^{*'}_{t2.} \otimes I_q \end{bmatrix} \Sigma^{-1}\epsilon'_{t.}.$$

From Eq. (5.40) and the discussion surrounding it in Chapter 5, we recall that if $D = (D_1, D_2)$ is the matrix of the characterisitic vectors of $\Pi(1)$ such that $D_1$ is $q \times r_0$, and corresponds to the **zero** roots (which correspond to the **unit roots** of the system), while $D_2$ corresponds to the nonzero roots (which correspond to the stationary roots of the system), and if $D^{-1} = (D^*_1, D^*_2)'$, $D^{*'}_1$ being $r_0 \times q$ then

$$D^{*'}_1 X'_{t.} \text{ is \textbf{nonstationary}, while } D^{*'}_2 X'_{t.} \text{ is \textbf{stationary}.}$$

Putting

$$H = \begin{bmatrix} D^{-1} \otimes I_q & 0 \\ 0 & I_{(n+k-1)q} \otimes I_q \end{bmatrix}$$

we may thus rewrite the relation above as

$$H\tilde{\Phi}H'H'^{-1}(\hat{\gamma} - \gamma^\circ) = \begin{bmatrix} \tilde{\Phi}^*_{11} & \tilde{\Phi}^*_{12} & \tilde{\Phi}^*_{13} \\ \tilde{\Phi}^*_{21} & \tilde{\Phi}^*_{22} & \tilde{\Phi}^*_{23} \\ \tilde{\Phi}^*_{31} & \tilde{\Phi}^*_{32} & \tilde{\Phi}^*_{33} \end{bmatrix} \begin{bmatrix} -(D'_1 \otimes I_q)(\hat{\pi}(1) - \pi(1)^\circ) \\ -(D'_2 \otimes I_q)(\hat{\pi}(1) - \pi(1)^\circ) \\ (\hat{\gamma}_{(2)} - \gamma^\circ) \end{bmatrix}$$

$$= -\sum_{t=1}^{T} \begin{bmatrix} D^{*'}_1 p'_{t-1.} \otimes A(1)'^{-1} \\ D^{*'}_2 p'_{t-1.} \otimes A(1)'^{-1} \\ V^{*'}_{t2.} \otimes I_q \end{bmatrix} \Sigma^{-1}\epsilon'_{t.}, \tag{6.117}$$

$$(D_1^{*'} \otimes I_q)\tilde{\Phi}_{11}(D_1^* \otimes I_q) \ = \ \tilde{\Phi}_{11}^*, \quad (D_1^{*'} \otimes I_q)\tilde{\Phi}_{11}(D_2^* \otimes I_q) = \tilde{\Phi}_{12}^*,$$

$$(D_2^{*'} \otimes I_q)\tilde{\Phi}_{11}(D_2^* \otimes I_q) \ = \ \tilde{\Phi}_{22}^*, \quad \tilde{\Phi}_{23}^* = (D_2^{*'} \otimes I_q)\tilde{\Phi}_{12}, \quad (D_1^{*'} \otimes I_q)\tilde{\Phi}_{12} = \tilde{\Phi}_{13}^*,$$

and $\tilde{\Phi}_{33}^* = \tilde{\Phi}_{22}$, $\tilde{\Phi}_{21}^* = \tilde{\Phi}_{12}^{*'}$, $\tilde{\Phi}_{31}^* = \tilde{\Phi}_{13}^{*'}$, $\tilde{\Phi}_{32}^* = \tilde{\Phi}_{23}^{*'}$. Define the matrix

$$I_* = \mathrm{diag}\left(\frac{1}{T}I_{r_0}, \frac{1}{\sqrt{T}}I_r, \frac{1}{\sqrt{T}}I_{(n+k-1)q}\right),$$

and note that, by the discussion in Chapter 3,

$$I_*H\tilde{\Phi}^*H'I_* \ = \ \begin{bmatrix} \tilde{\Phi}_{11}^*/T^2 & \tilde{\Phi}_{12}^*/T^{3/2} & \tilde{\Phi}_{13}^*/T^{3/2} \\ \tilde{\Phi}_{21}^*/T^{3/2} & \tilde{\Phi}_{22}^*/T & \tilde{\Phi}_{23}^*/T \\ \tilde{\Phi}_{31}^*/T^{3/2} & \tilde{\Phi}_{32}^*/T & \tilde{\Phi}_{33}^*/T \end{bmatrix},$$

$$\frac{1}{T^{3/2}}\tilde{\Phi}_{12}^* \ = \ \frac{1}{T^{3/2}}D_1^{*'}P_{-1}'P_{-1}D_2^* \otimes [A(1)\Sigma A(1)']^{-1} \xrightarrow{\mathrm{d}} 0,$$

$$\frac{1}{T^{3/2}}\tilde{\Phi}_{13}^* \ = \ \frac{1}{T^{3/2}}D_1^{*'}P_{-1}'V_2^* \otimes [\Sigma A(1)']^{-1} \xrightarrow{\mathrm{d}} 0,$$

$$\frac{1}{T^{3/2}}\tilde{\Phi}_{j1}^* \xrightarrow{\mathrm{d}} 0, \quad j = 1, 2, \quad \text{so that} \tag{6.118}$$

$$I_*H\tilde{\Phi}H'I_* \ \sim \ \begin{bmatrix} \tilde{\Phi}_{11}^*/T^2 & 0 & 0 \\ 0 & \tilde{\Phi}_{22}^*/T & \tilde{\Phi}_{23}^*/T \\ 0 & \tilde{\Phi}_{32}^*/T & \tilde{\Phi}_{33}^*/T \end{bmatrix} = \begin{bmatrix} \tilde{\Phi}_{11}^*/T^2 & 0 \\ 0 & \tilde{\Phi}_{(2)}^*/T \end{bmatrix}.$$

In the results above we have made use of the fact that $P_{-1}D_2^*$ is stationary, and have defined $P_{-1} = (p_{t-1\cdot})$, $V_2^* = (V_{t2\cdot}^*)$. It follows, therefore, that

$$T(D_1' \otimes I_q)(\hat{\pi}(1) - \pi(1)^\circ) \sim \left(\frac{1}{T^2}\tilde{\Phi}_{11}^*\right)^{-1}\frac{1}{T}\sum_{t=1}^{T}\left(D_1^{*'}p_{t-1\cdot}' \otimes A(1)'^{-1}\Sigma^{-1}\epsilon_{t\cdot}'\right)$$

$$= \ \left(\frac{1}{T^2}(D_1^{*'}P_{-1}'P_{-1}D_1^* \otimes [A(1)\Sigma A(1)']^{-1}\right)^{-1}\frac{1}{T}\sum_{t=1}^{T}\left(D_1^{*'}p_{t-1\cdot}' \otimes A(1)'^{-1}\Sigma^{-1}\epsilon_{t\cdot}'\right)$$

$$= \ \left[\left(\frac{D_1^{*'}P_{-1}'P_{-1}D_1^*}{T^2}\right)^{-1} \otimes I_q\right]\frac{1}{T}\sum_{t=1}^{T}\left(D_1^{*'}p_{t-1\cdot}' \otimes A(1)\epsilon_{t\cdot}'\right)$$

$$= \ \mathrm{vec}\left[\left(\frac{A(1)U'P_{-1}D_1^*}{T}\right)\left(\frac{D_1^{*'}P_{-1}'P_{-1}D_1^*}{T^2}\right)^{-1}\right], \tag{6.119}$$

or equivalently,

$$T(\hat{\Pi}(1) - \Pi(1)^\circ)D_1 \sim \left(\frac{A(1)U'P_{-1}D_1^*}{T}\right) \left(\frac{D_1^{*'}P_{-1}'P_{-1}D_1^*}{T^2}\right)^{-1}$$

$$\xrightarrow{\text{d}} A(1)T_1' \left(\int_0^1 dB(s)'B(s)\right) P_0 D_1^* \left[D_1^{*'}P_0' \left(\int_0^1 B(s)'B(s)ds\right) P_0 D_1^*\right]^{-1}$$

$$= A(1)T_1' \int_0^1 dB(s)'B_2(s) \left(\int_0^1 B_2(s)'B_2(s)ds\right)^{-1} T_2^{-1},$$

$$\Sigma = T_1'T_1, \quad B(s)P_0D_1^* = B_2(s)T_2, \quad T_2'T_2 = D_1^{*'}P_0'P_0D_1^*, \tag{6.120}$$

$P_0$ is the matrix of the triangular decomposition of the long-run covariance matrix of the process $X_{t \cdot}$, and $B(s)$, $B_2(s)$ are SMBM with $q$ and $r_0$ elements, respectively.

**Remark 9.** It might appear that the limiting distribution of the nonstationary component above differs from that given in Eq. (5.56) of Chapter 5 in the case of the $VAR(n)$ model. The difference, however, is more apparent than real. In Chapter 5, it was far more convenient to determine the limiting distribution of $TD_1'(\hat{\Pi}(1) - \Pi(1)^\circ)'$; when allowance is made for this, the limiting distribution above is **precisely the same**, *mutatis mutandis*, as that given in Proposition 2 of Chapter 5; for example, $B_2(s)$ above corresponds to $B_*(s)$ in Chapter 5; $B(s)$ above corresponds to $B_4(s)$ in Chapter 5; $T_1A(1)$ above, corresponds to $T_4$ in Chapter 5, which is the triangular decomposition of $\Sigma$, as is $T_1$; $T_2$ above corresponds to $T^*$ in Chapter 5 and is the decomposition of $D_1^{*'}C_*D_1^*$, where $C_*$ is the long run covariance matrix of the process $X_{t \cdot}$. The latter is computed, in the MARMA case as well, in accordance with Eq. (5.46), except that the matrices $H_j^*$ will have to reflect the MA parameter matrices $A_j$. Evidently, the entity $A(1)$ would be $I_q$ if we restrict ourselves to the $VAR(n)$ model. Thus, the limiting distribution of the conformity test statistic for the presence and rank of cointegration in $MARMA(n, m)$ models is *mutatis mutandis* of the same type as in the case of the $VAR(n)$ model. Moreover, the development in this section contains the (relevant) results of Chapter 5, as the special case resulting when we put $k = 0$.

We now examine the limiting distribution of the stationary components of the estimator. To this end, put

$$
\hat{\gamma}^* - \gamma^{*\circ} = \begin{bmatrix} -(D_2' \otimes I_q)(\hat{\pi}(1) - \pi(1)^\circ) \\ \hat{\gamma}_{\cdot 2} - \gamma_{\cdot 2}^\circ \end{bmatrix} \sim \tilde{\Phi}_{(2)}^{*-1} \sum_{t=1}^{T} \xi_t \epsilon_t',
$$

$$
\frac{1}{T} \tilde{\Phi}_{(2)}^* = \frac{1}{T} \begin{bmatrix} D_2^{*\prime} P_{-1}' P_{-1} D_2^* \otimes [A(1)\Sigma A(1)']^{-1} & D_2^{*\prime} P_{-1}' V_2^* \otimes A(1)'^{-1}\Sigma^{-1} \\ V_2^{*\prime} P_{-1} D_2^* \otimes \Sigma^{-1} A(1)^{-1} & V_2^{*\prime} V_2^* \otimes \Sigma^{-1} \end{bmatrix}
$$

$$
\overset{\text{a.c.}}{\to} \begin{bmatrix} M_{zz} \otimes [A(1)\Sigma A(1)']^{-1} & M_{zv} \otimes A(1)'^{-1}\Sigma^{-1} \\ M_{vz} \otimes \Sigma^{-1} A(1)^{-1} & M_{vv} \otimes \Sigma^{-1} \end{bmatrix} = \Phi,
$$

$$
\xi_t = \begin{bmatrix} D_2^{*\prime} p_{t-1\cdot}' \otimes A(1)'^{-1}\Sigma^{-1} \\ V_{t2\cdot}^{*\prime} \otimes \Sigma^{-1} \end{bmatrix}, \quad Z = P_{-1} D_2^*. \tag{6.121}
$$

The dimension of $\Phi$ is $(n+k-1)q^2 + rq$, and $\xi_t$ is $[(n+k-1)q^2 + rq] \times q$. Next, define the stochastic basis

$$
\mathcal{G} = \{\mathcal{G}_t : \mathcal{G}_t = \sigma(\epsilon_{s\cdot}, s \le t, t \in \mathcal{N}_0, \text{ with } \mathcal{G}_0 = (\emptyset, \Omega)\}, \tag{6.122}
$$

where $\Omega$ is the sample space of the underlying probability space $(\Omega, \mathcal{A}, \mathcal{P})$, and note that the sequence $\{\xi_t \epsilon_t' : t \in \mathcal{N}_0\}$ is a **martingale difference**. This is so since $\xi_t \epsilon_t'$ is $\mathcal{G}_t$-measurable, $E|\xi_t \epsilon_t'|^2 \le E \parallel \xi_t \parallel^2 E|\epsilon_t'|^2 < \infty$ $E(\xi_t \epsilon_t' | \mathcal{G}_{t-1}) = 0$. Moreover, since we confine ourselves to stationary roots, the sequence above obeys a **Lindeberg condition**. Thus, by Proposition 21, Chapter 5, in Dhrymes (1989), p. 337, $\zeta_{(T)}$ obeys a central limit theorem; more precisely Proposition 21 implies that for $\zeta_{(T)} = (1/\sqrt{T}) \sum_{t=1}^{T} \xi_t \epsilon_t'$,

$$
\zeta_{(T)} \overset{\text{d}}{\to} \zeta \sim N(0, \Psi), \quad \Psi = \plim_{T \to \infty} \frac{1}{T} E(\xi_t \epsilon_t' \cdot \epsilon_t \cdot \xi_t' | \mathcal{G}_{t-1}). \tag{6.123}
$$

It may be verified directly that $\Psi$ as defined above obeys $\Psi = \Phi$, the latter as defined in Eq. (6.121). Thus, we conclude

$$
\sqrt{T}(\hat{\gamma}^* - \gamma^{*\circ}) \overset{\text{d}}{\to} N(0, \Psi^{-1}). \tag{6.124}
$$

We have therefore proved

**Proposition 8.** Let $\{X_{t\cdot} : t \in \mathcal{N}_+\}$ be a cointegrated, $MARMA(n, k)$, stochastic process defined on the probability space $(\Omega, \mathcal{A}, \mathcal{P})$. More precisely

$$
\Pi(L)X_{t\cdot}' = A(L)\epsilon_{t\cdot}', \quad \Pi(L) = I_q + \sum_{i=1}^{n} \Pi_i X_{t-i\cdot}', \quad A(L) = I_q + \sum_{j=1}^{k} A_j \epsilon_{t-j\cdot}',
$$

such that $|\Pi(z)| = 0$ has $r_0$ **zero** roots and $r = q - r_0$ **stationary roots**, $|A(z)| = 0$ has all stationary roots, and $\epsilon_t.$ is a Gaussian $MWN(\Sigma)$, i.e. an i.i.d. normal sequence. Put the model in the ECM form

$$y'_{t.} = -\Pi(1)p'_{t-1.} + \sum_{i=1}^{n-1} \Pi_i^* y'_{t-i.} + \sum_{j=1}^{k} A_j \epsilon'_{t-j.} + \epsilon'_{t.}, \quad y'_{t.} = \Delta X'_{t.}, \quad p'_{t-1.} = X'_{t-1.},$$

where, as before, $\Pi_i^* = \sum_{j=1}^{n-i} \Pi_{i+j}$ and $\Pi(1) = \sum_{i=0}^{n} \Pi_i$. Note further that $\Pi(1)$ has $r_0$ **zero roots**, and $r = q - r_0$ roots that are **less than one in absolute value**. Let $D$ be the matrix of its characteristic roots, and partition

$$D = (D_1, D_2), \quad D^{-1} = (D_1^*, D_2^*)'$$

such that $D_1$ is $q \times r_0$, $D_1^{*'}$ is $r_0 \times q$, implying that $D_1^{*'} X'_{t.}$ is **nonstationary** and $D_2^{*'} X'_{t.}$ is **stationary**. Let

$$\gamma^* = \text{vec}(J^*), \quad J^* = (-\Pi(1)D_2, \Pi_1^*, \Pi_2^*, \ldots, \Pi_{n-1}^*)$$

$$\gamma_{\cdot 1} = -\text{vec}[\Pi(1)D_1], \quad \gamma = (\gamma'_{\cdot 1}, \gamma^{*'})'.$$

The following statements are true:

i. an ICE for the parameters of this model exists, as developed in the discussion surrounding Eq. (6.114);

ii. the ML estimator of the parameters $\gamma$ and $\Sigma$ is obtained as the converging iterate of the iteration procedure, as given in Eq. (6.113);

iii. the limiting distribution of the ML estimator of $\gamma$ has two components, one corresponding to $-T(D'_1 \otimes I_q)(\hat{\pi}(1) - \pi(1)^\circ)$, or equivalently, $-T(\hat{\Pi}(1) - \Pi(1)^\circ)D_1$, and another corresponding to $\sqrt{T}(\hat{\gamma}^* - \gamma^{*\circ})$, as indicated in Eqs. (6.120) and (6.121); moreover,

iv. the first (nonstationary) component obeys

$$T(\hat{\Pi}(1) - \Pi(1)^\circ)D_1 \overset{\text{d}}{\to} A(1)T'_1 \left( \int_0^1 dB(s)' B_2(s) \right) K^{-1} T_2^{-1},$$

$$\Upsilon' = K = \int_0^1 B_2(s)' B_2(s) ds,$$

where $T_1$, $T_2$ are as defined in the discussion surrounding Eq. (6.120), and $B$, $B_2$ are SMBM, of $m$ and $r_0$ elements, respectively, also as defined therein;

v. the second (stationary) component obeys

$$\sqrt{T}(\hat{\gamma}^* - \gamma^{*\circ}) \xrightarrow{\text{d}} N(0, \Psi^{-1}),$$

$$\Psi = \begin{bmatrix} M_{zz} \otimes [A(1)\Sigma A(1)']^{-1} & M_{zv} \otimes A(1)^{-1'}\Sigma^{-1} \\ M_{vz} \otimes \Sigma^{-1}A(1)^{-1} & M_{vv} \otimes \Sigma^{-1} \end{bmatrix}$$

as indicated in Eq. (6.124) and the discussion surrounding it.

vi. The two components are asymptotically mutually independent or more precisely "uncorrelated" in the sense that the covariance matrix of their (joint) limiting distribution is block diagonal.

**Remark 10.** Note that in the special case $k = 0$, the $MARMA(n, k)$ model reduces to the $VAR(n)$ and the covariance matrix of the limiting distribution reduces to

$$\Psi^{-1} = M^{*-1} \otimes \Sigma, \quad M^* = \begin{bmatrix} M_{zz} & M_{zx} \\ M_{xz} & M_{xx} \end{bmatrix},$$

because the vector $v_{t2}^*$ reduces to the vector $x_{t\cdot}$ as defined in Eq. (5.83) in Chapter 5. However, the covariance matrix of the limiting distribution in Chapter 5, as given in Eq. (5.56), is $\Sigma \otimes M^{*-1}$ and thus there appears to be a discrepancy. This **is not the case**, however, since in Chapter 5 we were working with a vector $\gamma^{**} = \text{vec}(J^{*'})$, while in the current discussion we are working with the vector $\gamma^* = \text{vec}(J^*)$, where $J^*$ is as defined in the statement of Proposition 8. Thus, even though the two vectors contain **the same elements**, they appear **in different order**. To verify that the two limiting distribution are indeed identical we note that if $F'$ is $m \times n$, by Lemma 9, in Dhrymes (1994) p. 215, there exists a permutation matrix $I_{(m,n)}$ such that $\text{vec}(F') = I_{(m,n)}\text{vec}(F)$. In this discussion $m = q$, $n = \nu = (n + k - 1)q + r$, $F = J^*$, and $I_{(q,\nu)} = (I_q \otimes e_{.1}, I_q \otimes e_{.2}, \ldots, I_q \otimes e_{.\nu})$, where $e_{.j}$ is a $\nu$-element **column** vector all of whose elements are zero, except the $j^{th}$, which is unity. Using the notation $\text{Cov}(\gamma^*)$ to denote the covariance of the limiting distribution in part v of Proposition 8 we have, as a consequence,

$$\text{Cov}(\gamma^{**}) = (I_q \otimes e_{.1}, \ldots, I_q \otimes e_{.\nu})\text{Cov}(\gamma^*)(I_q \otimes e_{.1}, \ldots, I_q \otimes e_{.\nu})'$$

$$= (I_q \otimes e_{.1}, \ldots, I_q \otimes e_{.\nu})(M^{*-1} \otimes \Sigma)(I_q \otimes e_{.1}, \ldots, I_q \otimes e_{.\nu})'$$

$$= \sum_{i=1}^{\nu} \sum_{j=1}^{\nu} \left( M^{*-1} \otimes e_{.i}\sigma_{ij}e_{.j}' \right) = \Sigma \otimes M^{*-1}.$$

The last equality is valid since $M^{*-1} \otimes e_{.i}\sigma_{ij}e_{.j}'$ is a **block matrix**, all of whose blocks are zero, save the $(i,j)^{th}$ which is $\sigma_{ij}M^{*-1}$. Thus, the two limiting distributions of Proposition 2, of Chapter 5, and Proposition 9, of this chapter, are **identical**.

We have therefore proved Corollary 5.

**Corollary 5.** The results of Proposition 8 contain, as a special case for $k = 0$, the limiting distribution results for the unrestricted estimators of the $VAR(n)$ model, established in Proposition 2 of Chapter 5.

## 6.4.2 Restricted Estimators

In the preceding section we did not made use of the cointegration hypothesis in the estimation phase, but we certainly did make use of it in deriving the limiting distribution of the resulting estimators. In this section we make use of it both at the estimation phase and in the derivation arguments. This requires additional notation. To this end write the ECM form as

$$y_{t.}' = -\Gamma B' p_{t-1.}' + \sum_{i=1}^{n-1} \Pi_i^* y_{t-i.}' + \sum_{j=1}^{k} A_j \epsilon_{t-j.}' + \epsilon_{t.}', \quad \Pi(1) = \Gamma' B, \quad B' = (I_r, \ B_0')' \tag{6.125}$$

where $\Gamma$, $B$ are (each) $q \times r$ of rank $r$, and $B_0$ is $r \times r_0$. The LF is

$$L(\Pi, A, \Sigma) = -\frac{Tq}{2}\ln(2\pi) - \frac{T}{2}\ln|\Sigma| - \frac{1}{2}\sum_{t=1}^{T} \epsilon_{t.}\Sigma^{-1}\epsilon_{t.}'.$$

Put

$$V_{t.} = (p_{t-1.}B, y_{t-1.}, y_{t-2.}, \ldots, y_{t-n+1.}, \epsilon_{t-1.}, \epsilon_{t-2.}, \ldots, \epsilon_{t-k.}),$$

$$J_* = (-\Gamma, \Pi_1^*, \ldots, \Pi_{n-1}^*, A_1, \ldots, A_k), \quad \gamma_* = \text{vec}(J_*), \tag{6.126}$$

so that all elements of $V_{t\cdot}$ are **stationary**, and write the ECM form of the model as

$$y_{t\cdot}' = (V_{t\cdot} \otimes I_q)\gamma_* + \epsilon_{t\cdot}', \quad \text{or} \quad \epsilon_{t\cdot}' = y_{t\cdot}' - (V_{t\cdot} \otimes I_q)\gamma_*. \tag{6.127}$$

As in the previous section, the LF and its derivatives are nonlinear functions; we also obtain the analog of Eq. (6.108)

$$\frac{\partial L}{\partial \gamma_*} = -\sum_{t=1}^{T} \epsilon_{t\cdot} \Sigma^{-1} \frac{\partial \epsilon_{t\cdot}'}{\partial \gamma_*}, \tag{6.128}$$

$$\frac{\partial \epsilon_{t\cdot}'}{\partial \gamma_*} = -(V_{t\cdot} \otimes I_q) - \sum_{j=1}^{k} A_j \frac{\partial \epsilon_{t\cdot}'}{\partial \gamma_*}, \quad \text{or} \quad \frac{\partial \epsilon_{t\cdot}'}{\partial \gamma_*} = -(V_{t\cdot}^* \otimes I_q), \quad V_{t\cdot}^* = [A(L)]^{-1} V_{t\cdot}.$$

Next, we need an expression for the derivative with respect to the unknown elements in the matrix $B$. To this end, note that in obtaining the derivative in Eq. (6.128), we have made use of the representation $\mathrm{vec}(\Gamma B' p_{t-1\cdot}') = (p_{t-1\cdot} B \otimes I_q)\mathrm{vec}(\Gamma)$. However, we also have the alternative representation

$$\mathrm{vec}(\Gamma B' p_{t-1\cdot}') = \mathrm{vec}[\Gamma p_{t-1\cdot}^{(1)'} + \Gamma B_0 p_{t-1\cdot}^{(2)'}] = \Gamma p_{t-1\cdot}^{(1)'} + (p_{t-1\cdot}^{(2)} \otimes \Gamma)b_0, \quad b_0 = \mathrm{vec}(B_0),$$

where $p_{t-1\cdot}^{(1)}$ contains the **first** $r = q - r_0$ elements of $p_{t-1\cdot}$, and $p_{t-1\cdot}^{(2)}$ contains the last $r_0$ elements. By the arguments given in the preceding discussion we obtain

$$\frac{\partial \epsilon_{t\cdot}'}{\partial b_0} = -(p_{t-1\cdot}^{(2)} \otimes \Gamma) - \sum_{j=1}^{k} A_j \frac{\partial \epsilon_{t-j\cdot}'}{\partial b_0} \quad \text{or} \tag{6.129}$$

$$\frac{\partial \epsilon_{t\cdot}'}{\partial b_0} = -[A(L)]^{-1}(p_{t-1\cdot}^{(2)} \otimes \Gamma) \sim -(p_{t-1\cdot}^{(2)} \otimes [A(1)]^{-1}\Gamma). \tag{6.130}$$

Finally, ignoring terms that converge to zero in probability when properly normalized, we find

$$\frac{\partial^2 L}{\partial \delta \partial \delta} = -\sum_{t=1}^{T} \begin{bmatrix} p_{t-1\cdot}^{(2)'} p_{t-1\cdot}^{(2)} \otimes \Gamma'[A(1)\Sigma A(1)']^{-1}\Gamma & p_{t-1\cdot}^{(2)'} V_{t\cdot}^* \otimes \Gamma'[\Sigma A(1)']^{-1} \\ V_{t\cdot}^{*'} p_{t-1\cdot}^{(2)} \otimes [\Sigma A(1)']^{-1}\Gamma & V_{t\cdot}^{*'} V_{t\cdot}^* \otimes \Sigma^{-1} \end{bmatrix}$$

$$= -\begin{bmatrix} P_{-1}^{(2)'} P_{-1}^{(2)} \otimes \Gamma'[A(1)\Sigma A(1)']^{-1}\Gamma & P_{-1}^{(2)'} V^* \otimes \Gamma'[\Sigma A(1)']^{-1} \\ V^{*'} P_{-1}^{(2)} \otimes [\Sigma A(1)']^{-1}\Gamma & V^{*'} V^* \otimes \Sigma^{-1} \end{bmatrix},$$

where $\delta = (b'_0, \gamma'_*)'$. The iteration procedure is given by

$$
\tilde{\delta}_{(s)} = \tilde{\delta}_{(s-1)} - \left[ \sum_{t=1}^{T} \left( \frac{\partial \epsilon'_{t \cdot}}{\partial \delta} \right)' \Sigma^{-1} \left( \frac{\partial \epsilon'_{t \cdot}}{\partial \delta} \right) \right]^{-1}_{(s-1)} \left[ \sum_{t=1}^{T} \left( \frac{\partial \epsilon'_{t \cdot}}{\partial \delta} \right)' \Sigma^{-1} \epsilon'_{t \cdot} \right]_{(s-1)},
$$

$$
\tilde{\Sigma}_{(s)} = \frac{1}{T} \tilde{U}'_{(s)} \tilde{U}_{(s)}, \quad \tilde{U}_{(s)} = (\tilde{\epsilon}_{t \cdot, (s)}), \quad \text{and} \tag{6.131}
$$

$$
\tilde{\epsilon}'_{t \cdot, (s)} = y'_{t \cdot} + \tilde{\Gamma}_{(s-1)} \tilde{B}'_{(s-1)} p'_{t-1 \cdot} - \sum_{i=1}^{n-1} \tilde{\Pi}^*_{i,(s-1)} y'_{t-i \cdot} - \sum_{j=1}^{k} \tilde{A}_{j,(s-1)} \tilde{\epsilon}'_{t-j \cdot, (s)},
$$

where the last equation is computed recursively given the initial conditions $\epsilon_{-t \cdot} = 0$, for $t \geq 0$. The order of iteration is as follows: Given an ICE, compute the $\epsilon$'s of the last equation of Eq. (6.131); then $\Sigma$ from the next to the last equation; then compute the derivative, recursively using the same initial conditions as above, from Eqs. (6.129) and (6.128). This enables the computation of the first iterate from the first equation in the equation set Eq. (6.131) and so on. The converging iterate, say $\hat{\delta}$, is the (restricted) ML estimator of $\delta$, and obeys $(\partial L / \partial \delta)(\hat{\delta}) = 0$.

An ICE may be obtained as in the preceding section, or through the **unrestricted** estimator obtained therein. The ICE for $\Gamma$ and $B_0$ may be obtained from the ICE of $\Pi(1)$, say $\tilde{\Pi}(1)$, by noting that $\Pi(1) = (\Gamma, \Gamma B_0)$; thus if $\tilde{\Pi}(1)_{(r)}$ denotes the first $r$ columns of $\tilde{\Pi}(1)$, **it is a consistent** estimator of $\Gamma$ and moreover

$$
\tilde{B}_0 = [\tilde{\Pi}(1)'_{(r)} \tilde{\Pi}(1)_{(r)}]^{-1} \tilde{\Pi}(1)'_{(r)} \tilde{\Pi}(1)_{(r_0)}
$$

is an ICE for $B_0$.

**Limiting Distribution**

Putting

$$
G_{11} = \Gamma'[A(1)\Sigma A(1)']^{-1}\Gamma, \quad G_{12} = \Gamma'[\Sigma A(1)']^{-1}, \quad G_{21} = G'_{12}, \quad G_{22} = \Sigma^{-1},
$$

we note that, as before, the limiting distribution is obtained from the relation

$$
I_*^{-1}(\hat{\delta} - \delta^\circ) = - \left[ I_* \left( \frac{\partial^2 L}{\partial \delta \partial \delta} \right) I_* \right]^{-1} I_* \left( \frac{\partial L}{\partial \delta} (\delta^\circ) \right)',
$$

$$I_* = \operatorname{diag}\left(\frac{1}{T}(I_{r_0} \otimes I_q),\ \frac{1}{\sqrt{T}}(I_{(n+k)q+r} \otimes I_q)\right), \qquad (6.132)$$

$$I_*\left(\frac{\partial^2 L}{\partial\delta\partial\delta}\right)I_* = -\begin{bmatrix} P_{-1}^{(2)'}P_{-1}^{(2)}/T^2 \otimes G_{11} & P_{-1}^{(2)'}V^*/T^{3/2} \otimes G_{12} \\ V^{*'}P_{-1}^{(2)}/T^{3/2} \otimes G_{21} & V^{*'}V^*/T \otimes \Sigma^{-1} \end{bmatrix}$$

$$= \begin{bmatrix} \tilde{\Phi}_{11}^* & \tilde{\Phi}_{12}^* \\ \tilde{\Phi}_{21}^* & \tilde{\Phi}_{22}^* \end{bmatrix} \xrightarrow{\mathrm{d}} \begin{bmatrix} P_{02}'K P_{02} \otimes G_{11} & 0 \\ 0 & M_{vv} \otimes \Sigma^{-1} \end{bmatrix},$$

where $K = \int_0^1 B_2(s)'B_2(s)ds$, $B_2(s)$ is an $r_0$-element SMBM, and $P_{02}$ is the matrix of the triangular decomposition of the long-run covariance matrix of $X_{t-1\cdot}^{(2)}$, i.e. the long-run covariance matrix of the last $r_0$ (nonstationary) elements of the underlying process. Thus, as in the discussion of the previous section, the estimator of the cointegrating vectors (i.e. the elements of the matrix $B_0$) is uncorrelated with the estimator of the coefficients of the stationary variables of the problem, i.e. the elements of the matrix $J_*$.

Since $\partial L/\partial\delta = (\partial L/\partial b_0,\ \partial L/\partial\gamma_*)$, we obtain

$$I_*\left(\frac{\partial L}{\partial\delta}(\delta^\circ)\right)' = \sum_{t=1}^T \begin{bmatrix} (1/T)(p_{t-1\cdot}^{(2)'} \otimes \Gamma'A(1)'^{-1}\Sigma^{-1}\epsilon_{t\cdot}') \\ (1/\sqrt{T})(V_{t\cdot}^{*'} \otimes \Sigma^{-1}\epsilon_{t\cdot}') \end{bmatrix}. \qquad (6.133)$$

The first element of the right member above obeys

$$\frac{1}{T}\operatorname{vec}(G_{12}U'P_{-1}^{(2)}) \xrightarrow{\mathrm{d}} \operatorname{vec}\left[G_{12}T_1'\left(\int_0^1 dB(s)'B_2(s)\right)P_{02}\right], \qquad (6.134)$$

where $B(s)$ is a $q$-element SMBM, and $B_2(s)$, $P_{02}$ are as in Eq. (6.132). To complete the derivation of the limiting distribution, we note that we have established

$$T(\hat{b}_0 - b_0^\circ) \sim \frac{1}{T}\sum_{t=1}^T [(P_{02}'K P_{02})^{-1} \otimes G_{11}^{-1}](p_{t-1\cdot}^{(2)'} \otimes \Gamma'A(1)'^{-1}\Sigma^{-1}\epsilon_{t\cdot}')$$

$$= [(P_{02}'K P_{02})^{-1} \otimes I_q]\left(\frac{1}{T}\operatorname{vec}(G_{11}^{-1}\Gamma'A(1)'^{-1}\Sigma^{-1}U'P_{-1}^{(2)})\right)$$

$$= \frac{1}{T}\operatorname{vec}[(G_{11}^{-1}\Gamma'A(1)'^{-1}\Sigma^{-1}U'P_{-1}^{(2)}(P_{02}'K P_{02})^{-1})]$$

$$\xrightarrow{\mathrm{d}} \operatorname{vec}[G_{11}^{-1}\Gamma'A(1)'^{-1}\Sigma^{-1}T_1'\left(\int_0^1 dB(s)'B_2(s)\right)P_{02}(P_{02}'K P_{02})^{-1}]$$

$$= \operatorname{vec}[(G_{11}^{-1}\Gamma'A(1)^{'-1}T_1^{-1}\left(\int_0^1 dB(s)'B_2(s)\right)K^{-1}P_{02}^{'-1}], \quad (6.135)$$

or, equivalently,

$$T(\hat{B}_0 - B_0^\circ) \xrightarrow{d} G_{11}^{-1}\Gamma'A(1)^{'-1}T_1^{-1}\left(\int_0^1 dB(s)'B_2(s)\right)K^{-1}P_{02}^{'-1}. \quad (6.136)$$

We note that $G_{11}^* = \Gamma'A(1)^{'-1}T_1^{-1}$ obeys $G_{11}^*G_{11}^{*'} = G_{11}^{-1}$. We may exploit this fact in order to simplify the representation in Eq. (6.136). To this end let $T_3$ be the $(r \times r)$ triangular matrix of the decomposition of $G_{11}^{-1}$ and define the $r$-element SMBM

$$B_3(s) = B(s)T_1^{'-1}A(1)^{-1}T_3^{-1}. \quad (6.137)$$

In this notation we may write the results in Eq. (6.136) as

$$T[(\hat{B}_0 - B_0^\circ)(P_{02}'KP_{02})^{1/2}] \xrightarrow{d} T_3'\left(\int_0^1 dB_3(s)'B_2(s)\right)P_{02}(P_{02}'KP_{02})^{-(1/2)}. \quad (6.138)$$

We note that $X_{t-1}^{(2)}$. behaves essentially like $D_1^{*'}X_{t-1}'$. The two BM, $B_3(s)T_3$ and $B_2(s)P_{02}(P_{02}'KP_{02})^{-(1/2)}$, may be shown to be mutually independent if traced back to their origins, as was done in the discussion surrounding Propositions 4 and 5 in Chapter 5. Let $\phi_{\cdot i}$ be the $i^{th}$ column of the right member of Eq. (6.138), $h_{\cdot i}$ the $i^{th}$ column of $P_{02}(P_{02}'KP_{02})^{-(1/2)}$, and note that

$$\phi_{\cdot i} = T_3'\int_0^1 dB_3(s)'h_i(s), \quad h_i(s) = B_2(s)h_{\cdot i}, \quad E(\phi_{\cdot i}|B_2) = 0,$$

$$\operatorname{Cov}(\phi_{\cdot i}, \phi_{\cdot j}|B_2) = T_3'\left(\int_0^1\int_0^1 E[dB_3(s)'dB_3(r)|B_2]h_i(s)h_j(r)\right)T_3$$

$$= T_3'\left(\int_0^1 I_{r_0}h_i(s)h_j(s)ds\right)T_3$$

$$= T_3'\left(\int_0^1 h_{\cdot i}'[B_2(s)'B_2(s)]h_{\cdot j}ds\right)T_3 = (h_{\cdot i}'Kh_{\cdot j})T_3'T_3.$$

Noting that $h_{\cdot i}'Kh_{\cdot j} = 1$ if $i = j$, and **zero** otherwise, we conclude that the conditional distribution of $\phi_{\cdot i}$, **does not depend** on $B_2$, and is $N(0, G_{11}^{-1})$. Thus, the same is true of the unconditional distribution, and we conclude

$$T[(P_{02}'KP_{02})^{1/2} \otimes I_r)(\hat{b}_0 - b_0^\circ)] \xrightarrow{d} \zeta_1 \sim N(0, I_{r_0} \otimes G_{11}^{-1}). \quad (6.139)$$

The second element of the right member, as in the discussion of unrestricted estimators, is a martingale difference (MD) obeying a Lindeberg condition; thus,

$$\frac{1}{\sqrt{T}}\sum_{t=1}^{T}(V_{t\cdot}^{*\prime}\otimes\Sigma^{-1}\epsilon_{t\cdot}^{\prime})\xrightarrow{d} N(0,M_{vv}\otimes\Sigma^{-1}).\qquad(6.140)$$

It follows immediately from Eq. (6.132) that

$$\sqrt{T}(\hat{\gamma}_{*}-\gamma_{*}^{\circ})\xrightarrow{d}\zeta_{2}\sim N(0,M_{vv}^{-1}\otimes\Sigma),\qquad(6.141)$$

which is **precisely of the same form as the one found in the unrestricted case**, if we make the association $\Gamma=\Pi(1)D_{2}$.

We have therefore proved Proposition 9.

**Proposition 9.** In the context of Proposition 8, the restricted ML estimator, i.e. the estimator resulting when the cointegration hypothesis, $\Pi(1)=\Gamma B'$, $B'=(I_{r},B_{0})$, is imposed, may be found by the iteration scheme of Eq. (6.131) given an ICE. Moreover, the following statements are true:

   i. an ICE exists by the methods given in Proposition 8, modified by $\tilde{\Gamma}=\tilde{\Pi}(1)_{(r)}$, $\tilde{B}_{0}=(\tilde{\Gamma}'\tilde{\Gamma})^{-1}\tilde{\Gamma}'\tilde{\Pi}(1)_{(r_{0})}$, where $\tilde{\Pi}(1)_{(r)}$ consists of the first $r$ columns of $\tilde{\Pi}(1)$, and $\tilde{\Pi}(1)_{(r_{0})}$ contains its last $r_{0}$ columns;

  ii. the ML estimator of the parameters $\gamma_{*}$ and $\Sigma$ is found as the converging iterate in the iteration scheme of Eq. (6.131);

 iii. the limiting distribution of the ML estimator has two components, one nonstationary and one stationary, in the sense that the former involves nonstationary entities, while the second does not.

 iv. the nonstationary component obeys

$$T(\hat{B}_{0}-B_{0}^{\circ})\xrightarrow{d} G_{11}^{-1}\Gamma'A(1)^{\prime-1}T_{1}^{-1}\left(\int_{0}^{1}dB(s)'B_{2}(s)\right)K^{-1}P_{02}^{\prime-1};$$

  v. if the entity above is suitably normalized it has a **standard** distribution; precisely,

$$T[(P_{02}'KP_{02})^{1/2}\otimes I_{r})(\hat{b}_{0}-b_{0}^{\circ})]\xrightarrow{d}\zeta_{1}\sim N(0,I_{r_{0}}\otimes G_{11}^{-1});$$

vi. the stationary component obeys

$$\sqrt{T}(\hat{\gamma}_* - \gamma_*^\circ) \xrightarrow{\mathrm{d}} \zeta_2 \sim N(0, M_{vv}^{-1} \otimes \Sigma);$$

vii. the two components are mutually uncorrelated;

viii. if we put $\Pi(1)D_2 = \Gamma$ and $D_2^{*\prime} P_{-1}' P_{-1} D_2^* = B_0 P_{-1}' P_{-1} B_0'$, the limiting distribution of the stationary part in this context is identical to the limiting distribution of the stationary part in the **unrestricted** case.

**Remark 11.** It is interesting to observe that in the special case $k = 0$, and thus $A(1) = I_q$, the covariance matrix of the limiting distribution in part vi reduces to

$$M_{vv} = \begin{bmatrix} M_{zz} & M_{zx} \\ M_{xz} & M_{xx} \end{bmatrix} = M^*.$$

This is so since from Eq. (6.126) and in the special case under consideration, the entity $V_t^*$ of this section reduces to $V_t. = (p_{t-1}.B, x_t.)$, in the notation of Chapter 5. Thus, the covariance matrix of the limiting distribution in part vi reduces to $M^{*-1} \otimes \Sigma$, while that in Proposition 2 of Chapter 5 is $\Sigma \otimes M^{*-1}$. The difference, as explained above, is that in this discussion we examine a vector of the form $\mathrm{vec}\,(C)$, while in Chapter 5 we examined a vector of the form $\mathrm{vec}\,(C')$.

**Test of Restrictions**

When we examined this particular approach to the estimation of parameters for a $VAR(n)$ model in Chapters 4 and 5, we did not provide for a test of the restrictions and cointegration rank, even though we discussed such tests in a variety of other contexts. The test procedure involved is really a likelihood ratio test (LRT) procedure, much as it is in the Johansen formulation. We begin by noting that, for the **restricted model**

$$\max_{H_0} L(\Pi, A, \Sigma) = -\frac{Tq}{2}[\ln(2\pi) + 1] - \frac{T}{2}\ln|\tilde{\Sigma}/T|, \quad \tilde{\Sigma} = \sum_{t=1}^{T} \tilde{\epsilon}_t. \tilde{\epsilon}_t.,$$

where $\tilde{\epsilon}_t.$ are the residuals obtained from the last equation set in Eq. (6.131), using the **converging iterate**, i.e. the ML estimator of the parameters in

$\Pi(L)$ and $A(L)$. Similarly, from the **unrestricted model** we obtain

$$\max_{H_1} L(\Pi, A, \Sigma) = -\frac{Tq}{2}[\ln(2\pi) + 1] - \frac{T}{2}\ln|\hat{\Sigma}/T|, \quad \hat{\Sigma} = \sum_{t=1}^{T} \hat{\epsilon}'_{t\cdot}\hat{\epsilon}_{t\cdot}.$$

The LRT statistic is thus defined as

$$\phi = \frac{e^{\max_{H_0} L}}{e^{\max_{H_1} L}} = \left(\frac{|\hat{\Sigma}|}{|\tilde{\Sigma}|}\right)^{T/2}, \quad \text{or } \phi^{2/T} = \lambda = \frac{|\hat{\Sigma}|}{|\tilde{\Sigma}|}. \tag{6.142}$$

If $\lambda$ is "large", i.e. "close" to one, we accept the restrictions; if $\lambda$ is "small", we reject the restrictions. Since the determinants in the fractions above occur repeatedly in our discussion, it is desirable to designate them more simply as $S_0 = \hat{\Sigma}$ and $S = \tilde{\Sigma}$, respectively. The objective of this section is to determine the limiting distribution of the log of the LRT statistic $\ln\lambda$.

In the discussion below we employ common notation to facilitate the derivations. To this end, let

$$V_{t\cdot} = (p_{t-1\cdot}, y_{t-1\cdot}, \ldots, y_{t-n+1\cdot}), \quad V_{t1\cdot} = (p_{-1\cdot}), \quad V_{t2\cdot} = (y_{t-1\cdot}, y_{t-2\cdot}, \ldots, y_{t-n+1\cdot}),$$

$$\gamma = (\gamma'_1, \gamma'_2)', \quad \gamma_1 = -\text{vec}[\Pi(1)], \quad \gamma_2 = \text{vec}(\Pi^*_1, \Pi^*_2, \ldots, \Pi^*_{n-1}), \tag{6.143}$$

and note that

$$\hat{\epsilon}'_{t\cdot} = [\hat{A}(L)]^{-1}(y'_{t\cdot} - (V_{t\cdot} \otimes I_q)\hat{\gamma}$$

$$= [\hat{A}(L)]^{-1}[\hat{A}(L)\epsilon'_{t\cdot} - (p_{t-1\cdot} \otimes I_q)(\hat{\gamma}_1 - \gamma^\circ_1) + r'_{t\cdot}],$$

$$r'_{t\cdot} = [A(L)^{-1} - \hat{A}(L)^{-1}]\epsilon'_{t\cdot} - (V_{t2\cdot} \otimes I_q)(\hat{\gamma}_2 - \gamma^\circ_2). \tag{6.144}$$

Note that all (random) variables in $r_{t\cdot}$ are **stationary**, and are attached to entities which **converge to zero**, at least in probability. Thus, we may safely ignore them. Consider now the matrix $g_t = [A(L)]^{-1}(p_{t-1\cdot} \otimes I_q)$ and note that its elements are **nonstationary**. Employing the usual decomposition for polynomial operators we find

$$A(L)g_t = A(1)g_t + \sum_{j=0}^{k-1} A^*_j(I - L)g_t \sim A(1)g_t,$$

because $(I - L)g_t$ is **stationary** and is thus dominated by the nonstationary component $A(1)g_t$. Consequently, we may rewrite

$$\hat{\tilde{\epsilon}}'_{t\cdot} = \hat{\epsilon}'_{t\cdot} - \hat{A}(1)^{-1}(p_{t-1\cdot} \otimes I_q)(\hat{\gamma}_1 - \gamma_1^o) + \hat{r}'_{t1\cdot},$$

$$\hat{r}'_{t1\cdot} = \hat{r}'_{t\cdot} - \hat{A}(1)^{-1} \sum_{j=0}^{k-1} \hat{A}_j^*(\Delta p_{t-1\cdot} \otimes I_q)(\hat{\gamma}_1 - \gamma_1^o). \tag{6.145}$$

Evidently, we may also write

$$\tilde{\epsilon}'_{t\cdot} = \epsilon'_{t\cdot} - \tilde{A}(1)^{-1}(p_{t-1\cdot} \otimes I_q)(\tilde{\gamma}_1 - \gamma_1^o) + \tilde{r}'_{t1\cdot}. \tag{6.146}$$

Using the results above we obtain

$$\tilde{\epsilon}'_{t\cdot} = \hat{\epsilon}'_{t\cdot} + (\tilde{\epsilon}'_{t\cdot} - \hat{\epsilon}'_{t\cdot}) = \hat{\epsilon}'_{t\cdot} + \hat{A}(1)^{-1}[\hat{\Pi}(1) - \tilde{\Pi}(1)]p'_{t-1\cdot} + \tilde{r}'_{t\cdot} - \hat{r}'_{t\cdot}$$

$$+ [\tilde{A}^{-1} - \hat{A}^{-1}]\Pi(1)^o p'_{t-1\cdot} = \hat{\epsilon}'_{t\cdot} + \hat{A}(1)^{-1}[\hat{\Pi}(1) - \tilde{\Pi}(1)]DD^{-1}p'_{t-1\cdot} + r_t^{*'}$$

$$= \hat{\epsilon}'_{t\cdot} + \hat{A}(1)^{-1}[\hat{\Pi}(1) - \tilde{\Pi}(1)]D_1 D_1^{*'} p'_{t-1\cdot} + r_{t1\cdot}^{*'},$$

$$r_{t1\cdot}^{*'} = \hat{r}_{t\cdot}^{*'} + \hat{A}(1)^{-1}[\hat{\Pi}(1) - \tilde{\Pi}(1)]D_2 D_2^{*'} p'_{t-1\cdot}. \tag{6.147}$$

The third equality follows by a simple rearrangement; the fourth serves to define the symbol $r_t^*$; note, also, that the term added to the residual $r_{t\cdot}^{*'}$, and appearing explicitly in the last equation of the set, is the difference in the stationary terms as between the unrestricted and the restricted estimators. As indicated in part viii of Proposition 9, this entity converges to zero in distribution and hence in probability.

In view of Eq. (6.147), and defining $\tilde{U} = (\tilde{\epsilon}_{t\cdot})$, etc., we may write

$$\tilde{U} = \hat{U} + P_{-1}D_1^* D_1'[\hat{\Pi}(1) - \tilde{\Pi}(1)]' A(1)^{'-1} + R_1.$$

It is shown in Reinsel and Ahn (1992) that

$$S_0 = S + S_1, \quad S_1 = \hat{A}(1)^{-1}[\hat{\Pi}(1) - \tilde{\Pi}(1)]D_1[D_1^{*'} P'_{-1} P_{-1} D_1^*]D_1'[\hat{\Pi}(1) - \tilde{\Pi}(1)]' A(1)^{'-1}$$

plus terms that converge to zero in probability.

The LRT statistic is the ratio of the determinants of $S_0$ and $S$. Consider now the characteristic roots of $S_1$ **in the metric of** $S$, and let $\Lambda$ be the diagonal matrix containing these roots. By Proposition 63 in Dhrymes (1984), p. 75, there exists a nonsingular matrix $Q$, such that [11]

$$S = Q'Q, \quad S_1 = Q'\Lambda Q, \quad S_0 = Q'(I_q + \Lambda)Q. \qquad (6.148)$$

Moreover the logarithm of the LRT statistic is given by

$$\lambda = \frac{|S_0|}{|S|} = |I_q + \Lambda|, \quad \ln\lambda = \sum_{j=1}^{q} \ln(1 + \lambda_j) \approx \sum_{j=1}^{q} \lambda_j. \qquad (6.149)$$

The preceding is a theoretically appealing determination of the precise nature of the test statistic, in that it makes clear its dependence on the **rank** of the matrix $S_1$; however, it is not useful in that it requires an additional computation. It is thus fortuitous that we may obtain exactly the same results by considering the roots of $S_0$ in the metric of $S$, or

$$0 = |\nu S - S_0| = |(\nu - 1)S - S_1|, \quad (\text{so that} \quad \nu_i = 1 + \lambda_i),$$

which are also the same as the roots of

$$|\nu I_q - \hat{T}_1'^{-1} S_0 \hat{T}_1^{-1}| = 0, \quad \text{or of} \quad |(\nu - 1)I_q - \hat{T}_1'^{-1} S_1 T_1^{-1}| = 0,$$

where $S = \hat{\Sigma} = \hat{T}_1'\hat{T}_1$. Thus, **to determine** the limiting distribution of the (log) of the LRT statistic we **need to determine** the limiting distribution of the entity $\hat{T}_1'^{-1} S_1 T_1^{-1}$. From Eqs. (6.120), (6.136), and interpreting $\tilde{\Pi}(1)D_1$ to be $\tilde{\Gamma}\tilde{B}_0$ in the restricted case, we find

$$[\hat{\Pi}(1) - \tilde{\Pi}(1)]D_1 \sim (A(1) - \Gamma G_{11}^{-1}\Gamma'A(1)'^{-1}\Sigma^{-1})(U'P_{-1}D_1^*(D_1^{*'}P_{-1}'P_{-1}D_1^*)^{-1}.$$

Let $F = A(1) - \Gamma G_{11}^{-1}\Gamma'A(1)'^{-1}\Sigma^{-1}$ and conclude that

$$T_1'^{-1} S_1 T_1^{-1} \sim F^{*'}\left(\frac{U'P_{-1}D_1^*}{T}\right)\left(D_1^{*'}\frac{P_{-1}'P_{-1}}{T^2}D_1^*\right)^{-1}\left(\frac{P_{-1}'UD_1^*}{T}\right)F^*,$$

$$\rightarrow F^{*'}T_1'\left(\int_0^1 dB(s)'B(s)\right)P_0 D_1^*\left[D_1^{*'}P_0'\left(\int_0^1 B(s)'B(s)\,ds\right)P_0 D_1^*\right]^{-1}$$

$$\times D_1^{*'}P_0'\left(\int_0^1 B(s)'B(s)\,ds\right)T_1 F^*, \quad F^* = F'A(1)'^{-1}T_1^{-1}.$$

---

[11] In the context of the current discussion of characteristic roots, we redefine $S = \hat{\Sigma}/T = \left(\sum_{t=1}^{T} \hat{\epsilon}_{t\cdot}'\hat{\epsilon}_{t\cdot}\right)/T$, and similarly for $S_0$. In this fashion, when we write $S = \hat{T}_1'\hat{T}_1$, we can be assured that $\hat{T}_1$ converges at least in probability to $T_1$, such that $\Sigma = T_1'T_1$.

From the rightmost member of the equations above it is clear that we are dealing with stochastic and Lebesgue integrals inolving the MBM $B(s)T_1F^*$ and $B(s)P_0D_1^*$, $B(s)$ being a $q$-element SMBM. Notice that

$$T_1F^* = T_1(A(1)' - T_1^{-1}T_1'^{-1}A(1)^{-1}\Gamma G_{11}^{-1}\Gamma')A(1)'^{-1}T_1^{-1}$$

$$= I_q - T_1'^{-1}A(1)^{-1}\Gamma G_{11}^{-1}\Gamma'A(1)'^{-1}T_1^{-1}, \qquad (6.150)$$

which is seen to be a **symmetric idempotent matrix of rank** $r_0$. Let $T_2$ be the matrix of the triangular decomposition of $D_1^{*'}P_0'P_0D_1^*$, and note that we may set $B(s)P_0D_1^* = B_2(s)T_2$, where $B_2(s)$ is an $r_0$-element SMBM. Similarly, $B(s)T_1F^*$ is a MBM with covariance matrix $C = T_1F^*$, which is also idempotent. Let $Q = (Q_1, Q_2)$ be the matrix of characteristic roots of $C$ such that $Q_2$ corresponds to the $r_0$ **unit roots** of $C$, and $Q_1$ corresponds to the $r$ zero roots. Then

$$Q'CQ = \begin{bmatrix} Q_1'CQ_1 & Q_1'CQ_2 \\ Q_2'CQ_1 & Q_2'CQ_2 \end{bmatrix} = \begin{bmatrix} 0 & 0 \\ 0 & I_{r_0} \end{bmatrix}, \quad \text{so that} \quad Q_1'CQ_1 = 0,$$

$$B(s)CQ = (0, \ B(s)CQ_2), \quad B_2(s) = B(s)CQ_2, \quad Q_2'CQ_2 = I_{r_0}.$$

We note that the covariance matrix of $B_2(s)$ is $Q_2'CQ_2 = I_{r_0}$, so that it is an $r_0$-element SMBM. Hence we obtain the result

$$\mathcal{I} = T_1'^{-1}S_1T_1^{-1} \xrightarrow{\text{d}} \left(\int_0^1 dB_2(s)'B_2(s)\right)\left(\int_0^1 B_2(s)'B_2(s)\,ds\right)^{-1}$$

$$\times \left(\int_0^1 B_2(s)'dB_2(s)\right). \qquad (6.151)$$

We summarize this discussion in Proposition 10.

**Proposition 10.** In the context of Propositions 8 and 9, a test of cointegrating restrictions imposed on the $MARMA(n,k)$ may be carried out in terms of the LRT statistic. The following statements are true:

i. The LRT tests the hypothesis that the restrictions imposed on the model are valid, so that the rank of $\Pi(1) = \Gamma B'$, as given in the restricted model is $r$. The alternative is that the rank is $q$. Thus, it is a test for the presence of cointegration of rank $r$, as against the alternative that the system is **stationary**.

ii. The LRT statistic is given by $\ln \lambda$ in Eq. (6.142) as amplified in Eq. (6.150), and is determined by the characteristic roots of

$$0 = |\nu I_q - \hat{T}_1'^{-1} S_0 \hat{T}_1^{-1}|, \quad S = \hat{T}_1' \hat{T}_1, \quad S = \frac{1}{T} \sum_{t=1}^{T} \hat{\epsilon}_{t\cdot}' \hat{\epsilon}_{t\cdot}, \quad S_0 = \frac{1}{T} \sum_{t=1}^{T} \tilde{\epsilon}_{t\cdot}' \tilde{\epsilon}_{t\cdot\cdot}$$

iii. The limiting distribution of the test statistic

$$\ln \lambda = \sum_{j=1}^{q} \ln \nu_j,$$

is given by the distribution of the **trace** of the $r_0 \times r_0$, stochastic integral matrix $\mathcal{I}$, as given in Eq. (6.151).

**Remark 12.** It bears repeating that the result above includes as a special case, for $k = 0$, the $VAR(n)$ model. In fact, if one examines the argument leading to Proposition 10 and the limiting distribution of the LRT statistic, the only reference to the MA part of the model is the term $A(1)$; in the VAR case this term reduces to $I_q$. The matrix $T_1 F^*$, however, continues to be symmetric and idempotent, even if we replace $A(1)$ by $I_q$. Consequently, it follows from Proposition 10 that the **same test** for the rank of cointegration applies to both the $VAR(n)$ and the $MARMA(n, k)$ models, and it involves the characteristic roots of the estimator of the covariance matrix of the white noise process **in the restricted case**, in the **metric** of the white noise covariance matrix estimator **in the unrestricted case**.

# Appendix I to Chapter 6

This appendix seeks to accomplish three things: to derive certain results that make the discussion in the main part of the chapter smoother; to unify the notation in the three cases considered therein, and to give the critical values for tests on the rank of cointegration based on estimators of the zero roots.

## No Constant in VAR, No Trend in Levels: Case i

Here we deal with the entities encountered in case i.

$$\frac{1}{T^2} P'_{-1} N_x P_{-1} = \frac{1}{T^2} \left( S'_{-1} N_x S_{-1} \right), \ N_x = I_T - X(X'X)^{-1}X', \ P_{-1} = e\mu_0 + S_{t-1\cdot},$$
$$(6.152)$$

where $\mu_0$ is a "constant of integration", $S_{t\cdot} = \sum_{s=1}^t \eta_{t\cdot}$ and [12]

$$\eta_{t\cdot} = \epsilon_{t\cdot} \frac{H(L)'}{\pi(L)},$$

$\eta$ being a **strictly stationary** process. Since, in the context of this discussion $S$ is a matrix containing "observations" on an $I(1)$ process, while $X$ is a matrix containing observations on a strictly stationary one it follows, from the discussion of Chapter 3, that $S'_{-1}X/T$ converges to an a.c. finite random element. Since $X'X/T$ converges, at least in probability, to a constant matrix it follows therefore that

$$\frac{1}{T^2} P'_{-1} N P_{-1} \sim \frac{1}{T^2} P'_{-1} P_{-1} \xrightarrow{d} P'_0 \left( \int_0^1 B(r)' B(r) dr \right) P_0 = P'_0 K P_0, \quad (6.153)$$

where $B$ is a $1 \times q$ (row) SMBM and $P_0$ is the matrix of the triangular decomposition of the "long-run" covariance matrix of the stationary process

---

[12] Note that whether we do or do not include the constant of integration, i.e. whether in solving $\Delta X_t = \eta_t$ we write $P_{-1} = S_{-1}$ or $P_{-1} = e\mu_0 + S_{-1}$, is completely irrelevant and has no impact on the derivation of the limiting distribution.

that gives rise to the $I(1)$ process, i.e.

$$P_0'P_0 = \Phi = E\eta_{1\cdot}'\eta_{1\cdot} + \sum_{\tau=1}^{\infty}\left(E\eta_{1\cdot}'\eta_{\tau+1\cdot} + E\eta_{\tau+1\cdot}'\eta_{1\cdot}\right).$$

Similarly, since $P_{-1}\Pi(1)'$ is strictly stationary it follows that $P_{-1}'P_{-1}\Pi(1)'/T$ converges to an a.c. finite random element and, moreover,

$$\frac{P_{-1}'NU}{T} \sim \frac{P_{-1}'U}{T} \xrightarrow{\mathrm{d}} P_0'\left(\int_0^1 B(r)' dB_1(r)\right)T_0 = P_0'\Upsilon P_0, \qquad (6.154)$$

where $P_0$ and $B$ are as above, but $T_0, B_1$ refer to the triangular decomposition and the SMBM, respectively, generated by the i.i.d. process $u$, whose covriance matrix is $\Sigma$. It is evident, from the preceding and the definitions in Eq. (6.2) that

$$\sqrt{T}A_{12} \xrightarrow{\mathrm{d}} 0,$$

and hence in probability as well. Finally, in view of the arguments given above

$$TA_{22} \xrightarrow{\mathrm{d}} T_0'\left(\Upsilon' K^{-1}\Upsilon\right)T_0. \qquad (6.155)$$

## Constant in VAR, No Trend in Levels: Case ii

In this context,

$$P_{-1} = e\mu_0 + S_{-1}, \quad \hat{M} = \frac{1}{T}\hat{\Pi}(1)'P_{-1}^{*'}P_{-1}^{*}\hat{\Pi}(1), \quad P_{-1}^{*} = \left(I - \frac{ee'}{T}\right)S_{-1}, \quad (6.156)$$

and

$$A_{11} = \frac{\Pi(1)P_{-1}^{*'}P_{-1}^{*}\Pi(1)'}{T}, \qquad (6.157)$$

$$A_{12} = -\frac{1}{T}\left(\frac{U^{*'}N_x P_{-1}^{*}}{T}\right)\left(\frac{P_{-1}^{*'}N_x P_{-1}^{*}}{T^2}\right)^{-1}\left(\frac{P_{-1}^{*'}P_{-1}^{*}\Pi(1)'}{T}\right),$$

$$A_{22} = \frac{1}{T}\left(\frac{U^{*'}N P_{-1}^{*}}{T}\right)\left(\frac{P_{-1}^{*'}N_x P_{-1}^{*}}{T^2}\right)^{-1}\left(\frac{P_{-1}^{*'}P_{-1}^{*}}{T^2}\right)$$

$$\times \left(\frac{P_{-1}^{*'}N_x P_{-1}^{*}}{T^2}\right)^{-1}\left(\frac{P_{-1}^{*'}N_x U^{*}}{T}\right),$$

where

$$(P_{-1}^*, X^*, U^*) = \left(I - \frac{ee'}{T}\right)(P_{-1}, X, U).$$

We note that

$$\frac{P_{-1}^{*'} N_x P_{-1}^*}{T^2} = \frac{P_{-1}^{*'} P_{-1}^*}{T^2} - \frac{P_{-1}^{*'} X^*}{T^2}\left(\frac{X^{*'} X^*}{T}\right)^{-1}\frac{X^{*'} P_{-1}^*}{T} \sim \frac{P_{-1}^{*'} P_{-1}^*}{T^2} \xrightarrow{\text{d}} P_0' K P_0.$$

In order to preserve notational unity in all three cases, put

$$K_0 = \int_0^1 B(r)' B(r)dr, \quad K_1 = \left(\int_0^1 B(r)dr\right)'\left(\int_0^1 B(r)dr\right), \quad K = K_0 - K_1.$$
$$(6.158)$$

Similarly,

$$\frac{P_{-1}^{*'} N_x U^*}{T} = \frac{P_{-1}^{*'} U}{T} - \frac{P_{-1}^{*'} X^*}{T}\left(\frac{X^{*'} X^*}{T}\right)^{-1}\frac{X^{*'} U}{T} \sim \frac{P_{-1}^{*'} U}{T} \xrightarrow{\text{d}} P_0' \Upsilon T_0, \quad (6.159)$$

where

$$\Upsilon_0 = \int_0^1 B(r)' dB_1(r), \quad \Upsilon_1 = \left(\int_0^1 B(r)' dr\right) B_1(1), \text{ so that } \Upsilon = \Upsilon_0 - \Upsilon_1.$$
$$(6.160)$$

Consequently,

$$T A_{22} \xrightarrow{\text{d}} T_0'\left(\Upsilon' K^{-1} \Upsilon\right) T_0. \quad (6.161)$$

**Remark A.1.** The difference in the definitions of $\Upsilon$ and $K$ between case i and case ii **does not result because** in case i we write $P_{-1} = S_{-1}$, while in case ii we introduce the constant of integration, $\mu_0$. It results, rather, **from the fact that the $VAR(n)$ is stated with a constant term**. When this is so it is well known, from least squares theory, that the estimators of the parameters other than the constant term involve **centered data**, centered about their respective sample means. In stationary processes this operation does not alter the limiting distribution of the estimators in question. In the case of $I(1)$ processes, however, this operation leaves behind a trace, in that it modifies $\Upsilon$ and $K$, as between cases i and ii. The introduction of the constant of integration in the levels is quite innocuous, in either case, and has no consequence relative to the form of the limiting distribution.

## Constant in VAR, Trend in Levels: Case iii

For this case

$$P_{-1}^* = N_g P_{-1} = N_g S_{-1}, \quad N_g = I_T - G(G'G)^{-1}G', \quad G = (e, t^*), \quad (6.162)$$

where $t^{*'} = (0, 1, 2, \dots, T-1)$, and we have made use of the model representation

$$P = e\mu_0 + t^*\mu_1 + S, \quad S = (S_{t\cdot}), \quad S_{t\cdot} = \sum_{s=1}^{t} \eta_{s\cdot}, \quad \eta_{t\cdot} = \epsilon_{t\cdot} \frac{H(L)'}{\pi(L)}. \quad (6.163)$$

We may verify that

$$(G'G)^{-1} = \frac{1}{T(T+1)} \begin{bmatrix} 2(2T-1) & -6 \\ -6 & 12/(T-1) \end{bmatrix}. \quad (6.164)$$

For purposes of estimation we write the error correction representation as

$$\Delta P = -P_{-1}^* \Pi(1)' + X\Pi^* + e\mu + U - G(G'G)^{-1}G'P_{-1}\Pi(1)', \quad (6.165)$$

and note that, by construction, the entity $P_{-1}\Pi(1)'$ represents the "data matrix" (observations) of the cointegral vector, and thus refers to a **strictly stationary process**, possibly with a nonzero mean. Obtaining the least squares estimator in the context of Eq. (6.117) we find

$$\hat{\Pi}(1)' = \Pi(1)' - (P_{-1}^{*'} N_x P_{-1}^*)^{-1}(P_{-1}^{*'} N_x) \left[ \left( I_T - \frac{ee'}{T} \right) U - G(G'G)^{-1}G'P_{-1}\Pi(1)' \right].$$
$$(6.166)$$

We note further that

$$\frac{P_{-1}^{*'} N_x P_{-1}^*}{T^2} \sim \frac{P_{-1}' N_g P_{-1}}{T^2}, \quad (6.167)$$

due to the fact that

$$\frac{1}{T^2} P_{-1}^{*'} X^* \left( \frac{X^{*'} X^*}{T} \right)^{-1} \frac{X^{*'} P_{-1}^*}{T} \xrightarrow{d} 0.$$

This is so since $P_{-1}^* = N_g S_{-1}$ represents the data matrix of an $I(1)$ process, while $X^*$ is the data matrix of a strictly stationary process, so that $P_{-1}^{*'} X^*/T$ converges to a well-defined random element.

Consider next

$$U_1^* + U_2^* = (P_{-1}^{*'} N_x) \left[ \left( I_T - \frac{ee'}{T} \right) U - G(G'G)^{-1} G' P_{-1} \Pi(1)' \right] \qquad (6.168)$$

and note

$$N_g N_x (I_T - N_g) = -N_g X^* (X^{*'} X^*)^{-1} X^{*'} G(G'G)^{-1} G',$$

so that

$$\frac{U_2^*}{T} = \frac{P_{-1}' N_g X^*}{T} \left( \frac{X^{*'} X^*}{T} \right)^{-1} \left[ \frac{1}{T} (X^{*'} G)(G'G)^{-1} G' \left( I_T - \frac{ee'}{T} \right) P_{-1} \Pi(1)' \right]. \qquad (6.169)$$

The matrix in square brackets converges in distribution to zero, while the matrix multiplying it on the left converges in distribution to an a.c. finite random element. Thus, we conclude

$$\frac{U_2^*}{T} \xrightarrow{\text{d}} 0, \text{ and hence } \frac{U_2^*}{T} \xrightarrow{\text{P}} 0,$$

so that

$$\frac{U_1^* + U_2^*}{T} \sim \frac{U_1^*}{T} \sim \frac{P_1' N_g U}{T}. \qquad (6.170)$$

The last relation holds because

$$N_g N_x \left( I_T - \frac{ee'}{T} \right) = N_g \left( I_T - \frac{ee'}{T} \right) N_x = N_g - N_g X^{*'} (X^{*'} X^*)^{-1} X^*,$$

and it is easily seen that

$$\frac{S_{-1}' N_g U}{T} \left( \frac{X^{*'} X^*}{T} \right)^{-1} \frac{X^{*'} U}{T} \xrightarrow{\text{d}} 0, \qquad (6.171)$$

because of the last factor.

Since the intuition behind the conformity test is to obtain an estimator of the covariance matrix of the cointegral vector and test for its rank, we are led to consider

$$\hat{M} = \frac{1}{T} \hat{\Pi}(1)' P_{-1}' N_g P_{-1} \hat{\Pi}(1) = A_{11} + A_{12} + A_{21} + A_{22}, \qquad (6.172)$$

where now

$$A_{11} = \frac{\Pi(1)' P_{-1}^{*'} P_{-1}^{*} \Pi(1)}{T}, \tag{6.173}$$

$$A_{12} = -\frac{1}{T} \left( \frac{U^{*'} N_x P_{-1}^{*}}{T} \right) \left( \frac{P_{-1}^{*'} N_x P_{-1}^{*}}{T^2} \right)^{-1} \left( \frac{P_{-1}^{*'} P_{-1}^{*} \Pi(1)}{T} \right),$$

$$A_{22} = \frac{1}{T} \left( \frac{U^{*'} N_x P_{-1}^{*}}{T} \right) \left( \frac{P_{-1}^{*'} N_x P_{-1}^{*}}{T^2} \right)^{-1} \left( \frac{P_{-1}^{*'} P_{-1}^{*}}{T^2} \right)$$

$$\times \left( \frac{P_{-1}^{*'} N_x P_{-1}^{*}}{T^2} \right)^{-1} \left( \frac{P_{-1}^{*'} N_x U^{*}}{T} \right),$$

and

$$P_{-1}^{*} = (I_T - G(G'G)^{-1}G')P_{-1}, \quad (X^*, U^*) = \left( I - \frac{ee'}{T} \right)(X, U).$$

As before,

$$\sqrt{T} A_{12} \xrightarrow{d} 0, \tag{6.174}$$

and

$$T A_{22} \sim \left( \frac{U' N_g P_{-1}}{T} \right) \left( \frac{P_{-1}' N_g P_{-1}}{T^2} \right)^{-1} \left( \frac{P_{-1}' N_g U}{T} \right)$$

$$= \left( \frac{U' N_g S_{-1}}{T} \right) \left( \frac{S_{-1}' N_g S_{-1}}{T^2} \right)^{-1} \left( \frac{S_{-1}' N_g U}{T} \right). \tag{6.175}$$

Put

$$S_{-1}' G = (T \bar{S}_{-1}', \sum_{t=1}^{T}(t-1)S_{t-1}') = (h_{\cdot 1}, h_{\cdot 2}),$$

$$U' G = (T \bar{u}', \sum_{t=1}^{T}(t-1)u_{t \cdot}') = (g_{\cdot 1}, g_{\cdot 2}), \tag{6.176}$$

and note that

$$T A_{22} \sim C_0' C_1^{-1} C_0, \quad C_0' = \left( \frac{\sum_{t=1}^{T} S_{t-1 \cdot}' u_{t \cdot}}{T} \right) - C_{01}',$$

$$C_1 = \left( \frac{\sum_{t=1}^{T} S_{t-1 \cdot}' S_{t-1 \cdot}}{T^2} \right) - C_{11}, \quad C_{11} = \frac{1}{T^2}(h_{\cdot 1}, h_{\cdot 2})(G'G)^{-1}(h_{\cdot 1}, h_{\cdot 2})',$$

$$C_{01}' = \frac{1}{T}(g_{\cdot 1}, g_{\cdot 2})(G'G)^{-1}(h_{\cdot 1}, h_{\cdot 2})'. \tag{6.177}$$

We now examine the second terms in $C_0'$ and $C_1$. Using the notation $(G'G)^{-1} = (b_{ij})$, we obtain

$$C_{01}' = \frac{1}{T}(g_{.1}, g_{.2})(G'G)^{-1}(h_{.1}, h_{.2})' = \frac{1}{T}[b_{11}g_{.1}h_{.1}' + b_{12}(g_{.1}h_{.2}' + g_{.2}h_{.1})'$$

$$+ b_{22}g_{.2}h_{.2}'],$$

$$C_{11} = \frac{1}{T^2}(h_{.1}, h_{.2})(G'G)^{-1}(h_{.1}, h_{.2})' = \frac{1}{T^2}[b_{11}h_{.1}h_{.1}' + b_{12}(h_{.1}h_{.2}' + h_{.2}h_{.1}')$$

$$+ b_{22}h_{.2}h_{.2}']. \tag{6.178}$$

From the results in Chapter 3, we conclude that

$$C_{01}' \xrightarrow{\text{d}} T_0'\Upsilon_1'P_0 = T_0'[4A_4'A_1 - 6(A_4'A_1 - A_3'A_1 + A_4'A_2) + 12(A_4'A_2 - A_3'A_2)]P_0,$$

$$C_{11} \xrightarrow{\text{d}} P_0'K_1P_0, \quad K_1 = 4A_1'A_1 - 6(A_1'A_2 + A_2'A_1) + 12A_2'A_2,$$

$$A_1 = \int_0^1 B(r)dr, \quad A_2 = \int_0^1 rB(r)dr, \quad A_3 = \int_0^1 B_1(r)dr, \quad A_4 = B_1(1).$$

From the discussion in the main text of this chapter, we conclude that the limiting distribution of the zero characteristic root estimators of the matrix $\hat{\Pi}(1)'P_{-1}^{*'}P_{-1}^*\hat{\Pi}(1)/T$ in the metric of $\hat{\Sigma}$ is the distribution of the roots of

$$|\rho I_{r_0} - J'D_2'T_0'\Upsilon'K^{-1}\Upsilon T_0D_2J| = 0, \quad \text{where} \quad J'^{-1}J^{-1} = D_2'\Sigma D_2. \tag{6.179}$$

The preceding discussion may be summarised as follows:

i. In case i, where the $VAR(n)$ has no constant term, the limiting distribution of the estimator of the zero roots is given by that of the characteristic roots of Eq. (6.131), where $K = K_0 - K_1$, $\Upsilon = \Upsilon_0 - \Upsilon_1$, with $\Upsilon_1 = K_1 = 0$.

ii. In case ii, where the $VAR(n)$ has a constant term, but it lies in $\mathcal{R}(\Gamma')$, so that the $X$-process **does not exhibit linear time trends**, the limiting distribution is as in case i, except that $\Upsilon_1 = A_1'A_4$, and $K_1 = A_1'A_1$.

iii. In case iii, where the $VAR(n)$ has a constant term and the $X$-process **does exhibit linear time trends**, the limiting distribution is as in case i, except that now

$$\Upsilon_1 = 4A_1' A_4 - 6(A_1' A_4 - A_1' A_3 + A_2' A_4) + 12(A_2' A_4 - A_2' A_3),$$

$$K_1 = 4A_1' A_1 - 6(A_1' A_2 + A_2' A_1) + 12A_2' A_2.$$

# Appendix II to Chapter 6

**Critical Values for the Hypotheses:** $H_0 : \sum_{i=q-r_0+1}^{q} \lambda_i = 0$ **(Trace)** and $H_0 : \lambda_{q-r_0+1} = 0$ **(Maximal Root)**.

The test statistics for these hypotheses are, respectively,

i. $T \sum_{i=q-r_0+1}^{q} \hat{\lambda}_i$ ; acceptance means the rank of cointegration is at most $q - r_0 = r$, while rejection means that the rank of cointegration **is greater than** $r$ ;

ii. $T \hat{\lambda}_{q-r_0+1}$ ; acceptance means that the rank of cointegration is at most $q - r_0 = r$, while rejection means that the rank of cointegration is greater that $r$ .

Thus these test statistics pertain to the same basic hypothesis, when applied to roots ordered in descending order by magnitude.

The term "trace" applied to the first test statistic is appropriate since, in the limit, it corresponds to the trace of the matrix $J' \tilde{M}_{22}^* J$ , in the discussion of the text, where $J' D_2' T_0' T_0 D_2 J = I_{r_0}$ .

The term "maximal root" is appropriate for the second test statistic since it is a test that the maximal ordered root of $\hat{M}$ that is zero is the $(q-r_0+1)^{th}$ ordered root.

All tabulations are generated by obtaining the characteristic roots of

$$|\rho I_{r_0} - C| = 0, \quad C = C_1' C_0^{-1} C_1, \tag{6.180}$$

where

$$C_1 = \frac{S_{-1}'(I - N)U}{T}, \quad C_0 = \frac{S_{-1}'(I - N)S_{-1}}{T^2}, \tag{6.181}$$

for $T = 2,000$ and 20,000 replications. In Eq. (6.181), $U = (u_{t.})$, $S = (S_{t.})$, $S_{t.} = \sum_{s=1}^{t} u_{s.}$, and $u_{t.}$ is an i.i.d. sequence of $N(0, I_q)$ **row** vectors. The particular specifications are as follows:

i. Case i, $N = 0$; this is the case where the VAR has no constant term.

ii. Case ii, $N = ee'/T$, where $e$ is a column vector of $T$ unities; this is the case where the VAR specification contains a constant term, but the underlying stochastic sequence ($X_{t.}$ has **no deterministic** time trends.

iii. Case iii, $N = G(G'G)^{-1}G'$, $G = (e, t^*)$, $t^* = (0, 1, 2, \ldots, T-1)'$; this is the case where the VAR specification contains a constant term and the underlying stochastic sequence ($X_{t.}$ has **(linear) deterministic** time trends.

Notice further that these tables contain all relevant information for carrying out tests in the Johansen (LRT) context as well. The critical values for such tests are contained in Tables 1 through 6 as the special case $q = i$, $r_0 = i$, $i = 1, 2, \ldots, 10$. Such critical values ought to be more accurate than those given in Osterwald-Lenum (1992) since the latter are obtained with $T = 400$ and 6,000 replications, while the entries in Tables 1 through 6 are obtained with sample size $T = 2,000$ and 20,000 replications, and are thus better approximations of the relevant limiting distributions.

Table 1 (CASE I)

**Trace**

Significance Points for Conformity
Cointegration Test.
Combinations of System Size ( $q$ ) and
Zero Roots Under the Null ( $r_0$ )

| | Level of Significance | | | | | | |
|---|---|---|---|---|---|---|---|
| q | .75 | .80 | .85 | .90 | .95 | .975 | .99 |
| | | | | $r_0 = 1$ | | | |
| 1 | 1.549 | 1.817 | 2.300 | 2.904 | 4.042 | 5.265 | 6.964 |
| 2 | 4.282 | 4.844 | 5.555 | 6.539 | 8.125 | 9.613 | 11.579 |
| 3 | 6.787 | 7.456 | 8.305 | 9.453 | 11.369 | 13.157 | 15.377 |
| 4 | 9.111 | 9.903 | 10.854 | 12.138 | 14.296 | 16.358 | 18.654 |
| 5 | 11.492 | 12.361 | 13.468 | 14.913 | 17.141 | 19.289 | 21.867 |
| 6 | 13.891 | 14.823 | 15.943 | 17.439 | 19.825 | 22.058 | 24.856 |
| 7 | 15.990 | 17.003 | 18.216 | 19.966 | 22.610 | 24.962 | 27.900 |
| 8 | 18.184 | 19.199 | 20.456 | 22.177 | 24.799 | 27.276 | 30.467 |
| 9 | 20.416 | 21.576 | 22.913 | 24.705 | 27.698 | 30.244 | 33.373 |
| 10 | 22.667 | 23.898 | 25.379 | 27.099 | 30.215 | 32.835 | 36.115 |
| | | | | $r_0 = 2$ | | | |
| 2 | 7.838 | 8.510 | 9.358 | 10.488 | 12.290 | 13.912 | 16.013 |
| 3 | 12.555 | 13.430 | 14.503 | 15.951 | 18.264 | 20.356 | 22.975 |
| 4 | 17.088 | 18.136 | 19.390 | 21.034 | 23.658 | 26.127 | 28.974 |
| 5 | 21.583 | 22.734 | 24.152 | 25.988 | 28.957 | 31.679 | 34.672 |
| 6 | 25.967 | 27.247 | 28.806 | 30.752 | 33.794 | 36.567 | 40.190 |
| 7 | 30.357 | 31.802 | 33.495 | 35.608 | 38.719 | 41.740 | 45.437 |
| 8 | 34.849 | 36.391 | 38.143 | 40.418 | 43.982 | 47.227 | 50.521 |
| 9 | 39.078 | 40.648 | 42.449 | 44.971 | 48.754 | 52.797 | 56.354 |
| 10 | 43.127 | 44.711 | 46.678 | 49.361 | 53.164 | 56.681 | 60.906 |
| | | | | $r_0 = 3$ | | | |
| 3 | 17.967 | 18.942 | 20.135 | 21.806 | 24.299 | 26.762 | 29.498 |
| 4 | 24.723 | 25.860 | 27.320 | 29.163 | 32.099 | 34.659 | 37.948 |
| 5 | 31.264 | 32.588 | 34.224 | 36.310 | 39.684 | 42.709 | 46.539 |
| 6 | 37.825 | 39.241 | 40.990 | 43.217 | 46.926 | 50.135 | 54.022 |
| 7 | 44.224 | 45.909 | 47.891 | 50.344 | 54.332 | 57.915 | 62.071 |
| 8 | 50.898 | 52.616 | 54.631 | 57.316 | 61.280 | 64.627 | 69.303 |
| 9 | 56.987 | 58.795 | 61.037 | 64.198 | 68.441 | 71.985 | 76.673 |
| 10 | 63.675 | 65.748 | 68.129 | 71.202 | 75.295 | 79.180 | 83.795 |

Table 1 (CASE I) continued

| q | .75 | .80 | .85 | .90 | .95 | .975 | .99 |
|---|-----|-----|-----|-----|-----|------|-----|
| | **Level of Significance** | | | | | | |
| | | | | $r_0 = 4$ | | | |
| 4 | 31.447 | 33.200 | 34.885 | 36.642 | 39.927 | 42.387 | 45.632 |
| 5 | 40.843 | 42.291 | 43.993 | 46.199 | 49.880 | 53.093 | 57.390 |
| 6 | 49.464 | 51.150 | 53.064 | 55.583 | 59.439 | 63.056 | 67.171 |
| 7 | 57.984 | 59.758 | 61.768 | 64.535 | 69.032 | 72.829 | 77.670 |
| 8 | 66.453 | 68.351 | 70.556 | 73.592 | 78.219 | 82.555 | 87.698 |
| 9 | 75.081 | 77.116 | 79.509 | 82.791 | 87.653 | 92.270 | 96.909 |
| 10 | 83.282 | 85.515 | 88.108 | 91.581 | 96.479 | 100.849 | 105.954 |
| | | | | $r_0 = 5$ | | | |
| 5 | 50.117 | 51.729 | 53.586 | 55.894 | 59.826 | 63.307 | 67.468 |
| 6 | 61.013 | 62.902 | 65.052 | 67.518 | 71.625 | 75.161 | 79.415 |
| 7 | 71.276 | 73.354 | 75.624 | 78.609 | 83.167 | 86.818 | 91.238 |
| 8 | 82.179 | 84.192 | 86.627 | 89.958 | 94.880 | 99.224 | 104.415 |
| 9 | 92.505 | 95.050 | 97.656 | 100.774 | 105.385 | 110.657 | 116.305 |
| 10 | 103.385 | 105.650 | 108.581 | 112.092 | 117.544 | 122.427 | 129.322 |
| | | | | $r_0 = 6$ | | | |
| 6 | 72.316 | 74.172 | 76.387 | 79.152 | 83.354 | 87.226 | 91.477 |
| 7 | 84.814 | 86.913 | 89.454 | 92.652 | 97.688 | 101.777 | 107.075 |
| 8 | 97.672 | 99.885 | 102.516 | 105.838 | 111.156 | 116.145 | 121.950 |
| 9 | 110.173 | 112.702 | 115.627 | 119.287 | 125.101 | 129.838 | 135.969 |
| 10 | 122.655 | 125.156 | 128.146 | 131.830 | 137.590 | 142.916 | 148.497 |
| | | | | $r_0 = 7$ | | | |
| 7 | 98.182 | 100.405 | 103.041 | 106.332 | 111.384 | 115.844 | 121.044 |
| 8 | 113.046 | 115.394 | 117.971 | 121.293 | 126.778 | 131.998 | 138.590 |
| 9 | 127.545 | 129.957 | 133.062 | 136.879 | 142.695 | 148.607 | 154.838 |
| 10 | 142.031 | 144.695 | 147.852 | 151.823 | 158.166 | 163.628 | 169.000 |
| | | | | $r_0 = 8$ | | | |
| 8 | 128.376 | 130.748 | 133.667 | 137.200 | 142.856 | 148.350 | 154.564 |
| 9 | 144.750 | 147.541 | 150.475 | 154.654 | 161.431 | 167.100 | 173.498 |
| 10 | 161.518 | 164.273 | 167.368 | 171.531 | 178.006 | 183.438 | 190.308 |
| | | | | $r_0 = 9$ | | | |
| 9 | 162.379 | 164.981 | 168.212 | 172.460 | 178.820 | 184.558 | 190.854 |
| 10 | 180.728 | 183.483 | 187.115 | 191.735 | 198.293 | 203.705 | 211.339 |
| | | | | $r_0 = 10$ | | | |
| 10 | 199.896 | 203.155 | 206.680 | 211.159 | 218.085 | 224.639 | 230.549 |

Table 2 (CASE I)

## Maximal Root

Significance Points for Conformity

Cointegration Test.

Combinations of System Size ( $q$ ) and

Zero Roots Under the Null ( $r_0$ )

| q | Level of Significance | | | | | | |
|---|---|---|---|---|---|---|---|
|  | .75 | .80 | .85 | .90 | .95 | .975 | .99 |
|  |  |  |  | $r_0 = 1$ |  |  |  |
| 1 | 1.549 | 1.817 | 2.300 | 2.904 | 4.042 | 5.265 | 6.964 |
| 2 | 4.282 | 4.844 | 5.555 | 6.539 | 8.125 | 9.613 | 11.579 |
| 3 | 6.787 | 7.456 | 8.305 | 9.453 | 11.369 | 13.157 | 15.377 |
| 4 | 9.111 | 9.903 | 10.854 | 12.138 | 14.296 | 16.358 | 18.654 |
| 5 | 11.492 | 12.361 | 13.468 | 14.913 | 17.141 | 19.289 | 21.867 |
| 6 | 13.891 | 14.823 | 15.943 | 17.439 | 19.825 | 22.058 | 24.856 |
| 7 | 15.990 | 17.003 | 18.216 | 19.966 | 22.610 | 24.962 | 27.900 |
| 8 | 18.184 | 19.199 | 20.456 | 22.177 | 24.799 | 27.276 | 30.467 |
| 9 | 20.416 | 21.576 | 22.913 | 24.705 | 27.698 | 30.244 | 33.373 |
| 10 | 22.667 | 23.898 | 25.379 | 27.099 | 30.215 | 32.835 | 36.115 |
|  |  |  |  | $r_0 = 2$ |  |  |  |
| 2 | 7.028 | 7.650 | 8.378 | 9.469 | 11.207 | 12.807 | 15.128 |
| 3 | 10.024 | 10.789 | 11.726 | 12.967 | 14.918 | 16.786 | 19.058 |
| 4 | 12.890 | 13.721 | 14.759 | 16.112 | 18.286 | 20.309 | 22.887 |
| 5 | 15.575 | 16.467 | 17.618 | 19.056 | 21.597 | 23.573 | 26.246 |
| 6 | 18.220 | 19.217 | 20.375 | 21.969 | 24.274 | 26.666 | 29.708 |
| 7 | 20.799 | 21.843 | 23.111 | 24.760 | 27.308 | 29.687 | 32.890 |
| 8 | 23.408 | 24.567 | 25.979 | 27.823 | 30.498 | 32.673 | 35.962 |
| 9 | 25.817 | 26.911 | 28.350 | 30.211 | 33.066 | 35.536 | 38.655 |
| 10 | 28.243 | 29.354 | 30.832 | 32.626 | 35.674 | 38.213 | 42.028 |
|  |  |  |  | $r_0 = 3$ |  |  |  |
| 3 | 12.632 | 13.468 | 14.401 | 15.748 | 17.727 | 19.676 | 21.981 |
| 4 | 15.799 | 16.666 | 17.772 | 19.169 | 21.425 | 23.516 | 26.174 |
| 5 | 18.706 | 19.650 | 20.872 | 22.456 | 24.817 | 26.995 | 29.773 |
| 6 | 21.651 | 22.605 | 23.797 | 25.317 | 27.946 | 30.306 | 33.105 |
| 7 | 24.307 | 25.341 | 26.635 | 28.302 | 31.044 | 33.557 | 36.798 |
| 8 | 27.316 | 28.375 | 29.667 | 31.396 | 34.310 | 37.074 | 40.671 |
| 9 | 29.804 | 31.039 | 32.364 | 34.233 | 37.036 | 39.960 | 43.135 |
| 10 | 32.573 | 33.705 | 35.153 | 36.974 | 40.077 | 42.777 | 46.168 |

Table 2 (CASE I), continued

| q | Level of Significance | | | | | | |
|---|---|---|---|---|---|---|---|
| | **.75** | **.80** | **.85** | **.90** | **.95** | **.975** | **.99** |
| | | | | $r_0 = 4$ | | | |
| 4 | 17.654 | 18.880 | 19.451 | 21.677 | 23.960 | 26.564 | 28.957 |
| 5 | 21.550 | 22.483 | 23.611 | 25.266 | 27.714 | 29.962 | 32.879 |
| 6 | 24.666 | 25.642 | 26.889 | 28.382 | 31.049 | 33.531 | 36.160 |
| 7 | 27.553 | 28.619 | 29.937 | 31.613 | 34.265 | 36.915 | 39.990 |
| 8 | 30.546 | 31.637 | 32.992 | 34.816 | 37.645 | 40.233 | 43.298 |
| 9 | 33.314 | 34.459 | 35.943 | 37.731 | 40.556 | 43.428 | 46.684 |
| 10 | 36.049 | 37.250 | 38.698 | 40.581 | 43.592 | 46.049 | 49.489 |
| | | | | $r_0 = 5$ | | | |
| 5 | 23.958 | 24.974 | 26.235 | 27.831 | 30.405 | 32.693 | 35.439 |
| 6 | 27.279 | 28.355 | 29.589 | 31.160 | 33.775 | 36.123 | 38.960 |
| 7 | 30.311 | 31.377 | 32.609 | 34.360 | 37.248 | 39.832 | 42.868 |
| 8 | 33.518 | 34.661 | 36.056 | 37.764 | 40.671 | 43.171 | 46.414 |
| 9 | 36.290 | 37.458 | 38.840 | 40.619 | 43.416 | 46.123 | 49.442 |
| 10 | 39.114 | 40.377 | 41.801 | 43.796 | 46.944 | 49.617 | 52.859 |
| | | | | $r_0 = 6$ | | | |
| 6 | 29.687 | 30.718 | 32.032 | 33.753 | 36.407 | 38.803 | 42.163 |
| 7 | 32.903 | 34.018 | 35.392 | 37.202 | 40.186 | 42.659 | 46.003 |
| 8 | 36.261 | 37.524 | 38.965 | 40.729 | 43.526 | 45.948 | 49.309 |
| 9 | 39.132 | 40.318 | 41.740 | 43.745 | 46.941 | 49.342 | 53.090 |
| 10 | 42.167 | 43.428 | 44.956 | 46.862 | 49.798 | 52.529 | 56.107 |
| | | | | $r_0 = 7$ | | | |
| 7 | 35.354 | 36.515 | 37.854 | 39.680 | 42.498 | 45.190 | 48.800 |
| 8 | 38.791 | 39.950 | 41.326 | 43.240 | 45.930 | 48.694 | 52.404 |
| 9 | 41.733 | 42.968 | 44.375 | 46.333 | 49.216 | 51.838 | 55.248 |
| 10 | 44.731 | 46.004 | 47.506 | 49.523 | 52.522 | 55.535 | 59.069 |
| | | | | $r_0 = 8$ | | | |
| 8 | 41.246 | 42.400 | 43.964 | 45.817 | 48.866 | 51.633 | 54.853 |
| 9 | 44.317 | 45.579 | 46.969 | 49.079 | 51.950 | 54.719 | 58.182 |
| 10 | 47.473 | 48.816 | 50.324 | 52.399 | 55.369 | 58.431 | 61.646 |
| | | | | $r_0 = 9$ | | | |
| 9 | 46.772 | 47.996 | 49.415 | 51.375 | 54.415 | 57.435 | 60.724 |
| 10 | 50.086 | 51.370 | 52.843 | 54.863 | 57.906 | 61.260 | 64.718 |
| | | | | $r_0 = 10$ | | | |
| 10 | 52.544 | 53.811 | 55.308 | 57.304 | 60.551 | 63.269 | 67.143 |

Table 3 (CASE II)

**Trace**

Significance Points for Conformity
Cointegration Test.
Combinations of System Size ($q$) and
Zero Roots Under the Null ($r_0$)

| q | **Level of Significance** | | | | | | |
|---|---|---|---|---|---|---|---|
|   | **.75** | **.80** | **.85** | **.90** | **.95** | **.975** | **.99** |
|   | | | | $r_0 = 1$ | | | |
| 1 | 4.371 | 4.916 | 5.618 | 6.574 | 8.185 | 9.756 | 11.667 |
| 2 | 6.857 | 7.542 | 8.392 | 9.506 | 11.405 | 13.235 | 15.464 |
| 3 | 9.184 | 9.987 | 10.950 | 12.203 | 14.231 | 16.188 | 18.582 |
| 4 | 11.562 | 12.470 | 13.604 | 14.985 | 17.267 | 19.337 | 22.041 |
| 5 | 13.820 | 14.776 | 15.902 | 17.429 | 19.962 | 22.254 | 25.207 |
| 6 | 16.072 | 17.080 | 18.338 | 19.926 | 22.469 | 24.730 | 28.056 |
| 7 | 18.242 | 19.376 | 20.722 | 22.508 | 25.184 | 27.574 | 30.598 |
| 8 | 20.527 | 21.695 | 23.170 | 24.935 | 27.799 | 30.447 | 33.367 |
| 9 | 22.764 | 24.015 | 25.505 | 27.398 | 30.320 | 33.024 | 36.473 |
| 10 | 24.871 | 26.121 | 27.789 | 29.794 | 32.870 | 35.773 | 38.798 |
|   | | | | $r_0 = 2$ | | | |
| 2 | 12.585 | 13.417 | 14.474 | 15.923 | 18.026 | 20.202 | 23.225 |
| 3 | 17.030 | 18.041 | 19.245 | 20.856 | 23.394 | 25.720 | 28.562 |
| 4 | 21.649 | 22.783 | 24.170 | 26.096 | 28.865 | 31.533 | 34.621 |
| 5 | 25.952 | 27.190 | 28.710 | 30.640 | 33.718 | 36.497 | 39.970 |
| 6 | 30.489 | 31.826 | 33.523 | 35.684 | 38.772 | 41.774 | 45.470 |
| 7 | 34.664 | 36.070 | 37.872 | 40.049 | 43.711 | 46.705 | 50.737 |
| 8 | 39.050 | 40.629 | 42.437 | 44.816 | 48.117 | 51.232 | 55.237 |
| 9 | 43.212 | 44.846 | 46.948 | 49.349 | 53.370 | 56.658 | 60.893 |
| 10 | 47.398 | 49.123 | 51.202 | 53.864 | 58.040 | 61.797 | 66.414 |
|   | | | | $r_0 = 3$ | | | |
| 3 | 24.514 | 25.651 | 27.005 | 28.701 | 31.444 | 34.190 | 37.202 |
| 4 | 31.327 | 32.654 | 34.199 | 36.265 | 39.584 | 42.690 | 46.364 |
| 5 | 37.834 | 39.211 | 41.015 | 43.315 | 47.029 | 49.964 | 54.100 |
| 6 | 44.204 | 45.793 | 47.590 | 50.254 | 53.966 | 57.686 | 61.761 |
| 7 | 50.766 | 52.448 | 54.405 | 57.052 | 61.085 | 64.658 | 68.352 |
| 8 | 57.130 | 58.918 | 60.972 | 63.844 | 67.794 | 71.433 | 76.234 |
| 9 | 63.440 | 65.433 | 67.811 | 70.651 | 75.213 | 79.348 | 84.685 |
| 10 | 69.577 | 71.635 | 74.098 | 77.138 | 81.887 | 85.907 | 90.604 |

Table 3 (CASE II) continued

| q | .75 | .80 | .85 | .90 | .95 | .975 | .99 |
|---|---|---|---|---|---|---|---|
| | | | | **Level of Significance** | | | |
| | | | | $r_0 = 4$ | | | |
| 4 | 42.689 | 45.741 | 47.324 | 49.705 | 53.487 | 56.146 | 60.238 |
| 5 | 49.335 | 50.932 | 52.933 | 55.470 | 59.333 | 62.881 | 67.126 |
| 6 | 58.031 | 59.762 | 61.864 | 64.610 | 68.743 | 72.384 | 76.883 |
| 7 | 66.278 | 68.207 | 70.481 | 73.356 | 77.948 | 82.062 | 87.064 |
| 8 | 74.675 | 76.900 | 79.473 | 82.657 | 87.360 | 91.734 | 96.771 |
| 9 | 83.210 | 85.305 | 88.013 | 91.235 | 96.779 | 101.030 | 106.327 |
| 10 | 91.723 | 94.012 | 96.841 | 100.375 | 105.285 | 109.922 | 115.031 |
| | | | | $r_0 = 5$ | | | |
| 5 | 60.715 | 62.439 | 64.486 | 67.175 | 71.350 | 74.982 | 79.545 |
| 6 | 71.334 | 73.236 | 75.613 | 78.554 | 83.118 | 87.482 | 91.994 |
| 7 | 81.875 | 83.984 | 86.558 | 89.673 | 94.096 | 98.242 | 103.545 |
| 8 | 92.571 | 94.665 | 97.391 | 100.758 | 106.018 | 110.707 | 116.037 |
| 9 | 103.214 | 105.487 | 108.139 | 111.514 | 116.928 | 121.459 | 127.458 |
| 10 | 113.233 | 115.807 | 118.773 | 122.476 | 127.889 | 133.392 | 139.122 |
| | | | | $r_0 = 6$ | | | |
| 6 | 84.745 | 86.735 | 89.268 | 92.279 | 97.176 | 101.267 | 106.379 |
| 7 | 97.305 | 99.362 | 101.922 | 105.322 | 110.479 | 114.997 | 120.367 |
| 8 | 110.074 | 112.430 | 115.175 | 118.550 | 123.768 | 129.046 | 134.589 |
| 9 | 122.512 | 125.064 | 127.936 | 131.704 | 137.670 | 142.731 | 148.737 |
| 10 | 134.850 | 137.501 | 140.754 | 144.963 | 151.081 | 156.410 | 162.339 |
| | | | | $r_0 = 7$ | | | |
| 7 | 112.564 | 114.926 | 117.660 | 121.211 | 126.282 | 130.977 | 136.825 |
| 8 | 127.363 | 129.823 | 132.772 | 136.386 | 142.500 | 147.480 | 153.442 |
| 9 | 141.816 | 144.689 | 148.008 | 151.950 | 157.551 | 163.035 | 169.421 |
| 10 | 156.186 | 159.073 | 162.497 | 166.663 | 173.394 | 178.952 | 186.074 |
| | | | | $r_0 = 8$ | | | |
| 8 | 144.741 | 147.433 | 150.432 | 153.755 | 159.762 | 165.084 | 171.371 |
| 9 | 161.355 | 164.277 | 167.491 | 171.624 | 177.892 | 183.464 | 190.394 |
| 10 | 177.521 | 180.525 | 184.081 | 189.077 | 195.771 | 201.622 | 208.308 |
| | | | | $r_0 = 9$ | | | |
| 9 | 180.486 | 183.477 | 187.107 | 191.582 | 198.316 | 203.732 | 210.388 |
| 10 | 198.757 | 201.907 | 205.307 | 209.927 | 216.775 | 223.344 | 230.632 |
| | | | | $r_0 = 10$ | | | |
| 10 | 220.148 | 223.474 | 227.140 | 231.415 | 238.791 | 245.344 | 253.906 |

Table 4 (CASE II)

## Maximal Root

Significance Points for Conformity
Cointegration Test.
Combinations of System Size ($q$) and
Zero Roots Under the Null ($r_0$)

| q | Level of Significance | | | | | | |
|---|---|---|---|---|---|---|---|
| | **.75** | **.80** | **.85** | **.90** | **.95** | **.975** | **.99** |
| | | | | $r_0 = 1$ | | | |
| 1 | 4.371 | 4.916 | 5.618 | 6.574 | 8.185 | 9.756 | 11.667 |
| 2 | 6.857 | 7.542 | 8.392 | 9.506 | 11.405 | 13.235 | 15.464 |
| 3 | 9.184 | 9.987 | 10.950 | 12.203 | 14.231 | 16.188 | 18.582 |
| 4 | 11.562 | 12.470 | 13.604 | 14.985 | 17.267 | 19.337 | 22.041 |
| 5 | 13.820 | 14.776 | 15.902 | 17.429 | 19.962 | 22.254 | 25.207 |
| 6 | 16.072 | 17.080 | 18.338 | 19.926 | 22.469 | 24.730 | 28.056 |
| 7 | 18.242 | 19.376 | 20.722 | 22.508 | 25.184 | 27.574 | 30.598 |
| 8 | 20.527 | 21.695 | 23.170 | 24.935 | 27.799 | 30.447 | 33.367 |
| 9 | 22.764 | 24.015 | 25.505 | 27.398 | 30.320 | 33.024 | 36.473 |
| 10 | 24.871 | 26.121 | 27.789 | 29.794 | 32.870 | 35.773 | 38.798 |
| | | | | $r_0 = 2$ | | | |
| 2 | 10.174 | 10.908 | 11.817 | 13.057 | 15.096 | 16.983 | 19.305 |
| 3 | 12.940 | 13.745 | 14.747 | 16.086 | 18.198 | 20.103 | 22.555 |
| 4 | 15.721 | 16.626 | 17.746 | 19.153 | 21.499 | 23.799 | 26.394 |
| 5 | 18.267 | 19.258 | 20.445 | 21.969 | 24.382 | 26.599 | 29.547 |
| 6 | 20.964 | 22.041 | 23.300 | 24.900 | 27.362 | 29.680 | 32.681 |
| 7 | 23.400 | 24.473 | 25.742 | 27.422 | 30.313 | 32.784 | 35.755 |
| 8 | 25.908 | 27.085 | 28.442 | 30.306 | 32.976 | 35.387 | 38.387 |
| 9 | 28.276 | 29.522 | 30.836 | 32.684 | 35.734 | 38.293 | 41.410 |
| 10 | 30.599 | 31.873 | 33.363 | 35.283 | 38.464 | 41.286 | 44.697 |
| | | | | $r_0 = 3$ | | | |
| 3 | 15.855 | 16.694 | 17.723 | 19.111 | 21.187 | 23.257 | 25.620 |
| 4 | 18.940 | 19.873 | 20.988 | 22.485 | 25.008 | 27.127 | 29.894 |
| 5 | 21.673 | 22.627 | 23.803 | 25.388 | 27.826 | 30.040 | 32.858 |
| 6 | 24.528 | 25.576 | 26.816 | 28.418 | 31.145 | 33.594 | 36.819 |
| 7 | 27.222 | 28.315 | 29.679 | 31.410 | 34.090 | 36.531 | 39.549 |
| 8 | 29.961 | 31.165 | 32.471 | 34.302 | 37.070 | 39.686 | 42.563 |
| 9 | 32.451 | 33.626 | 35.086 | 36.974 | 39.951 | 42.583 | 45.879 |
| 10 | 35.067 | 36.300 | 37.731 | 39.734 | 42.692 | 45.652 | 48.986 |

Table 4 (CASE II), continued

| q | .75 | .80 | .85 | .90 | .95 | .975 | .99 |
|---|---|---|---|---|---|---|---|
| | | | Level of Significance | | | | |
| | | | $r_0 = 4$ | | | | |
| 4 | 20.774 | 23.014 | 24.668 | 25.712 | 28.317 | 30.478 | 33.689 |
| 5 | 24.545 | 25.568 | 26.824 | 28.470 | 31.152 | 33.537 | 36.648 |
| 6 | 27.656 | 28.717 | 30.035 | 31.706 | 34.356 | 36.956 | 40.435 |
| 7 | 30.343 | 31.468 | 32.852 | 34.663 | 37.510 | 39.915 | 43.283 |
| 8 | 33.376 | 34.522 | 35.855 | 37.691 | 40.542 | 43.173 | 46.363 |
| 9 | 36.064 | 37.238 | 38.717 | 40.552 | 43.338 | 46.254 | 49.423 |
| 10 | 38.773 | 40.018 | 41.487 | 43.327 | 46.487 | 49.199 | 52.443 |
| | | | $r_0 = 5$ | | | | |
| 5 | 27.169 | 28.202 | 29.466 | 31.173 | 33.844 | 36.246 | 39.176 |
| 6 | 30.365 | 31.503 | 32.837 | 34.521 | 37.258 | 39.917 | 43.253 |
| 7 | 33.323 | 34.495 | 35.972 | 37.764 | 40.530 | 42.967 | 46.184 |
| 8 | 36.330 | 37.495 | 39.025 | 40.904 | 43.909 | 46.580 | 49.987 |
| 9 | 39.200 | 40.400 | 41.902 | 43.746 | 46.692 | 49.484 | 53.204 |
| 10 | 42.140 | 43.365 | 44.821 | 46.760 | 50.017 | 53.014 | 56.261 |
| | | | $r_0 = 6$ | | | | |
| 6 | 29.687 | 30.718 | 32.032 | 33.753 | 36.407 | 38.803 | 42.163 |
| 7 | 32.903 | 34.018 | 35.392 | 37.202 | 40.186 | 42.659 | 46.003 |
| 8 | 36.261 | 37.524 | 38.965 | 40.729 | 43.526 | 45.948 | 49.309 |
| 9 | 39.132 | 40.318 | 41.740 | 43.745 | 46.941 | 49.342 | 53.090 |
| 10 | 42.167 | 43.428 | 44.956 | 46.862 | 49.798 | 52.529 | 56.107 |
| | | | $r_0 = 7$ | | | | |
| 7 | 38.693 | 39.832 | 41.248 | 43.180 | 46.130 | 48.990 | 52.254 |
| 8 | 41.825 | 43.083 | 44.580 | 46.455 | 49.590 | 52.440 | 56.006 |
| 9 | 44.957 | 46.220 | 47.670 | 49.481 | 52.827 | 55.976 | 59.436 |
| 10 | 48.023 | 49.338 | 50.852 | 52.998 | 56.305 | 59.310 | 63.050 |
| | | | $r_0 = 8$ | | | | |
| 8 | 44.418 | 45.687 | 47.119 | 49.075 | 52.223 | 54.891 | 58.581 |
| 9 | 47.527 | 48.705 | 50.243 | 52.111 | 55.394 | 58.557 | 62.225 |
| 10 | 50.690 | 52.101 | 53.697 | 55.784 | 59.209 | 61.993 | 65.340 |
| | | | $r_0 = 9$ | | | | |
| 9 | 50.052 | 51.244 | 52.771 | 54.873 | 58.179 | 61.107 | 64.770 |
| 10 | 51.355 | 54.682 | 56.237 | 58.328 | 61.624 | 64.448 | 68.266 |
| | | | $r_0 = 10$ | | | | |
| 10 | 55.900 | 57.207 | 58.853 | 61.045 | 64.217 | 67.117 | 70.936 |

Table 5 (CASE III)

**Trace**

Significance Points for Conformity

Cointegration Test.

Combinations of System Size ($q$) and

Zero Roots Under the Null ($r_0$)

| q | .75 | .80 | .85 | .90 | .95 | .975 | .99 |
|---|-----|-----|-----|-----|-----|------|-----|
| | | | | **Level of Significance** | | | |
| | | | | $r_0 = 1$ | | | |
| 1 | 7.176 | 7.810 | 8.652 | 9.784 | 11.702 | 13.314 | 15.875 |
| 2 | 9.447 | 10.259 | 11.194 | 12.462 | 14.623 | 16.665 | 19.130 |
| 3 | 11.720 | 12.567 | 13.625 | 14.990 | 17.202 | 19.358 | 22.131 |
| 4 | 13.963 | 14.896 | 16.036 | 17.578 | 20.037 | 22.282 | 25.042 |
| 5 | 16.271 | 17.273 | 18.585 | 20.204 | 22.709 | 25.059 | 28.005 |
| 6 | 18.451 | 19.560 | 20.863 | 22.561 | 25.252 | 27.537 | 31.099 |
| 7 | 20.546 | 21.710 | 23.089 | 25.006 | 27.914 | 30.233 | 33.538 |
| 8 | 22.861 | 24.072 | 25.483 | 27.283 | 30.184 | 32.942 | 36.578 |
| 9 | 24.985 | 26.230 | 27.632 | 29.550 | 32.750 | 35.579 | 38.954 |
| 10 | 27.109 | 28.492 | 30.081 | 32.097 | 35.454 | 38.321 | 41.982 |
| | | | | $r_0 = 2$ | | | |
| 2 | 17.529 | 18.523 | 19.837 | 21.478 | 24.006 | 26.430 | 29.453 |
| 3 | 21.977 | 23.110 | 24.470 | 26.226 | 28.991 | 31.512 | 34.902 |
| 4 | 26.384 | 27.614 | 29.093 | 31.012 | 34.031 | 36.836 | 40.277 |
| 5 | 30.629 | 31.968 | 33.658 | 35.727 | 39.210 | 42.452 | 45.966 |
| 6 | 34.777 | 36.221 | 37.940 | 40.152 | 43.705 | 46.940 | 50.674 |
| 7 | 39.055 | 40.623 | 42.410 | 44.913 | 48.536 | 51.811 | 55.466 |
| 8 | 43.358 | 44.923 | 46.861 | 49.482 | 53.107 | 56.678 | 61.034 |
| 9 | 47.838 | 49.483 | 51.576 | 54.058 | 57.999 | 61.090 | 65.727 |
| 10 | 51.919 | 53.729 | 55.893 | 58.589 | 62.901 | 66.373 | 70.652 |
| | | | | $r_0 = 3$ | | | |
| 3 | 31.792 | 33.097 | 34.634 | 36.735 | 39.662 | 42.611 | 46.277 |
| 4 | 38.327 | 39.756 | 41.460 | 43.671 | 47.254 | 50.463 | 54.268 |
| 5 | 44.771 | 46.295 | 48.198 | 50.717 | 54.519 | 57.820 | 61.968 |
| 6 | 50.877 | 52.648 | 54.790 | 57.364 | 61.733 | 65.463 | 70.473 |
| 7 | 57.199 | 58.988 | 61.238 | 64.032 | 68.380 | 72.590 | 77.409 |
| 8 | 63.510 | 65.496 | 67.744 | 70.727 | 75.216 | 79.695 | 84.435 |
| 9 | 70.040 | 72.026 | 74.338 | 77.365 | 81.859 | 85.880 | 91.441 |
| 10 | 76.299 | 78.366 | 81.032 | 84.234 | 88.809 | 92.983 | 98.200 |

Table 5 (CASE III) continued

| q | Level of Significance | | | | | | |
|---|---|---|---|---|---|---|---|
| | .75 | .80 | .85 | .90 | .95 | .975 | .99 |
| | | | | $r_0 = 4$ | | | |
| 4 | 37.457 | 41.065 | 42.568 | 44.236 | 48.016 | 50.698 | 55.241 |
| 5 | 58.495 | 60.183 | 62.238 | 64.937 | 68.874 | 72.682 | 77.553 |
| 6 | 66.753 | 68.692 | 70.917 | 73.930 | 78.420 | 82.463 | 87.096 |
| 7 | 75.027 | 77.000 | 79.598 | 82.674 | 87.536 | 91.426 | 96.466 |
| 8 | 83.668 | 85.867 | 88.288 | 91.768 | 96.758 | 101.593 | 106.783 |
| 9 | 92.076 | 94.127 | 96.615 | 99.685 | 105.367 | 110.028 | 115.891 |
| 10 | 100.058 | 102.270 | 105.132 | 108.829 | 114.305 | 118.997 | 124.846 |
| | | | | $r_0 = 5$ | | | |
| 5 | 71.993 | 73.850 | 76.114 | 79.166 | 83.480 | 87.734 | 92.097 |
| 6 | 82.693 | 84.665 | 87.112 | 90.190 | 94.817 | 98.972 | 103.782 |
| 7 | 92.957 | 95.199 | 97.686 | 100.953 | 106.144 | 110.746 | 116.091 |
| 8 | 103.439 | 105.738 | 108.497 | 112.220 | 117.736 | 122.661 | 128.439 |
| 9 | 113.660 | 116.125 | 118.835 | 122.479 | 128.049 | 133.580 | 139.289 |
| 10 | 123.953 | 126.422 | 129.522 | 133.509 | 139.368 | 144.512 | 151.733 |
| | | | | $r_0 = 6$ | | | |
| 6 | 98.172 | 100.249 | 102.901 | 106.041 | 111.102 | 115.476 | 121.072 |
| 7 | 110.518 | 112.943 | 115.666 | 119.045 | 124.271 | 129.095 | 134.600 |
| 8 | 123.176 | 125.617 | 128.379 | 132.196 | 138.094 | 143.983 | 149.567 |
| 9 | 135.525 | 138.100 | 140.951 | 144.835 | 150.894 | 156.458 | 162.974 |
| 10 | 147.724 | 150.358 | 153.606 | 157.711 | 164.224 | 170.277 | 176.332 |
| | | | | $r_0 = 7$ | | | |
| 7 | 127.930 | 130.405 | 133.324 | 136.892 | 142.518 | 146.980 | 153.625 |
| 8 | 142.436 | 144.873 | 148.061 | 152.201 | 158.462 | 164.131 | 171.274 |
| 9 | 156.668 | 159.404 | 162.544 | 166.604 | 173.556 | 179.086 | 185.659 |
| 10 | 171.257 | 173.926 | 177.408 | 181.754 | 188.235 | 194.470 | 201.355 |
| | | | | $r_0 = 8$ | | | |
| 8 | 161.704 | 164.348 | 167.570 | 171.982 | 178.661 | 184.530 | 191.747 |
| 9 | 178.147 | 181.082 | 184.448 | 189.034 | 195.032 | 200.768 | 207.705 |
| 10 | 194.633 | 197.721 | 201.440 | 206.129 | 212.650 | 218.539 | 226.190 |
| | | | | $r_0 = 9$ | | | |
| 9 | 199.713 | 202.686 | 206.307 | 210.252 | 216.999 | 223.649 | 229.520 |
| 10 | 218.267 | 221.478 | 225.447 | 230.224 | 236.810 | 242.728 | 249.952 |
| | | | | $r_0 = 10$ | | | |
| 10 | 241.666 | 245.101 | 248.936 | 253.681 | 260.445 | 267.249 | 274.326 |

Table 6 (CASE III)

## Maximal Root

Significance Points for Conformity

Cointegration Test.

Combinations of System Size ($q$) and

Zero Roots Under the Null ($r_0$)

| q | Level of Significance | | | | | | |
|---|---|---|---|---|---|---|---|
| | .75 | .80 | .85 | .90 | .95 | .975 | .99 |
| | | | | $r_0 = 1$ | | | |
| 1 | 7.176 | 7.810 | 8.652 | 9.784 | 11.702 | 13.314 | 15.875 |
| 2 | 9.447 | 10.259 | 11.194 | 12.462 | 14.623 | 16.665 | 19.130 |
| 3 | 11.720 | 12.567 | 13.625 | 14.990 | 17.202 | 19.358 | 22.131 |
| 4 | 13.963 | 14.896 | 16.036 | 17.578 | 20.037 | 22.282 | 25.042 |
| 5 | 16.271 | 17.273 | 18.585 | 20.204 | 22.709 | 25.059 | 28.005 |
| 6 | 18.451 | 19.560 | 20.863 | 22.561 | 25.252 | 27.537 | 31.099 |
| 7 | 20.546 | 21.710 | 23.089 | 25.006 | 27.914 | 30.233 | 33.538 |
| 8 | 22.861 | 24.072 | 25.483 | 27.283 | 30.184 | 32.942 | 36.578 |
| 9 | 24.985 | 26.230 | 27.632 | 29.550 | 32.750 | 35.579 | 38.954 |
| 10 | 27.109 | 28.492 | 30.081 | 32.097 | 35.454 | 38.321 | 41.982 |
| | | | | $r_0 = 2$ | | | |
| 2 | 13.230 | 14.070 | 15.040 | 16.420 | 18.582 | 20.616 | 23.296 |
| 3 | 15.852 | 16.732 | 17.792 | 19.208 | 21.531 | 23.674 | 26.607 |
| 4 | 18.477 | 19.459 | 20.619 | 22.122 | 24.551 | 26.863 | 29.790 |
| 5 | 20.981 | 21.987 | 23.250 | 24.970 | 27.579 | 29.982 | 32.961 |
| 6 | 23.384 | 24.488 | 25.828 | 27.590 | 30.178 | 32.597 | 36.007 |
| 7 | 25.827 | 26.987 | 28.268 | 30.115 | 33.010 | 35.702 | 38.854 |
| 8 | 28.338 | 29.631 | 31.144 | 32.955 | 35.955 | 38.657 | 42.225 |
| 9 | 30.924 | 32.208 | 33.732 | 35.706 | 38.571 | 40.986 | 44.274 |
| 10 | 33.254 | 34.488 | 36.062 | 38.026 | 41.103 | 43.940 | 47.989 |
| | | | | $r_0 = 3$ | | | |
| 3 | 19.035 | 19.920 | 21.045 | 22.405 | 24.812 | 27.205 | 29.723 |
| 4 | 21.895 | 22.890 | 24.084 | 25.696 | 28.160 | 30.429 | 33.388 |
| 5 | 24.633 | 25.687 | 26.984 | 28.638 | 31.395 | 33.739 | 36.791 |
| 6 | 27.169 | 28.276 | 29.585 | 31.344 | 34.095 | 36.864 | 40.212 |
| 7 | 29.954 | 31.148 | 32.564 | 34.374 | 36.988 | 39.942 | 43.459 |
| 8 | 32.564 | 33.772 | 35.222 | 37.077 | 40.237 | 43.073 | 46.181 |
| 9 | 35.246 | 36.501 | 38.002 | 39.819 | 42.799 | 45.399 | 48.667 |
| 10 | 37.738 | 38.967 | 40.539 | 42.599 | 45.604 | 48.855 | 52.428 |

Table 6 (CASE III), continued

| | Level of Significance | | | | | | |
|---|---|---|---|---|---|---|---|
| q | .75 | .80 | .85 | .90 | .95 | .975 | .99 |
| | | | | $r_0 = 4$ | | | |
| 4 | 20.013 | 22.368 | 23.887 | 24.865 | 27.329 | 29.307 | 33.548 |
| 5 | 27.746 | 28.775 | 30.059 | 31.760 | 34.322 | 36.873 | 39.978 |
| 6 | 30.558 | 31.630 | 32.967 | 34.847 | 37.570 | 40.156 | 43.439 |
| 7 | 33.269 | 34.433 | 35.823 | 37.665 | 40.394 | 43.057 | 46.377 |
| 8 | 36.173 | 37.429 | 38.930 | 40.793 | 44.083 | 46.870 | 50.162 |
| 9 | 38.841 | 40.044 | 41.489 | 43.503 | 46.458 | 49.164 | 52.265 |
| 10 | 41.508 | 42.879 | 44.388 | 46.402 | 49.522 | 52.385 | 56.156 |
| | | | | $r_0 = 5$ | | | |
| 5 | 30.505 | 31.626 | 32.905 | 34.608 | 37.474 | 40.045 | 43.070 |
| 6 | 33.514 | 34.687 | 36.023 | 37.821 | 40.705 | 43.339 | 46.821 |
| 7 | 36.449 | 37.677 | 39.129 | 41.001 | 43.806 | 46.688 | 49.488 |
| 8 | 39.311 | 40.583 | 42.045 | 43.990 | 47.167 | 50.096 | 53.984 |
| 9 | 42.159 | 43.413 | 44.966 | 46.880 | 49.584 | 52.479 | 55.794 |
| 10 | 44.987 | 46.263 | 47.706 | 49.760 | 52.859 | 55.828 | 59.160 |
| | | | | $r_0 = 6$ | | | |
| 6 | 36.214 | 37.390 | 38.795 | 40.667 | 43.438 | 46.426 | 49.766 |
| 7 | 39.308 | 40.503 | 41.902 | 43.841 | 46.821 | 49.766 | 53.142 |
| 8 | 42.289 | 43.620 | 45.180 | 47.110 | 50.286 | 52.926 | 56.287 |
| 9 | 45.197 | 46.453 | 48.053 | 49.954 | 52.974 | 55.748 | 59.405 |
| 10 | 48.066 | 49.464 | 51.004 | 52.955 | 56.047 | 59.136 | 62.653 |
| | | | | $r_0 = 7$ | | | |
| 7 | 41.974 | 43.231 | 44.778 | 46.642 | 49.758 | 52.561 | 55.810 |
| 8 | 45.051 | 46.434 | 47.960 | 49.997 | 52.884 | 56.125 | 60.279 |
| 9 | 47.998 | 49.330 | 50.823 | 52.778 | 55.727 | 58.414 | 62.040 |
| 10 | 50.951 | 52.204 | 53.563 | 55.554 | 58.910 | 61.985 | 65.385 |
| | | | | $r_0 = 8$ | | | |
| 8 | 47.806 | 49.087 | 50.666 | 52.649 | 55.763 | 58.625 | 62.756 |
| 9 | 50.768 | 52.037 | 53.616 | 55.569 | 58.601 | 61.395 | 64.952 |
| 10 | 53.674 | 54.986 | 56.537 | 58.669 | 61.928 | 65.019 | 68.675 |
| | | | | $r_0 = 9$ | | | |
| 9 | 53.400 | 54.680 | 56.265 | 58.236 | 61.464 | 64.134 | 67.482 |
| 10 | 56.449 | 57.783 | 59.375 | 61.512 | 64.979 | 68.126 | 71.949 |
| | | | | $r_0 = 10$ | | | |
| 10 | 58.966 | 60.378 | 61.881 | 63.913 | 67.512 | 70.759 | 74.582 |

# Chapter 7

# Brownian Motion

## 7.1 Preliminaries

The origin of the theory of Brownian motion lies in an observation made by the English botanist Robert Brown in 1827; he reported that when particles of the pollen of plants were suspended in a liquid they seemed to execute complex movements in rather convoluted zig-zag fashion. Brown believed that he had discovered active molecules in organic and inorganic matter, while others disputed the validity of this claim. The matter remained unresolved until 1905 when Einstein revisited the subject, and modeled the movement of pollen grains by means of a set of differential equations, discovering in this fashion a central limit theorem! The topic was further explored by Paul Levy, a French mathematician and probabilist, and later by Norbert Wiener, an American mathematician who made important contributions in elucidating the mathematical properties of what we now term **Brownian motion**.

In a completely independent development, unrelated to the problem of characterizing the movement of particles suspended in a liquid, the French mathematician L. Bachelier (1900) in a doctoral thesis discovered the stochastic process now termed Brownian motion by studying and modeling the properties of fluctuations in financial markets. [1]

It is certainly not our intention here to provide a complete exposition of the origins and development of this subject. The interested reader is re-

---

[1] An English translation of his work appears as "Theory of Speculation" (Chapter 2), in Cootner (1964).

ferred to the excellent book by Hida (1980).[2] Our objective is to provide a modicum of information about the process and its properties. One may ask why an economist should study Brownian motion? Economists certainly do not study the movements of microscopic particles! Strange though it may seem, Brownian motion has found fruitful applications in financial theory, in explaining the price of risky assets such as stocks, options and similar securities. This received considerable impetus after the rediscovery of Bachelier's work in the fifties and other attempts to explain stock market fluctuations based on statistical mechanics which had their origin in the work of Einstein and Wiener. Of particular importance to us, Brownian motion (BM) has found extensive applications in econometrics: In the distribution of the least squares estimators of the GLM **when the regressors are** $I(1)$; in the distribution of tests statistics in the context of unit roots and cointegration problems, to mention but a few such applications examined in the preceding chapters.

## 7.2   Random Walk and Brownian Motion

The terms **random walk, Markov property** or **Markov chain**, and **martingale** appear frequently in the literature of econometrics and economic theory and it is important to have a clear understanding of their meaning. The imagery of the preceding section is an ideal context in which to develop the meaning of a random walk. Thus, suppose we idealize the observations above, and we conceive of the particle as beginning at the origin and moving, at each "step", a distance 1 (up) with probability $p$, or a distance minus 1 (down) with probability $q = 1 - p$; the movements at each step are assumed to represent a sequence of independent (Bernoulli) trials. The particle is then said to perform a random walk, and the plot of the position of the particle at each step is said to depict a sample path. Extracting the essential ideas from this description we arrive at a formal definition.

**Definition 1.** Let $X = \{X_n : n \in \mathcal{N}_0\}$ be a sequence of independent random variables defined on the probability space $(\Omega, \mathcal{A}, \mathcal{P})$. The entity $S_n = \sum_{j=1}^{n} X_j$, with the convention $S_0 = X_0 = 0$, $n \in \mathcal{N}_0$, is said to be a

---

[2] In fact some part of this exposition relies on Hida's exposition.

**random walk.**

**Remark 1.** Since the sequence $\{S_n : n \in \mathcal{N}_0\}$, where $\mathcal{N}_0 = \{0, 1, 2, \ldots\}$, is one of **independent increments**, i.e. $S_n - S_{n-1}$ is independent of $S_m - S_{m-1}$, for $n \neq m$, the definition states that a sequence with independent increments is a random walk. The typical usage in economics seems to define a random walk as a sequence whose increments are **independent** and **identically** distributed, but this is not the case in the general literature of mathematics.

If in the discussion above $p = q = (1/2)$ or, in the definition, if the distribution of the $X_j$ is **symmetric**, the stochastic sequence is said to represent a **symmetric random walk**.

The term **Markov chain** refers to the following context: We are given the particle of the previous examples and a collection, $S$, of "positions" that may be attained by the particle. The collection $S$ is at most **countable** and is often referred to as the **state space**; in this framework its elements are referred to as **states**. The probability of reaching state $j$ at step $n+1$, given that the particle is in state $i$ at step $n$, is denoted by $p_{ij,n}$, and the latter are collectively referred to as **transitional probabilities**. An application in economics is the modeling of total factor productivity (TFP) of firms or plants, in which the various "states" are simply bands of TFP attained by such plants or firms from year to year. Again extracting the basic concepts we are led to another formal definition.

**Definition 2.** Let $X = \{X_n : n \in \mathcal{N}_+\}$ be a sequence of r.v. defined on the probability space $(\Omega, \mathcal{G}, \mathcal{P})$ and taking values in the countable set $S$. Define the stochastic basis $\{\mathcal{G}_t : \mathcal{G}_t = \sigma(X_s, s \leq t), t \in \mathcal{N}_+\}$. The stochastic sequence $(X_n, \mathcal{G}_n)$ is said to have the **Markov property** if

$$p_{ij,n} = \mathcal{P}(X_{n+1} = j | \mathcal{G}_n) = \mathcal{P}(X_{n+1} = j | X_n = i);$$

It is said to be a **homogeneous Markov chain** if, in addition, $p_{ij,n} = p_{ij}$.

Finally, from Dhrymes (1989), p. 280, we have the following definition.

**Definition 3.** Let $X = \{X_n : n \in \mathcal{N}_0\}$ be a collection of r.v. defined on the probability space $(\Omega, \mathcal{G}, \mathcal{P})$ and define the stochastic basis (often called **filtration** in the literature) $\{\mathcal{G}_n : \mathcal{G}_n = \sigma(X_s, s \leq n, n \in \mathcal{N}_+)\}$ with

$\mathcal{G}_0 = (\emptyset, \Omega)$. The stochastic sequence $\{(X_n, \mathcal{G}_n), n \in \mathcal{N}_+\}$ is said to be a **martingale** if and only if $E|X_n| < \infty$ and $E(X_n|\mathcal{G}_{n-1}) = X_{n-1}$.

While the three concepts are similar, there are evidently differences. For example, after $n$ steps the position of the particle in the **symmetric random walk** above may be represented by

$$S_n = X_0 + \sum_{i=1}^{n} X_i, \quad \text{with} \quad X_0 = 0.$$

The sequence $\{S_n, \mathcal{G}_n : n \in \mathcal{N}_+\}$ is certainly a martingale since, for every $n \in \mathcal{N}_+$, $E|S_n| \leq n < \infty$ and $E(S_n|\mathcal{G}_{n-1}) = S_{n-1}$. A nonsymmetric random walk, however, is not necessarily a martingale. In the example given earlier, suppose $p = \alpha q$, $\alpha \neq 1$, so that $EX_n = (\alpha - 1)q = \mu \neq 0$. In this context

$$E(S_n|\mathcal{G}_{n-1}) = S_{n-1} + EX_n = S_{n-1} + \mu \neq S_{n-1}.$$

The behavior of many economic time series is modeled, in contemporary econometric practice, as $\Delta x_t = \epsilon_t$, where $x_t$ is the economic time series in question, and $\epsilon = \{\epsilon_t : t \geq 1\}$ is $WN(1)$. By recursion, and taking $x_0 = 0$, we find $x_t = \sum_{j=1}^{t} \epsilon_j$. This is, evidently, a martingale since $E|x_n| \leq nE|\epsilon_1|$ and, in the appropriate context, $E(x_n|\mathcal{G}_{n-1}) = x_{n-1}$. According to our definition of (general) random walks (Definition 1) it is also a random walk, even though the examples we had used earlier involved Bernoulli trials with a **finite number** of "states" (up or down, minus 1 or plus 1).

Since random walks are more familiar to economists it is perhaps best if we think of Brownian motion as the limit of a random walk process. The intuitive appeal of the motivation is strengthened if we keep the imagery conveyed by the particle movements of the previous section but **restrict its movement to the line**, although real particles evidently move in three-dimensional space. Suppose further that when the particle collides with molecules of the liquid, the particle is displaced by $\pm\sqrt{\delta}$ per collision. Let $\{X_n : n \in \mathcal{N}_+\}$ denote the impact sequence, and suppose its elements are i.i.d. with a (symmetric) Bernoulli distribution, i.e. $P(X_n = \sqrt{\delta}) = 1/2$, $P(X_n = -\sqrt{\delta}) = 1/2$, so that $EX_n = 0$, $\text{Var}(X_n) = \delta$. Assume the particle is initially at the origin, and denote by $S_n$ the position of the particle after $n$ collisions. Since we are dealing with microscopic particles, and infinitesimal displacements, we adjust the "time frame" accordingly. Thus, in a unit of time "$t$" we assume that there are $[t/\delta]$ collisions, where $[a]$ denotes the integer part of $a$. It follows immediately that the position of a particle, initially at zero, is given at time $t$ by

$$S_t = \sum_{j=1}^{[\frac{t}{\delta}]} X_j. \tag{7.1}$$

It is easy to verify that $ES_t = 0$, $\text{Var}(S_t) = [t/\delta]\delta = t$, for sufficiently small $\delta$, or more appropriately as $\delta \downarrow 0$.

One of the first central limit theorems (CLT), the De Moivre-Laplace CLT, states that for a normalized sum of Bernoulli variables with probabilities $p + q = 1$,

$$\lim_{n \to \infty} Pr\left\{ \left( \frac{S_n - np}{\sqrt{npq}} \right) \in (a, b) \right\} = \frac{1}{\sqrt{2\pi}} \int_a^b e^{-(1/2)\xi^2} \, d\xi.$$

Consequently, we have established that as $\delta \downarrow 0$, with $t$ fixed

$$\frac{S_t}{\sqrt{t}} \xrightarrow{\text{d}} N(0,1), \quad \text{or} \quad S_t \xrightarrow{\text{d}} N(0,t). \tag{7.2}$$

The entity in Eq. (7.1) is the prototype of the Brownian motion (BM). We shall now deal with BM more formally.

**Definition 4.** A collection of random elements $\xi = \{\xi_\tau. : \tau \in \mathcal{T}\}$, where each $\xi_\tau.$ is an $m$-element (row) vector, is said to be a **Gaussian system** if and only if for every $n \geq 1$ and $\tau_1, \tau_2, \ldots, \tau_n \in \mathcal{T}$ the random vector

$$\zeta_{(n)} = (\xi_{\tau_1}., \xi_{\tau_2}., \ldots \xi_{\tau_n}.)'$$

has a multivariate normal distribution. An equivalent requirement is that for $\tau_i$ as above, and $\alpha_i \in R$, $\xi = \sum_{i=1}^n \alpha_i \xi_{\tau_i}.$ is normally distributed.

An immediate consequence of the preceding is Proposition 1.

**Proposition 1.** Let $\xi$ be Gaussian system as in Definition 4, and put

$$C_{\tau,\tau'} = \text{Cov}(\xi_\tau., \xi_{\tau'}.).$$

The following statements are true.

i. $\xi_\tau.$ and $\xi_{\tau'}.$ are mutually independent, if and only if $C_{\tau,\tau'} = 0$.

ii. $\xi_{\tau_0}$. is independent of the Gaussian system $\xi^* = \{\xi_\tau. : \tau \in \mathcal{T}, \tau \neq \tau_0\}$ if and only if $C_{\tau,\tau_0} = 0$ for all $\tau \neq \tau_0$.

Proof: In view of Definition 4 this is obvious since independence between two jointly normal vectors is equivalent to zero covariance.

q.e.d.

The preceding then suggests the following.

**Definition 4a.** A Gaussian system as in Definition 4, is said to be an **independent system** if and only if $C_{\tau,\tau'} = 0$, for $\tau \neq \tau'$.

We now turn to the discussion of Brownian motion.

**Definition 5.** A stochastic process $B = \{B(t,\omega) : t \in R_+, \ R_+ = [0,\infty)\}$, defined on the probability space ($\Omega$, $\mathcal{A}$, $\mathcal{P}$) and taking values in the measurable space $(\mathcal{E}, \mathcal{B}(\mathcal{E}))$ [3] is said to be a one-dimensional **Brownian motion**, if and only if it is a Gaussian system and satisfies:

i. $m = 1$;

ii. $B(0,\omega) = 0$, a.c., i.e. if $A = \{\omega : B(0,\omega) \neq 0\}$ then $\mathcal{P}(A) = 0$;

iii. for every $t > 0$, and $s$ such that $t + s > 0$, $B(t + s, \cdot) - B(t, \cdot)$ is normally distributed on ($\Omega$, $\mathcal{A}$, $\mathcal{P}$), with mean zero and variance $|s|$;

iv. it follows immediately from ii and iii that for every $t > 0$, $B(t, \cdot) \sim N(0,t)$.

This will always be referred to as the **standard Brownian motion** and denoted by SBM.

It is said to be a $q$-dimensional **Brownian motion** if and only if it is a Gaussian system, and satisfies:

i. $m = q$;

ii. $B(0,\omega) = [B_1(0,\omega), B_2(0,\omega), \ldots, B_q(0,\omega)] = 0$, a.c.

---

[3] This is occasionally referred to as the **state space**; in nearly all of our discussions $\mathcal{E} = R^n$ and $\mathcal{B}(\mathcal{E}) = \mathcal{B}(R^n)$. Often $n = 1$, i.e. the state space is the ordinary one-dimensional Borel space.

iii. for every $t > 0$, and $s$ such that $t + s > 0$, $B(t+s, \cdot) - B(t, \cdot)$ is normally distributed on $(\Omega, \mathcal{A}, \mathcal{P})$, with mean zero and covariance matrix $|s|\Sigma$, $\Sigma \geq 0$;

iv. it follows immediately from ii and iii that for every $t > 0$, $B(t, \cdot) \sim N(0, t\Sigma)$; this stochastic process is referred to as the **multivariate Brownian motion** (MBM).

v. The **standard multivariate Brownian motion** (SMBM) obeys the requirements in i, ii and iii **but**, for every $t > 0$, $B(t, \cdot) \sim N(0, t I_q)$.

**Remark 2**. Notice that the SMBM, as defined above, simply consists of $q$ **independent** copies of the standard (univariate) BM.

Some authors denote the MBM by the notation $B(\Sigma)$. We do not follow this practice. In subsequent discussion, when we have occasion to deal with the $MBM(\Sigma)$, we will use the notation $BP$, where $B$ is the SMBM defined above, and $P$ is the (triangular) matrix of the decomposition $\Sigma = P'P$. Thus, in our usage the symbol $B$ will always denote the SMBM, which thus serves as the basic "unit" for such stochastic processes. This is analogous to writing a $N(0, \sigma^2)$ random variable as $\sigma N(0, 1)$. It is a useful device in that it eliminates ponderous, or obscure, notation but the reader should note that it does not address covariance or correlation issues. Thus, for example, if we deal with two MBM say $MBM(\Sigma_1)$ and $MBM(\Sigma_2$, we shall denote them by $BP_1$ and $BP_2$, respectively. This does **not** necessarily mean that the symbol $B$ refers to the **same** stochastic process (SMBM) in **both** MBM; it only means that it is a transformation of a SMBM and the matrix of the transformation is $P_i$, $i = 1, 2$. It is precisely the same situation we have when we state that $X$ and $Y$ are each $N(0, I)$. Evidently, the notation makes it clear that the two $N(0, I)$ variables are not the same! Nor is it the case that the notation $BP_i$, for $i = 1, 2$, tells us anything about the covariance relation between the two MBM, and precisely the same would be true even if we stated them as $MBM(\Sigma_1)$ and $MBM(\Sigma_2)$, respectively.

An immediate consequence of Definition 5 is [4] Proposition 2.

---

[4] Henceforth, to avoid repetition and ponderous notation, we deal only with one-dimensional BM unless indicated to the contrary, and we replace $B(t_i, \cdot)$ by $B_i$, and $B(t, \omega)$ by $B_t$ or $B(t)$. **We caution the reader**, however, that **in a multivariate**

**Proposition 2.** Let $B$ be a SBM, as in Definition 5. The following statements are true:

i. if $t_1, t_2 > 0$ are elements of $R_+$, $EB_1B_2 = \min(t_1, t_2)$;

ii. if $0 < t_1 < t_2 < \ldots < t_n$ are elements of $R_+$, the vector $\zeta_n = (B_1, B_2 - B_1, B_3 - B_2, \ldots, B_n - B_{n-1})'$ has the multivariate normal distribution with mean vector zero, and covariance matrix

$$\Sigma = \text{diag}(\sigma_{11}, \sigma_{22}, \sigma_{33}, \ldots, \sigma_{nn}),$$

where $\sigma_{11} = t_1$, $\sigma_{jj} = t_j - t_{j-1}$, $j = 1, 2, \ldots n - 1$;

iii. $B$ has independent increments, and given any partition of the interval $0 < a = t_1 < t_2 < t_3 < \ldots < t_n = b$, one can write

$$B(b, \cdot) = \sum_{j=1}^{n} (B_j - B_{j-1}), \quad B(0, \cdot) = 0, \text{ a.c.};$$

iv. the vector $\xi_n = (B_1, B_2, B_3, \ldots, B_n)'$ is normal with mean vector zero and covariance matrix $\Sigma = [\sigma_{ij}]$, such that $\sigma_{jj} = t_j$, and for $i \neq j$, $\sigma_{ij} = \min(t_i, t_j)$;

v. for $n = 3$

$$E(B_2|B_1, B_3) = \frac{(t_3 - t_2)B_1 + (t_2 - t_1)B_3}{t_3 - t_1}. \tag{7.3}$$

Proof: Since

$$EB_1B_2 = \frac{1}{2}\left[EB_2^2 + EB_1^2 - E(B_2 - B_1)^2\right] = \frac{1}{2}\left[t_2 + t_1 - |t_2 - t_1|\right],$$

if $t_2 > t_1$, the expectation in the left member is equal to $t_1$; if $t_2 < t_1$ the expectation in the left member is equal to $t_2$, which proves part i.

For part ii, we note that

$$E(B_i - B_{i-1})(B_j - B_{j-1}) = \min(t_i, t_j) - \min(t_i, t_{j-1}) - \min(t_{i-1}, t_j)$$

$$+ \min(t_{i-1}, t_{j-1}) = 0, \quad \text{for } i \neq j,$$

$$= t_i - t_{i-1}, \quad \text{for } i = j.$$

---

**context** the subscript, say $B_i$, will denote the $i^{th}$ component of the SMBM; in that context the notation $B_i(t_j)$ will indicate the ordinate of the $i^{th}$ component of the SMBM at "time" $t_j$. The context of the discussion will generally make clear the meaning of the notation.

This completes the proof of ii, since by property ii of Definition 4 each of the elements of the vector $\zeta_n$ is normally distributed.

For part iii, we note that the components of $\zeta_n$ are mutually independent and moreover, $B(b) = B(t_n) = \sum_{j=1}^{n}(B_j - B_{j-1})$, it being understood that $B_0 = B(0) = 0$.

For part iv we note that

$$\xi_n = \begin{bmatrix} 1 & 0 & 0 & 0 & \cdots & 0 \\ 1 & 1 & 0 & 0 & \cdots & 0 \\ 1 & 1 & 1 & 0 & \cdots & 0 \\ \vdots & \vdots & \vdots & \vdots & \cdots & \vdots \\ 1 & 1 & 1 & 1 & \cdots & 1 \end{bmatrix} \zeta_n.$$

The conclusion follows immediately from standard normal theory; similarly for part v.

q.e.d.

Another set of useful implications is readily derived from Definition 2.

**Proposition 3.** Let $B$ be a SBM, as in Definition 5. The following statments are true:

i. for arbitrary but fixed $a > 0$ and $\phi \neq 0$,

    1. $V(t), t \in R_+$ is a BM, where $V(t) = B(t + a) - B(a)$;

    2. $V(t), t \in R_+$ is a BM, where $V(t) = (1/\phi)B(\phi^2 t)$;

ii. $B(t)$ and $tB(1/t)$, $t > 0$, have the same distribution.

Proof: To prove i.1, we note that by ii of Proposition 2 $V(t)$ is a normal process. In addition, $V(0) = 0$ a.c. and for $t, t + s > 0$,

$$E[V(t + s) - V(t)]^2 = E[B(t + s + a) - B(t + a)]^2 = |s|.$$

To prove i.2 we note that if $t > 0$, $\phi^2 t > 0$, and both are elements of $R_+$; hence, $V$ is normally distributed for $t > 0$, with mean zero and variance $\phi^2 t / \phi^2 = t$. Moreover, for $t, s$ such that $t > 0$, $t + s > 0$,

$$E[V(t + s) - V(t)]^2 = \frac{1}{\phi^2} E[B(\phi^2(t + s)) - B(\phi^2 t)]^2 = \frac{\phi^2}{\phi^2}|s| = |s|.$$

Finally, to prove ii we note that for $0 < t < \infty$, $(1/t) \in R_+$, and $B(1/t)$ is also a BM, although in "reverse" time relative to $B(t)$; thus it is normally distributed, with mean zero and variance $(1/t)$. Consequently, $tB(1/t)$ is normally distributed with mean zero and variance $t$, as is $B(t)$.

q.e.d.

## 7.3   P. Levy's Construction

In the preceding we have established a number of properties for the BM, based on Definition 5. We have not, however, shown that such a process does, indeed, exist, since all our operations involved finite dimensional vectors. By the definition of BM we are essentially dealing with an infinity of random variables, since for each (fixed) $t$, $B(t, \omega)$ is a random variable and, for each $\omega$, BM is a function in $R^{[0,\infty]}$, the space of functions defined on $[0, \infty]$. We might characterize it, therefore, as an infinite dimensional vector; since for finite $n$, the distribution of $[B(t_1), B(t_2), \dots, B(t_n)]'$ is easily established, we may extend this distribution to $\mathcal{B}(R^{[0,\infty]})$ by invoking the Kolmogorov extension theorem [see Dhrymes (1989), p. 90]. Such an approach would ignore all properties of the sample paths of BM, other than their being real-valued functions.

It can be shown that the sample paths of the BM are **continuous**. To establish the **continuity of such sample paths**, we introduce a construction of the BM, due to P. Levy (1948), of which the following is the basic idea.

**Proposition 4.** Let $B$ be a SBM as in Proposition 2, and $(a, b) \subset [0, \infty)$; for every $t \in (a, b)$ we may represent

$$B(t) = \mu(t) + \sigma(t)X(t), \quad \sigma^2(t) = \frac{(t - a)(b - t)}{(b - a)}, \tag{7.4}$$

$$\mu(t) = \frac{(b - t)B(a) + (t - a)B(b)}{b - a}, \quad X_t \sim N(0, 1),$$

where $X(t)$ is independent of $B(s)$, $s \in [0, a] \cup [b, \infty)$.

Proof: From part v of Proposition 2, $\mu(t)$ is the conditional mean of $B(t)$, ($B_2$ in the proposition), **given** $B(a), B(b)(B_1, B_3$ in the proposition), and $X(t)$

is **independent of** $B(a)$ and $B(b)$; thus, it may be easily verified that the expectation of the representation above is zero, and its variance is $t$. Moreover, since $B$ is a Gaussian system and $\mu(t)$ is independent of $X(t)$ it follows that the right member of the representation is normal with mean zero and variance $t$.

q.e.d.

**Remark 3**. The proposition shows that if we know the value assumed by a BM at two points $a$ and $b$, we can determine its value **in the interval** $(a, b)$ by **interpolation**.

Let $Y = \{Y_n : n \geq 1\}$ be a sequence of unit normal variables. Let the sets $T_n$ be given by $T_n = \{2^{-(n-1)}k,\ k = 0, 1, 2, \ldots, 2^{n-1}\}$, $T_0 = \bigcup_{n=1}^{\infty} T_n$, and define the normal processes $X_n : n > 1$ and $X_1$ as follows:

$$X_{n+1}(t) = X_n(t), \text{ for } t \in T_n \tag{7.5}$$

$$= \frac{1}{2}\left[X_n(t + 2^{-n}) + X_n(t - 2^{-n})\right] + 2^{-\frac{n+1}{2}}Y_{k(t)}, \ t \in T_{n+1} \cap \overline{T}_n$$

$$= 2^n[(k+1)2^{-n} - t]X_{n+1}k2^{-n} + 2^n(t - k2^{-n})X_{n+1}[(k+1)2^{-n}],$$

$$\text{for} \quad t \in [k2^{-n},\ (k+1)2^{-n}],$$

$$X_1(0) = 0, \quad X_1(1) = Y_1, \quad X_1(t) = tY_1, \quad t \in (0, 1),$$

where $k(t) = 2^n + (1/2)(2^n t - 1)$, for $t \in T_{n+1} \cap \overline{T}_n$. Note that in the construction above, the normal process $X_1(t, \omega)$ is completely defined on $[0, 1]$. The expression for $X_{n+1}$ assumes that $X_n$ **has been defined** and specifies $X_{n+1}$, given $X_n$. Note further that the first two expressions give the definition of the process for the rational lattice points contained in $T_{n+1} \supset T_n$, while the third expression defines $X_{n+1}$ by **interpolation** between lattice points, thus completing the definition on $[0, 1]$. It is easy to verify that the definition of the process on the rational lattice points of $[0, 1]$, for each $n$, gives a sequence with independent normal increments having mean zero and appropriate variance. Indeed, if we confine $X_n(t)$ to $t \in T_n$ we see that we have a BM, **with discrete parameter**. Perhaps a few computations will make this transparent, and disclose the elegance and extreme simplicity of the scheme. We obtain, **ignoring the interpolation aspect of the** specification,

$$X_1(0) = 0, \quad X_1(1) = Y_1, \tag{7.6}$$

$$X_2(0) = 0, \quad X_2(1) = Y_1, \quad X_2(1/2) = \frac{1}{2}(Y_1 + Y_2),$$

$$X_3(0) = 0, \quad X_3(1) = Y_1, \quad X_3(1/2) = \frac{1}{2}(Y_1 + Y_2),$$

$$X_3(1/4) = \frac{1}{4}(Y_1 + Y_2) + \frac{1}{2^{3/2}}Y_4, \quad X_3(3/4) = \frac{3}{4}Y_1 + \frac{1}{4}Y_2 + \frac{1}{2^{3/2}}Y_5,$$

$$X_4(0) = 0, \quad X_4(1) = Y_1, \quad X_4(1/2) = \frac{1}{2}(Y_1 + Y_2),$$

$$X_4(1/4) = \frac{1}{4}(Y_1 + Y_2) + \frac{1}{2^{3/2}}Y_4, \quad X_4(3/4) = \frac{3}{4}Y_1 + \frac{1}{4}Y_2 + \frac{1}{2^{3/2}}Y_5,$$

$$X_4(1/8) = \frac{1}{8}(Y_1 + Y_2) + \frac{1}{2^{5/2}}Y_4 + \frac{1}{4}Y_8,$$

$$X_4(3/8) = \frac{3}{8}(Y_1 + Y_2) + \frac{1}{2^{5/2}}Y_4 + \frac{1}{4}Y_9,$$

$$X_4(5/8) = \frac{5}{8}Y_1 + \frac{3}{8}Y_2 + \frac{1}{2^{5/2}}Y_5 + \frac{1}{4}Y_{10},$$

$$X_4(7/8) = \frac{7}{8}Y_1 + \frac{1}{8}Y_2 + \frac{1}{2^{5/2}}Y_5 + \frac{1}{4}Y_{11}.$$

Note the essential simplicity of this scheme which defines the process by successive iterations; in the first iteration, we define the process for the points $\{0, 1\}$; in the second, **we retain the definitions of the first iteration, and we add a definition for the point** $\{1/2\}$; in the third iteration we add a definition for the points $\{1/4, 3/4\}$, and so on. It is intuitively clear that, proceeding in this ordered fashion, we shall "ultimately fill in" the definitions for all the rational lattice points in $[0, 1]$. The proof that the sequence of successive iterations converges, in the mean square sense, to the BM with continuous parameter over the unit interval is a technical mathematical argument we shall not produce here. The interested reader may consult the original source, Levy (1948) or Hida (1980) or Karatzas and Shreve (1988).

**Assertion 1.** The BM given in the preceding construction can be extended to $R_+ = [0, \infty)$ as follows: Let $B_1(t)$, $B_2(t)$ be **two independent BM, constructed as above, on the interval** $[0, 1]$. The desired BM on $[0, \infty)$

is then given by $B(t) = B_1(t)$, for $t \in [0, 1]$, and by $B(t) = B_1(1) + tB_2(1/t) - B_2(1)$, for $t > 1$.

Proof: It is evident by construction that $B(t)$ satisfies parts i and ii of Definition 3 and, for $t \in [0, 1]$, satisfies part iii as well. Thus, we need only check part iii for $t > 1$. But

$$EB^2(t) = EB_1(1)^2 + t^2 EB_2(1/t)^2 + EB_2(1)^2 - 2tEB_2(1/t)B_2(1) = t.$$

q.e.d.

## 7.4 Properties of Sample Paths

On the basis of the construction above, it is now relatively simple to show that almost all sample paths of a BM are **continuous**. In principle, this can be shown for any compact subset of $R_+$. However, for simplicity of exposition we continue to operate with the interval $[0, 1]$, and rely on Assertion 1 to justify the extension to $[0, t]$, $t \in R_+$. We begin with the following proposition.

**Proposition 5.** Let $B(t)$ be the $L^2$ limit of the sequence $\{X_n(t) : n \geq 1\}$, in the construction above, over the probability space $(\Omega, \mathcal{A}, \mathcal{P})$. Then

$$X_n(t) \overset{\text{a.c.}}{\to} B(t), \quad \text{uniformly in } t \in [0, 1], \tag{7.7}$$

and moreover $B(t)$ is continuous (in $t$) a.c.

Proof: Let

$$\xi_n(t, \omega) = X_{n+1}(t, \omega) - X_n(t, \omega),$$

and note that for $t \in T_n$, $\xi_n = 0$. For

$$t \in T_{n+1} \cap \overline{T}_n, \quad \text{i.e. for } t = 2^{-n}(2k - 1), \; k = 1, 2, \ldots, 2^{n-1},$$

we have

$$|\xi_n(t, \omega)| = 2^{-(n+1)/2}|Y_{k(t)}|, \quad k(t) = 2^n + \frac{1}{2}(2^n t - 1). \tag{7.8}$$

Choose a sequence $r_n \downarrow 0$; we shall now show that the event

$$A_n = \{\omega : \max_{0 \leq t \leq 1} |\xi_n(t, \omega)| \geq r_n\}$$

occurs for at most a finite number of indices, $n$, and hence that [5]

$$\mathcal{P}(A_n, \text{i.o.}) = 0.$$

Since the $X_n(t)$ are continuous in $t$, the result above establishes that $X_n(t)$ converges for almost all $\omega$, **uniformly in** $t$, to $B(t)$ and, moreover, almost all sample paths $B(t, \omega)$ are continuous. Since by construction, $|\xi_n(t, \omega)| = 2^{-(n+1)/2} |Y_{k(t)}(\omega)|$,

$$\max_{0 \leq t \leq 1} |\xi_n(t, \omega)| = 2^{-(n+1)/2} \max_{2^n \leq k \leq 2^n + 2^{n-1}} |Y_k(\omega)|. \tag{7.9}$$

Let

$$C_k = \{\omega : |Y_k(\omega)| > 2^{(n+1)/2} r_n\}, \quad C_n = \bigcup_{k=2^n}^{2^n + 2^{n-1}} C_k,$$

and note that $A_n \subseteq C_n$. We thus obtain, with $\alpha_n = 2^{(n+1)/2} r_n$, [6]

$$\mathcal{P}(A_n) \leq \mathcal{P}(C_n) = 2^{n-1} \frac{2}{\sqrt{2\pi}} \int_{\alpha_n}^{\infty} e^{-\frac{1}{2}\phi^2} \, d\phi$$

$$\leq \frac{2^n}{\sqrt{2\pi}} \frac{2^{n-1}}{2^{(n+1)/2} r_n} e^{-\frac{1}{2}(2^{(n+1)/2} r_n)^2}. \tag{7.10}$$

Set $r_n = [2nc(\ln 2)]^{1/2} 2^{-(n+1)/2}$, $c > 1$, and note that it obeys $r_n \downarrow 0$. With this substitution we obtain

$$\mathcal{P}(A_n) \leq \frac{1}{\sqrt{\pi}} 2^{-(c-1)n}. \tag{7.11}$$

By the Borel-Cantelli theorem [Dhrymes (1989), p. 136],

$$\mathcal{P}(A_n, \text{i.o.}) = 0, \quad \text{since} \quad \sum_{n=1}^{\infty} \mathcal{P}(A_n) < \infty.$$

But this means that $\max_{0 \leq t \leq 1} |\xi_n(t, \omega)| \xrightarrow{\text{a.c.}} 0$, which completes the proof of the proposition.

$$\text{q.e.d.}$$

---

[5] The notation i.o. means infinitely often; for a discussion of the concept see Dhrymes (1989), p. 9.

[6] The last inequality below is valid because

$$\frac{1}{\sqrt{2\pi}} \int_{\phi}^{\infty} e^{-\frac{1}{2}x^2} \, dx < \frac{1}{\sqrt{2\pi}} \frac{1}{\phi} e^{-\frac{1}{2}\phi^2}.$$

The continuity of the sample paths of BM is a very important property, and sets the stage for the derivation of the *Wiener measure*, which describes their distributional properties. It must not be thought, however, that the sample paths have other "nice" properties. To set off some of their "nasty" properties we introduce some additional concepts.

**Definition 6.** Let $f$ be a real-valued function defined on the interval, say $[0, t]$; a set of points $0 = t_1 < t_2 < t_3 \ldots < t_n = t$, denoted by $p_{nt}$, is said to be a **partition**.

The entity $\mu(p_n) = \sup_i |t_i - t_{i-1}|$ is said to be the **modulus of partition**.

We also use the notation $p_{nt}^k(f)$ to indicate

$$p_{nt}^k(f) = \sum_{i=0}^{n-1} |f(t_{i+1}) - f(t_i)|^k.$$

**Definition 7.** In the context of Definition 6, if the limit

$$\lim_{n \to \infty} p_n^k(f) = v^k(f)$$

exists, $v^k(f)$ is said to be the $k^{th}$ variation of $f$.

If $v(f) < \infty$ the function $f$ is said to be of **bounded variation** or of **finite variation**.

These concepts are easily extended to stochastic processes. Thus, we have

**Definition 8.** Let $\{X(t) : t \in R_+\}$ be a stochastic process and $\{\mathcal{G}_t : t \in R_+\}$ be a stochastic basis, i.e. a collection of nested $\sigma$-algebras such that $X(t)$ is $\mathcal{G}_t$-measurable. Let $p_{nt}$, $n \geq 1$ be a sequence of partitions of $[0, t]$, with modulus of partition $\mu(p_{nt})$. If as $\mu(p_{nt}) \to 0$, $p_{nt}^k(X) \to v_t^k(X)$ in some mode of convergence, and $v_t^k(X)$ is an a.c. finite random variable, $v_t^k(X)$ is said to be the $k^{th}$ **variation of the process**.

The connection between these concepts and the main topic of our discussion is reflected in the next proposition.

**Proposition 6.** In the context of Definitions 6 and 7, let $f$ be (at least right-) continuous and $\{p_n : n \geq 1\}$ a sequence of partitions of the interval $[0, 1]$, such that $\mu(p_n) \downarrow 0$. If the $k^{th}$ variation is bounded on $[0, 1]$, the following statements are true:

i. for $q > k$, $v^q(f) = 0$;

ii. for $q < k$, $v^q(f) = \infty$ if $v^k(f) > 0$.

Proof: For part i, we have

$$p_n^q(f) = \sum_{i=0}^{n-1} |f(t_{i+1}) - f(t_i)|^k |f(t_{i+1}) - f(t_i)|^{q-k},$$

$$\leq p_n^k(f) m_{nq}(f), \quad m_{nq} = \sup_s |f(t_s) - f(t_{s-1})|^{q-k}.$$

Since $v^k(f) < \infty$ and $m_{nq}(f) \to 0$ the proof of part i is complete.

As for part ii, suppose $v^k(f) > 0$, but $v^q(f) < \infty$. There exist constants, say $C > 0$ and $n_0$, such that for $n \geq n_0$, $p^q(f) \leq C$. Consequently, from

$$p_n^k(f) \leq p_n^q \, m_{nk}(f) \leq C \, m_{nk}(f), \quad m_{nk}(f) = \max_{1 \leq s \leq n} |f(t_s) - f(t_{s-1})|^{k-q},$$

we conclude that $v^k(f) = 0$, which is a contradiction.

q.e.d.

A similar result may be proved for stochastic processes, as is done in the following corollary.

**Corollary 1.** Let $\{(X(t), \mathcal{G}_t) : t \in R_+\}$ be (at least a right-) continuous stochastic process as in Definition 8, $\{p_{nt} : n \geq 1\}$ a nested sequence of partitions of the interval $[0, t]$, and suppose that

$$p_{nt}^k(X) \to v_t^k(X) \quad \text{as} \quad \mu(p_{nt}) \to 0,$$

in some mode of convergence (say, P or a.c.) where $v_t^k(X)$ is an a.c. finite random variable assuming values in $R_+$. The following statements are true:

i. for $q > k$, $p_{nt}^q(X) \to v_t^q(X) = 0$, in the same mode of convergence;

ii. for $q < k$, $p_{nt}^q(X) \to v_t^q(X) = \infty$ over the set $A = \{\omega : v_t^k(X) > 0\}$, in the same mode of convergence.

Proof: For part i, we have

$$p_{nt}^q(X) = \sum_{i=0}^{n-1} |X(t_{i+1}) - X(t_i)|^k |X(t_{i+1}) - X(t_i)|^{q-k}$$

$$\leq \sup_i |X(t_{i+1}) - X(t_i)|^{q-k} \, p_{nt}^k(X).$$

Since as $\mu(p_{nt}) \to 0$ with $q - k > 0$, $p_{nt}^k(X) \overset{\text{P, or a.c.}}{\longrightarrow} v_t^k(X)$, which is an a.c. finite random variable, the proof of part i is complete, by the (right) continuity of the process.

As for part ii, suppose $q < k$, $v_t^k(X)$ is an a.c. finite random variable, but $v_t^q(X) < \infty$ over some set, $A$, with $\mathcal{P}(A) > 0$. Thus, there exist constants $C$ and $n_0$ such that for all $n \geq n_0$

$$\mathcal{P}(A \cap B_n) \geq c > 0, \quad B_n = \{\omega : p_{nt}^q(X) \leq C\}.$$

Since $p_{nt}^k(X) \leq p_{nt}^q(X) \sup_i |X(t_{i+1}) - X(t_i)|^{k-q}$, this implies that on at least one proper subset of $A$ (of positive measure), $v_t^k(X) = 0$, which is a contradiction. Hence, over the set $A$, $v_t^q(X) = \infty$.

q.e.d.

The "nasty" aspect of the sample paths of BM is that, a.c., **they are not of bounded variation** and hence are nowhere differentiable. Perhaps this should not have been entirely unexpected, remembering the motivation and intuition given to BM. At any moment in "time" the particle may be "bumped" in any of an uncountable infinity of directions, and "time" here is **instantaneous**. Under the circumstances, continuity of paths is the most we can expect.

For ease of reference we shall cite a number of properties of BM paths without proof, although a proof of Proposition 6 will be given in a later section. The interested reader is referred for greater detail to Hida (1980), pp. 51-61, or Karatzas and Shreve (1988), pp. 103-116.

**Proposition 7.** Let $B$ be a BM defined over the probability space ($\Omega$, $\mathcal{A}$, $\mathcal{P}$) and consider its sample paths $B(t, \cdot) = \{B(t, \omega) : t \in R_+, \omega \in \Omega\}$. Let $p_{nt}$ be a sequence of successively finer partitions of $[0, t]$, i.e. $p_{nt} \subseteq p_{(n+1)t}$, and

$$p_{nt}^k(B)(\omega) = \sum_{i=0}^{n-1} |B(t_{i+1}, \omega) - B(t_i, \omega)|^k.$$

The following statements are true:

   i. $p_{nt}^2(B) \overset{\text{a.c.}}{\to} v^2(B)$, and $v^2(B)$ is an a.c. finite random variable;

   ii. if $\mu(p_n) \downarrow 0$, $v^2(B) = t$.

**Corollary 2.** Let $B$ be a BM defined as in Proposition 6; the following statements are true:

i. the sample paths of BM are of **unbounded variation**, a.c.;

ii. the sample paths of BM are nowhere differentiable a.c.

Proof: To prove i, we note that for $\mu(p_n) \downarrow 0$, $v^2(B) = 1$; hence, from part ii of Corollary 1, we conclude that $v(B) = \infty$; the proof of part ii, follows from part i, since a function of unbounded variation is nondifferentiable.

$$q.e.d.$$

## 7.5   Wiener Measure

The space $(R^{[0,\infty)}, \mathcal{B}[R^{[0,\infty)}])$ is inappropriate for dealing with BM, because $\mathcal{B}(R^{[0,\infty)})$ is "too small", or "too big" depending on one's point of view. The sample paths are **continuous functions** in $R^{[0,\infty)}$, i.e. for fixed $\omega$, $B(t, \omega)$ is a continuous function defined on $R_+$. Thus, it belongs to the space $C_{[0,\infty)}$, the space of continuous functions defined on $R_+$. It may be shown, however, **that the only subset of** $C_{[0,\infty)}$ which is $\mathcal{B}(R^{[0,\infty)})$-measurable is the null set. Hence, whatever probability measure may be established on $\left(R^{[0,\infty)}, \mathcal{B}(R^{[0,\infty)})\right)$ the resulting probability space will be irrelevant for the sample paths of BM, since subsets of $C_{[0,\infty)}$ **are not measurable** in that space.

A measurable space over $C_{[0,\infty)}$ may be built in the standard fashion, occasionally referred to as the **coordinate mapping process**. Let $D_n$ be an $n$-dimensional cylinder set;[7] for functions $x \in C_{[0,\infty)}$, and points $t_i \in R_+$, $i = 1, 2, \ldots, n$ define the set

$$A = \{x : [x(t_1), x(t_2), \ldots, x(t_n)] \in D_n\}. \tag{7.12}$$

The set $A$ is a collection of **functions**, $x \in C_{[0,\infty)}$. Do that for all possible cylinder sets of dimension $n \geq 1$. This will give rise to a collection of

---

[7] For a definition of cylinder sets and related issues see Dhrymes (1989), p. 81 ff. Notice that a cylinder set is a set of the form $T_1 \times T_2 \times \cdots \times T_n \times R \times R \times \cdots$, where $T_i = [a_i, b_i)$ are real intervals, specifically subsets of $R_+$ and, as such, $\mathcal{B}(R^{[0,\infty)})$-measurable.

sets like $A$, which will constitute an **algebra**, say $S$; let $\mathcal{B}(C_{[0,\infty)})$ be defined as $\sigma(S)$, i.e. the smallest $\sigma$-algebra containing $S$. The measurable space $(C_{[0,\infty)}, \mathcal{B}(C_{[0,\infty)}))$ admits of a probability measure, say $\mu_W$, termed the **Wiener measure** which describes the probability characteristics of the sample paths of BM. This may be done as follows. The probability space $(\Omega, \mathcal{A}, \mathcal{P})$, can accommodate a rich collection of random variables with which we can obtain the construction of the previous section. Thus, if $B(t)$ is a BM, for every cylinder set $D_n$ and points $t_i \in R_+$, $i = 1, 2, 3, \ldots, n$ we can determine

$$\mathcal{P}\{[B(t_1), B(t_2), B(t_3), \ldots, B(t_n)] \in D_n]\},$$

which is simply an assignment obtained from the multivariate normal. Since the inverse image of the cylinder sets $D_n$ are sets like $A$ of Eq. (6.12) and hence $\mu_W$-measurable [i.e. the inverse image $A$ of the cylinder set $D_n$ is an element of $\mathcal{B}(C_{[0,\infty)})$] we can determine the Wiener measure of the sample paths of BM by

$$\mu_W(A) = \mathcal{P}([B(t_1), B(t_2), B(t_3), \ldots, B(t_n)] \in D_n).$$

It may be further verified that $\mu_W$ **is a probability measure**, and in particular that $\mu_W(C_{[0,\infty)}) = 1$.

**Remark 4.** There seems to be some confusion in the econometrics literature regarding BM and Wiener measure, which is often erroneously referred to as a Wiener process. First, it is historically inappropriate to refer to BM as the Wiener process. On the other hand Wiener (1923) was the first to introduce a (normal) probability measure over $C_{[0,\infty)}$, and it is thus quite fitting that such a measure should be termed the **Wiener measure**. The second important aspect of the preceding discussion is that $\mathcal{P}[B(t_1), B(t_2), B(t_3), \ldots, B(t_n)] \in D_n]$ gives the probability assigned by $\mathcal{P}$ to a certain set, but **this set cannot be the set** $A$. If we wished to be more pedantic, or more complete depending on one's viewpoint, we should proceed as follows: Let $D = \{\omega : [B(t_1), B(t_2), B(t_3), \ldots, B(t_n)] \in D_n\}$. Note that $D \in \mathcal{A}$ and as such the entity $\mathcal{P}(D)$ is well defined, since $D \subset \Omega$ and is the inverse image of $D_n$ in $\Omega$. This shows that $D \in \mathcal{A}$. On the other hand, $A = \{x : [x(t_1), x(t_2), \ldots, x(t_n)] \in D_n\}$ **is the inverse image of** $D_n$

in $\mathcal{B}(C_{[0,\infty)})$. What is involved in the definition of the **Wiener measure is the assignment**

$$\mathcal{P}(D) = \mu_W(A). \tag{7.13}$$

Intuitively, if we are given that $n$ ordinates of BM lie in a set $D_n \subset R^n$, the Wiener measure gives us the probability to be assigned to the collection of sample paths (of the BM in question) that could give rise to ordinates in the set $D_n$. This is analogous to the case where if we are told that an $n$-element random vector lies in a set $D_n$, the usual probability measure gives the probability to be assigned to the collection of elements $\omega \in \Omega$ that could give rise to ordinates of the elements of the random vector located in the set $D_n$.

## 7.6  Brownian Bridge

We often require that the BM pass through a certain point, say $y$, at a given instant, say $t_0$; or we wish to deal with a BM that begins at a point other than zero, say $x$. It is relatively simple to accommodate such requirements. Thus, if $B^*(t,\omega)$ is a standard BM, the modified BM, $B(t,\omega) = B^*(t,\omega) + x$ satifies the last condition. To satisfy the first requirement, a little more care is required. First, we recall part v of Proposition 2, and Proposition 4, which imply that what we need is to represent $B(t), t \in (0, t_0)$ conditionally on the values at the end points. From standard normal theory, the conditional distribution of $B(t)|B(t_0)$ is normal with variance $t - (t^2/t_0)$. The entity $B^*(t,\omega) - (t/t_0)B^*(t_0,\omega)$ certainly has the required variance. Thus, if we define the modified BM by

$$B(t,\omega) = B^*(t,\omega) - \frac{t}{t_0}B^*(t_0,\omega) + a(t,t_0)x + b(t,t_0)y, \tag{7.14}$$

we may expect to find parameters $a$ and $b$ that satisfy these conditions. Since

$$B(0,\omega) = a(0,t_0)x + b(0,t_0)y, \quad B(t_0,\omega) = a(t_0,t_0)x + b(t_0,t_0)y, \tag{7.15}$$

the required conditions imply $a(0,t_0) = 1$, $a(t_0,t_0) = 0$; $b(0,t_0) = 0$, $b(t_0,t_0) = 1$. Thus, we may choose

$$a(t,t_0) = \frac{t_0 - t}{t_0}, \quad b(t,t_0) = \frac{t}{t_0}, \tag{7.16}$$

and the desired modification is given by

$$B(t,\omega) = B^*(t,\omega) - \frac{t}{t_0}B^*(t_0,\omega) + \frac{t_0 - t}{t_0}x + \frac{t}{t_0}y, \quad t \in [0, t_0]. \qquad (7.17)$$

We have, therefore, Definition 9.

**Definition 9.** A **Brownian bridge** or a **tied down BM** is a Gaussian system as in Eq. (6.17), where $B^*$ is a standard BM.

An immediate implication is stated in the following assertion.

**Assertion 2.** Let $B(t,\omega)$ be the Brownian bridge (BB) of Eq. (6.17). The following statements are true:

$$EB(t,\omega) = \mu(t) = \frac{t_0 - t}{t_0}x + \frac{t}{t_0}y;$$

$$\mathrm{Cov}[B(t,\omega), B(s,\omega)] = \frac{s(t_0 - t)}{t_0}, \quad \text{if } s < t,$$

$$\frac{t(t_0 - s)}{t_0}, \quad \text{if } s > t.$$

# Chapter 8

# Stochastic Integration

## 8.1 Motivation and Preliminaries

In this section we take up issues related to the integration of sample paths of Brownian motion (BM). Such issues arise in the econometrics of estimators and test statistics based on certain types of economic time series, for example, integrated processes of the form $\Delta^p x_t = \epsilon_t$, $p \geq 1$, $\epsilon = \{\epsilon_t : t \in \mathcal{N}\}$ being a $WN(\sigma^2)$ process.

The contemporary theory of stochastic integration [1] is cast in terms of semi-martingales, a term to be defined below. In our applications, on the other hand, we require only stochastic integrals that involve BM. In an attempt to present to the reader as much of the general theory as is possible, we frame our discussion in the context of semimartingales, but where specific examples or specific properties of semimartingales are required we fall back on BM. The reader interested in the more general aspects of stochastic integration is referred to Hida (1980), Karatzas and Shreve (1988), particularly Chapter 3, and Protter (1992), particularly Chapter 4.

The following elements make up the context in which we operate is: the probability space $(\Omega, \mathcal{A}, \mathcal{P})$ on which we define the stochastic processes $X$; the stochastic basis $\{\mathcal{A}_t : \mathcal{A}_t = \sigma(X_s, s \leq t), \mathcal{A}_0 = (\emptyset, \Omega), t \geq 0\}$; the space of square integrable functions defined [2] on the probability space above,

---

[1] If $f(t, \omega)$ is a suitable function and $X_t$ is a semimartingale, an integral of the form $\int_0^t f(s, \omega) dX_s$ is said to be a **stochastic integral**.

[2] For notational convenience we henceforth refer to this space as $\mathcal{H}^2$.

$\mathcal{H}^2(\Omega, \mathcal{A}, \mathcal{P})$, as defined in Chapter 2; the inner product defined, for any $x, y \in \mathcal{H}^2$, by $< x, y >= Exy$; and the norm defined by $\| x \|=< x, x >^{1/2}$.

Stochastic integration is the process by which we obtain the solution of the system

$$dY_t = a(t, \omega) \, dt + b(t, \omega) dX_t, \tag{8.1}$$

over the interval $[0, T)$; such a solution would be an integral (for $t \leq T$)

$$Y_t = Y_0 + \int_0^t a(s, \omega) \, ds + \int_0^t b(s, \omega) \, dX_s. \tag{8.2}$$

Until a meaning is assigned to these integrals, it is to be understood that Eq. (8.2) simply means that $\{Y_t : t \in [0, T)\}$ is a stochastic process (with continuous parameter) satisfying Eq. (8.1).

In summary, then, the problem of stochastic integration is simply to find meaning for integrals of the form

$$I_1 = \int b(t, \omega) \, dX_t, \quad I_2 = \int a(t, \omega) \, dt,$$

where $X_t = X(t, \omega)$ is a stochastic process with continuous parameter, or simply a continuous stochastic process.

## 8.2 The Ito Integral

One particular solution was given by the Japanese mathematician K. Ito, in a series of papers (1950), (1951a), (1951b), (1951c); others have given alternative solutions. It is the Ito solution that is of particular interest to us. He succeeded in providing a meaningful definition of the stochastic integral $\int_0^T f(t, \omega) dB_t$, $T \in [0, \infty)$, where $B$ is BM process, and established conditions under which such integrals exist in the context of the space $\mathcal{H}^2$.

### 8.2.1 Why Standard Integration Fails

Before we embark on the (rather complex) derivation of the Ito integral it is useful to ask why the standard approach to integration fails. For example, if we are presented with the entity $\int_0^t K_s \, dX_s$, the standard approach is as follows. Consider the partition $p_{nt}$, $0 = t_0 \leq t_1 \leq t_2 \leq \ldots \leq t_n = t$ and, for each $\omega \in \Omega$, form the sums

$$S_n = \sum_{j=0}^{n} K_{t_j} \Delta X_{t_{j+1}}, \quad X_{t_j} = X(t_j, \omega), \quad \Delta X_{t_{j+1}} = X_{t_{j+1}} - X_{j_j}.$$

The integral is then defined in terms of pointwise convergence, for almost all $\omega$, by

$$\int_0^t K_s \, dX_s = \lim_{n \to \infty} S_n.$$

To understand why this is not always possible, and particularly **not in the case where** $X$ **is Brownian motion**, we note the following important ressults.

**Assertion 1.** Let $f(t, \omega)$, defined on $R_+ \times \Omega$, where $R_+ = [0, \infty)$, be $\mathcal{B}(R) \times \mathcal{A}$-measurable and a.c. continuous in $t$ (i.e. $f$ is continuous in $t$ for all $\omega \in \Omega$, except possibly for a set of $\mathcal{P}$-measure zero). Let $p_{nt}$ be a partition as above and define $t_j \leq r_j \leq t_{j+1}$, $j = 0, 1, 2, \ldots n-1$. If $X(t, \omega)$ is a continuous process such that for each $\omega \in \Omega$ its paths are of bounded variation (on compact subsets of $R_+$) then

$$\lim_{n \to \infty} S_n = \int_0^t f(s, \omega) \, dX_s \quad a.c. \qquad S_n = \sum_{j=0}^{n-1} f(r_j, \omega) \Delta X_{t_{j+1}}.$$

Proof: By definition

$$|S_n| \leq \sum_{j=0}^{n-1} |f_j||X_{j+1} - X_j| \leq K \sum_{j=0}^{n-1} |X_{j+1} - X_j|,$$

where $K$ is a uniform bound for $f$ on $[0, t)$ and the rightmost sum converges at least in probability to the variation of $X$ which by assumption is finite.

$$\text{q.e.d.}$$

**Assertion 2.** If the sums $S_n$ of Assertion 1 converge to a limit for every **continuous** $f$ then the paths of the process $X$ are of bounded variation.

Proof: The proof of this result is rather involved and is not given here. The general idea, however, is evident from the inequality in the proof of Assertion 1. If, for every continuous function $f$, the sums $S_n$ converge as the modulus of partition goes to zero, we would expect the entity $\sum_{j=0}^{n-1} |X_{j+1} - X_j|$ to converge as well.

The reader who is interested in a formal proof may consult Protter (1992), p. 40.

A final point to note is given in Assertion 3.

**Assertion 3.** If the $X$-process in Assertion 1 is of bounded variation and has (a.c.) continuous paths, and if $f$ has continuous first-order derivative, then $f(X_t)$ is a bounded variation process, and moreover

$$f(X_t) - f(X_0) = \int_0^t f'(X_s)\, dX_s \quad \text{a.c.}$$

Proof: Since $f'$ is continuous on $[0, t]$ it is therefore bounded, and the integral in question exists. Let $p_{nt}$ be a partition such that $\mu(p_{nt})$ converges to zero with $n$; using telescoping sums we have, by the mean value theorem,

$$f(X_t) - f(X_0) = \sum_{j=0}^{n-1}[f(X_{t_{j+1}}) - f(X_{t_j})] = \sum_{j=0}^{n-1} f'(X_{r_j})(X_{t_{j+1}} - X_{t_j}),$$

with $t_j \leq r_j \leq t_{j+1}$. Upon taking the limit of the sums above, the conclusion follows from Assertion 1.

The import of the preceding discussion is that if we wish to obtain a standard Riemann-Stieltjes type integral on a path-by-path basis, the process in question **must have paths of bounded variation**. Since this is not so in the case of BM, we must seek a definition of the integral by using other means and other forms of convergence. The integral we seek to define can be obtained through $L^2$ convergence in the space $\mathcal{H}^2$ when certain **additional restrictions** are placed on the function $f$.

## 8.2.2 Derivation of the Integral

In the classic style of measure theory, we begin with simple functions and BM as follows: Let $[0, T) \subset [0, \infty)$, $B(t, \omega)$ be a BM with continuous parameter defined on the probability space above, $p_{nT}$ be a partition defined by the points $0 = t_0 \leq t_1 \leq t_2 \leq, \ldots, \leq t_n = T$,[3] and $e^n$ be a simple function of the form

$$e^n(t) = \sum_{j}^{n} a_j^n(\omega) I_{j,j+1}(t),$$

---

[3] The collection of points is to be understood *mutatis mutandis* when we consider the interval $[0, t)$; evidently, in this case $t_n = t$.

where $I_{j,j+1}$ is the indicator function of the interval $[t_j, t_{j+1})$.[4] For such a function it is natural to define

$$\int_0^T e^n(t,\omega)\,dt = \sum_{j=0}^n a_j^n(\omega)\Delta t_{j+1}^n, \quad \int_0^T e^n(t)\,dB_t = \sum_{j=0}^n a_j^n \Delta B_{t_{j+1}^n},$$

$$\Delta B_{t_{j+1}^n} = B(t_{j+1}^n,\omega) - B(t_j^n,\omega).$$

From Dhrymes (1989), Proposition 10, pp. 23-25, we recall that if $f$ is a **measurable function**, it can be approximated arbitrarily closely by a simple function and/or by a continuous function. Suppose again we wish to define the meaning of the integral $\int_0^t B_s\,dB_s$; since $B(t,\omega)$ is a.c. continuous in $t$ it may be approximated equally well by the simple functions [5]

$$e^n(t,\omega) = \sum_{j=0}^n B_j I_{j,j+1}, \quad \text{or} \quad g^n(t,\omega) = \sum_{j=0}^n B_{j+1} I_{j,j+1}.$$

The integral in question, therefore, may be approximated by the two expressions

$$I_t(e^n) = \int_0^t e^n(s,\omega)\,dB_s = \sum_{j=0}^{n-1} B_j[B_{j+1} - B_j] \approx I_t(B) \quad \text{and}$$

$$I_t(g^n) = \int_0^t g^n(s,\omega)\,dB_s = \sum_{j=0}^{n-1} B_{j+1}[B_{j+1} - B_j] \approx I_t(B).$$

Setting aside the question of whether such sums converge as $\mu(p_{nt}) \to 0$ and $n \to \infty$, we note that both approximations are equally reasonable and both entities $I_t(e^n)$ and $I_t(g^n)$ are stochastic processes. On the other hand, we observe that given the stochastic basis we had defined earlier,

$$EI_t(e^n) = \sum_{j=0}^{n-1} EE[B_j(B_{j+1} - B_j)|\mathcal{A}_{t_j}] = 0,$$

while

$$EI_t(g^n) = \sum_{j=0}^{n-1} EE[(B_{j+1} - B_j + B_j)(B_{j+1} - B_j)|\mathcal{A}_{t_j}] = \sum_{j=0}^{n-1}(t_{j+1}^n - t_j^n) = t.$$

---

[4] For greater notational precision such intervals should be referred to as $[t_j^n, t_{j+1}^n)$; in the interest of simplicity we use the two forms interchangeably, using the latter only when we wish to emphasize the connection of $n$ to the modulus of the partition.

[5] The symbols $B_j$, $B_{t_j}$, $B_{t_j^n}$ are used interchangeably and their common meaning is $B(t_j^n, \omega)$.

Evidently, if both approximations converge, so that $I_t(e^n)$ and $I_t(g^n)$ are both defined unambiguously, we find that the resulting integrals have **different** properties, even though both involve equally acceptable approximations to the **same** entity!

In particular, with the $e$-approximation $EI_t(B) = 0$, while with the $g$-approximation $EI_t(B) = t$. Thus, we are alerted to the fact that the functions that may be integrated (i.e. the potential integrands) must obey certain restrictions, in particular they should be **adapted to the stochastic basis we have constructed**.

Thus, in employing the approach above, the function $f$ cannot be left completely arbitrary. Since our primary focus is on stochastic integrals involving BM processes (with continuous parameter),[6] we may define the stochastic basis $\{\mathcal{A}_t : \mathcal{A}_t = \sigma(B_s, \ s \leq t), \ t \in [0, T), \ \mathcal{A}_0 = (\emptyset, \Omega)\}$ and the class of admissible functions, say $\mathcal{F}$, by

    i.  $f(t, \omega) : R_+ \times \Omega \to R$ is $\mathcal{B}(R) \times \mathcal{A}$-measurable;

    ii.  $f(t_j, \omega)$ is $\mathcal{A}_{t_j}$-measurable;

    iii.  $E\left[\int_0^\infty f^2(t, \omega)\, dt\right] < \infty$.

The conditions above are not the least stringent possible, but they accommodate a wide class of functions, certainly far more than are necessary for what we have in mind.

To define the integrals we employ the result of Proposition 10, in Dhrymes (1989), cited above.[7] Thus, we show the following:

    i.  The Ito integral is defined for (bounded) simple functions $e^n(t, \omega) \in \mathcal{F}$, and it obeys

$$E\left[\int_0^T e^n(t, \omega)\, dB_t\right]^2 = E\int_0^T [e^n(t, \omega)]^2\, dt;$$

---

[6] When a wider class of processes is employed, such as martingales or semimartingales, the stochastic basis should be redefined appropriately.

[7] All convergence results in this discussion are valid a.c., i.e. for all $\omega$, except possibly for a set of $\mathcal{P}$-measure zero; for convenience we shall often employ the phrase "for almost all" (a.a.) $\omega$, or even leave it out entirely to be understood as implicit in the context.

ii. If $f \in \mathcal{F}$ and, in addition, $f$ is bounded and for each $\omega$ it is continuous in $t$, there exist bounded simple functions $f^n(t, \omega)$ which for each $\omega$ are continuous in $t$,[8] such that for almost all (a.a.) $\omega$, $f^n \to f$.

iii. If $f \in \mathcal{F}$ is bounded, there exist bounded $g^n \in \mathcal{F}$ which for each $\omega$ are continuous in $t$ such that $g^n \to f$, for a.a. $\omega$.

iv. If $f \in \mathcal{F}$ there exist bounded functions $h^n$ such that $h^n \to f$.

Proof: For the first part (and $n \in \mathcal{N}_+$), let $\{t_j^n : j \geq 0\}$ be the points of the partition $p_{nT}$ and construct the simple functions $e^n(t, \omega)$ by $e^n(t, \omega) = \sum_{j=0}^{n-1} a_j^n(\omega) I_{j,j+1}(t)$, where $I_{j,j+1}$ is the indicator function of the interval $[t_j, t_{j+1})$, $a_j^n(\omega)$ is $\mathcal{A}_{t_j^n}$-measurable and $\sum_{j=0}^{n-1} E[a_j^n(\omega)]^2[t_{j+1} - t_j] < \infty$. Thus, for any $n$, the functions $e^n(t, \omega)$ are simple and obey $e^n \in \mathcal{F}$. The Ito integral is defined as

$$I(e^n) = \int_0^T e^n(\omega) dB_t = \sum_{j=0}^{n-1} a^j(\omega)[B_{j+1} - B_j], \qquad (8.3)$$

where $B_j = B(t_j^n, \omega)$. Consequently,

$$E[I(e^n)]^2 = \sum_{j=0}^{n-1} \sum_{k=0}^{n-1} E a_j^n a_k^n [B_{j+1} - B_j][B_{k+1} - B_k]$$

$$= \sum_{j=0}^{n-1} E[a_j^n(\omega)]^2[t_{j+1} - t_j] = E \int_0^T |a^n(\omega)|^2 \, dt. \qquad (8.4)$$

To prove part ii, let the partition $p_{nT}$ be given and let $f \in \mathcal{F}$ be bounded and, for each $\omega$, continuous in $t$. Consider now the simple functions

$$g^n(t, \omega) = \sum_{j=0}^{n-1} f(t_j, \omega) I_{j,j+1}(t).$$

For each $n$, it may be verified that $g^n$ is a simple bounded continuous function in $\mathcal{F}$ and, for each $\omega$,

$$\lim_{n \to \infty} \int_0^T [f(t, \omega) - g^n(t, \omega)]^2 \, dt = 0. \qquad (8.5)$$

---

[8] Continuous, in this discussion, means continuous for $t \in R_+$, except possibly for sets of $\mathcal{B}(R)$-measure zero.

Since the functions are bounded, it follows from Proposition 16 in Dhrymes (1989), pp. 43-45, that

$$\lim_{n\to\infty} E\left[\int_0^T [f(t,\omega) - g^n(t,\omega)]^2 \, dt\right] = 0. \tag{8.6}$$

To prove part iii, let $f \in \mathcal{F}$; we construct a sequence of functions $g^n \in \mathcal{F}$ which are bounded, simple, and, for each $\omega$, continuous in $t$; to this end, let $\alpha_n$ be a non-negative continuous function obeying [9]

$$\alpha_n(x) = 0, \quad x \notin (0, \frac{1}{n}),$$

$$\int_{-\infty}^{\infty} \alpha_n(x)dx = 1. \tag{8.7}$$

Define the approximating functions by

$$g^n(t,\omega) = \int_0^t \alpha_n(t - s)f(s,\omega) \, ds, \quad n \in \mathcal{N}_+.$$

It may be verified that $g^n$ is a (uniformly) bounded sequence of functions which, for each $\omega$, are continuous in $t$, and evidently converge to $f$. Again by Proposition 16 in Dhrymes (1989), pp. 43-45,

$$\lim_{n\to\infty} E\left[\int_0^T [f(t,\omega) - g^n(t,\omega)]^2 \, dt\right] = 0. \tag{8.8}$$

Finally for part iv, if $f \in \mathcal{F}$, define the sequence of bounded functions

$$g^n(t,\omega) = -n, \quad \text{if } f(t,\omega) \leq -n,$$

$$= f(t,\omega), \quad \text{if } -n \leq f(t,\omega) \leq n,$$

$$= n, \quad \text{if } f(t,\omega) > n. \tag{8.9}$$

Evidently, for each $\omega$, $g^n \to f$ and moreover, $|f - g^n| \leq |f|$. It follows then from Proposition 19, in Dhrymes (1989), p. 48, that

$$\lim_{n\to\infty} E\left[\int_0^T [f(t,\omega) - g^n(t,\omega)]^2 \, dt\right] = 0. \tag{8.10}$$

---

[9] One example of a function like that defined in Eq. (8.7) may be obtained by setting it equal to zero outside the interval $(0, 1/n)$, and assigning to it the value $\alpha_n(x) = -(n\pi/2)\sin n\pi(x - 1/n)$ in the interval $(0, 1/n)$.

The import of the preceding discussion is to establish a chain of approximations: For any $f \in \mathcal{F}$ we may approximate it by a sequence as in part iv;
any sequence as in part iv may be approximated by a sequence of the type
in part iii; any sequence in part iii may be approximated by a sequence of
the type in part ii, i.e. by a sequence of bounded simple functions that, for
each $\omega$, are continuous in $t$. But this means that any function $f \in \mathcal{F}$ may
be approximated by a sequence of bounded simple functions as in part ii. By
part i, this means that if $f \in \mathcal{F}$, and $g^n$ are the bounded simple functions
of part i, we must have

$$\lim_{n \to \infty} E \left[ \int_0^T |f - g^n|^2 \, dt \right] = 0.$$

We thus **define the Ito integral** as the $L^2$ limit

$$\int_0^T g^n(t, \omega) \, dB_t \xrightarrow{L^2} I_T(f) = \int_0^T f(t, \omega) \, dB_t, \tag{8.11}$$

in the sense that

$$\lim_{n \to \infty} E \left| I_T(f) - \int_0^T g^n(t, \omega) \, dB_t \right|^2 = 0.$$

Notice further that the preceding discussion establishes that the limit which
defines the Ito integral is **independent of the approximating** sequence in
the sense that

$$E[I_T(f)]^2 = E \left[ \int_0^T f^2 \, dt \right]. \tag{8.12}$$

Certain fundamental properties of the stochastic integral defined in Eq.
(8.11) are given below.

**Proposition 1.** Let $f \in \mathcal{F}$. Let $B(t, \omega)$ be a BM with continuous param-
eter defined on the space $\mathcal{H}^2(\Omega, \mathcal{A}, \mathcal{P})$, and $\{\mathcal{A}_t : \mathcal{A}_t = \sigma(B_s, s \leq t), \mathcal{A}_0 = (\emptyset, \Omega), t \in [0, \infty)\}$. The Ito integral defined in Eq. (8.11) has the following
properties:

    i. It is a.c. linear, in the sense that, for $f, g \in \mathcal{F}$, $0 \leq t \leq T < \infty$ and
    $c \in R$,

        1. $I_T(cf + g) = \int_0^T (cf + g) \, dB_t = c \int_0^T f \, dB_t + \int_0^T g \, dB_t = cI_T(f) + I_T(g)$;

2. $\int_0^T f \, dB_s = \int_0^t f \, dB_s + \int_t^T f \, dB_s$

for all $\omega \in \Omega$, except possibly for a set of $\mathcal{P}$-measure zero (or for almost all $\omega$).

ii. The Ito ingegral is a zero mean square integrable **martingale** relative to the stochastic basis defined above.

Proof: The assertions in part i are true for simple functions; since every $f \in \mathcal{F}$ can be derived as the limit of simple functions the assertions are, thus, equally true for any $f, g \in \mathcal{F}$.

For the proof of part ii, we note that for a simple function, $e^n(t, \omega) = \sum_{j=0}^{n-1} a_j^n(\omega) I_{j,j+1}(t)$, and $t \le T < \infty$

$$E I_t(e^n) = \sum_{j=0}^{n-1} E\{a_j(\omega) E[(B_{j+1} - B_j)|\mathcal{A}_{t_j}]\} = 0,$$

for all $n$. If $\{e^n : n \ge 1\}$ is a sequence of simple functions converging a.c. to $f \in \mathcal{F}$, we conclude that $E I_t(f) = 0$; to show that $\{(I_t(f), \mathcal{A}_t) : t \in [0, \infty)\}$ is a martingale we need to establish:

$$E|I_t(f)| < \infty, \quad \text{and} \quad E[I_t(f)|\mathcal{A}_r, \ r \le t] = I_r(f).$$

The first requirement is evidently satisfied since $E I_t(f) = 0$, for every $t$; as for the second, using part i.2 of this proposition we have

$$E[I_t(f)|\mathcal{A}_r] = \int_0^r f \, dB_s + E\left[\int_r^t f \, dB_s | \mathcal{A}_r\right] = \int_0^r f \, dB_s = I_r(f).$$

We need only show that $I_t(f)$ is square integrable; considering the approximation

$$f(t, \omega) \approx f_n(t, \omega) = \sum_{j=0}^{n-1} f(t_j, \omega) I_{j,j+1}(t),$$

we find

$$E[I_t(f_n)]^2 = E\left[\int_0^t f_n(t, \omega) \, dB_s\right]^2 = E\left[\sum_{j=0}^{n-1} f_n(t_j, \omega)(B_{j+1} - B_j)\right]^2$$

$$= E \sum_{j=0}^{n-1} f_n^2(t_j, \omega)(B_{j+1} - B_j)^2 = \sum_{j=0}^{n-1} E[f_n(t_j, \omega)]^2 (t_{j+1} - t_j)$$

$$\to \int_0^t E f^2(s, \omega) \, dt = E[I_t(f)]^2 < \infty.$$

The last inequality follows from part iii of the definition of the collection of admissible integrands, $\mathcal{F}$.[10]

q.e.d.

Two useful and important basic results may be mentioned at this time; the first provides an analog of Markov's inequality for martingales; the second, actually an implication of the first, provides a version of the **law of the iterated logarithm** for martingales in general, and for the BM and the Ito integral in particular.

**Proposition 2** (Doob's Martingale Inequality). Suppose $\{(X_t, \mathcal{A}_t) : t \geq 0\}$ is a martingale process (with continuous parameter), defined on the probability space $(\Omega, \mathcal{A}, \mathcal{P})$, and whose sample paths are at least right-continuous. Then for $p \geq 1$, $k > 0$,

$$\mathcal{P}(A^*) \leq \frac{1}{k^p} \sup_t E|X_t|^p, \quad A^* = \{\omega : \sup_t |X_t| \geq k\}.$$

Proof: See Doob (1953), Chapter 7, Theorem 3.2, or Karatzas and Shreve (1988), Chapter 1, Theorem 3.8.

**Corollary 1.** Let $\{B_t : t \geq 0\}$ be a BM process as above, and define

$$S_t = \sup_{s \leq t} B_s.\text{[11]} \tag{8.13}$$

In addition, define the sets

$$A_1 = \{\omega : \varlimsup_{t \to 0} \frac{B_t}{[2t \log_2(1/t)]^{1/2}} = 1\}, \quad A_2 = \{\omega : \varliminf_{t \to 0} \frac{B_t}{[2t \log_2(1/t)]^{1/2}} = -1\};$$

$$A_1^* = \{\omega : \varlimsup_{t \to \infty} \frac{B_t}{[2t \log_2 t]^{1/2}} = 1\}, \quad A_2^* = \{\omega : \varliminf_{t \to \infty} \frac{B_t}{[2t \log_2 t]^{1/2}} = -1\},$$

where $\log_2 t = \log(\log t)$. Then

$$\mathcal{P}(A_1) = \mathcal{P}(A_2) = \mathcal{P}(A_1 \cap A_2) = 1, \tag{8.14}$$

---

[10] The last **equality**, occasionally referred to as **Ito's isometry**, states that

$$\| I_t(f) \|^2 = \left\| \int_0^t f(s, \omega) \, ds \right\|^2 = E \int_0^t f^2(s, \omega) \, ds,$$

so that the two entities have the same norm.

[11] Notice that if $t$ is fixed, $S_t \sim |B_t|$.

and

$$P(A_1^*) = P(A_2^*) = P(A_1^* \cap A_2^*) = 1. \tag{8.15}$$

Proof: See Revuz and Yor (1991), Chapter 2.

**Remark 1**. Notice if the results in Eq. (8.14) are valid then so are the results of Eq. (8.15), and vice-versa. This is a consequence of the time inversion noted in Chapter 6, viz. if $B_t$ is a BM then so is $tB_{1/t}$.

**Corollary 2**. Consider the Ito integral of Eq. (8.11), and define the set

$$A_T = \{\omega : \sup_{0 \le t \le T} |I_t(f)| \ge k\}.$$

Then,

$$P(A_T) \le \frac{1}{k^2} E \left[ \int_0^T f^2(t, \omega) \, dt \right].$$

**Remark 2**. Note that Ito integrals do not obey the usual rules; in particular,

$$I_t(B) = \int_0^t B_s \, dB_s \ne \frac{1}{2} B_t^2.$$

The Ito integral **is not** a Riemann-Stieltjes (RS) integral; for example, the approximating sums for the Ito integral $I(f)$ are given by

$$\sum_{j=0}^{n-1} f(t_j^*, \omega) \Delta B_{j+1} \rightarrow I(f) = \int_0^t f(s, \omega) \, dB_s.$$

In the RS context, $t_j^*$ could be **any** point in the interval $[t_j^n, t_{j+1}^n)$ of the partition $p_{nt}$. In the Ito context **we must have** $t_j^* = t_j^n$. The choice $t_j^* = (1/2)(t_{j+1}^n + t_j^n)$, yields an entirely different integral, the **Stratonovich integral** whose properties are different from those of the Ito integral, a matter we not pursue in this volume. For a discussion of such issues see Protter (1992), pp. 215-234.

If a candidate for an Ito integral, $I_t(f)$, is presented we can check to see whether the claim is correct through the operation

$$\lim_{n \to \infty} E \left| I_t(f) - \sum_{j=0}^{n-1} f(t_j, \omega) \Delta B_{j+1} \right|^2 = 0.$$

If the condition above is satisfied, the proposed candidate for $I_t(f)$ is the proper one; if not, it is not. The preceding discussion, however, does not allow us to derive routinely an explicit closed-form representation of the Ito integral; we take up this problem in a later section. The following example illustrates the procedure.

**Example 1**. Here we examine the conjecture that

$$I_t(f) = \int_0^t B_s \, dB_s = \frac{1}{2} B_t^2.$$

To check this conjecture, and operating from first principles, we obtain the approximating sum $\sum_j B_j \Delta B_{j+1}$ and examine the limit of

$$
\begin{aligned}
\phi_n &= E \left| \frac{1}{2} B_t^2 - \sum_{j=0}^{n-1} B_j \Delta B_{j+1} \right|^2 \\
&= E \left[ \frac{1}{4} B_t^4 + (\sum_{j=0}^{n-1} B_j \Delta B_{j+1})^2 - B_t^2 \sum_{j=0}^{n-1} B_j \Delta B_{j+1} \right] \\
&= \frac{3}{4} t^2 + \sum_{j=0}^{n-1} t_j \Delta t_{j+1} - 2 \sum_{j=0}^{n-1} t_j \Delta t_{j+1} = \frac{3}{4} t^2 - \sum_{j=0}^{n-1} t_j \Delta t_{j+1} \\
&\to \frac{3}{4} t^2 - \frac{1}{2} t^2 = \frac{1}{4} t^2.
\end{aligned}
$$

In obtaining the third equality we have used the relation $B_t = B_t - B_{j+1} + \Delta B_{j+1} + B_j$. Since the limit of $\phi_n$ is $t^2/4 \neq 0$, the conjecture is not correct. In fact, it can be shown that the correct form of the integral is

$$\int_0^t B_s \, dB_s = \frac{1}{2} B_t^2 - \frac{1}{2} t. \tag{8.16}$$

To verify this claim, consider

$$
\begin{aligned}
\psi_n &= E \left[ \left( \frac{1}{2} B_t^2 - \sum_{j=0}^{n-1} B_j \Delta B_{j+1} \right) - \frac{1}{2} t \right]^2 = E \left( \frac{1}{2} B_t^2 - \sum_{j=0}^{n-1} B_j \Delta B_{j+1} \right)^2 \\
&\quad - t E \left[ \frac{1}{2} B_t^2 - \sum_{j=0}^{n-1} B_j \Delta B_{j+1} \right] + \frac{1}{4} t^2 \\
&\to \frac{1}{4} t^2 - \frac{1}{2} t^2 + \frac{1}{4} t^2 = 0.
\end{aligned}
$$

We conclude, therefore, that Eq. (8.16) represents the correct form of the integral in question.

## 8.3  Ito's Formula

To facilitate our presentation and to make clear the broad reach of Ito's formula, we introduce two additional concepts: That of conditional or **quadratic variation** [see Dhrymes (1989), pp. 287-289], and that of **semimartingales**.

**Definition 1.** Let $\{X_t, Y_t : t \in R_+\}$ be stochastic processes defined on the probability space $(\Omega, \mathcal{A}, \mathcal{P})$; on the set $[0, t)$, $t < \infty$ define the partition $p_{nt}$ and consider

$$QV_{nt}(X) = \sum_{j=0}^{n-1}(X_{t_{j+1}} - X_{t_j})^2 = \sum_{j=0}^{n-1}(\Delta X_{j+1})^2.$$

If as $\mu(p_{nt}) \to 0$ with $n \to \infty$, $QV_{nt}(X)$ converges at least in probability, we denote the limit by

$$[X, X]_t = \plim_{n \to \infty} QV_{nt}(X). \tag{8.17}$$

The term $[X, X]_t$ is said to be the quadratic variation of $X$ over $[0, t)$.

The quadratic **covariation**, occasionally also termed the cross quadratic variation, of the two processes $X$ and $Y$ over $[0, t)$ is given by

$$[X, Y]_t = \frac{1}{4}\{[X + Y, X + Y]_t - [X - Y, X - Y]_t\}, \tag{8.18}$$

and is evidently obtained as the limit (at least in probability)

$$[X, Y]_t = \plim_{n \to \infty} \sum_{j=0}^{n-1} \Delta X_{j+1} \Delta Y_{j+1}.$$

An immediate consequence of the preceding is Proposition 3.

**Proposition 3.** The quadratic variation of the SBM process $B = \{B(t, \omega) : t \in R_+\}$ over the interval $[0, t)$, $t < \infty$ is given by $[B, B]_t = t$.

Proof: By definition, and given a partition $p_{nt}$,

$$QV_{nt}(B) = \sum_{j=0}^{n-1}(\Delta B_{j+1})^2;$$

for every partition, the corresponding entitites $(\Delta B_{j+1})$ are independent normal variables with mean zero and variance $\Delta t_{j+1}^n$. Thus,

$$EQV_{nt}(B) = \sum_{j=0}^{n-1} \Delta t_{j+1}^n = t, \quad \text{and}$$

$$E\,|QV_{nt}(B) - t|^2 \;=\; \sum_{j=0}^{n-1} E\left[(\Delta B_{j+1})^2 - \Delta t_{j+1}^n\right]^2$$

$$= 3\sum_{j=0}^{n-1}(\Delta t_{j+1}^n)^2 - 2\sum_{j=0}^{n-1}(\Delta t_{j+1}^n)^2 + \sum_{j=0}^{n-1}(\Delta t_{j+1}^n)^2$$

$$= 2\sum_{j=0}^{n-1}(\Delta t_{j+1}^n)^2 \;\;\to\;\; 0, \tag{8.19}$$

which shows convergence in quadratic mean and hence in probability. Since we have convergence, and the summands are independent random variables, we have convergence a.c. as well, by Kolmogorov's zero-one law [see Proposition 3, in Dhrymes (1989), pp. 139-140].

q.e.d.

**Definition 2.** Let $X = \{X_t : t \in R_+\}$ be a continuous process defined on the probability space ($\Omega$, $\mathcal{A}$, $\mathcal{P}$); and let $\{\mathcal{A}_t : \mathcal{A}_t = \sigma(X_s, s \leq t,\ t \in R_+,\ \mathcal{A}_0 = (\emptyset, \Omega)\}$ be a stochastic basis. $X$ is said to be a **semimartingale** if and only if it can be represented as $X_t = M_t + A_t$, such that $M_t$ is a **local martingale** and $A_t$ is a process with **bounded variation**, both relative to (or adapted to) the stochastic basis above.

A continuous process $M = \{M_t : t \in R_+\}$ is said to be a **local martingale** if and only if

i. there exists an increasing sequence of stopping times $\{T_n : n \geq 1\}$ such that $T_n \overset{\text{a.c.}}{\to} \infty$;

ii. for every $n$, the stopped process $M^{T_n}$ is a **uniformly integrable** [12] **martingale**.

**Remark 3.** In connection with the definition, note the following:

i. A stopped process consists of $\{M_{\min_{t,T_n}} : n \geq 1\}$.

ii. An example of a stopping time is $T_n = \inf_{t \in R^+}\{t : M_t \geq n\}$.

Evidently, if $T_n$ is increasing for all $n \in \mathcal{N}_+$ it will obey $T_n \to \infty$ in one mode of convergence or another. Moreover, note that $T_n = n$ is

---

[12] For the concept of uniform integrability, see Dhrymes (1989), p. 167.

also a sequence of stopping times. For a fuller explanation of stopping times and stopped processes, see, for example, Dhrymes (1989), pp. 289-304.

iii. Any (minimally right-) continuous square integrable martingale is a continuous local martingale since it satisfies, automatically, conditions i and ii above.

iv. The entities $B_t$, $B_t^2 - t$, as well as $\exp\left(\alpha B_t - \frac{\alpha^2}{2} t\right)$, $\alpha \in R$, are right-continuous square integrable martingales and hence also local martingales. This may be seen as follows: To show that a process $X$ is a martingale process we need to show (a) $E|X_t| < \infty$ and (b) $E[X_t | \mathcal{A}_s, s < t] = X_s$ In the present case we have [13]

$$EB_t = 0, \quad E[B_t^2 - t] = 0, \quad E\left[e^{\left(\alpha B_t - \frac{\alpha^2}{2} t\right)}\right] = 1,$$

and hence that the condition $E|X_t| < \infty$ is satisfied for all three. One further verifies that

$$E[B_t | \mathcal{A}_s, s < t] = B_s + E[(B_t - B_s) | \mathcal{A}_s, s < t] = B_s,$$

$$E[(B_t^2 - t) | \mathcal{A}_s] = B_s^2 - s + E[\{[(B_t^2 - B_s^2) - (t - s)]| \mathcal{A}_s]$$

$$= B_s^2 - s + E\left[e^{\left(\alpha B_t - \frac{\alpha^2}{2} t\right)}\Big| \mathcal{A}_s\right]$$

$$= e^{\left(\alpha B_s - \frac{\alpha^2}{2} s\right)} + E\left[e^{\left(\alpha(B_t - B_s) - \frac{\alpha^2}{2}(t-s)\right)}\Big| \mathcal{A}_s\right]$$

$$= e^{\left(\alpha B_s - \frac{\alpha^2}{2} s\right)},$$

thus confirming that they are martingales; since they are evidently square integrable, all three are semimartingales as well.

---

[13] The reader should note certain aspects of the argument below: First if $|EX_t| < \infty$, it follows that $E|X_t| < \infty$. To see this, note that we may always write $X_t = X_t^+ - X_t^-$, where $X_t^+ = \max(X_t, 0)$ and $X_t^- = \max(-X_t, 0)$. Thus, $EX_t = EX_t^+ - EX_t^-$, and the expectation is said to exist if both terms are finite. Since $|X_t| = X_t^+ + X_t^-$ the conclusion follows immediately. The second aspect to be noted is that $Ee^{\alpha B_t - (\alpha^2/2)t} = e^{-(\alpha^2/2)t}G(1)$, where $G$ is the moment generating function of $\alpha B_t$, evaluated at one. Since $\alpha B_t \sim N(0, \alpha^2 t)$, its moment generating function with parameter $u$ is $G(u) = e^{(1/2)u^2\alpha^2 t}$, and the conclusion follows upon taking $u = 1$.

The class of semimartingales, however, is much larger than the class of (right-) continuous martingales. For example $B_t^2$ is a semimartingale, since it can be written as $B_t^2 = (B_t^2 - t) + t = M_t + A_t$, but is definitely not a martingale!

v. It is easy to find functions of bounded variation on compact subsets of $R_+$. In particular, the functions $f_k(t) = t^k$ for $k \in [0, \infty)$ are such functions.

Ito's formula is a particular answer to the following question: If $X$ is a continuous semimartingale[14] what is the class of functions, $h$, for which $h(X)$ is also a (minimally right-) continuous semimartingale? Ito's result identifies this to be the class of **twice continuously differentiable functions**, $C^2$. We begin with the following preliminary result.

**Proposition 4.** Let $\{X_t, Y_t : t \in R_+\}$ be continuous semimartingales on the probability space $(\Omega, \mathcal{A}, \mathcal{P})$; then over the interval $[0, t)$, $t < \infty$

$$X_t Y_t - X_0 Y_0 = \int_0^t X_s \, dY_s + \int_0^t Y_s \, dX_s + [X, Y]_t. \qquad (8.20)$$

Moreover, $XY$ is also a continuous (parameter) semimartingale process.

Proof: Consider the partition $p_{nt}$, let $Z_t = X_t Y_t$, and note that

$$Z_t - Z_0 = \sum_{j=0}^{n-1} \Delta Z_{j+1} = \sum_{j=0}^{n-1} (X_{j+1} Y_{j+1} - X_j Y_j) = \sum_{j=0}^{n-1} [X_j \Delta Y_{j+1} + Y_j \Delta X_{j+1}]$$

$$+ \sum_{j=0}^{n-1} \Delta X_{j+1} \Delta Y_{j+1}] \to \int_0^t X_s \, dY_s + \int_0^t Y_s \, dX_s + [X, Y]_t.$$

That $Z_t$ is a continuous parameter process is evident from its construction.

q.e.d.

**Remark 4.** The result of Proposition 4 is often referred in the literature as the **change in variable** or **integration by parts** formula. We also have the following useful corollary.

---

[14] The formal result requires only right-continuity; the assertion of continuity simplifies the argument leading to it, without putting appreciable restrictions on the applications we have in mind.

**Corollary 3.** In the context of Proposition 4 let

$$f : [0, t) \longrightarrow R, \ t < \infty,$$

be a continuous function of bounded variation, $X = f$ and $Y = B$. Then

$$f(t)B(t) = \int_0^t B_s \, df(s) + \int_0^t f(s) \, dB_s.$$

Proof: Proceeding in the same manner as in the proof of Proposition 4, we find

$$
\begin{aligned}
f(t)B_t &= \sum_{j=0}^{n-1} [f(t_{j+1})B_{j+1} - f(t_j)B_j] \\
&= \sum_{j=0}^{n-1} [f(t_j)\Delta B_{j+1} + B_j \Delta f_{j+1} + \Delta B_{j+1} \Delta f_{j+1}] \\
&\rightarrow \int_0^t f(s) \, dB_s + \int_0^t B_s \, df(s) + [X, f]_t.
\end{aligned}
$$

It remains only to show that $[X, f]_t = 0$. But this is evident since

$$\left( \sum_{j=0}^{n-1} \Delta B_{j+1} \Delta f_{j+1} \right) \leq \sup_{p_{nt}} |B_{j+1} - B_j| \sum_{j=0}^{n-1} |f_{j+1} - f_j| \rightarrow 0,$$

by the continuity of BM paths, and the fact that $f$ is of bounded variation.

$$\text{q.e.d.}$$

The general form of Ito's formula is basically an extension of the results above to the more general case where one shows that if $X$ is a (minimally right-) continuous semimartingale and

$$h : R_+ \times R \longrightarrow R, \quad h \in C^2,$$

i.e. $h$ is a continuously twice differentiable function, $h(X_t, t)$ is also a continuous semimartingale. Before we take up these issues, however, a few examples are in order.

**Example 2.** Consider the integral $\int_0^t B_s \, dB_s$, which we also examined in Example 1, and we verified that it is given by the expression in Eq. (8.16). At that time, however, we were unable to explain how it was derived. We shall do so now.

Using Proposition 4, we can certainly take $X = B$, $Y = B$, and thus obtain

$$B_t^2 = \int_0^t B_s \, dB_s + \int_0^t B_s \, dB_s + [B, B]_t = 2 \int_0^t B_s \, dB_s + t,$$

since $[B, B]_t$ is the limit of $\sum_{j=0}^{n-1} (\Delta B_{j+1})^2$ and by Proposition 7, of Chapter 7, this is simply $t$. Consequently, we have

$$\int_0^t B_s \, dB_s = \frac{1}{2} [B_t^2 - t],$$

which simply confirms the result obtained in Eq. (8.16). Incidentally, note that the right member above is a continuous square integrable martingale. This is implied by Proposition 4, and actually verified in Remark 3. **The Ito integral is the only integral that has this property.**

Even though Proposition 4 and its corollary enable us to obtain several particular Ito integrals, they apply only to a relatively restricted class of stochastic integrals that may be expressed as the product of two semimartingales. An extension to a much wider class of stochastic integrals is given in Proposition 5.

**Proposition 5** (Ito's Formula). Let $X$ be a continuous semimartingale relative to the stochastic basis $\{\mathcal{A}_t : \mathcal{A}_t = \sigma(X_s, \ s \leq t), \ t < \infty, \mathcal{A}_0 = (\emptyset, \Omega)\}$, and let $K_t$ be a continuous process of bounded variation, both relative to (i.e. adapted to) the stochastic basis above. If

$$h : R_+ \times R \longrightarrow R, \quad h \in C^2,$$

then

$$h(X_t, K_s) = h(X_0, K_0) + \int_0^t h_1(X_s, K_s) \, dX_s + \int_0^t h_2(X_s, K_s) \, dK_s$$

$$+ \frac{1}{2} \int_0^t h_{11}(X_s, K_s) \, d[X, X]_t,$$

where $h_{ij}$ indicates the (second or cross) derivative of $h$ with respect to its $i^{th}$ and $j^{th}$ arguments.

Proof: We note that $h(X_t, K_s) - h(X_0, K_0) = \sum_{j=0}^{n-1} \Delta h_{(j+1)}$, where $h_{(j)} = h(X_{t_j}, K_{t_j})$, and $t_j$ are the points of the partition $p_{nt}$. Now, using Taylor's theorem with remainder, up to second-order terms, we find

$$\Delta h_{(j+1)} \;=\; h_1 \Delta X_{j+1} + h_2 \Delta K_{j+1}$$

$$+\frac{1}{2}\left[h_{11}(\Delta X_{j+1})^2 + 2h_{12}\Delta X_{j+1}\Delta K_{j+1} + h_{22}(\Delta K_{j+1})^2)\right] + r_j,$$

$$r_j \;=\; o(|\Delta t_{j+1}|^2 + |\Delta K_{j+1}|^2 + |\Delta X_{j+1}|^2), \qquad\qquad (8.21)$$

where the notation $o(x)$ means $[o(x)/x] \to 0$ as $x \to 0$. Since $K_t$ and $X_t$ are continuous processes, $r_j \to 0$ with $n$, at least in probability. Hence, by the approximating arguments we had employed earlier, we conclude that

$$h(X_t, K_s) \;=\; h(X_0, K_0) + \int_0^t h_1(X_s, K_s)\, dX_s + \int_0^t h_2(X_s, K_s)\, dK_s$$

$$+\frac{1}{2}\int_0^t h_{11}(X_s, K_s)\, d[X, X]_t.$$

This argument is valid in view of the following:

i. In the quadratic term of Eq. (8.21)

$$\left|\sum_{j=0}^{n-1} h_{22}(\Delta K_{j+1})^2\right| \;\le\; c_{22} \sum_{j=0}^{n-1} (\Delta K_{j+1})^2 \to 0$$

by part i of Proposition 5, in Chapter 6, since $c_{22}$ is a uniform bound on $h_{22}$ and $K_t$ is of bounded variation.

ii. The term $\sum h_{12}\Delta X_{j+1}\Delta K_{j+1} \to 0$ since both processes are continuous and $K$ is of bounded variation.

iii. The term $\sum h_{11}(\Delta X_{j+1})^2 \to \int h_{11}\, d[X, X]_t$, since

$$\sum_{j=0}^{n-1}(\Delta X_{j+1})^2 \;\xrightarrow{\text{P}}\; [X, X]_t.$$

q.e.d.

**Remark 5.** Proposition 5 does not, unfortunately, provide an explicit formula for any stochastic integral one might wish to obtain; for example, it gives no **direct and explicit** representation of a closed-form expression of the integral $\int_0^t B_s^2\, dB_s$. What it does provide is **an explicit expression of**

**certain processes which may be represented as the sum of stochastic integrals**. Often, by manipulating such expressions it may be possible to obtain an explicit and closed form representation for a given stochastic integral.

An application of the preceding is given in Example 3.

**Example 3.** Let $Z_t = h(X_t, K_s) = (B_t^2 - t)^2$. In this context, $K_t = t$ and $X_t = B_t$. Evidently, $t$ is a function of bounded variation; noting that $B_0 = 0$ a.c., $h_1 = 4(B_t^2 - t)B_t$, $h_2 = -2(B_t^2 - t)$, and $h_{11} = 12B_t^2 - 4t$, we obtain

$$(B_t^2 - t)^2 = 4 \int_0^t (B_s^2 - s)B_s \, dB_s - 2 \int_0^t (B_s^2 - s) \, ds$$

$$+ \frac{1}{2} \int_0^t (12B_s^2 - 4s) \, d[B, B]_t$$

$$= 4 \int_0^t (B_s^2 - s)B_s \, dB_s + 4 \int_0^t B_s^2 \, ds,$$

which contains certain easily interpretable stochastic integrals.

Proposition 4 may also be used to derive the same result. To this end put $X_t = B_t^2 - t$, $Y_t = X_t$, and note that $[X, X]_t = [B^2, B^2]_t$. This is so since

$$\sum(\Delta B_{j+1}^2 - \Delta t_{j+1})^2 = \sum_{j=0}^{n-1} (\Delta B_{j+1}^2)^2 - 2 \sum_{j=0}^{n-1} (\Delta B_{j+1}^2 \Delta t_{j+1})$$

$$+ \sum_{j=0}^{n-1} (\Delta t_{j+1})^2 \rightarrow [B^2, B^2]_t - 2 \cdot 0 + 0,$$

and the last two terms of the expansion converge to zero by the continuity of BM paths, and the fact that $t$ is of bounded variation. We therefore find

$$X_t^2 = (B_t^2 - t)^2 = 2 \int_0^t X_s \, dX_s + [B^2, B^2]_t = 4 \int_0^t (B_s^2 - s)B_s \, dB_s + [B^2, B^2]_t.$$

To derive an expression for $[B^2, B^2]_t$, we begin by noting that it is simply the limit of $\sum_{j=0}^{n-1}(\Delta B_{j+1}^2)^2$. Expanding the square, we find

$$(B_{j+1}^2 - B_j^2)^2 = (B_{j+1} - B_j)^2(B_{j+1} + B_j)^2 = (\Delta B_{j+1})^2(\Delta B_{j+1} + 2B_j)^2.$$

It is easy to verify that all terms vanish in the limit, except for the term

$$4 \sum_{j=0}^{n-1} B_j^2 (\Delta B_{j+1})^2 \rightarrow 4 \int_0^t B_s^2 \, d[B, B]_s$$

so that

$$[B^2, B^2]_t = 4 \int_0^t B_s^2 \, d[B, B] = 4 \int_0^t B_s^2 \, ds.$$

Combining and collecting terms,

$$(B_t^2 - t)^2 = 4 \int_0^t (B_s^2 - s) B_s \, dB_s + 4 \int_0^t B_s^2 \, ds,$$

which is identical with the result obtained earlier by applying Proposition 5.

We complete this section by giving the Ito formula for the multidimensional case. Since no new concepts are involved, we simply present the result without proof.

**Proposition 6** (Multivariate Ito Formula). Let $X$ be a continuous $q$-element vector semimartingale process [i.e. $X_t = (X_t^{(1)}, X_t^{(2)}, \ldots, X_t^{(q)})$] defined on the probability space $(\Omega, \mathcal{A}, \mathcal{P})$ and

$$h : R_+ \times R^q \to R, \quad h \in \mathcal{C}^2.$$

Then

$$h(X_t) = h(X_0) + \sum_{i=1}^q \int_0^t h_i(X_s) \, dX_s^{(i)} + \frac{1}{2} \int_0^t \sum_{i,j=1}^q h_{ij}(X_s) \, d[X^{(i)}, X^{(j)}]_s.$$

**Corollary 4.** In the context of Proposition 6, if one of the processes, say the last, is of bounded variation then

$$h(X_t) = h(X_0) + \sum_{i=1}^{q-1} \int_0^t h_i(X_s) \, dX_s^{(i)} + \int_0^t h_q(X_s) \, dK_s$$

$$+ \frac{1}{2} \int_0^t \sum_{i,j=1}^{q-1} h_{ij}(X_s) \, d[X^{(i)}, X^{(j)}]_s,$$

where we have taken $X_t^{(q)} = K_t$ to be the process of bounded variation.

We conclude this section by deriving a number of stochastic integrals using Proposition 6.

**Example 4.** In this example we consider a function $h$ of the form

$$h(B_t, t) = B_t^r t^k, \quad h_1 = r B^{r-1} t^k, \quad h_2 = k B_t^r t^{k-1}, \quad \frac{1}{2} h_{11} = \frac{1}{2} r(r-1) B_t^{r-2} t^k,$$

and obtain the corresponding stochastic integrals by assigning various values
to the parameters $k$ and $r$.

1. For the case $k = 0$, $r = 1$, we find

$$B_t = \int_0^t dB_s,$$

as is to be expected.

2. for the case $k = 1$, $r = 0$, we find

$$t = \int_0^t ds,$$

again as is to be expected.

3. For the case $r = 1, k = 1$, we find

$$tB_t = \int_0^t s\, dB_s + \int_0^t B_s\, ds;$$

the first integral appears to be like a sum of independent normal variables,
and second is the area under a sample path of BM over the interval $[0, t)$.
It is evident that

$$\int_0^t s\, dB_s \sim N\left(0, \frac{1}{3}t^3\right), \quad \int_0^t B_s\, ds \sim N\left(0, \frac{1}{3}t^3\right),$$

which may be demonstrated either by using the approximation method, or
directly from the properties of SBM. Operating directly with the integral,
and noting that the SBM process is one with independent increments, we
find

$$I^* = \int_0^t s\, dB_s, \quad EI^* = 0, \quad \mathrm{Var}(I^*) = \int_0^t \int_0^t sr\, E(dB_s dB_r)$$

$$= \int_0^t \int_0^t \delta_{rs}\, s\, r\, ds = \int_0^t s^2\, ds = \frac{1}{3}t^3,$$

where $\delta_{sr}$ is the Kronecker delta, i.e. $\delta_{rs} = 1$ if $s = r$ and zero otherwise.
Thus,

$$\int_0^t s\, dB_s \sim N\left(0, \frac{1}{3}t^3\right), \quad \text{as claimed.}$$

For the second integral, using the arguments of Proposition 1, we obtain

$$I^{**} = \int_0^t B_s\, ds, \quad EI^{**} = \lim_{n \to \infty} \sum_{j=0}^{n-1} (EB_{s_j})\Delta s_{j+1} = 0,$$

$$\text{Var}(I^{**}) = \lim_{n\to\infty} \sum_{i=0}^{n-1}\sum_{j=0}^{n-1} \Delta s_{i+1}\Delta r_{j+1} E(B_s, B_{r_j})$$

$$= \lim_{n\to\infty} \sum_{i=0}^{n-1}\sum_{j=0}^{n-1} \min(s_i, r_j)\Delta s_{i+1}\Delta r_{j+1}$$

$$= \int_0^t \int_0^t \min(s,r)\,ds\,dr = \int_0^t \left(\int_0^r s\,ds\right) dr + \int_0^t r\left(\int_r^t ds\right) dr$$

$$= \frac{1}{3}t^3,$$

where $0 = s_0 \leq s_1 \leq s_2 \leq \cdots \leq s_n = t$, etc., is a partition of the interval $[0,t]$, so that

$$\int_0^t B_s\,ds \sim N\left(0, \frac{1}{3}t^3\right), \quad \text{as claimed.}$$

The development above implies, because of the representation of $tB_t$ above, that twice the covariance of the two integrals in the right member is $(1/3)t^3$. This can be shown directly from first principles. Denote the covariance by $\gamma$, use the partition above, and note that

$$\gamma = \lim_{n\to\infty} \sum_{i=0}^{n-1}\sum_{j=0}^{n-1} s_i \Delta r_{j+1} E B_{r_j}\Delta B_{s_{i+1}} = \lim_{n\to\infty} \sum_{i=0}^{n-1} \sum_{r_j \geq s_{i+1}} s_i \Delta s_{i+1}\Delta r_{j+1}$$

$$\to \int_0^t \left(s\int_s^t dr\right) ds = \int_0^t s(t-s)\,ds = \frac{t^3}{6}.$$

Thus, the sum of variances of the two integrals plus twice their covariance equals $t^3$, as implied by the fact that the two integrals sum to $tB_t$ whose variance is, evidently, $t^3$. In the argument above we have made use of the fact that $E B_{r_j}\Delta B_{s_{i+1}} = 0$, for $r_j < s_{i+1}$. For $r_j \geq s_{i+1}$, write

$$B_{r_j} = \Delta B_{s_{i+1}} + B_{s_i} + (B_{r_j} - B_{s_{i+1}});$$

and note that the last two terms in the right member above are independent of the first term, thus obtaining the desired result.

4. For the case $k = 0$, $r = 2$, we find

$$B_t^2 = 2\int_0^t B_s\,dB_s + \int_0^t d[B,B]_t, \quad \text{or}$$

$$\frac{1}{2}[B_t^2 - t] = \int_0^t B_s\,dB_s.$$

5. For the case $k = 1$, $r = 2$, we find

$$t B_t^2 - \frac{1}{2} t^2 = 2 \int_0^t s B_s \, dB_s + \int_0^t B_s^2 \, ds.$$

6. For $k = 2$, $r = 2$ we find

$$\frac{1}{2} [t^2 B_t^2 - (1/3) t^3] = \int_0^t s^2 B_s \, dB_s + \int_0^t s B_s^2 \, ds.$$

We may proceed in similar fashion with the same function $h$, or we may use other functional forms in order to derive expressions for stochastic integrals of interest.

7. Let us now examine the multivariate analog (SMBM) to item 4 above; in particular, let us establish the result

$$B(t)' B(t) = \int_0^t B(s)' \, dB(s) + \int_0^t B(s) \, dB(s)' + I_q t,$$

where $B(t) = [B_1(t), B_2(t), \ldots, B_q(t)]$. From Proposition 6, with $h(B) = B_i B_j$, we obtain

$$B_i(t) B_j(t) = \int_0^t B_i(s) \, dB_j(s) + \int_0^t B_j \, dB_i(s) + \int_0^t d[B_i, B_j]_s.$$

From the definition of quadratic covariation in Eq. (8.18), and because the components $B_i, B_j$ of the SMBM are **mutually independent**, we find

$$[B_i + B_j, B_i + B_j]_s = 2s, \quad [B_i - B_j, B_i - B_j]_s = 2s.$$

Consequently,

$$[B_i, B_j]_s = \delta_{ij} s,$$

where $\delta_{ij} = 1$ if $i = j$ and zero otherwise; it follows, therefore that

$$B_i(t) B_j(t) = \int_0^t B_i(s) \, dB_j(s) + \int_0^t B_j \, dB_i(s) + \delta_{ij} \, t,$$

or that

$$B(t)' B(t) = \int_0^t B(s)' \, dB(s) + \int_0^t B(s) \, dB(s)' + I_q t,$$

as claimed.

**Example 5.** In item 3 of the preceding example, we established the distribution of the stochastic integral $\int_0^t s \, dB(s)$; a natural extension of that result is to determine the distribution of the stochastic integral $\int_0^t B(s)' \, dB(s)$, where $B$ is a SMBM. We begin with the **scalar** case, and define

$$I_1 = \int_0^t B_1(s) \, dB_1(s),$$

where $B_1$ is a **scalar** SBM. Operating from first principles, the integral above is the limit of the sums

$$S_n = \sum_{j=0}^{n-1} B_1(s_j) \Delta B_1(s_{j+1}), \quad 0 = s_0 \le s_1 \le s_2, \cdots, \le s_n = t,$$

over all possible partitions $p_{nt}$ whose modulus converges to zero with $n$. We note that each term of the sum has mean zero since, by the properties of SBM, $B_1(s_j)$ and $\Delta B_1(s_{j+1})$ are **mutually independent**. The variance of each term is $s_j \Delta s_{j+1}$. Thus, we obtain

$$ES_n = 0, \quad \mathrm{Var}(S_n) = \sum_{j=0}^{n-1} s_j \Delta s_{j+1}, \quad \text{so that}$$

$$\lim_{n \to \infty} \mathrm{Var}(S_n) = \int_0^t s \, ds = \frac{1}{2} t^2.$$

We may also obtain the same result directly from the integral representation, if we recall that $B_1(s)$ is independent of $dB_1(s)$ and

$$E dB_1(s) dB_1(r) = \delta_{rs} \, ds.$$

Thus,

$$E I_1 = \int_0^t [E B_1(s)][E dB_1(s)] = 0,$$

$$E I_1^2 = \int_0^t \int_0^t E[B_1(s) B_1(r)] E[dB_1(s) dB_1(r)] = \int_0^t s \, ds = \frac{1}{2} t^2,$$

and the (first two) moments of the distribution of $I_1$ are easily determined; its distribution, however, is nonstandard since in $S_n$ we basically have the sum of **products of independent normals** and the product of two normals, unfortunately, does not have a standard distribution.

Next, let us consider integrals of the form

$$I_2 = \int_0^t B_1(s) \, dB_2(s),$$

where $B_i, i = 1, 2$, are distinct (and mutually independent) scalar SBM. We may certainly employ the same procedure as above, and define the approximating sums and so on. In the interest of brevity we consider only arguments based on the integral representation, thus obtaining

$$E I_2 = \int_0^1 [E B_1(s)][E dB_2(s)] = 0,$$

$$E I_2^2 = \int_0^t \int_0^1 [E B_1(s) B_1(r)][E dB_2(s) dB_2(r)] = \int_0^t \int_0^t \delta_{rs} s \, dr = \int_0^t s \, ds$$

$$= \frac{t^2}{2},$$

by the properties of standard (univariate) SBM. The distribution of $I_2$ may be easily obtained by the following convenient device.[15] Consider the **conditional distribution of** $I_2$ **given** $B_1$, and note that the integral reduces to the generic form

$$I_2 | B_1 = \int_0^t g(s) \, dB_2(s),$$

which is of the type dealt with in item 3 (case $k = 1$, $r = 1$) of the preceding example; an easy calculation shows that

$$E(I_2 | B_1) = 0, \quad \text{Var}(I_2 | B_1) = \int_0^t g^2(s) \, ds.$$

The fact that the **conditional distribution** of $I_2 | B_1$ is **normal** is quite evident from the approximating sum representation. Hence we conclude that

$$I_2 | B_1 \sim N(0, \ \phi), \quad \phi = \int_0^t B_1^2(s) \, ds,$$

which, evidently, depends on $B_1$ through its variance and, thus, the unconditional distribution of $I_2$ is nonstandard. However, if we consider the conditional distribution of the **normalized** integral $\phi^{-(1/2)} I_2 | B_1$, we obtain

$$\frac{1}{\sqrt{\phi}} I_2 | B_1 \sim N(0, \ 1),$$

which **does not depend** on $B_1$. We conclude therefore that the **unconditional distribution of the normalized integral** $(\phi^{-(1/2)} I_2)$ obeys

$$\frac{1}{\sqrt{\phi}} I_2 \sim N(0, \ 1).$$

---

[15] The same device cannot be applied to the integral $I_1$ since the latter involves only one distinct SBM.

The preceding discussion has established the following useful result. Define

$$\Psi = \int_0^t B'(s)\,dB(s) = \left( \int_0^t B_i(s)\,dB_j(s) \right), \quad i,j = 1, 2, \cdots, q,$$

where $B$ is the SMBM of Example 5. The off-diagonal elements of $\Psi$ have the following conditional distributions

$$\psi_{ij}|B_i \sim N(0, \phi_i), \quad \phi_i = \int_0^t B_i^2(s)\,ds, \quad t \in (0, 1], \quad i \neq j, \quad j = 1, 2, \cdots, q.$$

Moreover,

$$\frac{1}{\sqrt{\phi_i}}\psi_{ij} \sim N(0,\ 1), \quad \text{for all } i \neq j.$$

# Chapter 9

# Central Limit Theorems; Invariance

## 9.1   Introduction

All central limit theorems (CLT) discussed in this chapter are formulated as (weak) convergence of probability measures. The reader may be more familiar with CLT formulated as (weak) convergence of distribution functions. There is no conflict or difference between the two and we clarify this matter at a later stage in this discussion. For the moment, it will suffice to point out that the concept of convergence of distribution functions is tied to the real line and to note that the analytical methods employed in such CLT proofs cannot be readily adapted to more general spaces. The concept of convergence of probability measures (or probability distributions, as distinct from distribution functions), on the other hand, is equally at home on the real line or **on any other metric space**. For this reason it is a more powerful concept and is most appropriate for the topics of the ensuing discussion. The reader who desires a more extensive explanation may consult Billingsley (1968), pp. 2-6.

In dealing with the distributional aspects of estimators in the context of unit root or cointegration [1] models, we are dealing with very complex issues; these involve the nature of the stochastic entities under consideration. We

---

[1] This is a situation where it is claimed that even though a random element (vector), say $X$, is $I(1)$, there exists at least one **constant vector**, say $\beta$, such that $\beta' X$ is $I(0)$.

review this complexity first, before we embark on our main task.

In the preceding we dealt with scalar as well as multivariate stochastic processes, denoted by $X(t, \omega)$, for $t \in \mathcal{T}$ and $\omega \in \Omega$. In some of our discussions $\mathcal{T} = [0, 1]$, but in most $\mathcal{T} = R_+$, with the understanding that all arguments are carried out on compact subsets of $R_+$. The structure of such (scalar) processes is as follows:

$$X : R_+ \times \Omega \longrightarrow R,$$

such that for each $t \in \mathcal{T}$, $X_t$ is a random variable defined on the probability space $(\Omega, \mathcal{A}, \mathcal{P})$; thus a stochastic process is simply a collection of random variables defined on the probability space above.

If $X$ is a **multivariate process**, we are dealing with

$$X : R_+ \times \Omega \longrightarrow R^d,$$

where $d$ is the **dimension** of the vector. Thus, for each $t \in \mathcal{T}$ we are dealing with the random vector $X_t = (X_t^{(1)}, X_t^{(2)}, \ldots, X_t^{(d)})$. In this context the stochastic process is simply a collection of random vectors, defined on the probability space $(\Omega, \mathcal{A}, \mathcal{P})$ and taking values in $R^d$. Each component, say $X_t^{(j)}$, is thus a collection of random variables defined on that probability space.

If we fix $\omega$, and think of the entities $X(t, \omega)$ as **functions of** $t$ **alone**, then we are dealing with a collection of random functions, which are the **sample paths**, or "time paths" of the stochastic process. If this function is a.c. continuous, i.e. it is continuous for all $\omega$, except for possibly a set of $\mathcal{P}$-measure zero, then the sample paths are elements of the space $C(R_+, R^d)$, the space of continuous functions from $R_+$ to $R^d$.[2] If the sample paths are not continuous, **but are continuous on the right and have left limits,**

---

[2] The reader might wonder how a space like $C(R^+, R^d)$ can possibly contain the sample paths corresponding to the $\omega$'s in the set of $\mathcal{P}$-measure zero, referred to above. The answer is that it does not! The technical way of avoiding this is to produce a theorem that states that if a stochastic process has this property, there exists a **modification** of this process, i.e. another process that is equivalent to the given process in the sense that the two processes differ only on a set of $\mathcal{P}$-measure zero, and which has continuous sample paths, **without exception**. In all of our discussion we deal exclusively with this modification.

i.e. if $\lim_{s \downarrow t} X(s, \omega) = X(t, \omega)$ and $\lim_{s \uparrow t} X(s, \omega)$ exists (but it is **different from** $X(t, \omega)$), then the sample paths are elements of the space $D(R_+, R^d)$, the space of functions which are **right-continuous with left limits.** [3]

At this stage it is useful to highlight this dual perspective on stochastic processes, and make clear the connection between the two aspects. To do so we introduce the so-called **coordinate mapping** which operates on the stochastic process $X$ to produce, for each $\omega$, the ordinate of its sample path at "time" $t$. More precisely, the coordinate mapping is defined by

$$\phi_t(X) = X_t. \tag{9.1}$$

In a more revealing representation we would write

$$\phi_t \circ X(\omega) = \phi_t[X(\omega)] = X_t. \tag{9.2}$$

since we are really dealing with the **composition function** $\phi_t \circ X$. In this framework $\omega$ is treated as a parameter and is suppressed, but it is understood that $X_t$, treated solely as a function of $t$, is a **random function**.

Similarly, for $0 = t_1 < t_2 < \dots, t_m < \infty$ the mapping $\phi_{t_1, t_2, \dots, t_m}$ operating on the stochastic process $X$ produces

$$\phi_{t_1, t_2, \dots, t_m}(X) = (X_{t_1}, X_{t_2}, \dots, X_{t_m}). \tag{9.3}$$

We make these ideas concrete, first in the case of $d = 1$. Let $\mathcal{B}(C)$ be the Borel $\sigma$-algebra of the space $C(R_+, R^d)$ and consider the measurable space $(C, \mathcal{B}(C))$. What is a typical set in $\mathcal{B}(C)$? It is simply a collection of **functions** with particular characteristics. For example, by the continuity of the functions in $C$, $S \in \mathcal{B}(C)$ where

$$S = \{x : x(t) \leq s\}$$

is the collection of all sample paths of $X$ whose ordinate at $t$ does not exceed $s$. If $X$ is a stochastic process with continuous sample paths, what

---

[3] Such functions are occasionally called in the literature *càdlàg*, from the initials of the French expression *continu à droite, avec des limites à gauche*, which means in English right-continuous with left limits. Occasionally we use, for such functions, the designation rcll which is becoming increasingly popular. Incidentally, the reader should note that $C$ stands for continuous and $D$ for discontinuous.

is the connection between $X$ and $S$? To answer that question, consider the set $H = \{\omega : X(t, \omega) \leq s\}$ and note that $H \in \mathcal{A}$. A little reflection will convince the reader that $H = X^{-1}(S)$, i.e. it is the inverse image of $S$ under $X$; this is so since $H$ contains all the $\omega$ for which the corresponding sample paths have ordinates at $t$ which do not exceed $s$, while $S$ contains all the continuous sample paths (of $X$) whose ordinates at $t$ do not exceed $s$.

The situation is basically similar for $d > 1$. Thus, if

$$X : R_+ \times \Omega \longrightarrow R^d$$

with **continuous** sample paths, consider the measurable space $(C, \mathcal{B}(C))$. In the range space $(R^d, \mathcal{B}(R^d))$ let $A \in \mathcal{B}(R^d)$ and define $S = \{x : x(t) \in A\}$. Because the sample paths are continuous and thus measurable, $S \in \mathcal{B}(C)$. Consequently, $H = \{\omega : X(t, \omega) \in A\} = X^{-1}(S)$.

The preceding discussion discloses a dual perspective on stochastic processes. We can either view them as collections, or families, of random variables or vectors, defined on the probability space $(\Omega, \mathcal{A}, \mathcal{P})$ and indexed on some set $\mathcal{T}$,[4] or we can view them as a collection of (their) sample paths, obtained through coordinate mappings, and defined on the space $(C, \mathcal{B}(C))$. If a probability measure can be defined on that space, we can view the coordinate mappings as a stochastic process.

The situation is generally similar on the space $D(R_+, R^d)$, although certain properties of the space have to approached with considerably more care than is the case with $C(R_+, R^d)$. This is so because certain discontinuities (jumps) are allowed in the elements of $D(R_+, R^d)$, i.e. the functions representing the sample paths of the underlying stochastic process.

## 9.2 Convergence in $C(R_+, R^d)$

In discussing CLT, or other convergence issues, in the context of $C(R_+, R^d)$ or similar such spaces we deal with **sequences** of stochastic processes, say $X^* = \{X^{(n)} : n \in \mathcal{N}_+\}$. In this context, **each** $X^{(n)}$ is a **stochastic process**, $X^{(n)}(t, \omega)$, and if $d > 1$ we are dealing with a **vector** process, whose fuller

---

[4] In our discussions $\mathcal{T}$ is typically $[0, 1]$ but occasionally it may be $R_+$.

representation is

$$X^{(n)}(t,\omega) = (X^{(n1)}(t,\omega),\ X^{(n2)}(t,\omega),\ldots,X^{(ni)}(t,\omega),\ldots,X^{(nd)}(t,\omega)),$$

and where the second superscript, say $i$, corresponds to the $i^{th}$ component of the vector process.

If on $C(R_+, R^d)$ we introduce the metric

$$\rho(x,y) = \sum_{n=1}^{\infty} 2^{-n} \frac{\sup_{t \leq n}|x(t) - y(t)|}{1 + \sup_{t \leq n}|x(t) - y(t)|}, \tag{9.4}$$

and the topology of uniform convergence (on compact subsets of $R_+$), it may be shown that $(C,\rho)$ is a complete topological space with a countable dense subset [5] and we may define on it the Borel $\sigma$-algebra $\mathcal{B}(C)$, i.e. the $\sigma$-algebra generated by the open sets (balls) in $C$. It is easy to show that this is the $\sigma$-algebra generated by the coordinate mappings. To see that, let $\mathcal{G}$ be the $\sigma$-algebra of coordinate mappings; since the latter are continuous it follows that $\mathcal{G} \subset \mathcal{B}(C)$. Conversely, by the definition of the metric above, for **fixed** $y$ the set $\{x : \rho(x,y) < a\}$ is $\mathcal{G}$-measurable. Hence $\mathcal{G}$ contains all the open balls. It follows that $\mathcal{B}(C) \subset \mathcal{G}$ and thus, $\mathcal{G} = \mathcal{B}(C)$.

The measurable space $(C,\mathcal{B}(C))$ will be referred to as the **Wiener space**, in honor of the American mathematician Norbert Wiener, and will be denoted for notational simplicity by $\mathbf{W}$, or $\mathbf{W^d}$ when we wish to stress that we are dealing with vector processes. If a probability measure can be defined on $\mathbf{W}$, say $\mathcal{P}$, the space $(C,\mathcal{B}(C),\mathcal{P})$ shall be referred to as the **Wiener probability space**. We already know from Chapter 6 that if $C$ is the collection of (sample) paths of BM, a probability measure can indeed be defined, which we had termed the Wiener measure, $\mu_W$. Thus, the Wiener probability space $(C,\mathcal{B}(C),\mu_W)$ is the appropriate probability space when dealing with the sample paths of BM.

Convergence results on $\mathbf{W^d}$ are usually obtained in two steps:

   i. the relevant sequence of probability measures, say $\{P^{(n)} : n \in \mathcal{N}_+\}$ is shown to be (weakly) relatively compact;

---

[5] Occasionally in the literature, a metrizable complete topological space with a countable dense subset is said to be Polish; a space with a countable dense subset is said to be separable. Thus, a Polish space is a complete separable space.

ii. the finite dimensional distributions are shown to converge, i.e. if we look, for arbitrary $k$, at the probability distribution of

$$X_{(k)}^{(n)} = (X^{(n)}(t_1),\ X^{(n)}(t_2), \ldots, X^{(n)}(t_k)),$$

say $P^{nk}$, $n \geq 1$ then this sequence is shown to converge weakly, say to $P^k$.[6]

We make these concepts more precise in the next section, where we also indicate what modification is required in order to enable us to prove central limit theorems in the space $(D, \mathcal{B}(D))$, i.e. in the space of sample paths which are rcll (right continuous with left limits)).

## 9.2.1 Tightness and Relative Compactness on $C$

We begin with a definition.[7]

**Definition 1.** Let $\{P^n : n \in \mathcal{N}_+\}$, $P$ be probability measures defined on $\mathbf{W}^\mathbf{d}$, more generally on a metric space $(S, \mathcal{B}(S))$. The sequence is said to **converge weakly** to $P$, denoted by $P^n \overset{\text{w}}{\to} P$, if and only if any one of the following (equivalent) conditions is satisfied:

i. For every bounded continuous function $g$ on $S$,

$$\lim_{n \to \infty} \int g \, dP^n = \int g \, dP.$$

ii. For every closed set $A \subset S$

$$\limsup_{n \to \infty} P^n(A) \leq P(A).$$

iii. For every open set $B \subset S$

$$\liminf_{n \to \infty} P^n(B) \geq P(B).$$

iv. For every $P$-continuity set $F \subset S$ (i.e. one whose boundary $\partial F$ obeys $P(\partial F) = 0$)

$$\lim_{n \to \infty} P^n(F) = P(F).$$

---

[6] The notation, for example $P$ and $P^k$ is meant to highlight the fact that $P$ is a probability distribution (measure) that pertains to the stochastic process as an ensemble, while $P^k$ is meant to highlight the fact that it pertains only to the $k$-tuples $X_{(k)}^{(n)}$ above.

[7] This and the following section rely chiefly on the presentation in Billingsley (1968).

**Definition 2.** A family of probability measures, say $\Pi$, defined on $\mathbf{W^d}$, more generally on the metric space $(S, \mathcal{B}(S))$, is said to be **(weakly) relatively compact** if and only if **every** sequence in $\Pi$ contains a subsequence which converges weakly.

**Definition 3.** A family of probability measures $\Pi$, as in Definition 2, is said to be **tight** if and only if for every integer $q > 1$, there exists a compact set $K_q$ such that

$$P(K_q) \geq 1 - \frac{1}{q}, \quad \text{for every} \quad P \in \Pi.$$

The significance of the concepts introduced by the definitions above becomes quite evident from the two propositions below, which are given without proof. Their proofs are very technical in nature and will not materially aid the reader's understanding of the issues involved.

**Proposition 1** (Prohorov's Criterion). Let $\Pi$ be a family of probability measures defined on $\mathbf{W^d}$, more generally on a metric space $(S, \mathcal{B}(S))$;

i. if $\Pi$ is tight, then it is (weakly) relatively compact;

ii. if the space is, in addition, a metrizable complete topological space with a countable dense subset (i.e. complete and separable) then a weakly relatively compact family is tight.

Proof: See Billingsley (1968), p. 37-40.

**Proposition 2.** Let $\{P^{(n)} : n \in \mathcal{N}_+\}$ be a sequence of probability measures and $P$ a probability measure, both defined on $\mathbf{W^d}$; denote their finite dimensional distributions by $P^{(nk)}$ and $P^k$, respectively. If $P^{(nk)} \overset{w}{\to} P^k$, **and if the sequence is tight**, then $P^{(n)} \overset{w}{\to} P$.

Proof: See Billingsley (1968), p. 35ff. and p. 54.

**Remark 1.** In general, convergence of the sequence of finite dimensional distributions is not sufficient to ensure that the sequence converges weakly, **unless** the sequence in question is weakly relatively compact. By Prohorov's criterion (Proposition 1), weak compactness is equivalent to tightness, if the space in question is a metrizable complete topological space with a countable dense subset, i.e. complete and separable or Polish. The space $\mathbf{W^d}$ as defined above **is** separable; thus, for a sequence of probability measures de-fined on this space, weak convergence of the finite dimensional distributions **implies** the weak convergence of the sequence, **if the latter is tight**. But what does it mean, operationally, for a sequence to be tight in this context? We next obtain such a criterion.

Before we proceed, however, we need to define the relatively compact sets in $\mathbf{W^d}$. To this end we introduce, for every $N \in \mathcal{N}_+$ and $x \in \mathbf{W^d}$, the entity

$$v_N(x; \delta) = \sup_{|t-t'| \leq \delta \ t, t' \leq N} |x(t) - x(t')|, \qquad (9.5)$$

which measures the degree of fluctuation (variability) of the ordinates of sample paths. It is termed the **modulus of continuity** associated with an element $x \in \mathbf{W^d}$. The significance of this concept is underscored by the next proposition.

**Proposition 3** (Arzelà-Ascoli). A subset $A \subset C(R_+, R^d)$ has compact closure (on compact subsets of $R_+$) if and only if

   i. $\sup_{x \in A} |x(0)| < \infty$, i.e. if $|x(0)|$ is bounded on $R^d$;

   ii. for every $N$, $\lim_{\delta \to 0} \sup_{x \in A} v_N(x; \delta) = 0$.

Proof: See Billingsley (1968), p. 221.

Of more direct relevance is Proposition 4.

**Proposition 4.** A subset $A \in \mathbf{W^d}$ is relatively compact in $\mathbf{W^d}$ if and only if

   i. for every $x \in A$, $|x(0)|$ is bounded in $R^d$;

   ii. for every $N$, $\lim_{\delta \to 0} \sup_{x \in A} v_N(x; \delta) = 0$.

Proof: Since a relatively compact (sub)set is one in which every sequence contains a weakly converging subsequence, it is clear that every sequence in $A$ contains a subsequence that converges to an element of its closure; thus $A$ has a compact closure and the result follows from Proposition 3.

q.e.d.

We now come to the important problem of characterizing tightness, or equivalently weakly relative compactness, on $\mathbf{W^d}$.

**Proposition 5.** A sequence of probability measures $\{P^{(n)} : n \in \mathcal{N}_+\}$ defined on $\mathbf{W^d}$ is tight (equivalently, weakly relatively compact) if and only if the following conditions hold.

   i. For every integer $r \geq 1$, there exist an integer $n_0$ and a number $c > 0$ such that

$$P^{(n)}(B) \leq \frac{1}{r}, \quad \text{for all } n \geq n_0, \quad \text{where } B = \{x : |x(0)| > c\};$$

   ii. for any integers $q, r$, and $N \in \mathcal{N}_+$ there exists a number $\delta > 0$ such that

$$P^{(n)}(F) \leq \frac{1}{r}, \quad \text{for all } n \geq n_0, \quad \text{where } F = \{x : v_N(x; \delta) > \frac{1}{q}\}.$$

Proof: Necessity follows immediately from Propositions 1 and 4.

As for sufficiency, let us first show that conditions i and ii may be stated in terms $n_0 = 0$, so that their statement could be modified to read for all $n \geq 0$. Thus, suppose conditions i and ii hold and consider the finite sequence of probability measures $\{P^{(n)} : n \leq n_0\}$. The sequence is evidently tight (why?) and thus satisfies i and ii for some numbers $\delta' > 0$ and $c' > 0$. If we put $c^* = \max(c, c')$ and $\delta^* = \min(\delta, \delta')$ in their statement, conditions i and ii are satisfied for all $n > 0$. **We shall assume this in the argument to follow.** To prove sufficiency we must prove that given any integer $r$, however large, we can construct a relatively compact subset, say $G_r \in \mathbf{W^d}$, such that $P^{(n)}(G_r) > 1 - (1/r)$, **for every** $n \in \mathcal{N}_+$. Noting that $v_0(x; \delta) = 0$ for any $x$ and $\delta$, we need consider only $N \geq 1$. To this end define sequences $\{c_{N,r} : N \geq 1\}$ and $\{\delta_{N,r,q} : N \geq 1\}$ such that

$$\sup_n P^{(n)}(B_{N,r}) \leq 2^{-(N+1)} \left(\frac{1}{r}\right), \quad \text{where } B_{N,r} = \{x : |x(0)| > c_{N,r}\},$$

$$\sup_n P^{(n)}(F_{N,r,q}) \leq 2^{-(N+1)-q} \left(\frac{1}{r}\right), \quad \text{where}$$

$$F_{N,r,q} = \{x : v_N(x; \delta_{N,r,q}) > \frac{1}{q}\}.$$

Consider now the sets

$$G_{N,r} = \bar{B}_{N,r} \cap \bar{F}_{N,r,q}, \quad N \geq 1. \tag{9.6}$$

By Proposition 4, the set

$$G_r = \bigcap_{N=1}^{\infty} G_{N,r} \tag{9.7}$$

is relatively compact in $\mathbf{W^d}$. Since $\bar{G}_r = \bigcup_{N=1}^{\infty} \bar{G}_{N,r}$ , we conclude that for every $n \in \mathcal{N}_+$

$$P^{(n)}(\bar{G}_r) \leq \sum_{N=1}^{\infty} P^{(n)}(B_{N,r}) + \sum_{N=1}^{\infty} P^{(n)}(F_{N,r,q})$$

$$\leq \left( \sum_{N=1}^{\infty} 2^{-N} \right) \left( \frac{1}{2} + \frac{1}{2^{q+1}} \right) \frac{1}{r} \leq \frac{1}{r}, \quad \text{or that}$$

$$P^{(n)}(G_r) > 1 - \frac{1}{r}. \tag{9.8}$$

q.e.d.

**Remark 2.** In the preceding $\delta$ is not required to lie in $(0,1)$. A slightly stronger form of condition ii is: For any integers $q, r$ there exists a number $\delta \in (0,1)$ and an integer $n_0$, such that

$$P^{(n)}(F) \leq \frac{1}{r} \delta, \tag{9.9}$$

where $F$ is as defined in condition ii of Proposition 5. For details see Billingsley (1968), p. 53.

**Remark 3.** A sequence of stochastic processes $\{X^{(n)} : n \in \mathcal{N}_+\}$ is said to be tight if and only if the sequence of associated probability distributions (measures), $\{P^n : n \in \mathcal{N}_+\}$, is tight.

Since the remark above sets up an equivalence between tightness of probability measures and tightness of sequences one may conjecture that there exists an "independent" characterization of tightness for sequences of random functions. This is given, in a form relevant for our future discussion, by Proposition 5a.

**Proposition 5a.** Let $\{\xi_j : j \in \mathcal{N}_+\}$ be a sequence of random variables on the probability space $(\Omega, \mathcal{A}, \mathcal{P})$, let $S_n = \sum_{j=1}^{n} \xi_j$, $S_0 = 0$, and define,

$$X^{(n)}(r, \omega) = \frac{1}{\sigma\sqrt{n}} S_{j-1}(\omega) + n \left( r - \frac{j-1}{n} \right) \xi_j(\omega), \quad r = \left[ \frac{j-1}{n}, \frac{j}{n} \right),$$

so that for each $n$ the entity above is a (continuous) random function on $(0,1)$, where $\sigma$ is a suitable constant. The sequence of random functions $\{X^{(n)}(r,\omega) : n \in \mathcal{N}_+\}$ is tight if, given any (arbitrary) integer $q$ there exist a number $\lambda > 1$ and an integer $n_0$ such that for all $n \geq n_0$

$$A_k(n) = \{\omega : \max_{i \leq n} |S_{k+i} - S_k| \geq \lambda \sigma \sqrt{n}\}, \quad \text{obeys} \quad \mathcal{P}[A_k(n)] \leq \frac{1}{q} \frac{1}{\lambda^2},$$

for all $k$.

Proof: The proof of this is given in Billingsley (1968), pp. 59ff.

**Remark 4.** If $P$ has support $A$, i.e. if $P(A) = 1$, and $A$ is compact then evidently $P$ is tight. If for a **finite sequence** $\{P^j : j = 1, 2, \ldots, n\}$ the support for $P^j$ is $R^d$, and if there exist compact sets $K_j \subset R^d$ such that $P^j(K_j) \geq 1 - (1/r)$, for arbitrary integer $r$, the sequence $\{P^j : j = 1, 2, \ldots, n\}$ is evidently tight since the set $K = \bigcap_{j=1}^{n} K_j$ is compact, and for every $j$ $P^j(K) \geq 1 - (1/r)$. In fact, by Proposition 19 in Dhrymes (1989) p. 226 **any** probability measure defined on a metric space is tight; hence a finite family of probability measures is tight.

A simpler criterion for compactness, and one that is often easier to verify, is given in Proposition 6.

**Proposition 6** (Kolmogorov's Criterion). Let $\{X^{(n)} : n \in \mathcal{N}_+\}$ be a sequence of continuous stochastic processes defined on the probability space $(\Omega, \mathcal{A}, \mathcal{P})$ and taking values in $R^d$, such that

   i. the sequence $\{X_0^{(n)} : n \in \mathcal{N}_+\}$ is tight (or equivalently, the sequence of initial probability distributions, say $\{P_0^n : n \in \mathcal{N}_+\}$, is tight);

   ii. there exist three constants, $c_i > 0$, $i = 1, 2, 3$, such for all $t_1, t_2 \in R_+$ and every $n \in \mathcal{N}_+$

$$\int_{R^d} |X_{t_1}^{(n)} - X_{t_2}^{(n)}|^{c_1} \, dP^n \leq c_2 \, |t_1 - t_2|^{1+c_3}.$$

Then, the sequence of corresponding probability measures $\{P^n : n \in \mathcal{N}_+\}$ is weakly relatively compact.

Proof: The proof for a slightly weaker version is given in Billingsley (1968), pp. 95-96.

## 9.2.2 Donsker's Theorem; Invariance

We are now in a position to present the first of an important class of CLT that are indispensable in determining the limiting distributions of parameter estimators, or test statistics, in the context of models that involve unit roots.

We will make the proof more detailed than is necessary to illustrate some of the applications and implications of the preceding discussion. We begin by creating the framework in which we shall operate. Let $\{\xi_n : n \in \mathcal{N}_+\}$ be a sequence of i.i.d. random variables defined on the probability space ($\Omega$, $\mathcal{A}$, $\mathcal{P}$), with the properties $E\xi_n = 0$, $\mathrm{Var}(\xi_n) = \sigma^2 < \infty$. Set

$$S_n = \sum_{j=1}^{n} \xi_j, \quad n \in \mathcal{N}_+, \tag{9.10}$$

and with $[a]$ the **integer part** of a real number $a$ and $t \in [0,1]$, define the following stochastic processes

$$\begin{aligned}
X_t^{(n)} &= \frac{1}{\sqrt{n\sigma^2}} \, S_i(\omega), \quad \text{for } t = \frac{i}{n}, \quad i = 1,2,\ldots,n \tag{9.11}\\
&= \left[\frac{(i+1)/n - t}{1/n}\right] \frac{1}{\sqrt{n\sigma^2}} \, S_i(\omega) + \left[\frac{t - (i/n)}{1/n}\right] S_{i+1}(\omega)\\
&= \frac{1}{\sqrt{n\sigma^2}} S_{[nt]}(\omega) + \left[\frac{t - (i/n)}{1/n}\right] \frac{1}{\sqrt{n\sigma^2}} \xi_{[nt]+1}, \quad t \in \left(\frac{i}{n}, \frac{i+1}{n}\right).
\end{aligned}$$

It may be readily verified that $X^* = \{X_t^{(n)} : n \in \mathcal{N}_+\}$ is a sequence of continuous stochastic processes on $\mathcal{T} = [0,1]$; as such it is an element of $C(\mathcal{T}, R)$ and may thus be studied in the context of $(C, \mathcal{B}(C))$. If we introduce the metric

$$\rho_1(x, y) = \sup_{t \in \mathcal{T}} |x(t) - y(t)|, \tag{9.12}$$

the space $(C, \mathcal{B}(C))$ becomes a complete topological space with a countable dense subset. Thus, the propositions obtained earlier in this section are fully applicable, and they may be used to study issues of convergence for such stochastic processes. To anticipate further developments note that the "interpolation" component of the last equation in Eq. (9.11) obeys

$$\left|\left[\frac{t - (i/n)}{1/n}\right] \frac{1}{\sqrt{n\sigma^2}} \xi_{i+1}\right| \leq \left|\frac{1}{\sqrt{n\sigma^2}} \xi_{i+1}\right| \xrightarrow{\mathrm{P}} 0.$$

Thus, from the point of view of applying a CLT to the stochastic process we have defined, the "interpolation" component may be ignored. On the other hand, the reader should note that if *ab initio* we **defined** the stochastic process without the interpolation component, its sample paths would be **discontinuous** but **rcll** and, in principle, we should not be dealing with the space $(C, \mathcal{B}(C))$ but rather with $(D, \mathcal{B}(D))$, a topic to which we return below.

We take this opportunity to note that at the points of discontinuity **both** right and left limits exist, **but the left limit at** $t$ **is not equal to** the right limit which is $X_t^n$, i.e. there is a jump (saltus) at that point. Occasionally, a discontinuity of this type is referred to as a **discontinuity of the first kind**.

We now present one of the earliest examples of a class of CLT which came to be known as **functional central limit theorems**.

**Proposition 7** (Donsker's Theorem). Let $\{\xi_n : n \in \mathcal{N}_+\}$ be a sequence of i.i.d. random variables defined on the probability space $(\Omega, \mathcal{A}, \mathcal{P})$, with

$$E\xi_n = 0, \quad \mathrm{Var}(\xi_n) = \sigma^2 < \infty, \quad \text{for all } n,$$

and define the sequence of continuous stochastic processes as in Eq. (9.11). Then, the sequence in question converges in distribution to the standard Brownian motion[8] on $\mathcal{T} = [0, 1]$, i.e. the sequence defined in Eq. (9.11) obeys[9]

$$X_t^n \xrightarrow{\mathrm{d}} B_t.$$

Proof: The proof is structured in two steps; first, we show that the finite dimensional distributions converge; second, we show that the sequence of probability measures, or equivalently the sequence of stochastic processes $\{X_t^n : n \in \mathcal{N}_+\}$, is tight.

Let $k$ be an arbitrary integer and consider the $k$-tuple

$$X_{(k)}^{*n} = \phi_{t_1, t_2, \dots, t_k}(X^{(n)}) = (X_{t_1}^{(n)}, X_{t_2}^{(n)}, \dots, X_{t_k}^{(n)}).$$

---

[8] It might appear to the reader that we had already proved a convergence to BM result in Chapter 7 (Section 7.2), in connection with Eq. (7.2). Strictly speaking, however, all we did there was to show that we had convergence of the partial sums, for **fixed** $t$, i.e. that for any $t \in \mathcal{T}$, $S_t \xrightarrow{\mathrm{d}} N(0, t)$.

[9] The notation below is to be understood in the context of Remark 5.

We show that the sequence of probability distributions for such entities converges weakly. For notational convenience put

$$Y^{*n} = (Y_1^{(n)}, Y_2^{(n)}, \ldots, Y_k^{(n)}), \quad Y_1^{(n)} = X_{t_1}^{(n)}, \quad Y_i^{(n)} = X_{t_i}^{(n)} - X_{t_{i-1}}^{(n)}, \quad (9.13)$$

where $i = 2, 3, \ldots k$. By the argument given just before the statement of the proposition, we might as well think of the sequence $Y^{*n}$ as if it were defined by

$$Y_1^{(n)} = \left(\frac{[nt_1]}{n}\right)^{1/2} \frac{1}{\sqrt{j_1\sigma^2}} S_{j_1}, \quad \text{for } t_1 \in [(j_1/n), (j_2/n)),$$

$$Y_i^{(n)} = \left(\frac{[n(t_i - t_{i-1})]}{n}\right)^{1/2} \frac{1}{\sqrt{(j_i - j_{i-1})\sigma^2}}[S_{j_i} - S_{j_{i-1}}],$$

$$\text{for } t_i \in [(j_i/n), (j_{i+1}/n)), \text{ where}$$

$$[nt_i] = j_i, \quad 0 \le j_1 < j_2 < \ldots < j_k \le n. \quad (9.14)$$

Noting that $j_i = [nt_i]$ it follows that

$$\lim_{n\to\infty} j_i = \infty, \quad \lim_{n\to\infty} \frac{[nt_i]}{n} = t_i, \quad i = 1, 2, \ldots, k, \quad (9.15)$$

and

$$Y_1^{(n)} \sim t_1^{1/2} \frac{1}{\sqrt{j_1\sigma^2}} \sum_{s=1}^{j_1} \xi_s, \quad (9.16)$$

$$Y_i^{(n)} \sim (t_i - t_{i-1})^{1/2} \frac{1}{\sqrt{(j_i - j_{i-1})\sigma^2}} \sum_{s=j_{i-1}+1}^{j_i - j_{i-1}} \xi_s, \quad i = 2, 3, \ldots k.$$

From Eq. (9.16) it is evident that each component of $Y^{*n}$ contains a **distinct**, i.e. nonoverlapping, subset of the sequence $\{\xi_n : n \in \mathcal{N}_+\}$. Since the latter is one of i.i.d. random variables, it follows from a standard CLT that

$$Y^{*n} \xrightarrow{\text{d}} \zeta, \quad \zeta \sim N(0, \Phi),$$

where $\Phi = \text{diag}(t_1, t_2 - t_1, \ldots, t_k - t_{k-1})$. Since

$$X_{(k)}^{*n} = Y^{*n} + (0, Y_1^{(n)}, Y_1^{(n)} + Y_2^{(n)}, \ldots, \sum_{s=1}^{k-1} Y_s^{(n)}) = g(Y^{*n}), \quad (9.17)$$

it follows, from Proposition 28 and Corollary 5 in Dhrymes (1989), pp. 242-244, that $X^{*n}_{(k)} \xrightarrow{d} A\zeta$, where $A$ is a lower triangular matrix with the property

$$a_{ij} = 1, \quad i \geq j, \quad \text{and } a_{ij} = 0, \text{ for } i < j.$$

Consequently,

$$X^{*n} \xrightarrow{d} \zeta^* \sim N(0, \Sigma), \quad \Sigma = (\sigma_{ij}), \quad \sigma_{ij} = \min(t_i, t_j),$$

which is the distribution of $(B_{t_1}, B_{t_2}, \ldots, B_{t_k})$, $B$ being the standard BM stochastic process. This completes the first phase of the proof.

The second phase of the proof consists of verifying condition ii of Proposition 5, as modified in Remark 2, adapted to our context, or alternatively providing a proof for Proposition 5a; what we shall show is that for any integers $q, r$ there exists a number $\delta \in (0, 1)$ such that

$$\mathcal{P}(B_q) \leq \frac{1}{r}, \quad \text{where } B_q = \{\omega : \sup_{t \leq s \leq t + \delta \wedge 1} \left| X^{(n)}_s - X^{(n)}_t \right| > \frac{1}{q} \}, \qquad (9.18)$$

where $t + \delta \wedge 1 = \min(t + \delta, 1)$. From the definition of the process we obtain

$$X^{(n)}_s - X^{(n)}_t = \frac{1}{\sqrt{n\sigma^2}} \left[ S_{[ns]}(\omega) - S_{[nt]}(\omega) \right] + a(s, n)\xi_{[ns]+1} - b(t, n)\xi_{[nt]+1},$$

where $|a| < 1$, $|b| < 1$. To complete the second phase we must first translate the condition in Eq. (9.18) into one that involves the partial sums $S_j(\omega)$ and, second, verify that the "translated" condition is satisfied. Thus, if $\delta \in (0, 1)$ is given, we may assume without loss of generality that there exist two integers $k < m$ such that

$$\frac{k}{n} \leq t < \frac{k+1}{n}, \quad \frac{m}{n} \leq t + \delta < \frac{m+1}{n}, \quad m + 1 \leq n. \qquad (9.19)$$

In this context $[nt] = k$, $[ns] \leq i \leq k + n\delta < m + 1$. Since the behavior of $S_i - S_k$ over $k \leq i \leq k + n\delta$ is the same as the behavior of $S_{i+k} - S_k$ over the range $0 \leq i \leq n\delta$, and since $\xi_{i+1}$, $\xi_{k+1}$ are contained in $S_{k+i+1} - S_{k+1}$ by the interpolation nature of the definition of $X^{(n)}_t$, we have for $\alpha_n = (n\sigma^2)^{-(1/2)}$

$$\sup_{t \leq s \leq t + \delta \wedge 1} \left| X^{(n)}_s - X^{(n)}_t \right| \leq 2 \max_{0 \leq i \leq n\delta} \alpha_n \left| S_{k+i}(\omega) - S_k(\omega) \right|. \qquad (9.20)$$

Equivalently,

$$\max_{0 \leq i \leq n\delta} \alpha_n \left| S_{k+i}(\omega) - S_k(\omega) \right| \geq \frac{1}{2} \sup_{t \leq s \leq t + \delta \wedge 1} \left| X^{(n)}_s - X^{(n)}_t \right|,$$

which thus imposes the condition that $\delta \in (0,1)$. Next, we need to show that given any integer $r$ there exists an integer $q$ and a number $\delta \in (0,1)$ such that for all $n \geq n_0$

$$\mathcal{P}(B_q) \leq \frac{1}{r}\delta, \quad B_q = \{\omega : \sup_{t \leq s \leq t+\delta \wedge 1} |X_s^{(n)} - X_t^{(n)}| > \frac{1}{q}\},$$

by operating on the partial sums. To this end, define

$$C_q = \{\omega : \max_{0 \leq i \leq n\delta} \alpha_n |S_{k+i}(\omega) - S_k(\omega)| > \frac{1}{q}\}, \qquad (9.21)$$

and note that if $\omega \in B_q$ then $\omega \in C_{2q}$, so that $B_q \subset C_{2q}$. Next, suppose that there exists a number $\lambda > 1$ and an integer $n_1$ such that

$$\mathcal{P}(C_\lambda) \leq \frac{1}{r}\frac{1}{4q^2\lambda^2}, \quad \text{for all } n \geq n_1. \qquad (9.22)$$

Put $\delta = (1/4q^2\lambda^2)$ so that $\delta \in (0,1)$ and note that $\lambda^2 = (1/4q^2\delta)$; choose an integer $n_0$ such that $n_0 > (n_1/\delta)$. Thus, for all $n \geq n_0$ we have $n\delta \geq n_0\delta > n_1$. Hence for $n\delta$, we also have that the set $C_\lambda'$ obeys

$$\mathcal{P}(C_\lambda') \leq \frac{1}{r}\frac{1}{4q^2}\frac{1}{\lambda^2}, \quad C_\lambda' = \{\omega : \max_{0 \leq i \leq n\delta} \alpha_{n\delta} |S_{k+i}(\omega) - S_k(\omega)| > \frac{1}{q}\}. \qquad (9.23)$$

Since we have chosen

$$\delta = \frac{1}{4q^2}\frac{1}{\lambda^2}, \quad \text{it follows that } \delta \in (0,1) \text{ and } \lambda = \frac{1}{2q}\frac{1}{\delta},$$

and consequently that

$$C_\lambda' = \{\omega : \max_{0 \leq i \leq n\delta} |S_{k+i}(\omega) - S_k(\omega)| > \frac{1}{2q}\frac{1}{\delta^{(1/2)}}\sqrt{n\delta\sigma^2} = \frac{1}{2q}\frac{1}{\alpha_n}\} = C_{2q}.$$

Since $\lambda^2 = (1/4q^2\delta)$ we conclude that

$$\mathcal{P}(B_q) \leq \mathcal{P}(C_{2q}) < \frac{1}{r}\frac{1}{4q^2}\frac{1}{\lambda^2} = \frac{1}{r}\delta,$$

for all $n \geq n_0$ as required. To complete the proof, we need only verify that Eq. (9.22) is valid. From the lemma in Billingsley (1968), p. 69,

$$\mathcal{P}(C_\lambda) \leq 2\mathcal{P}(D_{\lambda-\sqrt{2}}), \qquad (9.24)$$

where $D_{(\lambda-\sqrt{2})} = \{\omega : \alpha_n|S_n| \geq (\lambda - \sqrt{2})\}$. By a standard CLT one can show that

$$\alpha_n S_n(\omega) = \frac{1}{\sqrt{n\sigma^2}}\sum_{i=1}^{n} \xi_i \xrightarrow{d} N(0,1). \qquad (9.25)$$

It follows then that

$$\mathcal{P}(C_\lambda) \leq 2\mathcal{P}(D_{\lambda-\sqrt{2}}) \longrightarrow 2 \int_{|\epsilon|>(\lambda-\sqrt{2})} d\Phi,$$

where $\Phi$ is the cumulative distribution function of the unit normal. By Chebyshev's inequality the right member above, involving the integral, is bounded by (it is less than or equal to) $[2/(\lambda - \sqrt{2})]$. This is not the tightest bound possible, but it will do since

$$\limsup_{n\to\infty} \mathcal{P}(C_\lambda) \leq \frac{2}{\lambda - \sqrt{2}},$$

which can be made arbitrarily small by taking $\lambda$ sufficiently large.

q.e.d.

**Remark 5.** Note that, by Proposition 28, Corollary 5, in Dhrymes (1989), pp. 242-244, if $g$ is a continuous function or, if not, its discontinuities are contained in a set of $\mathcal{B}(R^d)$-measure zero,

$$g(X_t^{(n)}) \xrightarrow{\text{d}} g(X_t) \quad \text{if } X_t^{(n)} \xrightarrow{\text{d}} X_t, \tag{9.26}$$

provided $g(X_t)$ is defined. Since the functions

$$g_1(X_t^{(n)}) = \sup_{t\in[0,1]} X_t^{(n)}, \quad g_2(X_t^{(n)}) = \inf_{t\in[0,1]} X_t^{(n)},$$

$$g_3(X_t^{(n)}) = \int_0^t X_s^{(n)} \, ds, \quad t \in [0,1],$$

and many others the reader may think of are continuous, it follows that **irrespective of any other properties the sequence** $\{\xi_n : n \in \mathcal{N}_+\}$ **may possess** (beyond or instead of those invoked in establishing convergence in Proposition 7),

$$g_1(X_t^{(n)}) \xrightarrow{\text{d}} \sup_{t\in[0,1]} B_t, \quad g_2(X_t^{(n)}) \xrightarrow{\text{d}} \inf_{t\in[0,1]} B_t,$$

$$g_3(X_t^{(n)}) \xrightarrow{\text{d}} \int_0^t B_s \, ds, \text{ for } t \in [0,1]. \tag{9.27}$$

**so long as convergence in distribution holds.**

**Remark 6.** Donsker's theorem is occasionally termed an **invariance principle**. This unusual term owes its origin to a standard practice of the day [see Donsker (1951)], which encompassed two steps: First, in proving a CLT one proved convergence **and** showed that the limiting distribution of the sequence in question was independent of the distribution of the individual terms of the sequence; second, one **then calculated the form** of the limiting distribution based on a convenient distribution for the individual terms of the sequence in question.

In the literature the result in Donsker's theorem is more typically referred to as a functional central limit theorem (FCLT). The justification for this term is not really due to the results discussed in Remark 5. These results are simply a consequence of the weak convergence of probability measures corresponding to **transformed random elements**, and have nothing to do with Donsker's theorem *per se*. For example, if $\{\xi_n : n \in \mathcal{N}_+\}$ is the sequence of i.i.d. random variables of Proposition 7,

$$\frac{1}{\sqrt{n\sigma^2}}S_n \xrightarrow{d} \epsilon \sim N(0,1), \quad \text{and thus} \quad \frac{1}{n\sigma^2}S_n^2 \xrightarrow{d} \epsilon^2 \sim \chi_1^2.$$

We do not refer to this fact as a FCLT!

The most compelling justification for the term is that we have convergence to a collection of "functions", viz. the paths of BM, or in other words we have **convergence in a function space**. But since the elements of the sequence whose limit is sought are themselves collections of "functions" it is not surprising that the limit is of the same nature. Nonetheless the terminology is useful in allowing us to distinguish between the convergence of $X_r^{(n)}$, **for fixed** $r$, and the convergence of the sequence of "random functions" $X_r^{(n)}$ to the function $B_r$, i.e. to $B$ as a **function of** $r \in [0,1]$. In any event, it bears repeating that in the case of a standard CLT, or a functional CLT, what we obtain is **not** convergence to a specific random element, but to an equivalence class of such elements whose probability distribution is given by the (weak) limit of the sequence of the measures in question.

It is useful to reiterate that, in this context, convergence means and refers to weak convergence of measures. Specifically, in the context of the stochastic processes defined in Eq. (9.11) convergence means the following: If $\{P_n : n \geq 1\}$ is the probability distribution (measure) induced by $X^{(n)}$ and defined on the space, say $(C, \mathcal{B}(C))$, or $(D, \mathcal{B}(D))$ when appropriate, then what we have proved is that $P_n \xrightarrow{w} P$ in the **sense of Definition 1**. If $P$ corresponds to the Wiener measure, $\mu_W$ introduced in Chapter 7, which is the probability measure corresponding to the sample paths of the standard BM, then what is being proved is that the sequence of probability distributions (measures) $\{P_n : n \in \mathcal{N}_+\}$ obeys $P_n \xrightarrow{w} \mu_W$. In such a case we (continue to) write

$$X_t^{(n)} \xrightarrow{d} B_t, \quad t \in \mathcal{T},$$

instead of the notation $X_r^{(n)} \Rightarrow B_r$, favored by some authors. This preserves the notational convention customarily employed in the statement of standard CLT. Notice further that in standard CLT we are also dealing with the weak convergence of measures, even though this aspect is not made clear in most discussions of the subject. For example, in the context of the probability space $(\Omega, \mathcal{A}, \mathcal{P})$ let $S_n/\sqrt{n}$ be an entity in whose limiting distribution we are interested, and suppose $P_n$ is the sequence of probability measures of that entity induced by $\mathcal{P}$; if $f_n$ is the density of $P_n$ in the sense that for any $\mathcal{B}(R)$-measurable set $B$

$$P_n(B) = \int_B f_n d\mu, \quad \text{where } \mu \text{ is simple Lebesgue measure,}$$

we may define the cumulative distribution function for the special sets $B = (-\infty, x]$, by $F_n(x) = P_n(B)$. One may show that if $f_n$ converges, $P_n$ converges as well, so that "weak convergence" and convergence in distribution **are not different concepts**, and a separate notation for weak convergence only serves to confuse rather than clarify things. The context invariably makes it clear whether we are dealing with a standard CLT or a FCLT. On this topic see also Billingsley (1968), pp. 47-50.

**Remark 7.** The reader may well have noticed that in stating Donsker's theorem we had only stated a **scalar** version, and the question arises as to whether we require another development when we are dealing with random **vectors** $\xi_j.$. The response to this question is negative, i.e. no separate multivariate CLT or FCLT theorems are required in this context. To understand why this is so, first note that the case of standard CLT is discussed in Dhrymes (1989), where it is shown that if the sequence $\zeta_t = \xi_t.\beta$ obeys a CLT, say it converges to $N(0, \beta'\Sigma\beta)$, for every $\beta \in R^d$ then $\{\xi_t. : t \in \mathcal{N}_+\}$ obeys a CLT, and converges to $N(0, \Sigma)$.

In the case of FCLT, we notice that there are two steps: first, we show the convergence of finite dimensional distributions, **then** we show **tightness** for the sequence of associated probability measures. Since the first part involves a standard CLT, nothing is further required if $\xi_j$. is a **vector**.

As for the second part, there is no difference in the properties, or definition, of probability distributions (measures) between the case where the random elements in question are scalars or vectors. Notice that generally the context is:

$$X_n : \Omega \to S$$

is a sequence of random elements on the probability space $(\Omega, \mathcal{A}, \mathcal{P})$, (the elements) assuming values in the metric space $(S, \rho)$, $\rho$ being the metric on $S$. If $S = R$ we may consider the space $(R, \mathcal{B}(R), \rho)$ and define the sequence of probability measures induced by the sequence $X_n$ and the probability measure $\mathcal{P}$ on the space $(R, \mathcal{B}(R), \rho)$ thus obtaining the probability space $(R, \mathcal{B}(R), P_n, \rho)$; tightness in this context means that the sequence $\{P_n : n \in \mathcal{N}_+\}$ satisfies the criterion of Proposition 5. Now if $S = R^d$, and proceeding in the same fashion, we may construct the probability space $(R^d, \mathcal{B}(R^d), P_n, \rho)$; the arguments about tightness, or the conditions stated in Propositions 5, or 5a, do not depend in any way on whether $d = 1$, as long as $d$ is **finite**. Thus, if suitable conditions are placed on the vector sequence $\{X_t. : t \in \mathcal{N}_+\}$, a multivariate FCLT may be proved in precisely the same fashion.

## 9.3 Convergence in $D(R_+, R^d)$

In our discussion of weak convergence, tightness, and other concepts in the context of the space $C(\mathcal{T}, R^d)$, where $\mathcal{T} = R_+$, we had made use of the fact that the sample paths considered therein were continuous. Since $D(\mathcal{T}, R^d)$ is the space of discontinuous sample paths, i.e. of stochastic processes whose sample paths exhibit discontinuities, it follows that the results obtained in $C$ are not necessarily valid in $D$, unless certain modifications are made. For example, in Proposition 7 define the stochastic processes by

$$X_t^{(n)} = \frac{1}{\sqrt{n\sigma^2}} S_i(\omega), \quad \text{for } t \in \left[\frac{i}{n}, \frac{i+1}{n}\right), \tag{9.28}$$

thus omitting the interpolation term in Eq. (9.11). When we do so, the processes have a left limit at $(i+1)/n$, but **it is not equal** to $X_t^{(n)}$ for

$t = (i+1)/n$, i.e.

$$\lim_{s \uparrow t} X^{(n)}(s) = X^{(n)}(t-) = \frac{1}{\sqrt{n\sigma^2}} S_i(\omega) \neq X^{(n)}(t) = \frac{1}{\sqrt{n\sigma^2}} S_{i+1}(\omega).$$

In the space $D(\mathcal{T}, R^d)$ we are dealing with sequences of stochastic processes whose paths are rcll. But exactly how "discontinuous" are such paths? This is answered by Proposition 8.

**Proposition 8.** Given any integer $r$ and $x \in D$ there exist points $t_i \in \mathcal{T}$, $i = 0, 1, 2, \ldots, m$ such that

    i.  $0 = t_0 < t_1 < \ldots, < t_m = 1$,    $\delta_i = [t_{i-1}, t_i)$, and

    ii.  $\sup_{t, t' \in \delta_i} |x(t) - x(t')| = v(x; \delta_i) < (1/r)$, $i = 1, 2, \ldots m$.

Proof: Let $s$ be the **supremum** of $t \in \mathcal{T}$ such that the interval $[0, t)$ can be decomposed into finitely many subintervals as in part ii above. Since by construction, for any $x \in D$, $x(0) = x(0+)$, evidently $s > 0$. But $s < 1$ is not feasible; for suppose it were, and $s < 1$; then there exists a number, say $t^* \leq 1$, such that $s < t^*$; consider the interval $[s, t^*)$ and note that **therein** $\lim_{t \downarrow s} = x(s+) = x(s)$, which is a contradiction. Thus, $s = 1$.

q.e.d.

**Remark 8.** The preceding establishes a useful characterization of the nature of discontinuities present in $D$.

    i. If a "limit of tolerance" is established, i.e. if an integer $r$ is specified, there are at most a finite number of discontinuities where the jump exceeds $(1/r)$. This is so since Proposition 8 states that we can find $m + 1$ points of partition of $\mathcal{T}$ such that the largest difference in ordinates therein is bounded by $(1/r)$. Thus, for the jump to exceed this magnitude the partition must have fewer than $m + 1$ partition points.

    ii. The number of discontinuities is at most countable; this is so essentially because the rationals are dense in $\mathcal{T}$.

    iii. As a consequence of the preceding, for any $x \in D$, $\sup_{t \in \mathcal{T}} |x(t)| < \infty$.

This shows that the discontinuities allowed for elements $x \in D$ are of the first kind, and are at most countable in number. Thus the problem we face is not severe. Moreover, it suggests a modulus of continuity for elements $x \in D$, analogous to that we established in Eq. (9.5) for elements of $C$.

For the points $0 = t_0 < t_1 < t_2 < \ldots, < t_m = 1$ and $\delta_i$ as above, let $\delta \in (0,1)$ such that $t_i - t_{i-1} > \delta$. Define

$$v_1(x; \delta) = \inf_{\mathcal{T}_m} \max_{0 \le i \le m} v(x; \delta_i), \qquad (9.29)$$

where $\mathcal{T}_m$ indicates the collection of the points of partition, for all possible partitions (involving $m + 1$ points). The entity $v_1(x; \delta)$ is said to be the **modulus of continuity** of elements $x \in D$. Proposition 8, thus, states that for every element $x \in D$, $\lim_{\delta \to 0} v_1(x; \delta) = 0$.

Before we discuss how to deal with problems introduced by the discontinuities noted above, we review the structure of the arguments presented in the context of the space $C$.

First, we introduced Prohorov's criterion in a metric space, i.e. if a family of probability measures is tight, then it is weakly relatively compact; if the space in question is also complete and separable, and if the family in question is weakly relatively compact, then it is tight (Proposition 1).

Second, if a sequence of probability measures $\Pi = \{P^n : n \in \mathcal{N}_+\}$ has the property that its finite dimensional distributions $P^{nk}$ converge weakly to a measure $P^k$, for arbitrary finite $k$, and if $\Pi$ is **tight**, then $P^{(n)} \overset{w}{\to} P$ (Proposition 2).

Third, compact sets were defined in terms of the Arzelà-Ascoli theorem (Proposition 3).

Fourth, a criterion for tight subsets in $\mathbf{W^d}$ was given (Proposition 4).

Fifth, a criterion for tightness (in the context) of a sequence of probability measures was given (Proposition 5) and modified (Remark 2).

In the context of the space $C$ convergence proofs, or central limit theorems, may be structured in two steps for a given sequence of probability measures (or a given sequence of stochastic processes): First, show that the finite dimensional distributions converge; second, show that the given sequence is tight.

Evidently, the same structure could be imparted on the space $D$; but the

tools used in the space $C$ cannot be employed in that context. This is
due to the fact that the arguments employed in the context of $C$ **use in
an essential fashion** the property that the sample paths of the stochastic
processes in question are **continuous**. If a topology and a metric can be
found so that $D$ becomes a metrizable topological space with a countable
dense set, the same results would apply to it. We should then be able to prove
the analog of Donsker's theorem, or any other such theorems, for stochastic
processes whose paths are rcll. [10] Thus, the transition to the space $D$ does
not entail any conceptual issues. The issues are essentially technical. Since
the mathematical arguments become overly complex we shall omit them and
refer the reader to Billingsley (1968) or other texts that deal with the space
$D$.

### 9.3.1   The Skorohod Topology

When dealing with the space $C(\mathcal{T}, R^d)$ (for $\mathcal{T} = [0,1]$), we introduced the
uniform topology based on the metric, which we shall here designate by

$$\rho^*(x,y) = \sup_{t \in \mathcal{T}} |x(t) - y(t)|, \quad \text{for elements} \quad x, y \in C(\mathcal{T}, R^d).$$

This topology is unsuitable for the space $D(\mathcal{T}, R^d)$. Instead we introduce
the **Skorohod topology**. The basic idea behind this topology is that the
graphs of two paths may be aligned (i.e. the graph of one may be moved onto
the graph of the other) by a deformation of both ordinates and abscissas. In
the case of the uniform topology of the space $C$ we only have perturbations
of the ordinates. The formal rendition of this idea, due to Skorohod, is as
follows. Let $F$ be a class of functions

$$f : \mathcal{T} \longrightarrow \mathcal{T},$$

which are continuous, strictly increasing and such that $f(0) = 0$ and $f(1) =
1$. The metric implied by this topology, say $\rho_1(x,y)$, for any two elements
of $D$ defines their distance as the **infimum** of $\eta$'s such that there exists
$f \in F$ with

$$\sup_{t \in \mathcal{T}} |f(t) - t| \leq \eta, \quad \text{and} \quad \sup_{t \in \mathcal{T}} |x(t) - y[f(t)]| \leq \eta. \qquad (9.30)$$

---

[10] Incidentally, in the applications of this framework in the context of unit root problems
we are dealing with stochastic processes whose paths are rcll. Thus, what we require are
convergence results in the space $D$, not in the space $C$. Evidently, $C \subset D$.

We verify that this is indeed a metric: It is non-negative, symmetric, obeys the triangle inequality, and it is zero if and only if $x = y$. The first and last requirements are evidently satisfied, so we concentrate on symmetry and the triangle inequality. By the properties of $F$, $f \in F$ implies $f^{-1} \in F$; moreover, if $f_1, f_2 \in F$ so does their composition $f = f_1 \circ f_2$. This is so since $f(t) = f_1[f_2(t)]$ is evidently continuous, strictly increasing (because both $f_1$ and $f_2$ are strictly increasing) and

$$f(0) = f_1[f_2(0)] = f_1(0) = 0, \quad f(1) = f_1[f_2(1)] = f_1(1) = 1.$$

As for symmetry note that putting $s = f^{-1}(t)$ we have

$$\sup_{s \in T} |f(s) - s| = \sup_{t \in T} |f^{-1}(t) - t|, \quad \text{and}$$

$$\sup_{s \in T} |x(s) - y[f(s)]| = \sup_{t \in T} |x[f^{-1}(t)] - y(t)|.$$

For the triangle inequality we note that with $s = f_1(t)$

$$|f_2 \circ f_1(t) - t| \leq |f_2(s) - s| + |f_1(t) - t|, \quad \text{and} \quad \sup_{s \in T} |f_2(s) - s| = \sup_{t \in T} |f_2(t) - t|,$$

or

$$\sup_{t \in T} |f_2 \circ f_1(t) - t| \leq \sup_{t \in T} |f_1(t) - t| + \sup_{t \in T} |f_2(t) - t|.$$

In a similar fashion we can show that

$$\sup_{t \in T} |x(t) - y[f_1 \circ f_2(t)]| \leq \sup_{t \in T} |x(t) - z[f_2(t)]| + \sup_{t \in T} |z(t) - y[f_1(t)]|.$$

This is accomplished by adding and subtracting $z[f_2(t)]$ and setting $s = f_2(t)$. Thus, the entity defined above is indeed a metric, and $D$ with this topology is a metric space and has a countable dense set, i.e. it is a separable space. [11] Unfortunately, however, it is **not complete**, i.e. fundamental sequences (the analog here of Cauchy sequences) do not necessarily converge. To rectify this problem we may alter the first condition in Eq. (9.30), so that now the requirement to be satisfied is

$$\sup_{t \in T} f^*(t) \leq \eta, \quad \text{where} \quad f^*(t) = \sup_{s \neq t} \left| \log \frac{f(t) - f(s)}{t - s} \right|, \tag{9.31}$$

---

[11] One such dense set consists of functions $x^m$, for some fixed integer $m$, and such that $x^m(i)$ is rational and otherwise assumes a **constant rational value** on the interval $[(i-1)/m, i/m)$ for $i = 1, 2, \ldots, m$.

where $f \in F$. We note that what Eq. (9.31) requires is that the slope of the chord from $f(t)$ to $f(s)$ (i.e. the slope of the deformation of the abscissas) be (maximally) nearly one, or that its logarithm be nearly zero, in other words that the function imposing the deformation be as close as possible to an identity operator. Note further that if the sup above is finite, then the function $f^*$ is continuous and strictly increasing and thus a member of $F$. We may thus define a new metric, $\rho(x, y)$, as the infimum of those $\eta$'s for which there exists a function $f \in F$ satisfying the requirements

$$\sup_{t \in \mathcal{T}} f^*(t) \le \eta, \quad \text{and} \quad \sup_{t \in \mathcal{T}} |x(t) - y[f^*(t)]| \le \eta. \tag{9.32}$$

It may be verified that this is also a metric and that the two metrics, $\rho_1$ introduced immediately above and $\rho$, are equivalent in the sense that they induce the same (Skorohod) topology, and that under $\rho$ the space $D$ is separable and complete. Thus, the metric space of the ensuing discussion is $(D, \rho)$.

It is evident that Skorohod convergence, to be denoted by $\overset{S}{\to}$, implies convergence at the points of continuity of the limit function, i.e. if $\{X_t^{(n)} : t \in \mathcal{T}, n \in \mathcal{N}_+\}$ is a sequence in $D(\mathcal{T}, R^d)$ such that $X^{(n)} \overset{S}{\to} X$, the latter implies that if $t \in \mathcal{T}$ is a point of continuity of $X$ then $X_t^{(n)} \to X_t$. If the sample paths of $\{X_t^{(n)} : n \in \mathcal{N}_+\}$ are continuous on $\mathcal{T}$, Skorohod convergence is uniform convergence. On $C$, the Skorohod topology coincides with the uniform topology, and if the requisite conditions for Skorohod tightness can be determined on $D$, we shall be able to establish roughly the **same criteria for convergence** on $D$, as we did earlier on $C$.

## 9.3.2   Tightness and Convergence on $D(\mathcal{T}, R^d)$

We do not give any proofs in this section, since the results we deal with have the same purpose in $D$ as the corresponding propositions have in $C$. Basically, we give the analogs, in $D$, of Propositions 1, 2, 3, 4 and 5 in $C$. Evidently, Proposition 1 (Prohorov's criterion) continues to be valid in $D$, since the latter, endowed with the Skorohod topology, is a metrizable, complete, topological space with a countable dense set. The analog of Propositions 3 and 4 is as follows:

**Proposition 9.** A set has compact closure in the Skorohod topology, if and only if

$$\sup_{x \in A} \sup_{t \in T} |x(t)| < \infty, \quad \text{and} \quad \lim_{\delta \to 0} \sup_{x \in A} v_1(x; \delta) = 0. \qquad (9.33)$$

Proof: See Billingsley (1968), pp. 116-118.

The analog of Proposition 5 is given by Proposition 10.

**Proposition 10.** A sequence of probability measures $\{P^n : n \in \mathcal{N}_+\}$ defined on $(D, \mathcal{B}(D))$ is tight if and only if the following conditions hold:

   i. For every integer $r$, there exists a number $c$ such that

$$P^n(B) \leq \frac{1}{r}, \quad \text{for all} \quad n \geq 1, \quad \text{where} \quad B = \{x : |x(0)| > c\};$$

   ii. for any integers $r, q$ there exists an integer $n_0$ and a number $\delta \in (0, 1)$ such that

$$P^n(F) \leq \frac{1}{r}, \quad \text{for all} \quad n \geq n_0, \quad \text{where} \quad F = \{x : v_1(x; \delta) > \frac{1}{q}\}.$$

Proof: See Billingsley (1968), p. 126-128. Note also that the condition $n \geq 1$ in i of Proposition 10, differs from the condition $n \geq n_0$ in the statement of Proposition 5. Recall, however, that we had shown therein that this condition can also be stated as $n > 0$.

The perceptive reader will note that we have not as yet stated an analog of the important Proposition 2, which enables us to structure proofs in the following manner: Show the convergence of finite dimensional distributions, and then show that the sequence of probability measures is tight. This was possible because the $\sigma$-algebra generated by the coordinate mappings coincided with the Borel $\sigma$-algebra, $\mathcal{B}(C)$. How does the situation differ in $(D, \mathcal{B}(D))$? We note that if $X^{(n)} \overset{S}{\to} X$, and if the paths of $X$ are continuous at $t$ then evidently $X^{(n)}(t) \overset{d}{\to} X(t)$. If $X$ has a **discontinuity** at $t$ there is convergence in the sense of Skorohod-convergence, but $X^{(n)}(t)$ **does not converge** to $X(t)$. To see this, let $f_n \in F$, such that $f_n(t) = t - (1/n)$ and note that $f^*(t) = 1$, so that it satisfies the relevant requirements. Now,

**define** $X^n(s) = X[f_n(s)]$, for $s = f_n(t)$. Since [12]

$$\sup_{s\in(0,1)} |X^{(n)}(s) - X\{f[f_n(s)]\}| = \sup_{t\in(0,1)} \left| X(t - \frac{1}{n}) - X(t - \frac{2}{n}) \right|$$

can be made arbitrarily small by taking $n$ large enough, it follows that $X^{(n)} \xrightarrow{S} X$. This shows, in addition, that the **coordinate mapping functions are continuous at $t$ if and only if the limit stochastic process has paths which are continuous at $t$**. [13] The measurability of the coordinate mappings (functions) is established in Billingsley (1968), p. 121, where it is shown that the $\phi_{t_1\ldots t_k}$ are $\mathcal{B}(D)$-measurable, for arbitrary $k$. We note that in the case of $C$ this is automatic, since the coordinate mappings are continuous.

Given that the (finite dimensional) coordinate mappings are measurable, let $B \in \mathcal{B}(R^{dk})$, for arbitrary $k$, and for $t_i \in \mathcal{T}_0$, $\mathcal{T}_0 \subset \mathcal{T} = [0, 1]$, $i = 1, 2, \ldots, k$, consider the collection

$$\mathcal{G} = \{A : A = \phi^{-1}_{t_1\ldots t_k}(B), \ B \in \mathcal{B}(R^{dk})\}.$$

It is clear that $\mathcal{G}$ is an algebra. Based on this construction we have Proposition 11.

**Proposition 11.** If $\mathcal{T}_0$ contains zero and one and is otherwise **dense** in $\mathcal{T}$, then $\sigma(\mathcal{T}_0) = \mathcal{B}(D)$.

Proof: See Billingsley (1968), pp. 121-123.

**Remark 9.** As a matter of notation, observe that if $P^n$ is the probability measure corresponding to the stochastic process $X^{(n)}$ then the finite dimensional probability distribution, or probability measure pertaining to the $k$-tuple $X^{(n)}_{(k)} = (X^{(n)}_{t_1}, X^{(n)}_{t_2}, \ldots, X^{(n)}_{t_k})$ is given by $P^{(n)}\phi^{-1}_{t_1\ldots t_k}$. This is because the probability measure of the $k$-tuples assigns a probability to the

---

[12] The range $(0, 1)$ in the ensuing argument is relevant because the coordinate mappings are continuous at $0$ and at $1$, by construction.

Notice also that $X[f_n(s)] = X\{f_n[f_n(t)]\} = X[f_n(t - (1/n))] = X[t - (2/n)]$.

[13] The reader may ask why we are content, in this context, to examine only the single mapping $\phi_t$ and not the multiple mapping $\phi_{t_1\ldots t_k}$. To answer this question, we note that $\phi_{t_1\ldots t_k}(X)$ is really shorthand for $(\phi_{t_1}(X), \phi_{t_2}(X), \ldots, \phi_{t_k}(X))$; thus if $X$ is continuous at $t_i$, $i = 1, 2, \ldots, k$, the statement above applies to $\phi_{t_1\ldots t_k}$ as well. The same is true when one examines the measurability of the coordinate mappings.

set ("event") [14]

$$X_{(k)}^{(n)} \in B \quad \text{for any } B \in \mathcal{B}(R^{dk}).$$

We may evaluate this by first finding the set

$$A = \{x : \phi_{t_1 \dots t_k}(X^{(n)}) \in B\} = \phi_{t_1 \dots t_k}^{-1}(B), \quad \text{and then evaluating } P^{(n)}(A).$$

A consequence of the preceding discussion is Proposition 12.

**Proposition 12.** Let $P$ be a measure and let $\mathcal{T}_P \subset \mathcal{T}$ be a set that contains all elements $t \in \mathcal{T}$ for which the coordinate mapping $\phi_t$ is continuous, except possibly on a set of $P$-measure zero. If the complement of $\mathcal{T}_P$ in $\mathcal{T}$ has at most a countable number of elements, and if $t_i \in \mathcal{T}_P$ for $i = 1, 2, \dots, k$, then $\phi_{t_1 \dots t_k}$ is continuous, except possibly on a set of $P$-measure zero.

Proof: Billingsley (1968), p. 124.

The result below establishes the connection between the finite dimensional distributions of a given sequence of measures $\{P^{(n)} : n \in \mathcal{N}_+\}$, tightness and the weak convergence of the sequence to some measure $P$, and is the analog of Proposition 2.

**Proposition 13.** Let $\{P^n : n \in \mathcal{N}_+\}$ be a **tight** sequence of measures, and $P$ a measure. If

$$P^n \phi_{t_1 \dots t_k}^{-1} \xrightarrow{w} P \phi_{t_1 \dots t_k}^{-1}, \quad \text{whenever } t_i \in \mathcal{T}_P, \quad i = 1, 2, \dots, k$$

then $P^n \xrightarrow{w} P$.

Proof: Billingsley (1968), p. 124-125.

This completes our discussion of the structure of the space $(D, \mathcal{B}(D))$.

We are now in a position to prove Donsker's theorem in $D$, which is really what is required for applications to unit root models.

**Proposition 14** (Donsker's Theorem). Let $\{\xi_n : n \in \mathcal{N}_+\}$ be a sequence of

---

[14] Note that, for each $t_i$, $X_{t_i}^n$ is a $d$-dimensional vector, and thus assumes values in $R^d$. The entity $(X_{t_1}^n, X_{t_2}^n, \dots, X_{t_k}^n)$ is a matrix, or a collection of $d$-element vectors, and as such it assumes values in $R^{dk}$. This fact explains the notation $B \in \mathcal{B}(R^{dk})$.

i.i.d. random variables [15] with mean zero and variance $\sigma^2 < \infty$ defined on some probability space ($\Omega$, $\mathcal{A}$, $\mathcal{P}$), and define the sequence of stochastic processes

$$X^{(n)}(t,\omega) = \frac{1}{\sqrt{n\sigma^2}} S_{i-1}(\omega), \quad \text{for } t \in \left[\frac{i-1}{n}, \frac{i}{n}\right), \quad i = 0, 1, 2, \ldots, n-1.$$

Then

$$X_t^{(n)} \xrightarrow{d} B_t, \quad t \in \mathcal{T}.$$

Proof: Since the sample paths of Brownian motion (BM) are continuous the argument in connection with verifying condition i of Proposition 10 is the same as the one given in Proposition 8. Thus, we need only show tightness for the associated sequence of probability measures. [16] Noting further that $v_1(x; \delta) \le v(x; 2\delta)$, for $\delta < (1/2)$ we see that the argument for verifying condition ii of Proposition 10 is precisely that employed in Proposition 8 and thus the sequence of stochastic processes defined above is tight.

q.e.d.

# 9.4   CLT for Stationary Sequences

A number of CLT for dependent sequences are given in Dhrymes (1989), Chapter 5. In this section we discuss systematically a number of other CLT, as well as FCLT for such sequences. Some have been employed extensively in the literature of econometric, and others may yet find important applications. The FCLT are evidently generalizations of Donsker's invariance principle. The proof of such theorems, however, is entirely outside the scope of this volume, and only references to their proofs are given.

We begin by presenting two rather general (and minimalist in terms of assumptions) functional central limit theorems.

---

[15] There is no difficulty in extending this result to the multivariate case, as we had indicated in Remark 7, above. But there is another aspect of the problem that the reader should note. **Provided** a CLT is applicable to a **dependent sequence** $\{\xi_n : n \in \mathcal{N}_+\}$, something like Donsker's theorem may be proved as well, as in fact we show later. Thus, the range of problems that may be solved by this result is very wide indeed!

[16] There is a minor technical problem in that the probability measure of BM is really a measure that is defined on $C$. It is a rather straightforward procedure to find a convention under which it is applicable to $D$.

**Proposition 15.** Let $Z = \{Z(t,\omega) : t \in R\}$ be a (zero mean), $R \times \mathcal{A}$-measurable real valued stationary process defined on the probability space $(\Omega, \mathcal{A}, \mathcal{P})$. Define $\mathcal{G}_t = \sigma(Z_s, s \leq t)$, for $t \in R$ and $\mathcal{G}_\infty = \mathcal{G}$; note that $(\Omega, \mathcal{G}, \mathcal{P})$ is the probability space with stochastic basis $\mathcal{G}_t$, $t \in R$ generated by the sequence $Z$. In this context let $\{T_s : s \in R\}$ be a group of measure preserving transformations which are ergodic;[17] If $p \in [2, \infty]$, $q \in [1, 2]$, such that $(1/p) + (1/q) = 1$ and[18]

$$\| Z_0 \|_p < \infty, \quad \int_0^\infty \| E(Z_t|\mathcal{G}_0) \|_q \, dt < \infty,$$

then the processes

$$X_t^{(n)} = \frac{1}{\sqrt{n}} \int_0^{nt} Z_s ds, \quad t \geq 0$$

converge in distribution (or weakly), for finite dimension (over $R_+$) to $\sqrt{c}B_t$, where $B_t$ is the standard BM and

$$c = 2 \int_0^\infty E(Z_0 Z_t) dt,$$

$c$ being a **finite** constant. If, in addition, $p = 2$ then over $R_+$

$$X_t^{(n)} \xrightarrow{\mathrm{d}} \sqrt{c}B_t.$$

Proof: See Jacod and Shiryaev (1987), pp. 449-452.

**Remark 10.** To make this discussion as self contained as possible we elucidate a few of the terms used in the proposition. Additional detail is given in Dhrymes (1989), pp. 349-352. We begin by noting that a **group** is analogous to the concept of an algebra in the context of a collection of sets and it is simply a collection that is closed with respect to unions. A collection of transformations, say $T^* = \{T_s : s \in R\}$ is said to be a group if for any $s, t \in R$, $T_t \circ T_s = T_{s+t} \in T^*$. In the context of Proposition 15 this means

---

[17] For a discussion of the meaning of measure preserving transformations and ergodicity see Dhrymes (1989), pp. 349-352.

[18] The notation $\| X \|_p$ for a random variable or a random element (vector), defined over a probability space $(\Omega, \mathcal{A}, \mathcal{P})$ means

$$\| X \|_p = \left( \int_\Omega |X|^p \, d\mathcal{P} \right)^{1/p}.$$

This is well defined if the random element in question has $p$-order moments, and is often termed the $L^p$ norm. Notice that if $\| X \|_p$ is well defined then so is $\| X \|_q$, for $q \leq p$.

that for any $s, t \in R$, $Z_{t+s}(\omega) = Z_t \circ T_s(\omega)$ and, consequently, for any set $B \in \mathcal{B}(R)$, $Z_{t+s}^{-1}(B) = T_t^{-1}[Z_s^{-1}(B)]$. Thus, $\mathcal{G}_{t+s} = T_t^{-1}(\mathcal{G}_s)$, or more elegantly, $\mathcal{G}_t = T_t^{-1}(\mathcal{G}_0)$. The transformations are measure preserving if for $A_0 \in \mathcal{G}_0$ and $A_t \in \mathcal{G}_t$, such that $A_t = T_t^{-1}(A_0)$, $\mathcal{P}(A_t) = \mathcal{P}(A_0)$. Ergodicity means that the invariant $\sigma$-algebra under the transformation group, i.e. the collection of sets

$$\mathcal{I} = \{A : A \in \mathcal{G}_0, \ T_t^{-1}(A) = A, \text{ for all } t \in R\}$$

is trivial in the sense that for **any** $A \in \mathcal{I}$ we have either $\mathcal{P}(A) = 0$, or $\mathcal{P}(A) = 1$.

**Remark 11**. In Proposition 15, what the term **finite dimension** means is the following: Given any $k$-tuplet $(t_1, t_2, \ldots, t_k)$, $k \in \mathcal{N}_+$, $t_i \in R_+$, the probability measures $P^{nk}$ of the collection $\{X_{t_1}^{n)}, \ldots, X_{t_k}^{(n)}\}$ converge (weakly) to the probability measure $\mu_W^k$, which is the probability measure corresponding to the $k$-tuplet $[\sqrt{c}B(t_1), \sqrt{c}B(t_2), \ldots, \sqrt{c}B(t_k)]$, and this holds for **any** $k \in \mathcal{N}_+$. However, this part of the conclusions **does not** involve a FCLT. Evidently, here we have an inability to prove that the sequence of probability measures in question is **tight**. On the other hand when $p = 2$, **and thus** $q = 2$, **tightness** may be established and consequently we have a FCLT!

The result in Proposition 15 is not extensively helpful in econometrics since it is not often that we are dealing with continuous (parameter) stochastic processes. However, Bergstrom (1990) and Merton (1990) constitute an important exception. Of greater potential applicability is the following result, which applies to the **discrete parameter** case.

**Proposition 16**. Let $\xi$ be a random variable (or element) defined on the probability space $(\Omega, \mathcal{A}, \mathcal{P})$, and $T$ be a measure preserving transformation

$$T : (\Omega, \mathcal{A}) \longrightarrow (\Omega, \mathcal{A})$$

which is ergodic and bijective. [19] Define $\xi_k = \xi \circ T^k$, where $T^{r+s} = T^r \circ T^s$

---

[19] Bijective means that the transformation is one to one and onto, i.e. for every $\omega_1 \in \Omega$ there exists $\omega_2 \in \Omega$ such that $T(\omega_1) = \omega_2$, and $T(\omega_1) = T(\omega_1^*)$ if and only if $\omega_1 = \omega_1^*$.

and let $p \in [2, \infty]$, $q \in [1, 2]$. Define

$$X_t^n = \frac{1}{\sqrt{n}} \sum_{r=1}^{[nt]} \xi \circ T^r, \quad \text{for } t \in R_+, \quad \text{and suppose further}$$

$$E\xi = 0, \quad \| \xi \|_p = [E|\xi|^p]^{1/p} < \infty, \quad \frac{1}{p} + \frac{1}{q} = 1, \tag{9.34}$$

$$\infty > \sum_{r=1}^{\infty} \| E(\xi \circ T^r | \mathcal{G}_0) \|_q; \quad \text{then} \tag{9.35}$$

$$X_t^n \xrightarrow{\text{d}} \sqrt{\sigma^2} B_t, \tag{9.36}$$

where the convergence is finite dimensional, [20] $B$ is the standard BM and

$$\sigma^2 = E\xi^2 + 2 \sum_{r=1}^{\infty} E(\xi)(\xi \circ T^r), \quad \sigma^2 \in (0, \infty). \tag{9.37}$$

If $p = 2$ then, unambiguously (i.e. there is **convergence in measure**), $X_t^n \xrightarrow{\text{d}} \sqrt{\sigma^2} B_t$, $t \in R_+$.

Proof: See Jacod and Shiryaev (1987), pp. 456.

**Remark 12.** The result above is stated rather abstractly; it gives, in the context of stationary ergodic sequences, a sufficient set of conditions for the processes in the first equation of the equation set Eq. (9.34) to converge in finite dimension or in function space. When $p > 2$ (and thus $q < 1$) there is convergence in finite dimension only; as we had noted in Remark 11 this is not enough to produce **tightness** in the sequence of (the relevant) probability measures. When $p = 2$, and **thus** $q = 2$, we note that the condition in Eq. (9.35) involves the conditional second moments of $\xi \circ T^s$ given $\xi$, $s \geq 1$; specifically, it states that

$$\text{if } \sum_{r=1}^{\infty} \| E(\xi \circ T^r | \mathcal{G}_0) \|_2 < \infty \text{ then } \sigma^2 \in (0, \infty).$$

Renaming $\xi = \xi_0$, and taking $\mathcal{G}_0 = \sigma(\xi_0)$, the relation above may be stated

---

[20] This term means, as we had also noted earlier, that the finite dimensional distributions, i.e. the distributions of $(X_{t_1}^n, X_{t_2}^n, \ldots, X_{t_k}^n)$, for arbitrary $k$, converge (weakly) as $n \to \infty$ on compact subsets of $R_+ \cup \{0\}$.

in the more familiar **scalar** terms, [21]

$$\text{if} \ \sum_{j=1}^{\infty} |E(\xi_k|\xi_0)| < \infty \ \text{then} \ \sigma^2 \in (0, \infty).$$

We next connect the conditional moments above to the "covariance" function of the sequence. Since

$$\zeta_k(\xi_0) = E(\xi_k|\xi_0), \quad E\xi_0\xi_k = \int_{\Omega} \xi_0\zeta_k(\xi_0)d\mathcal{P} = E\xi_0\zeta_k(\xi_0),$$

it follows from Hölder's inequality [see Proposition 16 in Dhrymes (1989), pp. 108-109], that with $p_1 = p_2 = 2$

$$|E\xi_0\xi_k| \leq \kappa(0)^{1/2} \ \| \ \zeta_k(\xi_0) \ \|_2 \ .$$

where $\kappa(\cdot)$ is the autocovariance function of the sequence. We conclude from Eq. (9.35)

$$\left| \sum_{k=1}^{\infty} E(\xi_0\xi_k) \right| < \infty.$$

In turn, this implies that $\sigma^2$, as defined in Eq. (8.37), is finite; moreover, since the sequence in question is stationary, $\sigma^2 > 0$ by the results of Chapter 1.

We take this opportunity to note another important consequence of stationarity. Define $S_n = \sum_{j=1}^{n} \xi_j$ and note that

$$ES_n^2 = \sum_{j=1}^{n} E\xi_j^2 + 2 \sum_{\tau=1}^{n} \sum_{j=1}^{n-\tau} E\xi_j\xi_{j+\tau} = n\kappa(0) + 2 \sum_{\tau=1}^{n} (n - \tau)\kappa(\tau).$$

It follows therefore that

$$\lim_{n\to\infty} \frac{1}{n} ES_n^2 = \sigma^2 = \lim_{n\to\infty} \left( \kappa(0) + 2 \sum_{\tau=1}^{n} (1 - \frac{\tau}{n})\kappa(\tau) \right) = \kappa(0) + 2 \sum_{\tau=1}^{\infty} \kappa(\tau),$$

---

[21] Strictly speaking the popular notation $E(X|Y)$ is not very meaningful; it is commonly used in statistics as a mnemonic to aid integrations involved in the computation of moments; the popular description of the operation is that in the integration operation we treat $Y$ as a constant. Such statements have no well-defined meaning in the context of formal probability theory; we can only condition on a $\sigma$-algebra. Thus, the proper meaning of the notation above is $E(X|\mathcal{F})$, where $\mathcal{F} = \sigma(Y)$. For a more extended discussion of conditioning in probability theory see Chapter 2 in Dhrymes (1989).

which is well defined by the condition in Eq. (9.35). In fact, if we put

$$\sigma_n^2 = ES_n^2, \quad \text{we obtain} \quad \sigma_n^2 = n\sigma^2[1 + o(1)].\ ^{22}$$

Notice further that the condition in Eq. (9.35) has another important implication, viz. that

$$\lim_{n\to\infty} E(\xi_n|\xi_0) = 0 = (\lim_{n\to\infty})E\xi_n.$$

Consequently, the $\sigma$-algebras $\sigma(\xi_n)$ and $\sigma(\xi_0)$ are asymptotically independent. But this implies the ergodic theory **mixing condition**, since with $\xi_n = \xi_0 \circ T^n$ and suitable sets $A, B$

$$\lim_{n\to\infty} \mathcal{P}(A \cap T^{-n}(B)) = \mathcal{P}(A)\mathcal{P}(B).$$

Thus, in the case of $p = q = 2$ special properties accrue to the stationary ergodic sequence of Proposition 16, which allow a FCLT to be proved.

We return to this issue below, after having completed the examination of various **mixing properties** of, or mixing conditions placed on, stochastic sequences.

**Remark 13.** In Chapter 1 we had given, for square integrable stationary sequences, the definition of the spectral density, or spectral matrix, and their connection to the autocovariance function or kernel of the sequence in question. These relations are given by

$$f(\lambda) = \frac{1}{2\pi} \sum_{\tau=-\infty}^{\infty} e^{-i\lambda\tau} K(\tau), \quad K(\tau) = \int_{-\pi}^{\pi} e^{i\lambda\tau} f(\lambda)\, d\lambda,$$

where $f$ is the spectral matrix and $K(\cdot)$ the autocovariance kernel for a vector process, and the spectral density and autocovariance function for a scalar process, respectively.

---

[22] The notation $x_n = O(c_n)$, for $c_n \geq 0$ and $c_n \uparrow \infty$, means that $x_n/c_n$ remains bounded as $n \to \infty$; the notation $x_n = o(c_n)$ means that $x_n/c_n$ converges to zero as $n \to \infty$, or more formally, for a sequence $\{c_n : n \in \mathcal{N}_+,\ c_n \geq 0,\ c_n \uparrow \infty\}$

$$x_n = O(c_n) \text{ means } \lim_{n\to\infty} \frac{x_n}{c_n} \leq K < \infty, \quad x_n = o(c_n) \text{ means } \lim_{n\to\infty} \frac{x_n}{c_n} = 0.$$

Thus $o(1)$ in this instance means that the term in question, viz.

$$-\frac{2}{\sigma^2}\left(\sum_{\tau=n+1}^{\infty} \kappa(\tau) + \frac{1}{n}\sum_{\tau=1}^{n} \tau\kappa(\tau)\right) \to 0$$

with $n$.

In Proposition 16, the condition stated in Eq. (9.37) requires the sum of autocovariances to converge. In terms of the relationships stated above, we have therefore

$$\sigma^2 = \sum_{\tau=-\infty}^{\infty} \kappa(\tau) = 2\pi f(0), \quad \Sigma = \sum_{\tau=-\infty}^{\infty} K(\tau) = 2\pi f(0),$$

for scalar and vector processes, respectively.

**Remark 14.** Since the concepts of strict stationarity and ergodicity are frequently invoked in many convergence discussions in the literature of econometrics, this is an opportune moment to clarify their relation. In fact it may be shown that given any strictly stationary sequence, $X$, on a probability space $(\Omega,\ \mathcal{A},\ \mathcal{P})$ we may define, on perhaps another probability space $(\Omega_1,\ \mathcal{A}_1,\ \mathcal{P}_1)$, a random element, say $Y$, and a measure preserving transformation, say $T$, such that the sequences $X = \{X_t : t \in \mathcal{N}_0\}$ and $Y^* = \{Y \circ T^k : k \in \mathcal{N}_0\}$ have a.c. the same distribution. The transformation is said to be **ergodic** if all invariant sets, under the transformation, have measure either zero or one. Thus, $X$ is said to be stationary and ergodic, if there exists an ergodic measure preserving transformation such that the sequences $X$ and $Y^*$ have the same distribution.

The relevance of the preceding, for our discussion here and in Chapter 6, is found in Proposition 27, Chapter 5, Dhrymes (1989), the so called mean ergodic theorem. Specifically, if $T$ is a measure preserving transformation and $\mathcal{I}$ is the $\sigma$-algebra of invariant sets then

$$\frac{1}{n}\sum_{j=0}^{n-1} Y \circ T^j \overset{\text{a.c.}}{\to} E(Y|\mathcal{I}), \quad \text{and if } T \text{ is \textbf{also ergodic}} \quad \frac{1}{n}\sum_{j=0}^{n-1} Y \circ T^j \overset{\text{a.c.}}{\to} E(Y).$$

Thus, by far, the most important of the two properties is stationarity; ergodicity is mainly useful by adding the property that all invariant sets are trivial in the sense that they have measure either zero or one, and by ruling out certain anomalies.

Actually, the notion of ergodicity has been gradually abandoned in this field of research and has been supplanted by various mixing conditions, a subject we develop below.

We conclude this segment of our discussion by giving a standard CLT result for stationary ergodic processes.

**Proposition 16a.** Let $X = \{X_t. : t \in \mathcal{N}_0\}$ be a zero mean stationary ergodic sequence defined on the probability space $(\Omega, \mathcal{A}, \mathcal{P})$, and suppose $E|X_0|^2 < \infty$, and that $X_0$ is $\mathcal{G}_0$-measurable, $\mathcal{G}_0 \subset \mathcal{A}$. Let [23]

$$S_{n.} = \sum_{j=0}^{n} X_{j.}.$$

If

$$\left\| \sum_{r=1}^{\infty} E[X_{r.}' E(X_{n.}|\mathcal{G}_0)] \right\| \qquad \text{converges for each } n \geq 0$$

$$\lim_{n \to \infty} \left\| \sum_{j=k}^{\infty} E[X_{j.}' E(X_{n.}|\mathcal{G}_0)] \right\| = 0, \text{ uniformly for } k \geq 1,$$

then

$$\lim_{n \to \infty} \frac{1}{n} E S_{n.}' S_{n.} = \lim_{n \to \infty} \frac{1}{n} \Sigma_n = \Sigma \geq 0$$

$$S_n P_n \stackrel{\mathrm{d}}{\to} N(0, I), \quad \text{provided } \Sigma > 0, \text{ where } \Sigma_n = P_n'^{-1} P_n^{-1}.$$

Proof: For a proof of the univariate version of this result see Heyde (1974); see also Hall and Heyde (henceforth HH) (1980), pp. 133-134.

**Example 1.** Consider the general linear process (GLP)

$$Z_k = \sum_{j=0}^{\infty} a_j \epsilon_{k-j}, \quad \sum_{j=0}^{\infty} |a_j| < \infty,$$

where $\epsilon$ is $WN(1)$. The condition on the coefficients above is necessary for the GLP to be defined, in terms of a.c. convergence; if only $L^2$ convergence is employed, the relevant condition is $\sum_{j=0}^{\infty} |a_j|^2 < \infty$.

Introduce the measure preserving transformation, $T$, and redefine the process by

$$Z_k = Z_0 \circ T^k, \quad k \geq 1, \quad \text{where } Z_0 = \sum_{j=0}^{\infty} a_j \epsilon_{-j}, \quad \sum_{j=0}^{\infty} |a_j| < \infty, \qquad (9.38)$$

---

[23] The result is stated purposefully for a **vector sequence**. As we illustrate in Remark 15 below, the following is true for all FCLT or standard CLT results we consider in this discussion: If they hold for scalar sequences they hold for vector sequences as well, *mutatis mutandis*.

so that it conforms to the framework in Proposition 16. This means that $T$ is defined by the operation

$$Z_k(\omega) = Z_0[T^k(\omega)] = \sum_{j=0}^{\infty} a_j \epsilon_{k-j}, \quad k \geq 1.$$

The sequence $Z = \{Z_k : k \in \mathcal{N}_0\}$ is stationary by construction. To show that it is ergodic, we need to show that all invariant sets are $\mathcal{P}$-trivial, i.e. they have probability either zero or one. Thus, let $B \in \mathcal{B}(R)$, assuming we are dealing with random variables, or $B \in \mathcal{B}(R^d)$ if we are dealing with $d$-element vectors, and consider the sets

$$A_k = Z_k^{-1}(B) = T^{-k}[Z_0^{-1}(B)] = T^{-k}(A_0), \quad A_0 = Z_0^{-1}(B).^{24}$$

The transformation above is indeed measure preserving since the probability that $Z_k$ will assume a value in $B$ is exactly the same as the probability that $Z_0$ will assume a value in $B$, and thus $\mathcal{P}(A_k) = \mathcal{P}[T^{-k}(A_0)] = \mathcal{P}(A_0)$; however, the sets $A_k$, $k \geq 0$ need not all be the same. Consider now the collection of all invariant sets, i.e. sets such that

$$A_k = T^{-k}(A_0) = A_0, \quad k \geq 1, \quad \text{for some } B \in \mathcal{B}(R^d).$$

The mirror image, $B \in \mathcal{B}(R^d)$, of such an invariant set say $A_k = A_0$, obeys $Z_k(A_0) = B$, **for all** $k \geq 0$. Since the sequence involves a $WN(1)$, or $MWN(I)$ process, this is not possible, unless either $B = \emptyset$ or $B = R^d$, in which case, respectively, $A_k = \emptyset$, or $A_k = \Omega$, for all $k \in \mathcal{N}_0$. This means that all invariant sets have measure either zero or one and are thus $\mathcal{P}$-trivial. Consequently, the sequence $Z$ is stationary and ergodic, and otherwise conforms to the requirement of Proposition 16.

The condition in Eq. (9.35) is satisfied with $p = 2$, since

$$\| Z_0 \|_2^2 = \sum_{j=0}^{\infty} \sum_{s=0}^{\infty} a_j a_s \left[ E\epsilon_{-j}\epsilon_{-s} \right] = \sum_{j=0}^{\infty} a_j^2 < \infty. \tag{9.39}$$

This is implied by the condition defining the sequence in view of the fact that

$$\sum_{j=0}^{\infty} a_j^2 \leq \left( \sum_{j=0}^{\infty} |a_j| \right)^2. \tag{9.40}$$

---

[24] Note that if we define $\mathcal{G}_k = \sigma(Z_s : s \leq k)$, for $k \geq 1$ and $\mathcal{G}_0 = \sigma(Z_0)$, this relation implies $\mathcal{G}_k = T^{-k}(\mathcal{G}_0)$, which is a property of the sequences considered in Proposition 16.

Now define the $\sigma$-algebra generated by $Z_0$, viz. $\mathcal{G}_0 = \sigma(\epsilon_r, \ r \leq 0)$. Write

$$Z_k = \sum_{j=0}^{k-1} a_j \epsilon_{k-j} + Z_{(k|0)}, \quad Z_{(k|0)} = \sum_{j=k}^{\infty} a_j \epsilon_{k-j} = \sum_{s=0}^{\infty} a_{s+k} \epsilon_{-s},$$

and note that

$$E(Z_k | \mathcal{G}_0) = Z_{(k|0)} + E\left( \sum_{j=0}^{k-1} a_j \epsilon_{k-j} | \mathcal{G}_0 \right) = Z_{(k|0)}.$$

With $q = 2$, we find

$$\| Z_{(k|0)} \|_2^2 = E \sum_{j=0}^{\infty} \sum_{s=0}^{\infty} a_{j+k} a_{s+k} \epsilon_{-j} \epsilon_{-s} = \sum_{j=0}^{\infty} a_{j+k}^2 = |c_k|^2. \tag{9.41}$$

The condition in Eq. (9.35), requires that

$$\sum_{k=1}^{\infty} |c_k| < \infty.$$

To investigate the conditions under which the series, above, converges we note that

$$|c_k| = \left( \sum_{j=0}^{\infty} a_{j+k}^2 \right)^{1/2} \leq \sum_{j=0}^{\infty} |a_{j+k}|, \quad \text{so that}$$

$$\sum_{k=1}^{\infty} |c_k| \leq \sum_{k=1}^{\infty} \sum_{j=0}^{\infty} |a_{j+k}| = \sum_{j=0}^{\infty} j |a_j| = C_1. \tag{9.42}$$

Evidently, we require that $C_1 < \infty$, but this is not a condition required for the definition of GLP; hence this involves an **added restriction**.

Next, we note that

$$\sigma^2 = EZ_0^2 + 2 \sum_{k=1}^{\infty} E Z_k Z_0 = \sum_{j=0}^{\infty} a_j^2 + 2 \sum_{k=1}^{\infty} E Z_{(k|0)} Z_0$$

$$= \sum_{j=0}^{\infty} a_j^2 + 2 \sum_{k=1}^{\infty} \sum_{j=0}^{\infty} a_{j+k} a_j \leq \left( \sum_{j=0}^{\infty} |a_j| \right)^2 < \infty, \tag{9.43}$$

by the condition defining the GLP.

The discussion of the example above yields a very useful result for econometric applications, which is given in Corollary 1.

**Corollary 1.** In the context of Proposition 16, consider the GLP

$$Z_k = \sum_{j=0}^{\infty} a_j \epsilon_{k-j}, \quad k \in \mathcal{N}_0, \quad \sum_{j=1}^{\infty} j |a_j| < \infty, \qquad (9.44)$$

where $\epsilon$ is $WN(1)$. Then,

$$X_t^n = \frac{1}{\sqrt{n}} S_{[nt]} \overset{d}{\to} \sqrt{\sigma^2} B_t, \quad \text{for} \ \ t \in R_+.$$

**Remark 15.** The corollary above applies to the multivariate case as well. In such a case we would write

$$Z_k = \sum_{j=0}^{\infty} A_j \epsilon_{k-j}', \quad \sum_{j=0}^{\infty} j \parallel A_j \parallel < \infty,$$

where $\parallel \cdot \parallel$ is the norm of a matrix defined for example as $[\text{tr}(A_j' A_j)]^{1/2}$, and $\epsilon$ is $MWN(I_q)$. The conclusion is then

$$X_t^n = \frac{1}{\sqrt{n}} S_{[nt]} \overset{d}{\to} B_t H, \quad S_n = \sum_{k=1}^{n} Z_k',$$

where $H$ is a matrix such that $\Sigma = H'H$, $B_t$ is the SMBM examined in Chapter 7, and $\Sigma > 0$, defined by

$$\Sigma = E Z_0 Z_0' + \sum_{k=1}^{\infty} E[Z_0 Z_k' + Z_k' Z_0]$$

$$= \sum_{j=0}^{\infty} A_j A_j' + \sum_{k=1}^{\infty} \sum_{j=0}^{\infty} \left( A_j A_{j+k}' + A_{j+k} A_j' \right)$$

$$\parallel \Sigma \parallel \ \leq \ \sum_{j=0}^{\infty} \parallel A_j \parallel^2 + 2 \sum_{k=1}^{\infty} \sum_{j=0}^{\infty} \parallel A_{j+k} \parallel \parallel A_j \parallel \ \leq \ \left( \sum_{j=0}^{\infty} \parallel A_j \parallel \right)^2 < \infty,$$

by the condition defining the multivariate GLP.

Moreover, the sequence $\{Z_k : k \geq 1\}$ is stationary and ergodic; it evidently obeys the condition in Eq. (9.34); we need only verify the condition in Eq. (9.35). By definition

$$Z_{(k|0)} = \sum_{j=0}^{\infty} A_{j+k} \epsilon_{-j}, \quad \text{and}$$

$$\parallel Z_{(k|0)} \parallel_2^2 = E|Z_{(k|0)}|^2 = \sum_{j=0}^{\infty} \sum_{s=0}^{\infty} E \left( \epsilon_{-j}' A_{j+k}' A_{s+k} \epsilon_{-s} \right)$$

$$= \sum_{j=0}^{\infty} \operatorname{tr} A'_{j+k} A_{j+k} = \sum_{j=0}^{\infty} \| A_{j+k} \|^2; \quad \text{thus,}$$

$$\| Z_{(k|0)} \|_2 = \left( \sum_{j=0}^{\infty} \| A_j \|^2 \right)^{1/2} \leq \sum_{j=0}^{\infty} \| A_{j+k} \| .$$

It follows, therefore, that

$$\sum_{k=1}^{\infty} \| Z_{(k|0)} \| \leq \sum_{k=1}^{\infty} \sum_{j=0}^{\infty} \| A_{j+k} \| = \sum_{j=1}^{\infty} j \| A_j \| < \infty,$$

by the condition we have placed on the GLP. Thus, the condition in Eq. (9.35) is satisfied, and the conclusions of Proposition 16 hold with respect to **multivariate sequences**.

We now give a standard multivariate CLT for the GLP.

**Corollary 1a.** Consider the GLP $Z = \{Z_t : t \in \mathcal{N}_0\}$ of the example above, satisfying the condition

$$\sum_{j=0}^{\infty} \| A_j \| < \infty, \quad \text{and let} \quad S_T = \frac{1}{\sqrt{T}} \sum_{t=1}^{T} Z'_t.$$

Then

$$S_T \xrightarrow{d} N(0, \Phi), \quad \Phi = \sum_{j=0}^{\infty} A_j A'_j.$$

Proof: See Chapter 7 in Anderson (1971) .

A less restricted version due to Hannan (1970), p. 221, states that

$$S_T \xrightarrow{d} N(0, \Phi), \quad \Phi = 2\pi f(0),$$

where $f(\lambda)$ is the spectral matrix, provided $f$ is uniformly bounded, **continuous** at $\lambda = 0$, and $f(0) > 0$.

As the preceding example (and Remark 14) makes quite clear, in dealing with dependent sequences, the question of whether the processes defined above admit of a Donsker-type CLT depends on whether or not they exhibit diminishing dependence as the "distance" between any two random elements is increased. To make these ideas more precise we need to formalize the measurement of the degree of dependence among the elements of a stochastic sequence.

## 9.5   Mixing Sequences

The "mixing" property is a concept that was invented as a suitable characterization of the magnitude of serial, or "time", dependence amongst the elements of a sequence, or as a bound on that dependence; in the evolution of this literature it has supplanted the assumption of ergodicity. Its significance was highlighted in the discussion immediately above, as well as in Remark 12.

We begin with the basic notion of dependence between sets in $\sigma$-fields. [25]

**Definition 4**. Let $(\Omega, \mathcal{A}, \mathcal{P})$ be a probability space and $\mathcal{D}, \mathcal{F}$ be $\sigma$-fields [or $\sigma$-(sub)algebras] contained in $\mathcal{A}$. A number of numerical measures of the dependence of $\mathcal{D}$ and $\mathcal{F}$ have been proposed in the literature; the most common are given below. [26]

   i.

$$\alpha(\mathcal{D}, \mathcal{F}) = \sup_{D \in \mathcal{D}, \ F \in \mathcal{F}} |\mathcal{P}(F \cap D) - \mathcal{P}(F)\mathcal{P}(D)|;$$

  ii. for sets $D$ such that $\mathcal{P}(D) > 0$,

$$\phi(\mathcal{D}, \mathcal{F}) = \sup_{D \in \mathcal{D}, \ F \in \mathcal{F}} |\mathcal{P}(F|D) - \mathcal{P}(F)|;$$

 iii. for any sets $D$, $F$ such that $\mathcal{P}(D) > 0$, $\mathcal{P}(F) > 0$,

$$\psi(\mathcal{D}, \mathcal{F}) = \sup_{D \in \mathcal{D}, \ F \in \mathcal{F}} \frac{|\mathcal{P}(F \cap D) - \mathcal{P}(F)\mathcal{P}(D)|}{\mathcal{P}(F)\mathcal{P}(D)}$$

$$= \sup_{D \in \mathcal{D}, \ F \in \mathcal{F}} \left| \frac{|\mathcal{P}(F \cap D)}{\mathcal{P}(F)\mathcal{P}(D)} - 1 \right|;$$

 iv. for **real** random variables $X, Y$ which are respectively $\mathcal{D}$- and $\mathcal{F}$-measurable, such that $EX = 0$, $EY = 0$, $\| X \|_2 < \infty$, and $\| Y \|_2 < \infty$,

$$\rho(\mathcal{D}, \mathcal{F}) = \sup_{X,Y} \frac{|E(XY)|}{\| X \|_2 \| Y \|_2} = \sup_{X,Y} |\text{Corr}(X, Y)|.$$

---

[25] The terms $\sigma$-algebra and $\sigma$-field are synonymous in this context.

[26] What is involved in all of them is a measure of the maximal departure between the probability of the intersection of sets $D \cap F$ and the product of the probabilities assigned to the two sets, such that $D \in \mathcal{D}$ and $F \in \mathcal{F}$; or, of the (conditional) probability of the "event" $D$ **given** $F$, and the (unconditional) probability of the "event" $D$.

For an extensive review of this literature see Bradley (1986).

v.

$$\beta(\mathcal{D}, \mathcal{F}) = \sup_{D, F} \left( \frac{1}{2} \sum_{i=1}^{n} \sum_{j=1}^{m} |\mathcal{P}(D_i \cap F_j) - \mathcal{P}(D_i)\mathcal{P}(F_j)| \right),$$

where the sup is taken over all possible pairs of partitions of $\Omega$, $D = \{D_i : i = 1, 2, \ldots, n\}$, $F = \{F_j : j = 1, 2, \ldots, m\}$.

The relations above are all symmetric with respect to $\mathcal{D}$ and $\mathcal{F}$, except for the one in ii. This is so since $\mathcal{P}(F|D) = \mathcal{P}(F \cap D)/\mathcal{P}(D)$; consequently, for any sets $F \in \mathcal{F}$, $D \in \mathcal{D}$

$$|\mathcal{P}(F|D) - \mathcal{P}(F)| = \frac{1}{\mathcal{P}(D)} |\mathcal{P}(F \cap D) - \mathcal{P}(F)\mathcal{P}(D)|,$$

and the nonsymmetric nature of this definition becomes transparent.

**Remark 16.** In general, the measures of dependency above are utilized for the purpose of enforcing conditions of "asymptotic independence" on the elements of a sequence; this may be done by requiring such measures to converge to zero with $n$, where the latter is the "distance" between two elements of a sequence; for example, one may define $\mathcal{D}$ to be the $\sigma$-algebra $\sigma(\xi_r, r \leq m)$ and $\mathcal{F}$ to be the $\sigma$-algebra $\sigma(\xi_r, r \geq m + n)$. If, in this context, we denote the measure in i by $\alpha(\mathcal{D}, \mathcal{F}) = \alpha(n)$, asymptotic independence is enforced by the condition $\alpha(n) \to \infty$. Sequences that obey such a condition in the context of i are said to be $\alpha$-mixing; in the context of ii $\phi$-mixing; in the context of iii $\psi$-mixing, although in HH (1980) p. 40, this measure of dependency is termed *-mixing; in the context of iv $\rho$-mixing, and in the context of v **absolutely regular**. It may be shown, as a consequence of the discussion above, that all measures of dependence are **non-negative** and, moreover, obey the bounds

$$\alpha(\mathcal{D}, \mathcal{F}) \leq \frac{1}{4}, \ \beta(\mathcal{D}, \mathcal{F}) \leq 1, \ \phi(\mathcal{D}, \mathcal{F}) \leq 1, \ \rho(\mathcal{D}, \mathcal{F}) \leq 1.$$

For $\rho$-mixing the bound is obvious since a correlation coefficient is bounded by one; for $\phi$-mixing note that if the two $\sigma$-algebras contain a common set of arbitrarily small measure, say $D = F$, then $\mathcal{P}(D|D) - \mathcal{P}(D) \geq 1 - \delta$, for arbitrarily small $\delta$. Finally, for $\alpha$-mixing we note that for $D = F$,

$$|\mathcal{P}(D \cap D) - \mathcal{P}(D)^2| = \mathcal{P}(D)[1 - \mathcal{P}(D)] \leq \frac{1}{4},$$

and so on. In general, the equalities may be shown to hold in the case where $\mathcal{D} = \mathcal{F} = (\Omega, A, \bar{A}, \emptyset)$, with $\mathcal{P}(A) = 1/2$. For example, in the absolutely regular case consider the **identical** partitions $D_1 = A$, $D_2 = \bar{A}$ and $F_i = D_i, i = 1, 2$. Then

$$\mathcal{P}(D_i \cap F_i) - \mathcal{P}(D_i)\mathcal{P}(F_i) = \frac{1}{2} - \frac{1}{4}, \quad i = 1, 2$$

$$\mathcal{P}(D_i \cap F_j) - \mathcal{P}(D_i)\mathcal{P}(F_j) = \frac{1}{4}, \quad i, j = 1, 2, \text{ and } i \neq j,$$

which shows that in this case $\beta(\mathcal{D}, \mathcal{F}) = 1$.[27]

Our discussion focuses mainly on $\alpha$-, $\phi$-, and $\rho$-mixing which appear extensively in the literature; absolute regularity is rather seldom encountered. The usefulness of $\psi$-mixing derives from the fact that if a sequence has this property then it is a **mixingale** (difference) which is, basically, a sequence that asymptotically behaves like a martingale. For a discussion of these issues see HH (1980), pp. 18-23. A mixingale is defined below.

**Definition 5.** Let $\{\xi_n : n \in \mathcal{N}\}$ be a collection of random elements defined on the probability space $(\Omega, \mathcal{A}, \mathcal{P})$ and define the nested sequence of $\sigma$-algebras $\mathcal{G}_k = \sigma(\xi_n : n \leq k)$, $k \in \mathcal{N}$. The sequence $\{(\xi_n, \mathcal{G}_n) : n \in \mathcal{N}\}$ is said to be a **mixingale** (sequence) if and only if there exist sequences of non-negative constants, $c_n$, $\gamma_m$ with $\gamma_m \downarrow 0$ such that

$$\| E(\xi_n | \mathcal{G}_{n-m}) \|_2 \leq \gamma_m c_n,.$$

$$\| \xi_n - E(\xi_n | \mathcal{G}_{n+m}) \|_2 \leq \gamma_{m+1} c_n, \quad \text{for all } n \geq 1, \ m \geq 0.$$

**Remark 17.** It may be shown that if a zero mean (square integrable) sequence $\{\xi_n : n \in \mathcal{N}_+\}$ is $\phi$-mixing then it is a **mixingale** with $c_n = 2 \| \xi_n \|_2$ and $\gamma_m = \phi(m)^{1/2}$, where $\phi(m) = \phi(\mathcal{D}, \mathcal{F})$ for $\mathcal{D} = \sigma(\xi_s : s \leq k)$ and $\mathcal{F} = \sigma(\xi_s : s \geq k + m)$.

Evidently, the GLP of Example 1 is a mixingale with $c_n = \sigma^2$, and $\gamma_m = \sum_{j=m}^{\infty} a_j^2$ which clearly converges to zero with $m$.

We return to the various mixing concepts above with the additional purpose of making them more useful in the context of Proposition 16. Thus, in

---
[27] Why is the expression for $\beta(\mathcal{D}, \mathcal{F})$ multiplied by $1/2$? Hint: Is $\mathcal{P}(D_i \cap F_j) = \mathcal{P}(F_j \cap D_i)$?

addition to showing the relation among these measures of dependence, we also seek to connect the various mixing properties to moments of sequences. This will allow us to understand the nature of the interplay between the order and magnitude of the moments, on the one hand, and the "correlation" among elements of a sequence on the other, that are compatible with the conclusions of that proposition.

## 9.5.1 Covariance or Correlation Inequalities

We begin by clarifying the connection between mixing and certain conditional and unconditional moments of a random variable (or vector). [28]

**Proposition 17.** Let $X$ be a random element (vector) defined on the probability space $(\Omega, \mathcal{A}, \mathcal{P})$, which is $\mathcal{D}$-measurable, $\mathcal{D} \subset \mathcal{A}$, and possesses $p$-order moments. Let $\mathcal{F} \subset \mathcal{A}$ be another $\sigma$-(sub)algebra. Then the following statements are true:

   i. $\| E(X|\mathcal{F}) - E(X) \|_q \leq 2^{p-1} \phi(\mathcal{F}, \mathcal{D})^{1/p} \| X \|_p$,
     provided $(1/p) + (1/q) = 1$;

  ii. $\| E(X|\mathcal{F}) - E(X) \|_q \leq 2(2^{1/q} + 1)\alpha(\mathcal{F}, \mathcal{D})^{(1/q)-(1/r)} \| X \|_r$,
     where $1 \leq q \leq r \leq \infty$;

 iii. $\| E(X|\mathcal{F}) - E(X) \|_2 \leq \rho(\mathcal{F}, \mathcal{D}) \| X - E(X) \|_2$.

Proof: See Jacod and Shiryaev (1987), pp. 457.

The result above is still somewhat abstract, and does not convey immediately the import of the measures of dependence we had introduced; in particular, it is not clear how it measures or restricts the magnitude of the "correlation" exhibited by the elements of a given sequence. To make the results of Proposition 17 more easily accessible we state Corollary 2.

---

[28] In the following discussion of various FCLT it is to be understood that convergence on $R_+$ means convergence on compact subsets, i.e. convergence may be shown on any subset of $R_+ \cup \{0\}$ of the form $[t_1, t_2]$, with $t_1 \geq 0$, $t_2 < \infty$. Of direct interest in econometrics is the compact subset $[0, 1]$, and all econometric applications deal specifically with convergence on $[0, 1]$. The topology of the space is the Skorohod topology, discussed in a previous section. Also, of particular relevance for such applications is the space $D$, i.e. the space of discontinuous functions which are rcll. We do not repeat this in each instance.

**Corollary 2.** Let $X$ be as in Proposition 17, $Y$ be $\mathcal{F}$-measurable and suppose $\| X \|_p < \infty$, $\| Y \|_q < \infty$. The following statements are true:

i. $|E(XY) - E(X)E(Y)| \leq 2\phi(\mathcal{D}, \mathcal{F})^{1/p} \| X \|_p \| Y \|_q$,
   for $(1/p) + (1/q) = 1$;

ii. $|E(XY) - E(X)E(Y)| \leq 8\alpha(\mathcal{D}, \mathcal{F})^{1/r} \| X \|_p \| Y \|_q$,
    for $(1/p) + (1/q) + (1/r) = 1$;[29]

iii. $|E(XY) - E(X)E(Y)| \leq \rho(\mathcal{D}, \mathcal{F}) \| X \|_2 \| Y \|_2$.

iv. $|E(XY) - E(X)E(Y)| \leq \psi(\mathcal{D}, \mathcal{F}) \| X \|_1 \| Y \|_1$.

Proof: The proof of i is given in Ibragimov (1962) and may also be found in Billingsley (1968); the proof of ii was given by Wolkonski and Rozanov (1959) and Davydov (1968); both are given in the appendix of HH (1980); the proof of iii is immediate from the definition of $\rho(\mathcal{D}, \mathcal{F})$, and the proof of iv was given by Blum, Hanson, and Koopmans (1963).

Before we proceed with other aspects it is useful to note that, for any sets $D \in \mathcal{D}$ and $F \in \mathcal{F}$, we have the general relations

$$\left| \frac{\mathcal{P}(D \cap F)}{\mathcal{P}(D)\mathcal{P}(F)} - 1 \right|$$

$$= \frac{1}{\mathcal{P}(D)} |\mathcal{P}(F|D) - \mathcal{P}(F)| = \frac{1}{\mathcal{P}(F)\mathcal{P}(D)} |\mathcal{P}(F \cap D) - \mathcal{P}(F)\mathcal{P}(D)|;$$

or, alternatively

$$|\mathcal{P}(F \cap D) - \mathcal{P}(F)\mathcal{P}(D)|$$

$$= \mathcal{P}(F)|\mathcal{P}(F|D) - \mathcal{P}(F)| = \mathcal{P}(F)\mathcal{P}(D) \left| \frac{\mathcal{P}(D \cap F)}{\mathcal{P}(D)\mathcal{P}(F)} - 1 \right|.$$

If we now employ the special case $\mathcal{D} = \mathcal{F} = (\Omega, A, \bar{A}, \emptyset)$, with $\mathcal{P}(A) = 1/2$, we easily see that

$$\alpha(\mathcal{D}, \mathcal{F}) \leq \frac{1}{2}\phi(\mathcal{D}, \mathcal{F}) \leq \frac{1}{4}\psi(\mathcal{D}, \mathcal{P}), \quad \text{or} \quad 4\alpha(\mathcal{D}, \mathcal{F}) \leq 2\phi(\mathcal{D}, \mathcal{F}) \leq \psi(\mathcal{D}, \mathcal{F}).$$

---

[29] Note that in this case we must evidently have $p > 1$, $q > 1$ and moreover, if $q = 2$ then we must have $p = 2 + \delta$, $\delta > 0$.

The relation among the measures of dependency above and the sort of correlation structure they impose on the elements of the sequence is given in Corollary 3.

**Corollary 3.** Under the conditions of Corollary 2,

$$4\alpha(\mathcal{D},\mathcal{F}) \leq \rho(\mathcal{D},\mathcal{F}) \leq 2\phi(\mathcal{D},\mathcal{F})^{1/2};$$

$$\rho(\mathcal{D},\mathcal{F}) \leq \psi(\mathcal{D},\mathcal{F}).$$

Proof: Since $\rho(\mathcal{D},\mathcal{F})$ measures the **maximal** correlation between $\mathcal{D}$- and $\mathcal{F}$-measurable random elements, consider the sets $A_1 \in \mathcal{D}$, $A_2 \in \mathcal{F}$, and define the random elements

$$X_1 = \chi(A_1), \quad X_2 = \chi(A_2),$$

where $\chi(\cdot)$ stands, generically, for the indicator function of a set.[30] Note that

$$EX_i = \mathcal{P}(A_i), \quad \operatorname{Var}(X_i) = \mathcal{P}(A_i)[1 - \mathcal{P}(A_i)] = s_i, \quad i = 1, 2,$$

and consequently,

$$\rho(\mathcal{D},\mathcal{F}) \geq \sup_{A_1 \in \mathcal{D}, A_2 \in \mathcal{F}} \frac{|E[(X_1 - EX_1)(X_2 - EX_2)]|}{(s_1 s_2)^{1/2}}$$

$$= \sup_{A_1 \in \mathcal{D}, A_2 \in \mathcal{F}} \frac{|\mathcal{P}(A_1 \cap A_2) - \mathcal{P}(A_1)\mathcal{P}(A_2)|}{(s_1 s_2)^{1/2}}$$

$$\geq 4\alpha(\mathcal{D},\mathcal{F}).$$

For any zero mean (square integrable) elements we have

$$\rho(\mathcal{D},\mathcal{F}) = \sup_{X \in \mathcal{D}, Y \in \mathcal{F}} \frac{|EXY|}{\| X \|_2 \| Y \|_2} \leq 2\phi(\mathcal{D},\mathcal{F})^{1/2},$$

the last inequality by i of Corollary 2. It follows, therefore that

$$4\alpha(\mathcal{D},\mathcal{F}) \leq \rho(\mathcal{D},\mathcal{F}) \leq 2\phi(\mathcal{D},\mathcal{F})^{1/2}.$$

---

[30] This means that $\chi_i(\omega) = 1$, if $\omega \in A_i$, and zero otherwise.

To prove the second part of the corollary let $X$, $Y$ be, respectively, $\mathcal{D}$- and $\mathcal{F}$-measurable zero mean **simple elements**, i.e.

$$X = \sum_{i=1}^{n} x_i \chi(D_i), \quad Y = \sum_{j=1}^{m} y_j \chi(F_j), \quad D_i \in \mathcal{D}, \quad F_j \in \mathcal{F}.$$

Consequently,

$$
\begin{aligned}
\mathrm{Cov}(X,Y) &= EXY - (EX)(EY) \\[2mm]
&= \sum_{i=1}^{n} \sum_{j=1}^{m} x_i y_j \left[ \mathcal{P}(D_i \cap F_j) - \mathcal{P}(D_i)\mathcal{P}(F_j) \right] \\[2mm]
&= \sum_{i=1}^{n} \sum_{j=1}^{m} x_i y_j \mathcal{P}(D_i)\mathcal{P}(F_j) \left( \frac{\mathcal{P}(D_i \cap F_j)}{\mathcal{P}(D_i)\mathcal{P}(F_j)} - 1 \right);
\end{aligned}
$$

$$|\mathrm{Cov}(X,Y)| \leq \sum_{i=1}^{n} \sum_{j=1}^{m} |x_i| \, |y_j| \mathcal{P}(D_i)\mathcal{P}(F_j)\psi_{D,F}, \quad \text{where}$$

$$\psi_{D,F} = \max_{i,j} \left| \frac{\mathcal{P}(D_i \cap F_j)}{\mathcal{P}(D_i)\mathcal{P}(F_j)} - 1 \right|; \quad \text{thus}$$

$$|\mathrm{Cov}(X,Y)| \leq (E|X|)(E|Y|)\psi_{D,F} \leq \| X \|_2 \| Y \|_2 \, \psi_{D,F};$$

Taking suprema we establish $\rho(\mathcal{D},\mathcal{F}) \leq \psi(\mathcal{D},\mathcal{F})$.

q.e.d.

**Remark 18.** Before we proceed, it is useful to clarify the connection among the various types of mixing developed above, and their connection to the property of ergodicity. The fact that $\alpha$-, $\phi$-, and $\rho$-mixing all imply ergodicity is formally shown in Proposition 18, below. Moreover, the preceding discussion clearly shows the following: [31]

 i. $\psi$-mixing implies $\phi$-mixing;

 ii. $\phi$-mixing implies $\rho$-mixing and absolute regularity;

 iii. $\rho$-mixing implies $\alpha$-mixing (or strong mixing);

 iv. absolute regularity implies $\alpha$-mixing.

---

[31] For examples illustrating the relations below see Bradley (1980) and (1981).

v. absolute regularity is neither implied by, nor does it imply, $\rho$-mixing.

It is clear, thus, that among the five mixing conditions we have considered in this discussion (sometimes referred to as strong mixing conditions), $\alpha$-mixing is the **weakest** in the sense that it is implied by the other conditions, but **it does not imply any of them**.

**Definition 6.** Let $Y = \{Y_t : t \in \mathcal{N}_0\}$ be a (stationary) sequence of random elements defined on the probability space ($\Omega$, $\mathcal{A}$, $\mathcal{P}$), consider the sequence of $\sigma$-(sub)algebras

$$\mathcal{G}_0 = \sigma(Y_0), \ \mathcal{G}^n = \sigma(Y_s, s \geq n), \ \mathcal{G}^{n-1} \supseteq \mathcal{G}^n, \ n \geq 1,$$

and note that $Y_n$ is $\mathcal{G}_n$-measurable. Moreover, let

$$\alpha_n = \alpha(\mathcal{G}_0, \mathcal{G}^n), \ \ \phi_n = \phi(\mathcal{G}_0, \mathcal{G}^n), \psi_n = \psi(\mathcal{G}_0, \mathcal{G}^n), \ \ \rho_n = \rho(\mathcal{G}_0, \mathcal{G}^n).$$

If

  i. $\alpha_n \to 0$ (as $n \to \infty$), the sequence $Y$ is said to be $\alpha$-mixing, occasionally also termed strong mixing;

  ii. $\phi_n \to 0$ (as $n \to \infty$), the sequence $Y$ is said to be $\phi$-mixing, sometimes also termed uniform mixing;

  iii. $\psi_n \to 0$ (as $n \to \infty$), the sequence $Y$ is said to be $\psi$-mixing;

  iv. $\rho_n \to 0$ (as $n \to \infty$), the sequence $Y$ is said to be $\rho$-mixing.

It is interesting to note that if a **stationary** sequence satisfies any of the mixing conditions above then **it is ergodic**, a result given in Proposition 18.

**Proposition 18.** Let $X = \{X_t : t \in \mathcal{N}_0\}$ be a stationary sequence defined on the probability space ($\Omega$, $\mathcal{A}$, $\mathcal{P}$), such that it possesses moments of order $r \geq 2$. If it satisfies any of the conditions of Definition 6 then it is ergodic.

Proof: Without loss of generality, we may proceed by creating an equivalent stationary sequence as follows: Let $\xi$ be a random variable (or vector), such that $\| \xi \|_r < \infty$ and define the measure preserving transformation $T$ such that

$$X_0 = \xi, \ \ X_k = \xi \circ T^k.$$

This is an equivalent sequence, and all argumentation may be carried out in terms of the representation above. To this end, define the $\sigma$-algebras

$$\mathcal{G}_0 = \sigma(\xi), \quad \mathcal{G}^k = \sigma(\xi \circ T^n, \ n \geq k)$$

and consider the $\sigma$-algebra of invariant sets $\mathcal{I}$. We need to prove that all sets $S \in \mathcal{I}$ are $\mathcal{P}$-trivial, i.e. they have measure either zero or one or, put another way, all invariants are constants. Evidently, for $\phi$-mixing, we need only consider sets of positive measure. Now, under uniform mixing, if a set $S \in \mathcal{I}$ then $S \in \mathcal{G}_0 \cap \mathcal{G}^k$, for all $k \geq 1$ and consequently,

$$|\mathcal{P}(S|S) - \mathcal{P}(S)| \leq \phi_k, \quad \text{for all } k.$$

By condition ii of Definition 6 we conclude that $1 - \mathcal{P}(S) = 0$, which implies $\mathcal{P}(S) = 1$.

Next let $S \in \mathcal{I}$, not necessarily of positive measure, and obtain, for strong mixing,

$$|\mathcal{P}(S \cap S) - P(S)^2| \leq \alpha_k.$$

By condition i of Definition 6, we conclude

$$P(S) - P(S)^2 = 0,$$

which implies either $\mathcal{P}(S) = 0$ or $\mathcal{P}(S) = 1$; in either case $S$ is $\mathcal{P}$-trivial.

For the case of $\rho$-mixing, let $\chi$ be the indicator function of the set $S$, i.e. $\chi(\omega) = 1$ if $\omega \in S$ and zero otherwise. Define $Z = \chi - \mathcal{P}(S)$ and note that

$$EZ^2 = \mathcal{P}(S) - \mathcal{P}(S)^2 \leq \rho_k.$$

By condition iv of Definition 6 we conclude that

$$\mathcal{P}(S) - \mathcal{P}(S)^2 = 0,$$

which again implies either $\mathcal{P}(S) = 0$ or $\mathcal{P}(S) = 1$. Since, by Remark 18, $\psi$-mixing implies $\phi$-mixing, the former implies ergodicity by the discussion above.

q.e.d.

**Remark 19**. The import of the mixing conditions is that the elements of the sequence in question are **asymptotically independent**, in the sense that for $A_0 \in \mathcal{G}_0$ and $A_k \in \mathcal{G}^k$, we have

$$\lim_{k \to \infty} \mathcal{P}(A_k | A_0) = \mathcal{P}(A), \quad or \quad \lim_{k \to \infty} \mathcal{P}(A_0 \cap A_k) = \mathcal{P}(A_0)\,(\mathcal{P}(A)),$$

where $A = \lim_{k \to \infty} A_k$. It is this feature that makes mixing conditions attractive for proving CLT.

## 9.6 Limit Theorems for Mixing Sequences

### 9.6.1 Laws of Large Numbers

The reader should bear in mind that it is not difficult to prove laws of large numbers (LLN) if one **makes sufficiently strong assumptions** regarding the existence of higher moments. Thus, for example, if the fourth moment is assumed to be uniformly bounded for some sequence of independent zero mean random variables $\{X_n : n \in \mathcal{N}_+\}$ it is easily obtained from Chebyshev's inequality that

$$\frac{S_n}{n} \xrightarrow{\mathrm{P}} \lim_{n \to \infty} \frac{ES_n}{n}, \quad where \quad S_n = \sum_{j=1}^{n} X_j^2.$$

The point of the mathematical theory regarding LLN as well as CLT is to obtain results with the least restrictive moment assumptions.

First, we note that in the case of $\psi$-mixing sequences McLeish (1975) proves Proposition 19.

**Proposition 19**. Let $\{\xi_k : k \in \mathcal{N}_+\}$ be a mixingale sequence with characteristics $c_n$ and $\gamma_m$, defined on the probability space $(\Omega, \mathcal{A}, \mathcal{P})$; if there exists a sequence $\{b_n : b_n > 0, \ b_n \uparrow \infty\}$ such that

$$\sum_{n=1}^{\infty} \frac{c_n^2}{b_n^2} < \infty, \quad \gamma_n = O(n^{-(1/2)}(\ln n)^{-2}), \quad or \quad \gamma_n \sim \frac{1}{n^{1/2}(\ln n)^2},$$

then

$$\frac{1}{b_n} \sum_{j=1}^{n} \xi_n \xrightarrow{\text{a.c.}} 0.$$

Proof: See McLeish (1975).

More generally, for (weakly) dependent sequences proving WLLN proceeds by finding bounds of the form $E|S_n|^\tau = O(n^\gamma)$, where $S_n = \sum_{s=1}^n X_s$, and $\gamma < 1$, $\tau \geq 2$, or in a broader context $\tau > 0$; similarly, proving SLLN entails finding bounds of the form

$$EM(n)^\tau = O(n^\gamma), \quad M(n) = \max_{1 \leq j \leq n} |S_j|.$$

For example, if $X = \{X_j : j \in \mathcal{N}_+\}$ is an $\alpha$-mixing sequence obeying, see Doukhan (1994), Chapter 1.4,

$$\sum_{j=1}^\infty (j+1)^{k-2}\alpha(j)^{\frac{\delta}{k+\delta}} < \infty, \quad k = 2r, \; r \geq 1, \; k \geq \tau$$

and if $E|X_j|^{\tau+\delta} < \infty$, for some $\delta > 0$, then

$$E|S_n|^\tau \leq K \left( \sum_{j=1}^n \| X_j \|_{\tau+\delta}^\delta \right)^{\frac{\tau}{2}},$$

for some constant $K$. Consequently, under the simple condition that

$$\sup_{j \in \mathcal{N}_+} \| X_j \|_{\tau+\delta} \leq K_1 < \infty,$$

we find, using Markov's inequality, [32]

$$Pr\{|S_n| > \epsilon n\} \leq \frac{E|S_n|^{\tau+\delta}}{\epsilon^{\tau+\delta}n^{\tau+\delta}} \leq \frac{K_2}{\epsilon_1 n^{\frac{\tau}{2}}}.$$

In the context above we may thus say that the sequence $S_n/n$ converges in probability to zero **at the rate** $n^{-(\tau/2)}$. Similar results may be shown for the entities $EM(n)^{\tau+\delta}$. The preceding discussion leads to Proposition 20.

**Proposition 20.** Let $\{\xi_n : n \in \mathcal{N}_+\}$ be a (zero mean) stationary sequence defined on the probability space ($\Omega$, $\mathcal{A}$, $\mathcal{P}$), such that $E|\xi_1|^{2k+\delta} < \infty$, for $\delta > 0$, $k \geq 1$. If, in addition, the sequence is $\alpha$-mixing and

$$\sum_{r=1}^\infty (r+1)^{2(k-1)}\alpha(r)^{\frac{\delta}{2k+\delta}} < \infty,$$

---

[32] For a zero mean random variable $\xi$, it is a general custom to call the statement

$$Pr\{|\xi| > \epsilon\} \leq \frac{E\xi^2}{\epsilon^2}, \quad \textbf{Chebyshev's inequality}$$

and the statment, for $k \neq 2$,

$$Pr\{|\xi| > \epsilon\} \leq \frac{E|\xi|^k}{\epsilon^k}, \quad \textbf{Markov's inequality}.$$

then

$$\text{i.} \quad \frac{S_n}{n} \xrightarrow{\text{P}} 0, \quad \text{where} \quad S_n = \sum_{j=1}^{n} \xi_j;$$

$$\text{ii.} \quad \frac{S_n}{n} \xrightarrow{\text{a.c.}} 0;$$

convergence is at the rate $n^{-k}$.

Proof: See Doukhan (1994), Chapter 1.4.

## 9.6.2 Law of the Iterated Logarithm (LIL)

The LIL is essentially an alternative characterization of the rate at which an entity converges to its limit. [33] The classic LIL states:

**LIL1.** If $X = \{X_t : t \in \mathcal{N}_+\}$ is a sequence of square integrable, zero mean, i.i.d. random variables and $S_n = \sum_{t=1}^{n} X_t$,

$$\frac{S_n}{n} \xrightarrow{\text{a.c.}} 0, \qquad \frac{S_n}{\sqrt{n}} \xrightarrow{\text{d}} \sigma N(0,1),$$

where $\sigma^2$ is the common variance. An equally valid representation of the limiting process is to note that the second entity converges a.c. on the real line, in the sense that its limit points cover the real line. The LIL reflects this point of view by stating that taking $\sigma^2 = 1$ for simplicity and defining

$$\zeta_n(\omega) = (2n \log_2 n)^{-(1/2)} S_n(\omega), \quad \log_2 n = \log \log n,$$

we obtain

$$\overline{\lim_{n \to \infty}} \, \zeta_n = 1, \text{ a.c.,} \quad \underline{\lim_{n \to \infty}} \, \zeta_n = -1, \text{ a.c.}$$

Somewhat more precisely what these statements mean is that for the set sequences

$$A_n = \{\omega : \zeta_n = 1\}, \quad B_n = \{\omega : \zeta_n = -1\}$$

$$\mathcal{P}(A^*) = 1, \quad \mathcal{P}(B_*) = 1, \quad \text{where} \quad A^* = \overline{\lim_{n \to \infty}} \, A_n, \quad B_* = \underline{\lim_{n \to \infty}} \, B_n.$$

---

[33] Recall that the basic concept of the LIL was introduced in Chapter 7, in connection with the SBM.

In terms of the classical LIL obtained above it makes perfectly good sense to say that $S_n$ "grows at the rate" $(2n \log_2 n)^{(1/2)}$; notice further that since in the context above $S_n/\sqrt{n}$ converges in distribution to a $N(0,1)$ and since

$$\zeta_n = (2 \log_2 n)^{-(1/2)} \frac{S_n}{\sqrt{n}},$$

it, similarly, makes perfectly good sense to say that the convergence of the CLT is at the rate $(2 \log_2 n)^{1/2}$. Moreover, writing

$$\zeta_n = (2n^{-1} \log_2 n)^{-(1/2)} \frac{S_n}{n}$$

we may similarly say that the convergence of the SLLN is, in the present case, at the rate $(2n^{-1} \log_2 n)^{1/2}$.

The LIL that is relevant for our topics, however, is following one.

**LIL2.**     Let $X = \{X_t : t \in \mathcal{N}_+\}$ be a zero mean, strictly stationary, $\phi$-mixing (uniformly mixing) sequence, obeying

$$E|X_1|^{2+\delta} < \infty, \quad \sum_{n=1}^{\infty} \phi_n^{\frac{1+\delta}{2+\delta}} < \infty.$$

Put $S_n = \sum_{j=1}^{n} X_j$, $\sigma_n^2 = ES_n^2$, and

$$Z_n(t) = \frac{1}{\sigma_n}[S_k + (nt - k)X_{k+1}],$$

$$\zeta_n(t) = (2\sigma_n^2 \log_2 \sigma_n^2)^{-(1/2)}[S_k + (nt - k)X_{k+1}],$$

for $k \leq (nt - k) \leq k + 1$, and $k = 0, 1, 2, \ldots, n - 1$. Then

$$\frac{\sigma_n^2}{n} \to \sigma^2, \quad \sigma^2 \in [0, \infty); \quad \text{if } \sigma^2 > 0, \quad \text{then}$$

$$Z_n(t) \overset{\mathrm{d}}{\to} B(t), \ t \in [0, 1], \ \textbf{which is the SBM}; \quad \text{moreover,}$$

$$\zeta_n(t) \qquad \text{is relatively compact for } n > K_1,$$

where $K_1$ is a finite constant; the set of its $[\zeta_n(t)]$ limit points is given by $P = \{x : x \in C([0, 1], R)\}$, such that $x(0) = 0$ and

$$\int_0^1 \left(\frac{dx(t)}{dt}\right)^2 dt \leq 1.$$

Proof: See HH (1980), pp. 141-146.

Note that since $\sigma_n^2/n \to \sigma^2$, if $\sigma^2 > 0$, we may write

$$\zeta_n(t) = (2\log_2 n\sigma^2)^{-(1/2)} Z_n(t),$$

and we see again that the rate of convergence of the FCLT is $(2\log_2 n\sigma^2)^{(1/2)}$ which, if we put $\sigma^2 = 1$, is identical to what we obtained earlier for the standard CLT applied to i.i.d. zero mean, finite variance, random variables.

### 9.6.3   CLT and FCLT for Mixing Sequences

In this section we give a number of CLT and FCLT for stationary mixing sequences. To avoid repetitive statements regarding the nature of the sequence in question, the nature of the probability space and whether we are dealing with CLT or FCLT we set the general context now and retain it for the entire section.

The general context is as follows: Let $\xi = \{\xi_t : t \in \mathcal{N}_+\}$ be a zero mean, at least square integrable, strictly stationary (unless otherwise indicated), mixing sequence and define

$$S_n = \sum_{j=1}^{n} \xi_j, \quad \sigma_n^2 = ES_n^2,$$

such that $\sigma_n^2 \to \infty$. Moreover, let

$$Z_n = \frac{S_n}{\sigma_n}, \quad X^n(r) = \frac{1}{\sigma_n}S_{[nr]}, \quad r \in \left[\frac{j-1}{n}, \frac{j}{n}\right), \quad j = 1, 2, \ldots, n,$$

$$X^n(0) = 0, \text{ for all } n.$$

We shall say that $\xi$ obeys a (standard) CLT if and only if

$$Z_n \xrightarrow{d} N(0,1);$$

if $\sigma_n^2/n$ converges to a constant, say $\sigma^2$, we shall say that $\xi$ obeys a (standard) CLT if and only if

$$Z_n^* = \frac{S_n}{\sqrt{n}} \xrightarrow{d} \sigma N(0,1), \quad \text{where } \sigma = \sqrt{\sigma^2}.$$

We shall say that $\xi$ obeys a FCLT if and only if the element $X^n$, i.e. the function $X^n(r)$ for $r \in [0,1]$, obeys

$$X^n \xrightarrow{\text{d}} B,$$

where $B$ is the standard Brownian motion; if $\sigma_n^2/n$ converges to a constant, say $\sigma^2$, we shall say that $\xi$ obeys a FCLT if and only if

$$X^{*n} \xrightarrow{\text{d}} \sigma B \quad \text{on} \quad [0,1],$$

where $X^{*n}$ is the function $X^{*n}(r)$ below, for $r \in [0,1]$, and

$$X^{*n} = \frac{1}{\sqrt{n}} S_{[nr]}, \quad r \in \left[\frac{j-1}{n}, \frac{j}{n}\right), \quad j = 1, 2, \ldots, n, \; X^{*n}(0) = 0, \; \text{for all } n.$$

We shall deal only with three types of dependency measures (1) $\alpha$-mixing, (2) $\phi$-mixing and (3) $\rho$-mixing, since these are the most commonly utilized conditions; $\rho$-mixing is a particularly useful concept when dealing with covariance stationary, rather than strictly stationary, sequences.

For $\alpha$-mixing (strong mixing) we have Proposition 21.

**Proposition 21.** Let $\xi$ be the process above and suppose it is strictly stationary and strong mixing, then it obeys the standard CLT if and only if

$$\frac{S_n^2}{\sigma_n^2} \text{ is } \textbf{uniformly integrable.}$$

The condition above is satisfied if

$$E|S_n|^{2+\delta} = o(\sigma_n^{2+\delta}), \text{ for } \delta > 0.$$

Proof: See Denker (1986).

The corresponding FCLT is then given in Proposition 22.

**Proposition 22.** In the context of Proposition 21 a necessary and sufficient condition for the sequence $\xi$ to obey a FCLT is the following: Given any integer $r$, however large, there exist a number $\lambda > 1$ and an integer $n_0$ such that for all $n \geq n_0$

$$\mathcal{P}(A_n) < \frac{1}{r}\frac{1}{\lambda^2}, \quad \text{where } A_n = \{\omega : \max_{j \leq n} |S_j| \geq \lambda \sigma_n\}, \quad \text{and}$$

$$\frac{S_n^2}{\sigma_n^2} \quad \text{is } \textbf{uniformly integrable.}$$

Proof: Proposition 21 and Proposition 5a. This becomes clear if we note that for **strictly stationary** processes

$$A_n = A_k(n) = \{\omega : \max_{i \leq n} |S_{k+i} - S_k| \geq \lambda \sigma \sqrt{n}\},$$

by the construction of $A_k(n)$ in Proposition 5a.

**Remark 20**. In connection with Proposition 22 above, note that the second condition ensures that $\xi$ obeys a standard CLT which enables us to assert the convergence of the finite dimensional distributions; the first condition implies, by Proposition 5a, that the sequence of elements $\{X^n\}$ is **tight**; the desired result is then obtained because of Prohorov's criterion (Proposition 1) and Proposition 2.

**Proposition 23**. Let $\xi$ be the standard (strictly stationary) sequence and suppose

$$E|\xi_1|^{2+\delta} < \infty, \quad \text{for arbitrary } \delta > 0;$$

then $\xi$ obeys a standard CLT.

If, in addition,

$$\sum_{n=1}^{\infty} \alpha_n^{\delta/2+\delta} < \infty,$$

then $\xi$ obeys a FCLT.

Proof: For the first part see Proposition 21 as well as Ibragimov and Linnik (1971); for the second part see Oodaira and Yoshihara (1972).

For $\phi$-mixing (uniform mixing) sequences we have Proposition 24.

**Proposition 24**. Let $\xi$ be the standard (uniform mixing) sequence and suppose in addition it obeys a **Lindeberg** condition, i.e.

$$\frac{1}{\sigma_n^2} \sum_{j=1}^{n} \int_{A_{jn}} |\xi_j(\omega)|^2 \, d\mathcal{P} = \frac{n}{\sigma_n^2} \int_{A_{1n}} |\xi_1|^2 d\mathcal{P} \to 0,$$

where $A_{jn} = \{\omega : |\xi_j(\omega)| \geq \epsilon \sigma_n\}$, for every $\epsilon > 0$. Then $\xi$ obeys a FCLT.

Proof: See Peligrad (1985). Note, for example, that if the first (left) version of the condition above holds then $\xi$ would obey a FCLT, even if it were **only covariance stationary** instead of **strictly** stationary.

A somewhat weaker form is given in Proposition 25.

**Proposition 25.** Suppose the sequence $\xi$ satisfies

$$\sum_{n=1}^{\infty} \phi_n^{1/2} < \infty;$$

then the series

$$\sigma^2 = E\xi_1 + 2\sum_{k=1}^{\infty} E\xi_1\xi_{1+k}$$

converges absolutely and $\xi$ obeys a FCLT.

Proof: See Billingsley (1968), pp. 174-177.

For $\rho$-mixing sequences, the strongest results pertain to **covariance stationary** sequences. We first give a result that pertains to the behavior of $\sigma_n^2$ and the spectral density of the process.

**Proposition 26.** Let $\xi$ be covariance stationary and $\rho$-mixing;

  i. if $\sigma_n^2 \to \infty$ then $\sigma_n^2 = nh(n)$, where $h$ is a slowly varying [34] (positive) function on $R_+$;

  ii. if $\sum_{j=1}^{\infty} \rho(2^j) < \infty$, $\xi$ has a **continuous** spectral density, say $f$, and if $f(0) \neq 0$, then

$$\sigma_n^2 = 2\pi f(0)n[1 + o(1)].$$

Proof: See Ibragimov (1975) and Ibragimov and Rozanov (1978).

Convergence results are given in the two propositions below.

**Proposition 27.** Let $\xi$ be a $\rho$-mixing covariance stationary sequence.

---

[34] A function, $f$, is said to be **slowly varying** if and only if it is monotone increasing (nondecreasing), measurable, continuous or, if discontinuous its discontinuities are confined to a set of measure zero, and there exists $t \in R_+$ such that

$$\sup_{x>0} \frac{f(tx)}{f(x)} < \infty.$$

For example, in terms of the definition above, the polynomial function $f(x) = x^m$, with finite $m > 0$ is slowly varying, but $f(x) = e^{\alpha x}$, for finite $\alpha > 0$ is not.

   i. If

$$\sigma_n^2 \to \infty \quad \text{and} \quad \sum_{j=1}^{\infty} \rho(2^j) < \infty$$

then $\xi$ obeys a standard CLT;

  ii. if, in addition,

$$\sum_{j=1}^{\infty} [\rho(2^j)]^{1/2} < \infty$$

then $\xi$ obeys a FCLT.

Proof: For the proof of part i, see Ibragimov (1975); for part ii see Peligrad (1982).

**Proposition 28.** Let $\xi$ be strictly stationary, $\rho$-mixing, and suppose

$$E|\xi_1|^r < \infty, \quad \text{for} \quad r > 2.$$

The following are true:

   i. $\xi$ obeys a FCLT;

  ii. for every $k \in [1, r]$

$$E \left| \frac{S_n}{\sigma_n} \right|^k \to m_k$$

where $m_k$ is the $k^{th}$ absolute moment of the standard normal distribution.

Proof: See Ibragimov (1975) for part i; for part ii see Peligrad (1985).

We illustrate some aspects of the preceding discussion in the example below.

**Example 2.** Consider again the GLP of Example 1,

$$Z_k = \sum_{j=0}^{\infty} a_j \epsilon_{k-j}, \quad k \in \mathcal{N}_0,$$

recall that

$$Z_k = \sum_{j=0}^{k-1} a_j \epsilon_{k-j} + Z_{(k|0)}, \quad Z_{(k|0)} = \sum_{j=0}^{\infty} a_{j+k} \epsilon_{-j},$$

and define

$$\mathcal{G}_0 = \sigma(Z_0), \quad \mathcal{G}^k = \sigma(Z_n : n \geq k), \quad k \in \mathcal{N}_+.$$

In the context of the $\sigma$-algebras above, a set in $\mathcal{G}_0$ refers to an "event" relative to the random variable $Z_0$ and a set in $\mathcal{G}^k$ refers to an "event" relative to the random variables $Z_n, n \geq k$. In view of Definition 6, Proposition 17, and Corollary 2 note that, for the case of $p = 2$, $\phi$-mixing, and $\rho$-mixing, we have

$$\phi(\mathcal{G}_0, \mathcal{G}^k) = \sup_{A_0 \in \mathcal{G}_0, \ A_k \in \mathcal{G}^k} |\mathcal{P}(A_k|A_0) - \mathcal{P}(A_k)|$$

$$\geq \frac{1}{4} \left( \frac{|E Z_0 Z_k|}{\| Z_0 \|_2 \| Z_k \|_2} \right)^2,$$

$$\rho(\mathcal{G}_0, \mathcal{G}^k) = \sup_{X \in \mathcal{G}_0, \ Y \in \mathcal{G}^k} \frac{|E(XY)|}{\| X \|_2 \| Y \|_2}.$$

For the GLP, normalize first through division by $c = (\sum_{j=0}^{\infty} a_j^2)^{1/2}$, thus interpreting the coefficients $a_j$ as $a_j/c$, and observe that it is sufficient to deal with $Z_0$ and $Z_n, n \geq 1$. Since

$$Z_0 Z_n = Z_0 \left( \sum_{j=0}^{n-1} a_j \epsilon_{n-j} \right) + Z_0 Z_{(n|0)},$$

we find

$$|E Z_0 Z_n| = |E Z_0 Z_{(n|0)}| = \left| \sum_{j=0}^{\infty} a_j a_{j+n} \right| \leq \sum_{j=0}^{\infty} |a_{j+n}| \, |a_j|.$$

For $\phi$-mixing we must have

$$\phi_n \to \infty, \quad \text{so that a necessary condition is} \quad \lim_{n \to \infty} \left( \sum_{j=0}^{\infty} a_j a_{j+n} \right) = 0,$$

which does not really help us very much beyond what we shall obtain in the $\rho$-mixing case. In particular, we find

$$\rho_k \leq \sup_{n \geq k} \sum_{j=0}^{\infty} |a_{j+n}| \, |a_j| = |a_{j_k}| \sum_{j=0}^{\infty} |a_j|.$$

Since

$$|a_{j_k}| \leq \sup_{n \geq k} |a_n|,$$

it follows that $|a_{j_1}| \geq |a_{j_2}| \geq |a_{j_3}| \ldots$, and $j_1 \geq j_2 \geq j_3, \ldots$; thus, if there exists an index $N$ such that, for $r \geq N$, and $\gamma > 1$, $|a_r| \sim c_0 r^{-\gamma}$, we shall be able to conclude that, for $r \geq N$, $|a_{j_k}| \leq c_0 j_k^{-\gamma}$, and consequently that

$$\rho_k \leq c_1 \frac{1}{j_k^{\gamma}} \to 0, \qquad \sum_{k=1}^{\infty} \rho_k \leq M_0 + M_1 \sum_{j_k=N}^{\infty} \frac{1}{j_k^2} < \infty,$$

for suitable constants $M_0, M_1$. In view of the fact that the GLP obeys $\sum_{j=0}^{\infty} j|a_j| < \infty$, which in turn implies that $\lim_{j \to \infty} j|a_j| = 0$, there exists an integer $N$ such that for $r \geq N$, $|a_r| \leq c_0 r^{-\gamma}$, for some (positive) constants $c_0$, and $\gamma > 1$.

We conclude, therefore, that the GLP with the condition $\sum_{j=0}^{\infty} j|a_j| < \infty$, is $\rho$-mixing and thus, by Proposition 28,

$$\frac{1}{\sqrt{n}} S_{[nt]} \xrightarrow{d} \sqrt{\sigma^2} B_t.$$

## 9.6.4   CLT for Nonstationary Mixing Sequences

In the discussion of the previous section we have established very powerful tools for showing that stochastic processes of the type first defined in Eq. (9.11) converge in distribution to the standard BM, **provided** the underlying sequence is **stationary** and obeys certain mixing conditions. When that sequence, however, is **not** stationary, the results above do not necessarily apply. Nonstationarity, however, need not imply anything "sinister" like unbounded moments, as is the case with $I(1)$ processes. It may simply mean "heterogeneity" in the sense that the distribution of the elements of the sequence differ, or that only moments of certain orders differ. We give below a proper definition of a measure of the degree of dependence among elements of a sequence, which is not confined to stationary processes alone. This is the definition implicit in the results to be cited below, even though the statement of the theorem and its conditions are rather complex.

To reader may ask what the difference is between these two situations and why is it simpler to state conditions and convergence results for stationary (and ergodic or mixing) sequences. The basic answer is that stationarity puts a great deal of structure in the probabilistic "behavior" of, and ergodicity ensures availability of certain other properties for, the stochastic sequence to which we wish to apply a Donsker-like result. If we eliminate one or both of these types of conditions, we need to replace it (them) by some assumption(s) regarding the magnitude of the moments, and the "speed" with which the dependence among elements of the sequence decays (mixing conditions). These alternative assumptions, or conditions, however, cannot be neatly categorized in a set that characterizes an interesting class of stochastic sequences; instead, they **have to be asserted initially, and individually,** and exhibit certain "tradeoffs". Thus, a more stringent requirement on moments, i.e. one that requires moments of elements of the sequence to grow "slowly", may be given up for a weaker form of dependence, i.e. one that requires the correlation between two elements to vanish more quickly, when their "distance" from each other increases (and vice-versa). We shall see an example of this in Proposition 29 and its Corollaries.

When the sequence $\{Y_t : t \in \mathcal{N}\}$ is not stationary, the definitions of the various measures of dependence need to be altered, although the differences are rather minor. For the sake of completeness and precision we now give the formal definition of these properties for sequences that need not be strictly stationary.

**Definition 7.** Let

$$\mathcal{G}_k = \sigma(Y_t : t \leq k), \quad \mathcal{G}^{k+n} = \sigma(Y_t : t \geq k + n),$$

and define the mixing coefficients by

$$\alpha(n) = \sup_{k \in \mathcal{N}} \alpha_n(k), \quad \alpha_n(k) = \alpha(\mathcal{G}_k, \mathcal{G}^{k+n});$$

$$\phi(n) = \sup_{k \in \mathcal{N}} \phi_n(k), \quad \phi_n(k) = \phi(\mathcal{G}_k, \mathcal{G}^{k+n});$$

$$\psi(n) = \sup_{k \in \mathcal{N}} \psi_n(k), \quad \psi_n(k) = \psi(\mathcal{G}_k, \mathcal{G}^{k+n});$$

$$\rho(n) = \sup_{k \in \mathcal{N}} \rho_n(k), \quad \rho_n(k) = \rho(\mathcal{G}_k, \mathcal{G}^{k+n});$$

and similarly with the other dependence measures. If

 i. $\alpha(n) \to 0$ (as $n \to \infty$), the sequence $Y$ is said to be $\alpha$-mixing;

 ii. $\phi(n) \to 0$ (as $n \to \infty$), the sequence $Y$ is said to be $\phi$-mixing;

 iii. $\psi(n) \to 0$ (as $n \to \infty$), the sequence $Y$ is said to be $\psi$-mixing;

iv. $\rho_n \to 0$ (as $n \to \infty$), the sequence $Y$ is said to be $\rho$-mixing.

We now present a more general result than the one(s) given above, which applies to stationary as well as nonstationary processes, but is also far more complex in its statement and the conditions it imposes on the underlying sequence.

**Proposition 29.** Let $\xi = \{\xi_n : n \in \mathcal{N}_+\}$ be a sequence of random variables defined on the probability space $(\Omega, \mathcal{A}, \mathcal{P})$. Let $\mathcal{G}_m = \sigma(\xi_i, 1 \le i \le m)$, $\mathcal{G}^{n,k} = \sigma(\xi_i, m + k \le i \le n)$, and define the mixing coefficients

$$\alpha_n(k) = \sup_{m \le n-k} \alpha(\mathcal{G}_m, \mathcal{G}^{n,k}),$$

for $k \le n - 1$, and zero otherwise. Further, define [35]

$$S_n = \sum_{i=1}^{n} \xi_i, \quad \alpha(k) = \sup_{n \in \mathcal{N}_+} \alpha_n(k), \tag{9.45}$$

$$X_t^n = \frac{S_{[nt]}}{\sqrt{n}}, \quad t \in [0,1]; \quad \text{assume} \tag{9.46}$$

$$E\xi_n = 0, \quad E\xi_n^2 < \infty, \quad n \in \mathcal{N}; \tag{9.47}$$

$$\frac{ES_n^2}{n} \to \sigma^2 > 0; \tag{9.48}$$

$$\infty > \sup_{m,n \in \mathcal{N}_+} \frac{E(S_{m+n} - S_m)^2}{n}, \tag{9.49}$$

and let $\beta \in (2, \infty]$, $\gamma = (2/\beta)$. If the conditions in Eqs. (9.47), (9.48), (9.49) are satisfied and, for a sequence $a = \{a_n : n \in \mathcal{N}_+, a_n \in [1, \infty]\}$,

$$\lim_{n \to \infty} \left[ \sup_{i \le n} \| \xi_i \|_\beta^2 \left( \sum_{i \ge a_n} \alpha(i)^{1-\gamma} + \frac{a_n^{2-\gamma}}{n^{1-\gamma}} \right) \right] = 0,$$

then the processes $X_t^n$ converge weakly to the standard BM, i.e.

$$X_t^n \overset{\mathrm{d}}{\to} \sqrt{\sigma^2} B_t, \quad t \in [0,1].$$

Proof: See Herrndorf (1984b).

---

[35] Note that the notation $[a]$ means the integer part of $a$; thus $[nt]$ means the **largest** integer equal to or less than $nt$.

**Corollary 4.** Let $\beta \in (2, \infty]$, $\gamma = (2/\beta)$. If the conditions in Eqs. (9.47) and (9.48) are satisfied and, moreover,

$$\sum_{i=1}^{\infty} \alpha(i)^{1-\gamma} < \infty, \quad \limsup_{n \to \infty} \| \xi_n \|_\beta < \infty,$$

then

$$X_t^n \overset{\mathrm{d}}{\to} \sqrt{\sigma^2} B_t, \quad t \in [0, 1].$$

Proof: Herrndorf (1984b).

**Corollary 5.** Let $\beta \in (2, \infty]$, $\gamma = (2/\beta)$. If Eqs. (9.47) and (9.48) hold and, moreover, one of the following equations is satisfied with $a > 1/(1-\gamma)$,

$$\alpha(k) = O(k^{-a}), \ \| \xi_n \|_\beta = o(n^c), \ c = \frac{1-\gamma}{2} \frac{a}{1+a} - \frac{1}{2} \frac{1}{1+a}, \quad (9.50)$$

$$\alpha(k) = O(b^{-k}), \ b > 1, \ \| \xi_n \|_\beta = o(d_n), \ d_n = \frac{n^{(1-\gamma)/2}}{(\ln n)^{1-(\gamma/2)}}, \quad (9.51)$$

then

$$X_t^n \overset{\mathrm{d}}{\to} \sqrt{\sigma^2} B_t, \quad t \in [0, 1].$$

Proof: Herrndorf (1984b).

# Frequently Used Symbols

| Symbol | Defined | Meaning |
|---|---|---|
| AIC | Chapter 2 | Akaike information criterion |
| AICC | Chapter 2 | Akaike information criterion, corrected |
| $AR(k)$ | Chapter 1 | autoregressive process of order $k \in \mathcal{N}_0$ |
| $ARMA(m,n)$ | Chapter 1 | autoregressive moving average process of order $m,n \in \mathcal{N}_0$, |
| BIC | Chapter 2 | Bayesian information criterion |
| BM | Chapter 7 | Brownian motion |
| CLT | Chapter 9 | central limit theorem |
| FCLT | Chapter 9 | functional central limit theorem |
| FPE | Chapter 2 | final prediction error |
| FPEM | Chapter 2 | final prediction error matrix |
| $\mathcal{G}_t$ | Chapter 2 | (sub) $\sigma$-algebra |
| $I(0)$ | Chapter 1 | a strictly stationary process |
| $I(k)$ | Chapter 1 | process whose $k^{th}$ difference is strictly stationary, i.e. $X_t \sim I(k)$ if and only if $(I-L)^k X_t$ is $I(0)$ |
| IV | Chapter 3 | instrumental variable(s) |
| $K(\cdot)$ | Chapter 1 | autocovariance kernel of multivariate AR process |
| $MA(k)$ | Chapter 1 | moving average process of order $k \in \mathcal{N}_0$ |
| $MAR(k)$ | Chapter 1 | **multivariate** autoregressive process of order $k \in \mathcal{N}_0$ |
| $MARMA(m,n)$ | Chapter 1 | **multivariate** autoregressive moving average process of order $m,n \in \mathcal{N}_0$ |
| $MBM$ | Chapter 7 | multivariate Brownian motion |
| $MMA(k)$ | Chapter 1 | **multivariate** moving average process of order $k \in \mathcal{N}_0$ |

| Symbol | Defined | Meaning |
|---|---|---|
| $MWN(\Sigma)$ | Chapter 1 | multivariate white noise process with covariance matrix $\Sigma > 0$ |
| $SBM$ | Chapter 7 | standard Brownian motion |
| $SMBM$ | Chapter 7 | standard multivariate Brownian motion |
| $SSF$ | Chapter 2 | state space form (of a model) |
| $VAR(k)$ | Chapter 1 | vector autoregressive process of order $k \in \mathcal{N}_0$ |
| $WN(\sigma^2)$ | Chapter 1 | white noise process with variance $\sigma^2$, i.e. a sequence of independent identically distributed random variables with mean zero, variance $\sigma^2$ |
| $(\Omega, \mathcal{A}, \mathcal{P})$ | Chapter 1 | probability space, where $\Omega$ is the sample space, $\mathcal{A}$ is a $\sigma$-algebra, and $\mathcal{P}$ a probability measure |
| $\mathcal{N}$ | Chapter 1 | integer lattice on the real line, i.e. the set $\{0, \pm 1, \pm 2, \ldots\}$ |
| $\mathcal{N}_0$ | Chapter 1 | the set $\{0, 1, 2, \ldots\}$ |
| $\mathcal{N}_+$ | Chapter 1 | the set $\{1, 2, 3, \ldots\}$ |
| $\mathcal{N}_-$ | Chapter 1 | the set $\{-1, -2, -3, \ldots\}$ |
| $\mathcal{C}$ | Chapter 1 | the field of complex numbers |
| $\mathcal{C}^m$ | Chapter 1 | the $m$-dimensional field of complex numbers, i.e. the product of $\mathcal{C}$ with itself $m$ times |
| $\kappa(\cdot)$ | Chapter 1 | the autocovariance function of a stationary process |
| $\sigma(X_{s\cdot} : 1 \le s \le t)$ | Chapter 2 | the $\sigma$-algebra generated by the r.v. $X_{s\cdot}$, $1 \le s \le t$ usually denoted, generically, by $\mathcal{G}_t$. |

Comment: The term multivariate process is more flexible than the term vector process, which is preferred in the literature of econometrics; multivariate is a term that may refer to vectors as well as more complex arrays, such as matrices for instance. Thus, it is the term of preference in this volume, except for the multivariate AR process, where the term VAR is so well established a usage that it would be counterproductive to refer to it by any other term.

# Graphs of Sequences of Various Types

In the pages following we present a series of graphs for sequences of various types. The purpose is to acquaint the reader with the somewhat distinctive profiles of various types of time series, for example, i.i.d. sequences, linear time trends with errors, stationary sequences, nonstationary sequences, cointegrated sequences of various ranks, and the cointegral vectors that arise from them.

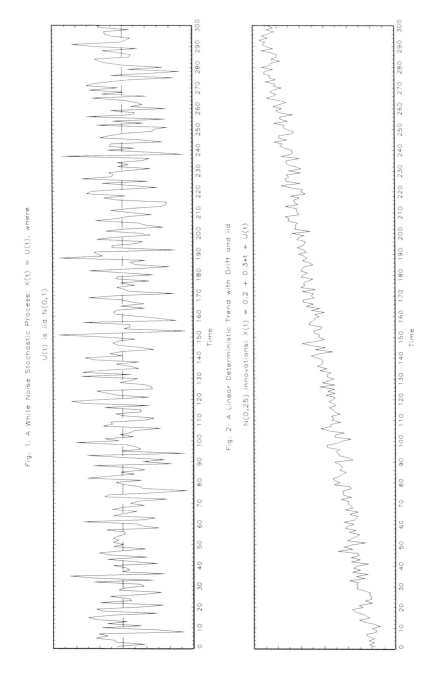

Fig. 1: A White Noise Stochastic Process: X(t) = U(t), where

U(t) is iid N(0,1)

Fig. 2: A Linear Deterministic Trend with Drift and iid

N(0,25) Innovations: X(t) = 0.2 + 0.3*t + U(t)

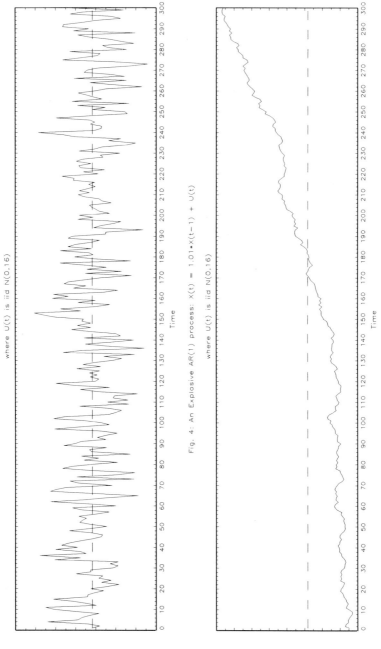

Fig. 3: A Stable AR(1) process: X(t) = 0.6 + 0.7*X(t-1) + U(t)

where U(t) is iid N(0,16)

Fig. 4: An Explosive AR(1) process: X(t) = 1.01*X(t-1) + U(t)

where U(t) is iid N(0,16)

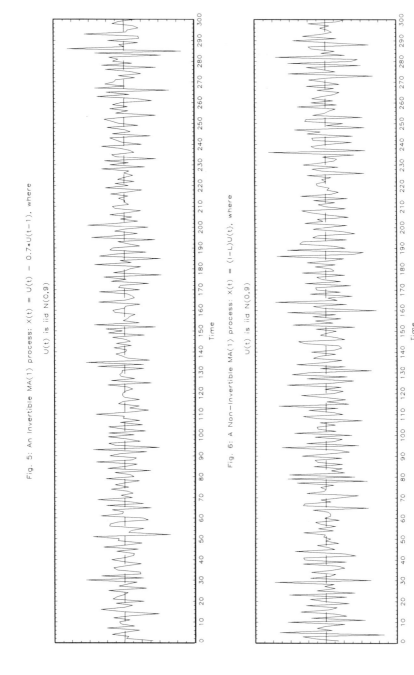

Fig. 5: An invertible MA(1) process: X(t) = U(t) − 0.7*U(t−1), where
U(t) is iid N(0,9)

Fig. 6: A Non-invertible MA(1) process: X(t) = (1−L)U(t), where
U(t) is iid N(0,9)

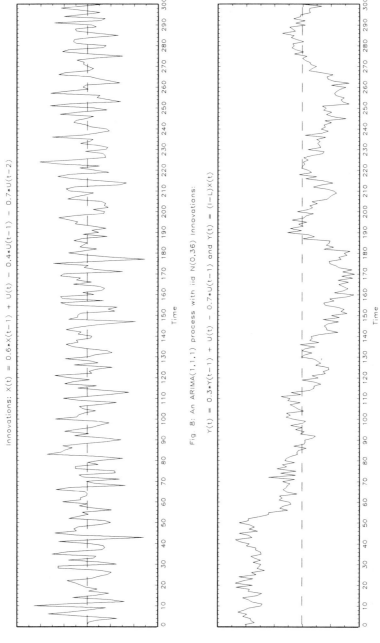

Fig. 7: A Stable and Invertible ARMA(1,2) process with iid N(0,36)

Innovations: X(t) = 0.6*X(t−1) + U(t) − 0.4*U(t−1) − 0.7*U(t−2)

Fig 8: An ARIMA(1,1,1) process with iid N(0,36) Innovations:

Y(t) = 0.3*Y(t−1) + U(t) − 0.7*U((t−1) and Y(t) = (I−L)X(t)

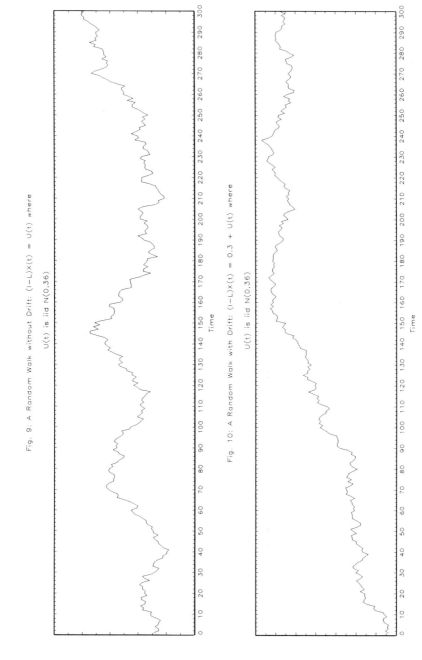

Fig. 9: A Random Walk without Drift: $(I-L)X(t) = U(t)$ where

U(t) is iid N(0,36)

Fig. 10: A Random Walk with Drift: $(I-L)X(t) = 0.3 + U(t)$ where

U(t) is iid N(0,36)

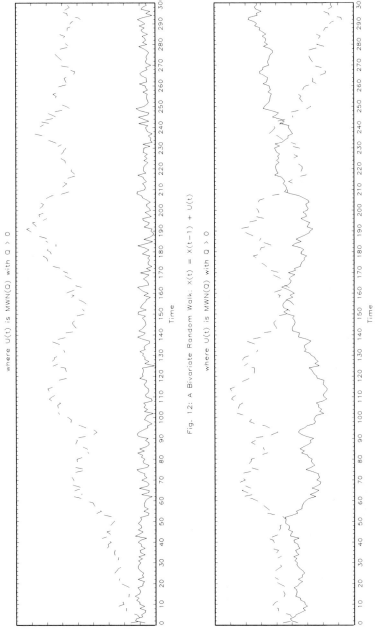

Fig. 11: A Bivariate Cointegrated VAR(1) process X(t) = X(t−1)*A + U(t)
where U(t) is MWN(Q) with Q > 0

Fig. 12: A Bivariate Random Walk: X(t) = X(t−1) + U(t)
where U(t) is MWN(Q) with Q > 0

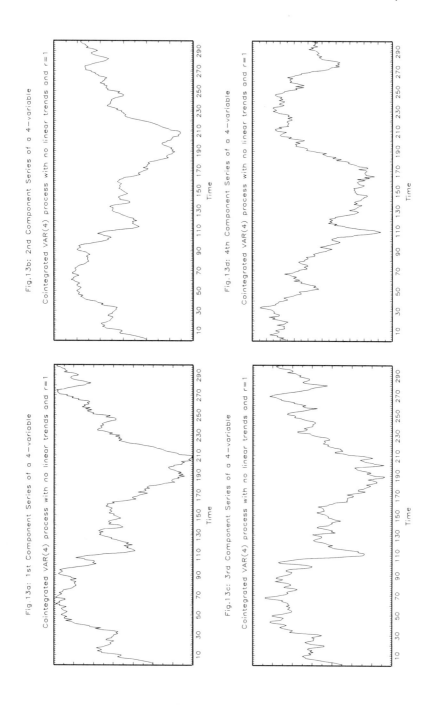

Fig.13a: 1st Component Series of a 4-variable

Cointegrated VAR(4) process with no linear trends and r=1

Fig.13b: 2nd Component Series of a 4-variable

Cointegrated VAR(4) process with no linear trends and r=1

Fig.13c: 3rd Component Series of a 4-variable

Cointegrated VAR(4) process with no linear trends and r=1

Fig.13d: 4th Component Series of a 4-variable

Cointegrated VAR(4) process with no linear trends and r=1

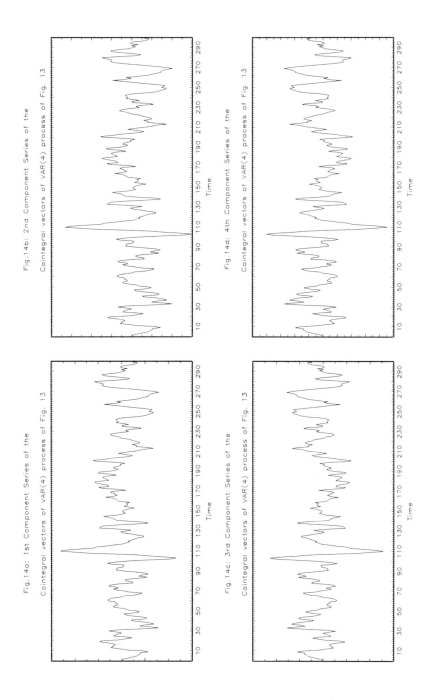

Fig.14a: 1st Component Series of the
Cointegral vectors of VAR(4) process of Fig. 13

Fig.14b: 2nd Component Series of the
Cointegral vectors of VAR(4) process of Fig. 13

Fig.14c: 3rd Component Series of the
Cointegral vectors of VAR(4) process of Fig. 13

Fig.14d: 4th Component Series of the
Cointegral vectors of VAR(4) process of Fig. 13

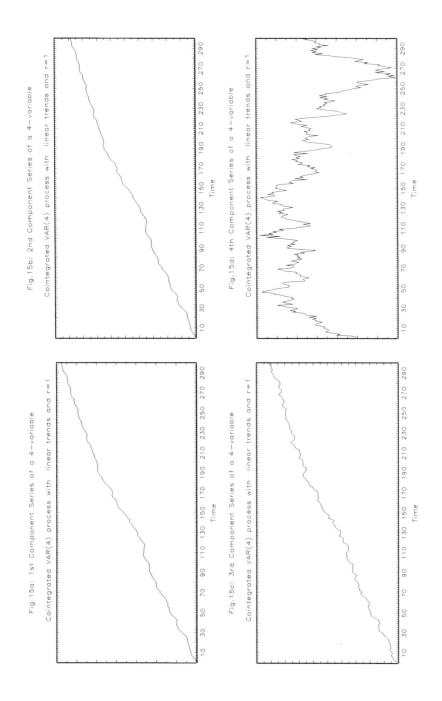

Fig.15a: 1st Component Series of a 4-variable

Cointegrated VAR(4) process with linear trends and r=1

Fig.15b: 2nd Component Series of a 4-variable

Cointegrated VAR(4) process with linear trends and r=1

Fig.15c: 3rd Component Series of a 4-variable

Cointegrated VAR(4) process with linear trends and r=1

Fig.15d: 4th Component Series of a 4-variable

Cointegrated VAR(4) process with linear trends and r=1

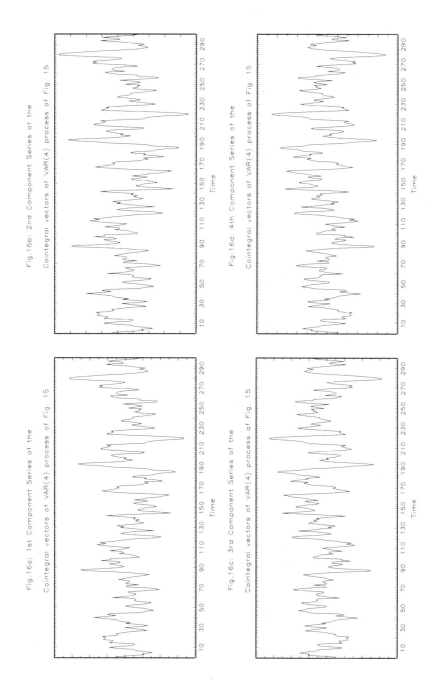

Fig.16a: 1st Component Series of the
Cointegral vectors of VAR(4) process of Fig. 15

Fig.16b: 2nd Component Series of the
Cointegral vectors of VAR(4) process of Fig. 15

Fig.16c: 3rd Component Series of the
Cointegral vectors of VAR(4) process of Fig. 15

Fig.16d: 4th Component Series of the
Cointegral vectors of VAR(4) process of Fig. 15

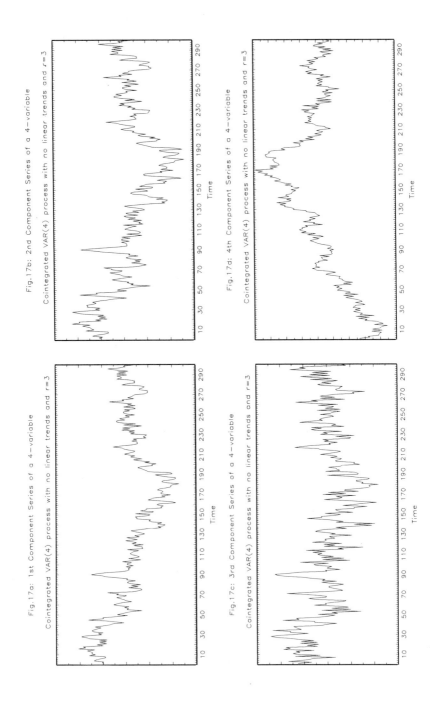

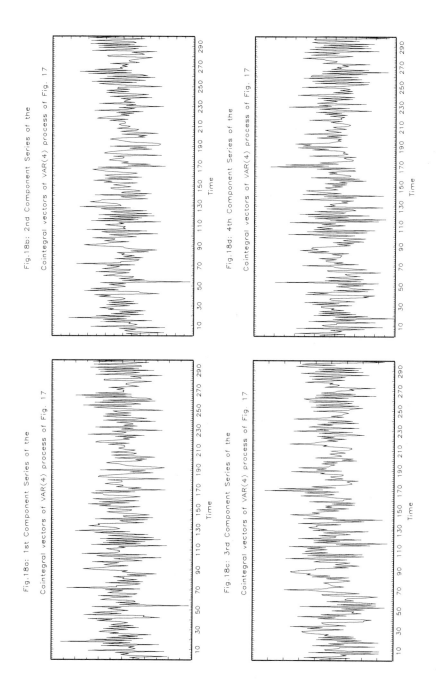

Fig.18a: 1st Component Series of the

Cointegral vectors of VAR(4) process of Fig. 17

Fig.18b: 2nd Component Series of the

Cointegral vectors of VAR(4) process of Fig. 17

Fig.18c: 3rd Component Series of the

Cointegral vectors of VAR(4) process of Fig. 17

Fig.18d: 4th Component Series of the

Cointegral vectors of VAR(4) process of Fig. 17

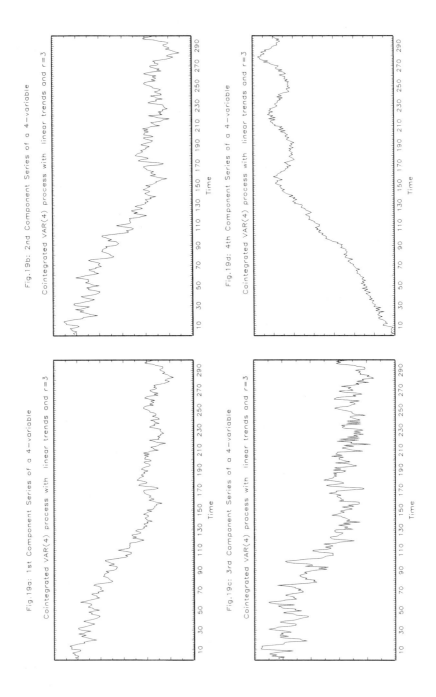

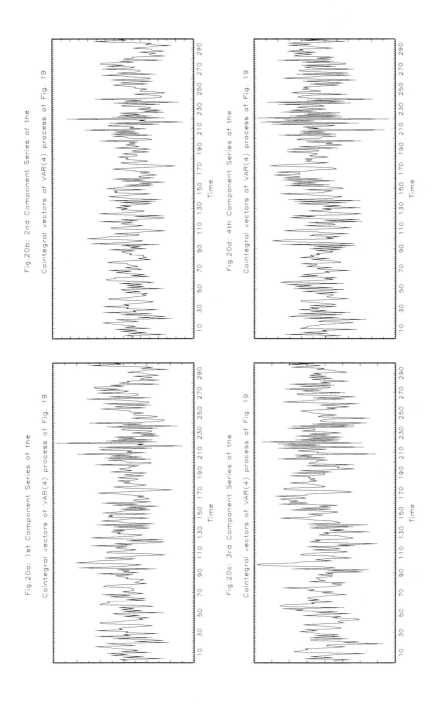

Fig.20a: 1st Component Series of the

Cointegral vectors of VAR(4) process of Fig. 19

Fig.20b: 2nd Component Series of the

Cointegral vectors of VAR(4) process of Fig. 19

Fig.20c: 3rd Component Series of the

Cointegral vectors of VAR(4) process of Fig. 19

Fig.20d: 4th Component Series of the

Cointegral vectors of VAR(4) process of Fig. 19

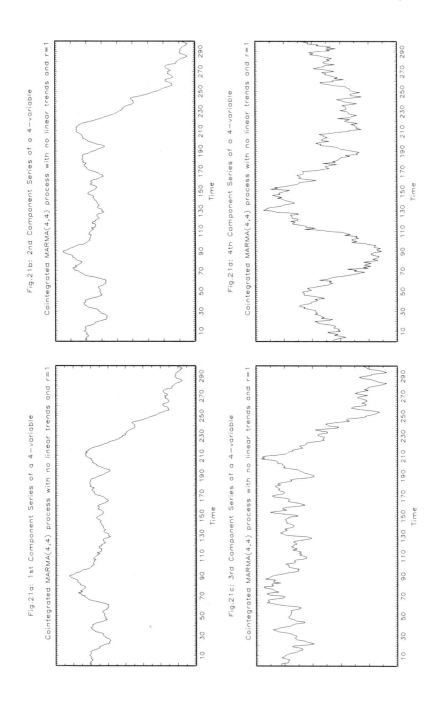

Fig.21a: 1st Component Series of a 4-variable

Cointegrated MARMA(4,4) process with no linear trends and r=1

Fig.21b: 2nd Component Series of a 4-variable

Cointegrated MARMA(4,4) process with no linear trends and r=1

Fig.21c: 3rd Component Series of a 4-variable

Cointegrated MARMA(4,4) process with no linear trends and r=1

Fig.21d: 4th Component Series of a 4-variable

Cointegrated MARMA(4,4) process with no linear trends and r=1

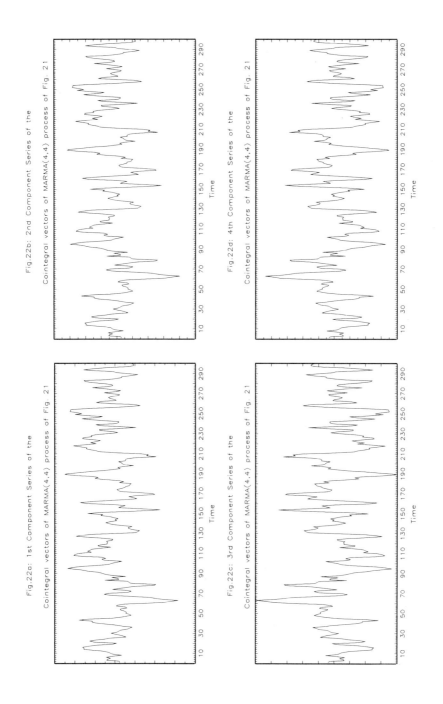

Fig.22a: 1st Component Series of the
Cointegral vectors of MARMA(4,4) process of Fig. 21

Fig.22b: 2nd Component Series of the
Cointegral vectors of MARMA(4,4) process of Fig. 21

Fig.22c: 3rd Component Series of the
Cointegral vectors of MARMA(4,4) process of Fig. 21

Fig.22d: 4th Component Series of the
Cointegral vectors of MARMA(4,4) process of Fig. 21

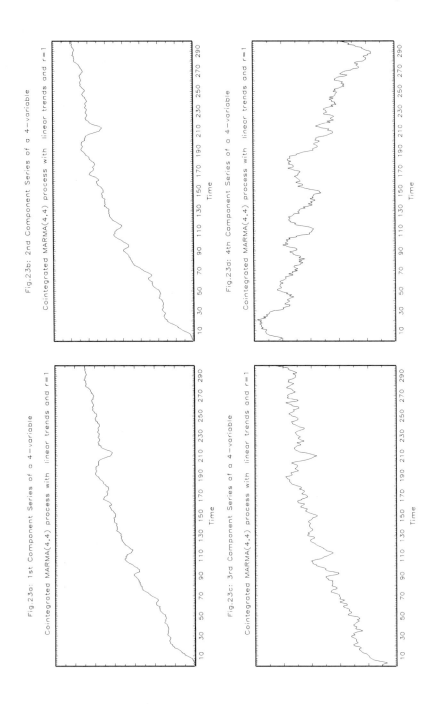

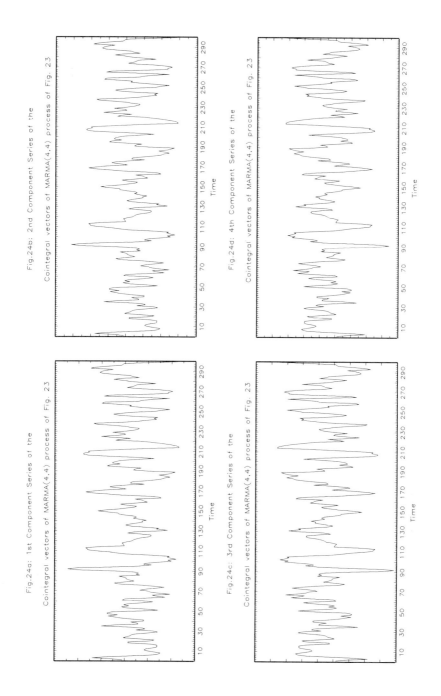

Fig.24a: 1st Component Series of the
Cointegral vectors of MARMA(4,4) process of Fig. 23

Fig.24b: 2nd Component Series of the
Cointegral vectors of MARMA(4,4) process of Fig. 23

Fig.24c: 3rd Component Series of the
Cointegral vectors of MARMA(4,4) process of Fig. 23

Fig.24d: 4th Component Series of the
Cointegral vectors of MARMA(4,4) process of Fig. 23

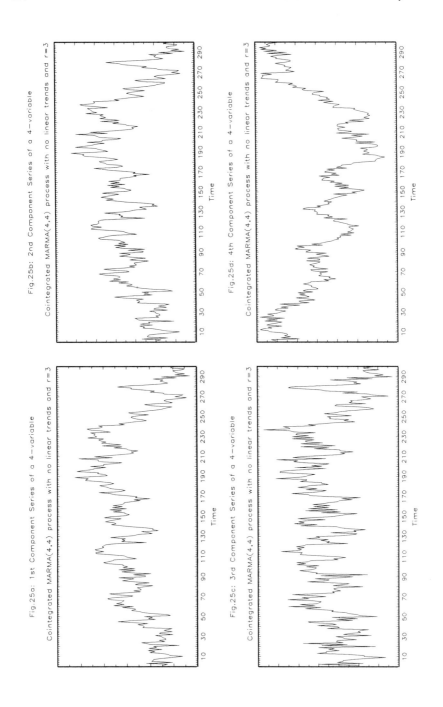

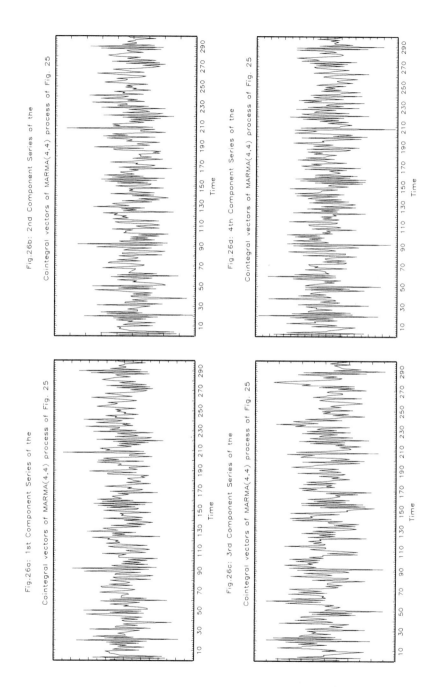

Fig.26a: 1st Component Series of the

Cointegral vectors of MARMA(4,4) process of Fig. 25

Fig.26b: 2nd Component Series of the

Cointegral vectors of MARMA(4,4) process of Fig. 25

Fig.26c: 3rd Component Series of the

Cointegral vectors of MARMA(4,4) process of Fig. 25

Fig.26d: 4th Component Series of the

Cointegral vectors of MARMA(4,4) process of Fig. 25

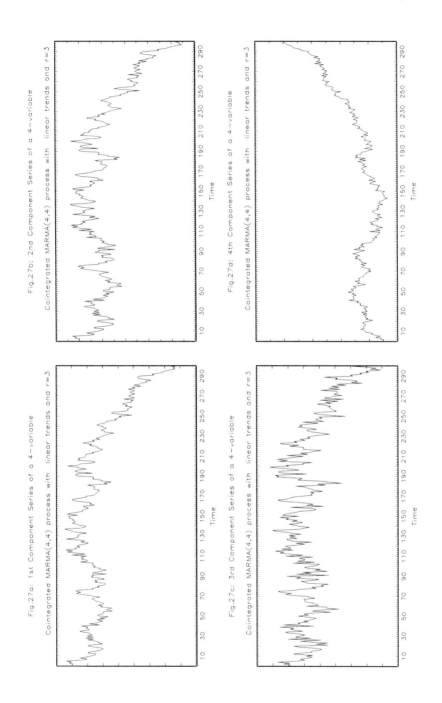

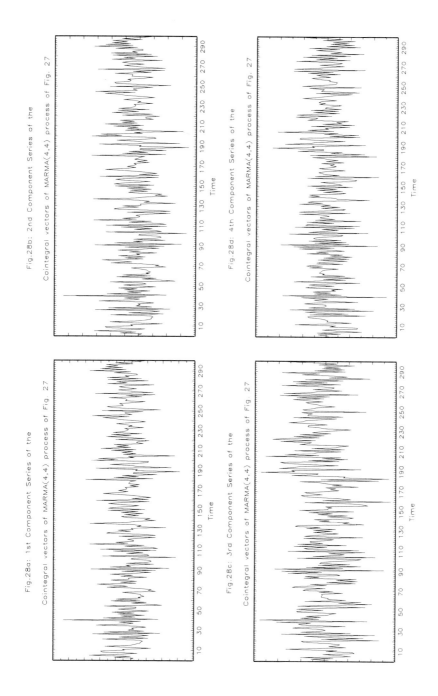

Fig.28b: 2nd Component Series of the
Cointegral vectors of MARMA(4,4) process of Fig. 27

Fig.28d: 4th Component Series of the
Cointegral vectors of MARMA(4,4) process of Fig. 27

Fig.28a: 1st Component Series of the
Cointegral vectors of MARMA(4,4) process of Fig. 27

Fig.28c: 3rd Component Series of the
Cointegral vectors of MARMA(4,4) process of Fig. 27

# Bibliography

Ahn, S. K. and G. C. Reinsel (1990), "Estimation for Partially Nonstationary Multivariate Autoregressive Models", *Journal of the American Statistical Association*, vol. 85 (Theory and Methods), pp. 813-823.

Akaike, H. (1969), "Fitting Autoregressive Models for Prediction", *Annals of the Institute of Statistical Mathematics* (Tokyo), vol. 21, pp. 243-247.

Akaike, H. (1973), "Information Theory and an Extension of the Maximum Likelihood Principle" in *Second International Symposium on Information Theory*, B. F. Petrov and F. Csaki (eds.), Budapest: Akademiai Kiado, pp. 267-281.

Akaike, H. (1978), "Time Series Analysis and Control through Parameteric Models", in *Applied Time Series Analysis*, D. F. Findley (ed.), New York: Academic Press.

An, H. Z., Z. G. Chen and E. J. Hannan (1983), "A Note on ARMA Estimation", *Journal of Time Series Analysis*, vol. 4, pp. 9-17.

Anderson, T. W. (1951), "Estimating Linear Restrictions on Regression Coefficients for Multivariate Normal Distributions", *Annals of Mathematical Statistics*, vol. 22, pp. 327-351.

Anderson, T. W. (1958), *Introduction to Multivariate Statistical Analysis*, New York: Wiley, second edition 1984.

Anderson, T. W. (1971), *The Statistical Analysis of Time Series*, New York: Wiley.

Anderson, T. W. (1976), "Estimation of Linear Functional Relationships", *Journal of the Royal Statistical Society, Series B*, vol. 38, pp. 1-36.

Andrews, D. W. K. (1991), "Heteroskedasticity and Autocorrelation Consistent Covariance Matrix Estimation", *Econometrica*, vol. 59, pp. 817-858.

Ansley, C. F. (1979) "An Algorithm for the Exact Likelihood of a Mixed Autoregressive-Moving Average Process", *Biometrika*, vol. 66, pp. 59-65.

Bachelier, Louis (1900), "Théorie de la Spéculation", *Annales de l' Ecole Normale Superieure*, series 3, vol. 17, pp. 21-86, reprinted in English translation in Cootner (1964).

Bannerjee, A., J.J. Dolado, D.F. Hendry and G.W. Smith (1986), "Exploring Equilibrium Relatinships in Econometrics through Static Models: Some Monte Carlo Results", *Oxford Bulletin of Economics and Statistics*, vol. 28, pp. 253-277.

Bellman, R. (1960), *Matrix Analysis*, New York: McGraw-Hill.

Bergstrom, A.R. (1990), *Continuous Time Econometric Modelling*, Oxford: Oxford University Press.

Beveridge, S. and C. R. Nelson (1981), "A New Approach to Decomposition of Economic Time Series into Permanent and Transitory Components with Particular Attention to the Measurement of the Business Cycle", *Journal of Monetary Economics*, vol. 7, pp. 151-174.

Billingsley, P. (1968), *Convergence of Probability Measures*, New York: Wiley.

Blanchard, O. J. (1989), "A Traditional Interpretation of Macroeconomic Fluctuations", *American Economic Review*, vol. 79, pp. 1146-1164.

Blanchard, O. J. and D. Quah (1989), "The Dynamic Effects of Aggregate Demand and Supply Disturbances", *American Economic Review*, vol. 79, pp. 655-673.

Blangiewicz M. and W. Charemza (1990), "Cointegration in Small Samples: Empirical Percentiles, Drifting Moments and Customized Testing", *Oxford Bulletin of Economics and Statistics*, vol. 52, pp.303-315.

Blum, J. R., D. L. Hanson and L. H. Koopmans (1963), "On the Strong Law of Large Numbers for a Class of Stochastic Processes", *Z. Wahrsch. Verw. Gebiete*, vol. 2, pp. 1-11.

Box. G. E. P. (1954a), "Some Theorems on Quadratic Forms Applied in the

Study of Analysis of Variance Problems, I: Effect of Inequality of Variance in the One-Way Classification", *Annals of Mathematical Statistics*, vol. 25, pp. 290-302.

Box. G. E. P. (1954b), "Some Theorems on Quadratic Forms Applied in the Study of Analysis of Variance Problems, II: Effect of Inequality of Variance in the Two-Way Classification", *Annals of Mathematical Statistics*, vol. 25, pp. 484-498.

Box, G. E. P. and G. M. Jenkins (1970), *Time Series Analysis, Forecasting and Control*, San Fransisco: Holden-Day.

Bradley, R. C. (1980), "On the Strong Mixing and Weak Bernoulli Conditions", *Z. Wahrsch. Verw. Gebiete*, vol. 51, pp. 49-54.

Bradley, R. C. (1981), "Central Limit Theorems under Weak Dependence", *Journal of Multivariate Analysis*, vol. 11, pp. 1-16.

Bradley, Richard C. (1986), "Basic Properties of Strong Mixing Conditions", in Eberlein and Taqqu (1986).

Brockwell, P. J. and R. A.Davis (1987), *Time Series: Theory and Mathods*, New York: Springer Verlag.

Brockwell, P. J. and R. A. Davis (1988), "Simple Consistent Estimation of the Coefficients of a Linear Filter", *Stochastic Processes and their Applications*, vol. 22, pp. 47-59.

Brockwell, P. J. and R. A.Davis (1991), *Time Series: Theory and Mathods*, New York: Springer Verlag, second edition.

Brown, B. M. (1971), "Martingale Central Limit Theorems", *Annals of Mathematical Statistics*, vol. 42, pp. 59-66.

Cheung, Y. and K. S. Lai (1993), "Finite Sample Sizes for Johansen's Likelihood Ratio Tests for Cointegration", *Oxford Bulletin of Economics and Statistics*, vol. 55, pp. 313-328.

Cootner, P. H. (ed.) (1964), *The Random Character of Stock Prices*, Cambridge, MA: MIT Press, revised edition.

Davydov, Y. A. (1968), "Convergence of Distributions Generated by Stationary Stochastic Processes", *Theory of Probability and Applications*, vol. 13, pp. 691-696.

Denker, M. (1986), "Uniform Integrability and the Central Limit Theorem for Strongly Mixing Processes", in Eberlein and Taqqu (1986).

Dhrymes, P. J. (1970), *Econometrics: Statistical Foundations and Applications*, New York: Harper and Row.

Dhrymes, P. J. (1971), *Distributed Lags: Problems of Formulation and Estimation*, San Francisco: Holden-Day.

Dhrymes, P. J. (1974) *Econometrics: Statistical Foundations and Applications*, New York: Springer Verlag, second edition.

Dhrymes, P. J. (1981), *Distributed Lags: Problems of Formulation and Estimation*, Amsterdam: North Holland, second edition.

Dhrymes, P. J. (1984), *Mathematics for Econometrics*, New York: Springer Verlag, second edition.

Dhrymes, P. J. (1989), *Topics in Advanced Econometrics: vol. I, Probability Foundations*, New York: Springer Verlag.

Dhrymes, P. J. (1994a), "Specification Tests in Simultaneous Equations Systems", (1994), *Journal of Econometrics*, vol. 64, pp. 45-76.

Dhrymes, P. J. (1994b), *Topics in Advanced Econometrics: vol. II, Linear and Nonlinear Simultaneous Equations*, New York: Springer Verlag.

Dickey, D. A. and W. A. Fuller (1979), "Distribution of the Estimators for Autoregressive Time Series with a Unit Root", *Journal of the American Statistical Association*, vol. 74, pp. 42-431.

Dickey, D. A. and W. A. Fuller (1981), "Likelihood Ratio Statistics for Autoregressive Time Series with a Unit Root", *Econometrica*, vol. 49, pp. 1057-1072.

Donsker, M. D. (1951), "An Invariance Principle for Certain Probability Limit Theorems", *Memoirs of the American Mathematical Society*, vol. 6, pp. 1-11.

Doob, J.L. (1953), *Stochastic Processes*, New York: Wiley.

Doukhan, P. (1994), *Mixing: Properties and Examples*, Lecture Notes in Statistics, New York: Springer Verlag.

Eaton, Morris L. (1983), *Multivariate Statistics: A Vector Space Approach*, New York: Wiley.

Eberlein, E. and M. S. Taqqu (eds.) (1986), *Dependence in Probability and Statistics*, Boston: Birkhäuser.

Engle, R.F. and C.W.J. Granger (1987), "Cointegration and Error Correction: Representation, Estimation and Testing", *Econometrica*, vol. 55, pp. 251-276.

Epanechnikov, V. A. (1969), "Nonparaetric Estimation of a Multivariate Probability Density", *Theory of Probability and Its Applications*, vol. 14, pp. 153-158.

Fama, E. F. (1965), "Behavior of Stock Market Prices", *Journal of Business*, vol. 38, pp. 34-105.

Fama, E.F. (1970), "Efficient Capital Markets: A Review of Theory and Empirical Work", *Journal of Finance*, vol. 25, pp. 383-417.

Gantmacher, F.R. (1959), *The Theory of Matrices*, vols. I and II, New York: Chelsea.

Gardeazabal, J. and M. Regulez (1992), *The Monetary Model of Exchange Rates and Cointegration*, Berlin: Springer-Verlag.

Gonzalo J. (1994), "Comparison of Five Alternative Methods of Estimating Long-Run Equilibrium Relationships", *Journal of Econometrics*, vol. 60, pp. 203-233.

Granger, C.W.J. (1981), "Some Properties of Time Series Data and Their Use in Econometric Specification", *Journal of Econometrics*, vol. 16, pp. 121-130.

Granger, C. W. J. and P. Newbold (1974), "Spurious Regressions in Econometrics", *Journal of Econometrics*, vol. 2, pp. 111-120.

Hall, P. and C. C. Heyde (1980), *Martingale Limit Theory and Its Application*, New York: Academic Press.

Hall, R. E. (1978), "Stochastic Implications of the Life Cycle Permanent Income Hypothesis", *Journal of Political Economy*, vol. 86, pp. 971-986.

Hannan, E. J. (1970), *Multiple Time Series*, New York: Wiley.

Hannan, E. J. (1973), "Central Limit Theorems for Time Series Regression", *Z. Wahr. verw. Geb.*, vol. 26, pp. 157-170.

Hannan, E. J. and M. Deistler (1988), *The Statistical Theory of Linear Systems*, New York: Wiley.

Haug, A. (1996), "Tests for Cointegration; A Monte Carlo Comparison", *Journal of Econometrics*, vol. 71, pp.89-115.

Henderson, H. V. and S. R. Searle (1981), "The Vec-Permutation Matrix, the Vec Operator and Kronecker Products: A Review", *Linear and Multilinear Algebra*, vol. 9, pp.271-288.

Herrndorf, N. (1983), "The Invariance Principle for $\phi$-mixing Sequences", *Z. Wahr. Verw. Geb.*, vol. 63, pp. 97-108.

Herrndorf, N. (1984a), "A Functional Central Limit Theorem for $\rho$-mixing Sequences", *Journal of Multivariate Analysis*, vol. 15, pp. 141-146.

Herrndorf, N. (1984b), "A Functional Central Limit Theorem for Weakly Dependent Sequences of Random Variables", *The Annals of Probability* vol. 12, pp. 141-153.

Heyde, C.C. (1974), "On a Central Limit Theorem for Stationary Processes", *Z. Wahrsch. Verw. Gebiete*, vol. 30, pp. 315-320.

Hida, T. (1980), *Brownian Motion*, New York: Springer Verlag.

Hooker M.A. (1993), "Testing for Cointegration: Power versus Frequency of Observation", *Economics Letters*, vol. 41, pp.359-362.

Hotelling, H (1936), "Relations between Two Sets of Variates", *Biometrika*, vol. 28, pp. 321-77.

Hurvich, C. M. and C. L. Tsai (1989), "Regression and Time Series Model Selection in Small Samples", *Biometrika*, vol. 76, pp. 297-307.

Hwang, J. and P. Schmidt (1993), "On the Power of Point Optimal Tests of the Trend Stationary Hypothesis", *Economics Letters*, vol. 43, pp. 143-147.

Ibragimov, I. A. (1962), "Some Limit Theorems for Stationary Processes", *Probability Theory and Applications*, vol. 7, pp. 349-82.

Ibragimov, I. A. (1975), "A Note on the Central Limit Theorem for Dependent Random Variables", *Probability Theory and Applications*, vol. 20, pp. 135-141.

Ibragimov, I. A. and Y. Y. Linnik (1971), *Independent and Stationary Sequences of Random Variables*, Groningen: Wolters-Noordhoff.

Ibragimov, I. A. and Y. A. Rozanov (1978), *Gaussian Random Processes*, Berlin: Springer Verlag.

Ito, K. (1950), "Stochastic Differential Equations in a Differentiable Manifold", *Nagoya Mathematical Journal*, vol. 2, pp. 35-47.

Ito, K. (1951a), "On a Formula Concerning Stochastic Differentials", *Nagoya Mathematical Journal*, vol. 3, pp. 55-65.

Ito, K. (1951b), "On Stochastic Differential Equations", *Memoirs of the American Mathematical Society*, vol. 4, pp. 1-51.

Ito, K. (1951c), "Multiple Wiener Integral", *Journal of the Mathematical Society of Japan*, vol. 3, 157-169.

Jacod, J. and A. N. Shiryaev (1987), *Limit Theorems for Stochastic Processes*, Berlin Heidelberg: Springer Verlag.

Johansen, S. (1988), "Statistical Analysis of Cointegration Vectors", *Journal of Economic Dynamics and Control*, vol. 12, pp. 231-254.

Johansen, S. (1991), "Estimation and Hypothesis Testing of Cointegration Vectors in Gaussian Vector Autoregressive Models", *Econometrica*, vol. 59, pp. 1551-1580.

Johansen, S and K. Juselius (1990), "Maximum Likelihood Estimation and Inference on Cointegration, with Applications to the Demand for Money", *Oxford Bulletin of Economics and Statistics*, vol. 52, pp. 169-210.

Johnson, N. L. and S. Kotz (1972), *Distributions in Statistics: Continuous Multivariate Distributions*, New York: Wiley.

Karatzas, I. and S.E. Shreve (1988), *Brownian Motion and Stochastic Cal-*

*culus*, New York: Spriner-Verlag.

Karatzas, I. and S.E. Shreve (1991), *Brownian Motion and Stochastic Calculus*, New York: Spriner-Verlag, second edition.

Kremers, J.J.M., N.R. Ericsson and J.J. Dolado (1992), "The power of Cointegration Tests", *Oxford Bulletin of Economics and Statistics*, vol. 54, pp. 325-348.

Levy, P. (1948), *Processus Stochastiques et Mouvement Brownien*, Paris: Gauthier-Villars.

Mandelbrot, B. (1966), "Forecasts of Futures Prices, Unbiased Markets and Martingale Models", *Journal of Business*, vol. 39, pp. 242-255.

Mann, H. B. and A. Wald (1943), "On the Statistical Treatment of Linear Stochastic Difference Equations", *Econometrica*, vol. 11, pp.173-220.

McLeish, D. L. (1975), "A Maximal Inequality for Dependent Strong Laws", *Annals of Probability*, vol. 3, pp. 829-839.

McLeish, D. L. (1977), "On the Invariance Principle for Nonstationary Mixingales", *Annals of Probability*, vol. 5, pp. 616-621.

Merton, R. (1990), *Continuous-time Finance*, Cambridge, MA: Basil Blackwell.

Nelson, C.R. and C. I. Plosser (1982), "Trends versus Random Walks in Macroeconomic Time Series: Some Evidence and Implications", *Journal of Monetary Economics*, vol. 10, pp. 139-162.

Oodaira, H and K. Yoshihara (1972), "Functional Central Limit Theorem for Strictly Stationary Processes Satisfying the Strong Mixing Condition", *Kodai Math. Sem. Rep.*, vol. 24, pp. 259-69.

Osterwald-Lenum, M. (1992), "A Note with Quantiles on the Asymptotic Distribution of the Maximum Likelihood Cointegration Rank Test Statistics", *Oxford Bulletin of Economics and Statistics*, vol. 54, pp. 461-471.

Park, J. Y. and P.C.B. Phillips (1988), "Statistical Inference in Regressions with Integrated Processes: Part 1", *Econometric Theory*, vol. 4, pp. 469-497.

Park, J. Y. and P.C.B. Phillips (1989), "Statistical Inference in Regressions with Integrated Processes: Part 2", *Econometric Theory*, vol. 5, pp. 95-131.

Peligrad, M (1982), "Invariance Principles for Mixing Sequences of Random Variables", *Annals of Probability*, vol. 10, pp. 968-81.

Peligrad, M. (1985), "An Invariance Principle for $\phi$-Mixing Sequences", *Annals of Probability*, vol. 13, pp. 1304-1313.

Phillips, P. C. B. (1986), "Understanding Spurious Regressions in Econometrics", *Journal of Econometrics*, vol. 33, pp. 311-340.

Phillips, P. C. B. (1987), "Time Series Regression with a Unit Root", *Econometrica*, vol. 55, pp. 277-301.

Phillips, P. C. B. (1988), "Weak Convergence to the Matrix Stochastic Integral $\int_0^1 B \, dB'$ ", *Journal of Multivariate Analysis*, vol. 24, pp. 252-264.

Phillips, P.C.B. (1991), "Optimal Inference in Cointegrated Systems", *Econometrica*, vol. 59, pp. 283-306.

Phillips, P.C.B. and S. Durlauf (1986), "Multiple Time Series Regression with Integrated Time Series", *Review of Economic Studies*, vol. 53, pp. 473-496.

Phillips, P.C.B. and B.E. Hansen (1990). "Statistical Inference in Instrumental Variables Regression with $I(1)$ Processes", *Review of Economic Studies*, vol. 57, pp. 99-125.

Phillips, P.C.B. and M. Loretan (1991), "Estimating Long-run Economic Equilibria", *Review of Economic Studies*, vol. 58, pp. 407-436.

Phillips, P.C.B. and S. Ouliaris (1990), "Asymptotic Properties of Residual Based Tests for Cointegration", *Econometrica*, vol. 58, pp. 165-93.

Phillips, P.C.B. and P. Perron (1987), "Testing for a Unit Root in Time Series Regression", *Biometrika*, vol. 75, pp. 335-346.

Phillips, P. C. B. and V. Solo (1992), "Asymptotics for Linear Processes", *Annals of Statistics*, vol. 20, pp. 971-1001.

Priestley, M. B. (1962), *Spectral Analysis and Time Series*, vols. I and II, New York: Academic Press.

Protter, P. (1992), *Stochastic Integration and Differential Equations* (second Corrected Printing), New York: Springer Verlag.

Rao, M. M. (1961), "Consistency and Limit Distributions of Estimators in Explosive Stochastic Difference Equations", *Annals of Mathematical Statistics*, vol. 32, pp. 195-218.

Reinsel, G.C. (1993), *Multiple Time Series*, New York: Springer Verlag.

Reinsel, G. C. and H. Ahn (1992), Reinsel, G. C., S. Basu and F. S. Yap (1992), "Maximum Likelihood Estimation in the Multivariate Autoregressive-Moving Average Model from a Generalized Least Squares Viewpoint", *Journal of Time Series Analysis*, vol. 13, pp. 133-145.

Revuz, D. and M. Yor (1991), *Continuous Martingales and Brownian Motion*, Berlin, Heidelberg: Springer Verlag.

Rosenblatt, M. (1956), "A Central Limit Theorem and a Strong Mixing Condition", *Proceedings of the National Academy of Sciences (USA)*, vol. 42, pp. 43-47.

Rubin, H. (1950), "Consistency of Maximum Likelihood Estimates in the Explosive Case", in *Statistical Inference in Dynamic Economic Models*, T.C. Koopmans (ed.), New York: Wiley.

Said, S. E. and D. A. Fuller (1984), "Testing for Unit Roots in Autoregressive-Moving Average of Unknown Order", *Biometrika*, vol. 71, pp. 599-607.

Saikkonen, P. (1992), "Estimation and Testing of Cointegrated Systems by an Autoregressive Approximation", *Econometric Theory*, vol. 8, pp. 1-27.

Samuelson, P. A. (1965), "Proof that Properly Anticipated Prices Fluctuate Randomly", *Industrial Management Review*, vol. 6, pp. 41-50.

Samuelson, P. A. (1973), "Proof that Properly Discounted Present Values of Assets Fluctuate Randomly", *Bell Journal of Economics and Management Science*, vol. 4, pp. 369-374.

Schwarz, G. (1978), "Estimating the Dimension of a Model", *The Annals of Statistics*, vol. 6, pp. 461-464.

Schwert, W.G. (1989), "Tests for Unit Roots: A Monte Carlo Investigation", *Journal of Business and Economic Statistics*, vol. 7, pp.147-159.

Scott, D. W. (1992), *Multivariate Density Estimation*, New York: Wiley.

Siddiqui, M. M. (1965), "Approximations to the Distribution of Quadratic Forms", *Annals of Mathematical Statistics*, vol. 36, pp. 677-682.

Sims, C. A., J. H. Stock and M. W. Watson (1990), "Inference in Linear Time Series with some Unit Roots", *Econometrica*, vol. 58, pp. 113-144.

Stock, J. H. (1987), "Asymptotic Properties of Least Squares Estimators of Cointegrating Vectors", *Econometrica*, vol. 55, pp. 1035-1056.

Stock, J. H. and M. W. Watson (1988), "Testing for Common Trends", *Journal of the American Statistical Association*, vol. 83, pp. 1097-1107.

Stone, R. A. (1947), "On the Interdependence of Blocks of Transactions", *Journal of the Royal Statistical Society*, Supplement, vol. 9, pp. 1-32.

Strasser, H. (1986), "Martingale Difference Arrays and Stochastic Integrals", *Probability Theory and Related Fields*, vol. 72, pp. 83-98.

Toda, H. (1995), "Finite Sample Performance of Likelihood Ratio Tests for Cointegrating Ranks In Vector Autoregressions", *Econometric Theory*, vol. 11, pp. 1015-1032.

Tsay, S. R. and G. C. Tiao (1990), "Asymptotic Properties of Multivariate Nonstationary Processes with Applications to Autoregressions", *The Annals of Statistics*, vol. 18, pp. 220-50.

Tso, M. K-S. (1981), "Reduced Rank Regression and Canonical Analysis", *Royal Statistical Society, Series B*, vol. 43, pp. 183-189.

Tunnicliffe-Wilson, G. (1972) "The Factorization of Matricial Spectral Densities", *SIAM Journal of Applied Mathematics*, vol. 23, pp. 420-426.

Wald, A. (1950), "A Note on the Identification of Economic Relations", Chapter 3, pp. 238-234, in Koopmans T.C. (ed.) (1950), *Statistical Inference in Dynamic Economic Models*, Monograph 10, Cowles Foundation for Research in Economics, New York: Wiley; also reprinted in Anderson, T. W. (ed.), (1955), *Selected Papers in Statistics and Probability by Abraham Wald*, New York: McGraw-Hill.

White, J. S. (1957), "The Limiting Distribution of the Serial Correlation Coefficient in the Explosive Case", *Annals of Mathematical Statistics*, vol. 29, pp. 1188-1197.

Whittle, P. (1963), "On the Fitting of Multivariate Autoregression and the Approxiate Canonical Factorization of Special Density Matrix", *Biometrika*, vol. 40, pp. 129-134.

Wolkonski, V. A. and Y. A. Rozanov (1959), "Some Limit Theorems for Random Functions, Part I", *Theory of Probability and Applications*, vol. 4, pp. 178-97.

Yap, S. F. and G. C. Reinsel (1995), "Estimation and Testing for Unit Roots in a Partially Nonstationary Autoregressive Moving Average Model", *Journal of the American Statistical Association*, vol. 90, pp. 253-267.

# Index

516